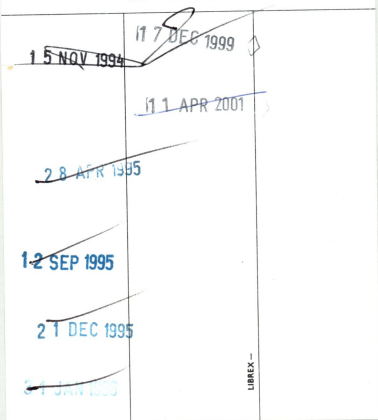

LASERS AND HOLOGRAPHY

LASERS AND HOLOGRAPHY

P C MEHTA
*Instruments Research and Development Establishment
Defence Research and Development Organisation
Dehra Dun, India*

V V RAMPAL
*Department of Science and Technology
New Delhi, India*

World Scientific
Singapore • New Jersey • London • Hong Kong

Published by

World Scientific Publishing Co. Pte. Ltd.
P O Box 128, Farrer Road, Singapore 9128
USA office: Suite 1B, 1060 Main Street, River Edge, NJ 07661
UK office: 73 Lynton Mead, Totteridge, London N20 8DH

621.367
MEH

LASERS AND HOLOGRAPHY

Copyright © 1993 by World Scientific Publishing Co. Pte. Ltd.

All rights reserved. This book, or parts thereof, may not be reproduced in any form or by any means, electronic or mechanical, including photocopying, recording or any information storage and retrieval system now known or to be invented, without written permission from the Publisher.

For photocopying of material in this volume, please pay a copying fee through the Copyright Clearance Center, Inc., 27 Congress Street, Salem, MA 01970, USA.

ISBN 981-02-1214-3

Printed in Singapore.

Dedicated to

BHAPPO

*"The Candle burns
and brightens those
that cause it to burn"*

PREFACE

The coming of lasers and their use for holography marked an important era in the advancement of physical sciences. This generated diverse fields of activity and gave new life to well researched areas like Raman spectroscopy. Holography provided an excited medium for three dimensional display. Lasers and holography continue to find new areas of applications. Their use in industry, medicine, communication and instrumentation for analysis and measurement are now well established. Individually, the two subjects, Lasers and Holography, have been treated at length in the literature and excellent reviews are available. The present book covers both the areas in a comprehensive manner. It is not always easy to get an overall view of a modern discipline in the presence of vast literature scattered in journals, books, reports and conference proceedings. Our effort of presenting both lasers and holography in a single volume may therefore appear rather ambitious but we hope it will achieve its purpose of providing the essentials of the topics and application update in a fairly reasonable extent.

The subject of holography is still evolving and a number of new applications such as computer interconnects and in medicine are emerging fast. It simply depends on ones imagination to exploit the unique features of holograms. A complete treatment of such applications and their thorough understanding therefore need further developments in architectures, techniques and materials.

The first three chapters discuss the physics and technology of lasers, while chapter 4 deals with the principles and techniques of holography. The chapters 5 to 9 describe the applications. The applications in the areas of medicine, information processing, nondestructive testing and interconnections have been included for their relevance in the context of recent developments.

In writing the book a large number of original publications have been used. However, the references cited do not represent the complete bibliography on the subject. Only those references are listed which have been used for writing the book. It is natural in such a book that many important publications may have been left unintentionally, for which

PREFACE

the authors regret.

The book aims for a readership of those studying for a Master's degree in Applied Optics, Lasers and Holography and requiring an understanding of the generation and use of coherent optical radiation. The book will be useful to students, researchers and professionals actively involved in the development of lasers, holography and their applications. It is hoped that the students and researchers would find it useful and interesting.

One of the authors (PCM) is grateful to Dr O.P. Nijhawan, Director, IRDE, Dehra Dun for helpful suggestions. The authors are also thankful to Mr. Devendra Mohan, Mr. K.S.S. Rao, Mr. Chandra Bhan and Mr. Pritam Lal for providing some of the photographs used in the book. We take great pleasure in acknowledging the valuable help of Dr R.K. Tyagi and Mr. A.K. Musla for their critical reading of the manuscript. We wish to specially thank Mr. R.K. Kukreti for his patience and skill in typing. We are obliged to Mr. A.K. Musla for his painstaking efforts in type setting all the equations and tables, and formatting the manuscript in the computer for laser printing. Finally, we express our appreciation to the World Scientific Publishing Company for the high quality of publication.

The authors acknowledge the patience and moral support of their spouses, Dr. Mrs. P. Mehta (PCM) and Mrs. S. Rampal (VVR) during the writing of the book.

May 19, 1993.
P.C. Mehta
V.V. Rampal

CONTENTS

PREFACE — vii

1. OPTICAL RADIATION AND PHOTONS — 1

1.1. Nature of light — 1
    1.1.1. Quantum description of radiation — 4
    1.1.2. Fluctuation properties — 8
    1.1.3. Power flow of electromagnetic radiation — 9

1.2. Interaction of radiation with matter—Emission and absorption of radiation — 10
    1.2.1. Transition probability — 11
    1.2.2. Emission and absorption of radiation by bound electrons — 14

1.3. Spontaneous and stimulated radiation — 20
1.4. Einstein coefficients — 24
1.5. Optical gain — 26
1.6. Gain saturation — 29
1.7. Optical resonators — 31
    1.7.1. Resonant modes of optical cavity — 33
    1.7.2. Theoretical methods for analyzing the modes — 38
    1.7.3. Gaussian beams — 39
    1.7.4. Design of open resonator cavity — 45
    1.7.5. 'Q' of optical resonator — 52
    1.7.6. Unstable resonators — 54

1.8. Threshold condition for laser oscillation — 57
1.9. Cavity coupling — 59
1.10. Frequency of resonant modes — 60
1.11. Frequency selection of laser oscillation — 61
1.12. Transverse modes selection — 62
1.13. Longitudinal mode selection — 63
1.14. Mode competition — 65
1.15. Hole burning — 66
1.16. Mode pulling — 67
1.17. Frequency stability of laser output — 72
1.18. Single mode operation of a laser — 73
1.19. Coherence of laser radiation — 74
    1.19.1. Time coherence — 75

	1.19.2. Spatial coherence	78
	1.19.3. Time and space coherence	79
	1.19.4. Transient coherence	81
	1.19.5. Higher order coherence functions	82
	1.19.6. Factors responsible for imparting coherence to laser radiation	82
1.20.	Laser noise	84
1.21.	General treatment of laser oscillation	89

2. SPATIAL, TEMPORAL AND SPECTRAL CHARACTERISTICS OF LASER — 92

2.1.	Introduction	92
2.2.	Mode locking	93
2.3.	Methods of mode locking	98
	2.3.1. Mode locking with saturable absorber- Passive mode locking	99
	2.3.2. Active mode locking	102
	2.3.3 Acousto-optic and electro-optic modulators as mode locking devices	105
	2.3.4. Self mode locking	107
	2.3.5. Experimental arrangement of an active mode locked laser	108
	2.3.6. Stabilization of mode locked lasers	108
2.4.	Measurement on mode locked pulses	110
2.5.	Generation and measurement of ultrashort pulses	115
	2.5.1. Synchronous mode locking	116
	2.5.2. Colliding pulse mode locking	116
	2.5.3. Self phase modulation and pulse compression	118
	2.5.4. Measurement of ultrafast pulses	121
2.6.	Mode locking of transverse modes	121
2.7.	Q-switching and cavity dumping	123
	2.7.1. Cavity dumping	123
	2.7.2. Q-switching	125
2.8.	Relaxation oscillation	136

3. SPECIFIC LASER SYSTEMS — 143

3.1.	Solid state lasers	144
	3.1.1. The Ruby laser	145
	3.1.2. Neodymium lasers	150
	3.1.3. Tunable solid state lasers	159
	3.1.4. Sensitized solid state laser materials	163
	3.1.5. Eye safe solid state lasers	164

	3.1.6. Diode laser pumping of solid state lasers	165
	3.1.7. The slab lasers	167
3.2.	Colour centre lasers	168
3.3.	Semiconductor lasers	173
	3.3.1. The p-n junction laser diode	178
	3.3.2. Heterojunction lasers	180
	3.3.3. Recent advances	184
	3.3.4. Quantum well lasers	184
	3.3.5. Distributed feedback lasers	188
3.4.	Dye lasers	193
	3.4.1. The dye as laser medium	193
	3.4.2. Spectra of organic dyes	195
	3.4.3. Requirements for starting oscillation	199
	3.4.4. Cavity arrangements	200
	3.4.5. Output characteristics	203
	3.4.6. Specific purpose developments	205
3.5.	Gas lasers	206
	3.5.1. He-Ne laser	206
	3.5.2. Argon ion laser	209
	3.5.3. Carbon dioxide laser	210
	3.5.4. TEA CO_2 laser	212
	3.5.5. Gas dynamic CO_2 laser	215
3.6.	Chemical lasers	218
	3.6.1. HF/DF laser	219
3.7.	Carbon monoxide (CO) laser	220
3.8.	Excimer lasers	221
3.9.	Nitrogen laser	224
3.10.	Metal vapour lasers	226
	3.10.1. He-Cd laser	228
	3.10.2. Hg-Br laser	228
	3.10.3. Copper vapour laser	229
	3.10.4. Plasma recombination laser	231
3.11.	Far infrared (FIR) lasers	234
3.12.	Free electron laser (FEL)	238
3.13.	Harmonic generation of laser radiation through nonlinear processes	241
	3.13.1. Second order effects in nonlinear crystals- Generation of second harmonics	243
	3.13.2. Third order nonlinear processes in gaseous media- Generation of tunable UV and IR	245

4.	**HOLOGRAPHY: PRINCIPLES AND TECHNIQUES**	**251**
4.1.	Introduction	251
4.2.	Characteristics of a hologram	254
4.3.	In-line Holography: Gabor holography	256
4.4.	Off-axis Holography: Leith-Upatnieks holography	258
4.5.	Holographic imaging equations	263
4.6.	Image magnification	267
4.7.	Hologram aberrations	268
4.8.	Orthoscopic and pseudoscopic images	269
4.9.	Classification of holograms	272
	4.9.1. Amplitude and phase holograms	272
	4.9.2. Classification based on hologram thickness	274
	4.9.3. Classification based on direction of reconstructed image	281
	4.9.4. Classification according to recording arrangement	285
4.10.	Practical holography	294
	4.10.1. Laser	294
	4.10.2. Reference-to-object intensity ratio	296
	4.10.3. Angle between reference and object beams	297
	4.10.4. Polarization of light beams	297
	4.10.5. Vibration isolation table	298
	4.10.6. Optical components and mounts	299
	4.10.7. Hologram recording geometries	301
	4.10.8. Hologram of a moving object	306
	4.10.9. Efficiency of a hologram	309
	4.10.10 Refractive index modulation	311
	4.10.11. Signal-to-Noise Ratio (SNR)	312
4.11.	Holographic recording materials	314
	4.11.1. Modulation transfer function	315
	4.11.2. Nonlinear recording	316
	4.11.3. Silver halide emulsion	318
	4.11.4. Hardened dichromated gelatin (DCG)	338
	4.11.5. Photopolymers	352
	4.11.6. Photoresists	358
	4.11.7. Photothermoplastics	361
	4.11.8. Photochromic materials	364
	4.11.9. Photorefractive crystals (Electro-optic materials)	367
	4.11.10. Summary of recording materials	371
	4.11.11. Health hazards of hologram processing chemicals	373
4.12.	Display holography	378
	4.12.1. Requirements of a display hologram	378

	4.12.2. 360° hologram	380
	4.12.3. Rainbow hologram	385
	4.12.4. Holographic stereogram	398
	4.12.5. Viewing zone	402
	4.12.6. Change of size	403
	4.12.7. Dispersion compensation	404
	4.12.8. Sources for reconstruction	408
	4.12.9. Hologram display systems	408
	4.12.10. Holographic 3D printer	412
	4.12.11. Holographic television	414
	4.12.12. Holographic cinematography	414
4.13.	Colour holography	416
	4.13.1. Chromaticity diagram	416
	4.13.2. Recording of colour holograms	418
	4.13.3. Volume colour holograms	418
	4.13.4. Recording geometry	423
	4.13.5. Pseudocolouring	426
4.14.	Special techniques	429
	4.14.1. Local reference beam hologram	429
	4.14.2. Multiple-exposure holography (Scanning object beam holography)	430
	4.14.3. Multiplexed hologram	432
	4.14.4. Multifaceted hologram	436
	4.14.5. Pinhole hologram	439
	4.14.6. Edge-lit hologram	443
4.15.	Hologram replication	445
	4.15.1. Optical interferometric techniques	445
	4.15.2. Mechanical replication technique (Embossed holograms)	449
4.16.	Polarization holography	450
	4.16.1. Depolarization effects	450
	4.16.2. Polarization recording	452
4.17.	Evanescent wave holography (Waveguide holography)	453
	4.17.1. Holographic lithography	459
5.	**HOLOGRAPHIC INTERFEROMETRY**	**479**
5.1.	Introduction	479
5.2.	Double-exposure holographic interferometry	479
5.3.	Single-exposure real-time holographic interferometry	482
5.4.	Time-average holographic interferometry	484
5.5.	Stroboscopic holographic interferometry	487
5.6.	Temporally modulated holography	488

5.7.	Fringe linearization holographic interferometry	489
5.8.	Desensitized holographic interferometry	489
5.9.	Digital holographic interferometry	490
5.10.	Fringe localization	491
	5.10.1. Pure translation	492
	5.10.2. Pure rotation about an axis in the surface	493
5.11.	Sandwich hologram interferometry	495
5.12.	Applications	497
	5.12.1. Holographic nondestructive testing (HNDT)	497
5.13.	Holographic contouring	510
	5.13.1. Two wavelength holographic contouring	510
	5.13.2. Two refractive index holographic contouring	511
	5.13.3. Contouring by change in the illuminating angle	511
5.14.	Holographic interferometry with fibre optics	511

6. HOLOGRAPHIC OPTICAL ELEMENTS — 517

6.1.	What is a HOE?	517
6.2.	Hologram of a point	519
6.3.	Resolution of a HOE	520
6.4.	Design aspects	522
6.5.	Fabrication	525
6.6.	Holographic gratings/mirrors	527
6.7.	Applications of HOEs	533
	6.7.1. Spectral filters	534
	6.7.2. Applications in optical communication	535
	6.7.3. HOEs in compact disks	537
	6.7.4. Holographic laser beam attenuator	537
	6.7.5. HOE based fibre optic gyroscope	537
	6.7.6. Holographic scanner	539
	6.7.7. Diffractive-refractive telescope	540
	6.7.8. Applications in architecture	541
	6.7.9. Beam combiners	543
	6.7.10. Fingerprint sensor	547
	6.7.11. HOEs in Art	549

7. INTERCONNECTS — 552

7.1.	Introduction	552
7.2.	Optical interconnects	553
7.3.	Classification of holographic interconnects	554
7.4.	HOE size	555
7.5.	Desirable characteristics	558

7.6.	Configurations	559
	7.6.1. Free-space focussed spot interconnects	559
	7.6.2. Waveguide hologram interconnects	565
	7.6.3. Perfect shuffle interconnects	572
	7.6.4. Dynamic holographic interconnects	572
7.7.	Challenges	579
8.	**INFORMATION PROCESSING**	583
8.1.	Optical data processor	583
8.2.	Image sharpening (Deblurring)	593
	8.2.1. Fourier transform division filter method (Inverse filter method)	593
	8.2.2. Correlative holographic decoding method	596
8.3.	Image subtraction	599
	8.3.1. Single-exposure method	599
	8.3.2. Double-exposure method	600
	8.3.3. Holographic beam splitter method	600
	8.3.4. Cross-polarization method	600
	8.3.5. Holographic shear lens method	602
	8.3.6. Rainbow holographic method	602
	8.3.7. Dynamic hologram method	603
	8.3.8. Colour coding method	604
8.4.	Character and pattern recognition	605
	8.4.1. Vander Lugt frequency plane correlator	605
	8.4.2. Experimental technique	607
	8.4.3. Applications	609
	8.4.4. Fresnel hologram correlator	612
8.5.	Holographic information storage	616
	8.5.1. Holographic associative memory	621
8.6.	Holographic image multiplication	626
8.7.	Coherent noise reduction	629
9.	**MEDICAL APPLICATIONS OF LASERS AND HOLOGRAPHY**	634
9.1.	Medical applications of lasers	634
	9.1.1. Introduction	634
	9.1.2. Laser-tissue interaction mechanisms	636
	9.1.3. Delivery of laser radiation through fibres	636
	9.1.4. Study of tissue fluorescence as a diagnostic aid	638
	9.1.5. Medical applications of solid state lasers	641
	9.1.6. Use of lasers in ophthalmology	645

	9.1.7. Angioplasty with lasers	648
	9.1.8. Other medical applications	650
9.2.	Medical applications of holography	652
	9.2.1. Holographic endoscopy	653
	9.2.2. Holography in ophthalmology	654
	9.2.3. Holography in dentistry	656
	9.2.4. Holography in otology	657
	9.2.5. Miscellaneous studies	658
	9.2.6. Biological holography with X-ray laser	659

NAME INDEX   664

SUBJECT INDEX   677

CHAPTER 1

OPTICAL RADIATION AND PHOTONS

1.1. Nature of Light

Light has dominated our lives for unknown years. Nearly eighty percent of sensor input to brain is through eyes. The nature of light, consequently, has been a subject of intense curiosity and a debate about its being wave like or corpuscular in nature has continued since early seventeenth century. Phenomena like diffraction, interference and polarization, supported wave theory while rectilinear propagation lent credence to the corpuscular nature of light. Names like Grimaldi, Hooke, Newton, Huygens, Young, Fresnel and Kirchoff have been associated with one or the other theory and kept alive the debate about the true nature of light.

In the later part of nineteenth century, the work of Maxwell and Poynting unified the understanding of electricity and magnetism and electromagnetic fields were shown to carry energy and momentum. The classical theory of electromagnetism found it difficult to explain photoelectric emission, since it was experimentally shown that kinetic energy of photoelectrons depend on frequency of incident radiation and not on its intensity. The difficulty of classical electrodynamics in explaining the spectral distribution of light emitted by a black body at temperature T is now history. The work of Stefan, Boltzmann, Wien, Rayleigh and Jeans presented limitations in fully explaining the experimental observations.

The classical Rayleigh Jeans Law predicts that the energy per unit frequency interval, $E(\nu)\,d\nu$ is determined by

$$E(\nu)\,d\nu = (8\pi\nu^2 kT/c^3)\,d\nu \qquad (1.1)$$

and is unbounded as ν increases. This leads to ultraviolet catastrophe since total emitted energy E, given by

$$E = \int_0^\infty E(\nu)\,d\nu \qquad (1.2)$$

is ultraviolet divergent. Wien's law, which has validity at high frequencies, is again unable to explain observations at low frequencies since

$$\rho(\nu) = \alpha\nu^3 \exp(-\beta\nu/T), \qquad (1.3)$$

where $\rho(\nu)$ is energy density in the frequency interval $d\nu$ such that

$$E(\nu)d\nu = V \rho(\nu) d\nu, \qquad (1.4)$$

V is cavity volume and β is Wien's constant.

The difficulties of classical representations were overcome by Planck who proposed the energy distribution formula

$$E(\nu)d\nu = (8\pi^2\nu^2/c^3)h\nu \, d\nu \, [\exp(h\nu/kT)-1]^{-1}, \qquad (1.5)$$

where h is Planck's constant. He assumed that oscillators in thermal equilibrium have discrete energy levels and not the continuous range associated with classical oscillators. It is easy to see from Eq. 1.5 that for small ν, since

$$\exp(h\nu/kT) \to (1 + h\nu/kT),$$

Eq. 1.5 reduces to Eq. 1.1; for large ν, however, it follows Eq. 1.3. The ultraviolet divergence is also removed, since integration of Eq. 1.5 over all frequencies, provides

$$E = \int_0^\infty E(\nu)d\nu = (8\pi^5/15h^3c^3)(kT)^4, \qquad (1.6)$$

which is finite and agrees with Stefan's law

$$E = \sigma T^4, \qquad (1.7)$$

where σ is a constant.

Planck's postulation, an early attempt towards quantum concepts, quantizes the oscillators (emitters) but not the radiation as such. It established that classical ideas were not enough to explain certain phenomena. Another limitation of classical approach was the inability to explain photo-electric effect. It was observed experimentally that kinetic energy of photoelectrons depended on the frequency of incident radiation and not on its intensity. Einstein applied quantum concept to provide a theoretical explanation of this phenomenon and later extended the idea of quantized oscillators to the theory of specific heat of solids.

Planck's quantum ideas have also been applied to Bohr Sommerfield theory of atom, Compton scattering of X-rays and Einstein's studies on atoms in thermal equilibrium with black body radiation, leading to the postulates on the existence of both spontaneous and stimulated transitions. The idea of radiation as quanta, nurtured by the work of Planck, Einstein and Compton, was however in sharp conflict with wave representation of interference and diffraction. The quality of wave and particle nature in matter was brought out by de Broglie with the suggestion that any moving particle with momentum p has a wavelength associated with it, such that

$$p = h/\lambda, \tag{1.8}$$

which is consistent with Compton's value of $p=h\nu/c$ for light quantum.

The quantum concept assumed special significance for atomic theory. The Bohr Sommerfield condition for circular atomic electron orbit of radius r and with momentum p gives

$$(2\pi r)p = nh, \tag{1.9}$$

which on applying Eq. 1.8 shows that the circumference of circular orbit is an integral number of de Broglie wavelengths. This suggests wave-like properties of atomic particles, and that electrons in stationary orbits are represented by standing waves.

With wave particle duality, the presence of a particle was interpreted by Born in terms of probability waves. The amplitude of the wave at a point gives a measure of the probability of the particle being there. This quantum approach used a probabilistic view.

The insufficiency of wave model or the particle model alone is now accepted. The wave particle duality, however, leads to a definite uncertainty in the simultaneous measurement of parameters. Heisenberg's uncertainty principle provides quantitative estimate of the uncertainty in the position Δx of a wave like object of uncertain momentum Δp, such that

$$\Delta p \, \Delta x \geq \hbar/2. \tag{1.10}$$

Similar relation exists for energy and time, i.e.

$$\Delta E \, \Delta t \geq \hbar/2 \tag{1.11}$$

This is related to the basic premise of quantum mechanics that measuring process may affect the physical system and that it is impossible to measure simultaneously certain pairs of variable with arbitrary precision. The quantum approach takes a probabilistic view.

In the quantum terminology for radiation field in a cavity, quanta of energy $h\nu$ are the photons of the system, and as such are not localized particles but characteristic excitations of the cavity modes spread over the mode volume. Unlike old quantum theory, quantum mechanics introduces non localized corpuscular photons which are mode excitations of the system. In the quantum explanation, photon is an occupation of one of the normal modes of the system and detector measures the occupation of normal modes. Photon division is precluded by quantum theory, but is allowed by semiclassical radiation theory in which the photon is viewed as a packet of electromagnetic energy.

According to Dirac, the radiation field is dynamically equivalent to an infinite set of uncoupled harmonic oscillators, one for each mode of the electromagnetic field. This enables systematic quantization of the free field as an independent dynamical system by representation of each field mode as a harmonic oscillator. The single excitation of a mode, or state, of the radiation field represents the presence of a photon in that mode.

It is to be noted that a great many interactions of light and matter may be solved without invoking the concept of photon or field quantization. The semiclassical theory, where radiation is represented by a classical wave and the atom is quantized, is fully sufficient for a great many purposes. In two main areas, however, the photon behaviour of light needs to be considered explicitly, viz. (i) when spontaneous emission is involved and (ii) when understanding of intensity correlations in or between light beams is needed or when the behaviour of an atom is dictated by the intensity correlations of a source of radiation.

Photoelectric effect and experiments on electron, proton and neutron diffraction have shown the elementary particles to exhibit both particle and wave properties. The phenomenon of electromagnetic radiation is no exception to this rule. Most conspicuous property of electromagnetic radiation, particularly at low frequencies, is its wave behaviour. However, it also possesses particle properties which become more and more pronounced at higher frequencies. The wave particle dualism of all elementary physical phenomenon is hard to visualize, though its mathematical description is unambiguously provided by the quantum theory. The particle associated with electromagnetic field is said to have zero rest mass so that photons always move with the speed of light. A photon can never be observed at rest like other elementary particles.

1.1.1. Quantum Description of Radiation

Considerations of theoretical consistency and experimental evidence demand quantization of electromagnetic field [1.1]. Concepts of photon and spontaneous emission of radiation need this quantization. The first instance of the need of photon as quantum of radiation was in the explanation of black body radiation (Planck's relation Eq. 1.5). The word photon was coined two decades later by Lewis in 1926 for light quantum. For many applications, classical description of electromagnetic waves is satisfactory even up to optical range. However, at optical frequencies, there are effects which need quantum description. Some of these effects are photoelectric effect, photon counting statistics, spontaneous emission and noise in lasers and waves in crystal lattices (phonons).

According to quantum theory, photons are described as quantum states of the radiation modes of the field. Since the choice of radiation modes is quite arbitrary, the assignment of photons to plane waves, spherical waves or cylindrical waves etc. is equally arbitrary.

Mechanical lattice vibrations of a single crystal have much in common with electromagnetic radiation field. Quantum theory leads to the quantum states of the normal modes of these vibrations which are termed phonons. These phonons have very similar properties to the photons of the electromagnetic radiation field, but with the difference that the electromagnetic field propagates in a continuum while the lattice vibrations are confined to the discrete points of the atoms of crystal lattice.

Quantized description of the electromagnetic field results in a representation that decomposes the field into infinitely many uncoupled harmonic oscillators. Each radiation oscillation of the field is represented by one mode of that field. The decomposition into modes is completely arbitrary (as long as they form a complete orthogonal set of functions). For reasons of convenience and simplicity, the plane travelling wave representation is often adopted, though by no means the only possible way.

Consider the Maxwell's equations in free space

$$\nabla \times \mathbf{H} = \epsilon \, \partial \mathbf{E}/\partial t$$
$$\nabla \times \mathbf{E} = -\mu \, \partial \mathbf{H}/\partial t$$
$$\nabla \cdot \mathbf{E} = 0$$
$$\nabla \cdot \mathbf{H} = 0 \qquad (1.12)$$

If we introduce a vector potential $\mathbf{A}$ such that

$$\mathbf{E} = -\partial \mathbf{A}/\partial t,$$
$$\mu \mathbf{H} = \nabla \times \mathbf{A}$$

and

$$\nabla \cdot \mathbf{A} = 0,$$

we can obtain the wave equation

$$\nabla^2 \mathbf{A} = \epsilon\mu \, \partial^2 \mathbf{A}/\partial t^2 \qquad (1.13)$$

Eqs. 1.12 or 1.13 can be regarded as a quantum theory describing individual photons. A second quantization however makes the particle properties apparent and is capable of describing many photons simultaneously. The process of second quantization changes a single particle wave theory into a theory describing many noninteracting particles of the same kind exhibiting their particle properties.

To see this, we put

$$\mathbf{A} = (1/\epsilon^{1/2}) \sum_n q_n(t) \, \mathbf{U}_n(x,y,z), \qquad (1.14)$$

which, on applying Eq. 1.13, gives

$$\nabla^2 \mathbf{U}_n + \epsilon\mu\omega_n^2 \, \mathbf{U}_n = 0 \qquad (1.15)$$

$$\partial^2 q_n(t)/\partial t^2 + \omega_n^2 \, q_n = 0. \qquad (1.16)$$

This involves introduction of a set of orthogonal functions which corresponds to the expression of electromagnetic field in terms of modes. For a cavity, the choice of $\mathbf{U}_n$ functions denotes the modes of the cavity. For interaction between atoms and the radiation field in free space, we have a great deal of freedom in the choice of $\mathbf{U}_n$ functions. A simple choice is the plane standing waves of free space. The boundary conditions on $\mathbf{U}_n$ restrict the possible values of the radian frequency ω_n. While Eq. 1.14 decomposes the electromagnetic field into modes, Eq. 1.16 has the form of an infinite set of uncoupled harmonic oscillators so that each mode behaves mathematically like a harmonic oscillator.

A single mode of frequency ω, which satisfies Maxwell's equations and the periodic cavity boundary condition, can be written as [1.2- 1.4]

$$E_x(z,t) = (2\omega^2 m/V\epsilon_o)^{1/2} q(t) \, \sin Kz$$

$$H_y(z,t) = (\epsilon_o/K) \, (2\omega^2 m/V\epsilon_o)^{1/2} \, \dot{q}(t) \, \cos Kz, \qquad (1.17)$$

where V is cavity volume, $K=(\omega/c)z$, $q(t)$ is a time dependent envelope with dimension of length and m is a constant with dimension of mass.

The field Hamiltonian or energy of the single mode field is

$$\mathbb{H} = (1/2) \int [\epsilon_0 E_x^2(z,t) + \mu_0 H_y^2(z,t)] dV, \qquad (1.18)$$

which on using Eq. 1.17 for E_x and H_y become

$$\mathbb{H} = (1/2) m\omega^2 q^2 + p^2/2m \qquad (1.19)$$

with $p = m\dot{q}$.

Eq. 1.19 identifies $\mathbb{H}$ as a harmonic oscillator of mass m, frequency ω and position coordinate $q(t)$. A single mode of the radiation field thus behaves like a simple harmonic oscillator.

In the terminology of quantum representation, the canonical position and momentum variables p and q are replaced by the non commuting Hermitian operators p and q which obey

$$[\hat{q},\hat{p}] = \hat{q}\hat{p} - \hat{p}\hat{q} = j\hbar \qquad (1.20)$$

and the oscillator is described by the Hamiltonian

$$\hat{\mathbb{H}} = (1/2) m\omega^2 \hat{q}^2 + \hat{p}^2/2m. \qquad (1.21)$$

The electric and magnetic field operators are now given by

$$\hat{E}_x(z,t) = (2\omega^2 m/V\epsilon_0)^{1/2} \hat{q}(t) \sin Kz$$

$$\hat{H}_y(z,t) = (\epsilon_0/K)(2m\omega^2/V\epsilon_0)^{1/2} (\hat{p}/m) \cos Kz \qquad (1.22)$$

Comparing Eqs. 1.17 and 1.22 one finds that formally very little is altered on quantization. Quantum representation essentially involves noncommutability of the operators and the laws of probability determine the mode amplitude.

Expressing $p_1(t)$, $q_1(t)$ for the ıth mode in terms of the creation and anihilation operators $a_1^+(t)$ and $a_1(t)$ [1.2], one can express the Hamiltonian $\mathbb{H}$ for the radiation field as

$$\mathbb{H} = \sum_1 \hbar\omega_1 (a_1^+ a_1 + 1/2), \qquad (1.23)$$

where the expectation value of the operator $a_1^+ a_1$ equals the number of quanta n_1 in the ıth mode of the resonator. The electromagnetic field inside a resonator is considered as an ensemble of independent harmonic oscillators.

In the quantum representation, the average energy per mode becomes

$$\overline{E} = (h\nu/2) + h\nu/[\exp(h\nu/kT)-1] \qquad (1.24)$$

and the black body (thermal radiation) energy density per unit frequency width as

$$\rho(\nu) = p(\nu)\overline{E} = (8\pi h\nu^3/c^3)[(1/2) + 1/\{\exp(h\nu/kT)-1\}], \qquad (1.25)$$

where $p(\nu)$ is the mode density per unit frequency given by

$$p(\nu) = 8\pi h\nu^3/c^3, \qquad (1.26)$$

c being the velocity of light in the resonator medium.

In Eq. 1.24 the factor $h\nu/2$ represents the zero point energy. Quantum theory provides the concept of zero point energy $H_i = (1/2)\hbar\omega_i$ for the ith mode as a result of the zero point fluctuations of the variables p_i and q_i. Since these variables cannot simultaneously become zero, the energy of harmonic oscillators cannot become zero. This zero point energy is not available for energy exchange.

Analogous to Eq. 1.25 the vibrational energy density for lattice vibrations per unit volume per unit frequency at temperature T is

$$\rho(\nu) = (12\pi\nu^2/v_a^3) [h\nu/2 + h\nu/\{\exp(h\nu/kT)-1\}], \qquad (1.27)$$

where v_a denotes the acoustic velocity.

Regarding the need for quantum description, one may note that if the difference between quantum states is very small (i.e. difference in energy states of a system is negligibly small) there is very little difference between the classical treatment and the quantum description. Quantization of an LC circuit is an example where either of the two descriptions (classical or quantum) is equally valid. However, in the visible region of optical spectrum, the frequency is high enough to make quantum effects observable.

1.1.2. Fluctuation Properties

The fluctuation properties of black body radiation are shown to be [1.4]

$$<[\Delta E(\nu)]^2> = h\nu\, \rho(\nu)\, [1 + 1/\{\exp(h\nu/kT)-1\}]V, \qquad (1.28)$$

where $\rho(\nu)$ is energy density at frequency ν and V is the volume containing the radiation modes.

In the high frequency Wien limit, $h\nu/kT \gg 1$

$$\langle[\Delta E(\nu)]^2\rangle \simeq h\nu\, \rho(\nu)V = \langle[\Delta E(\nu)]^2\rangle_p \qquad (1.29)$$

and black body radiation behaves apparently as a collection of independent particles.
In the low frequency Rayleigh Jeans limit, $h\nu/kT \ll 1$,

$$\langle[\Delta E(\nu)]^2\rangle = [\Delta E(\nu)]^2\rangle_w, \qquad (1.30)$$

the black body radiation behaves as a superposition of classical waves. The classical limit $h \to 0$, gives only the wave fluctuations. The simultaneous presence of wave and particle properties is a quantum effect and requires a non-vanishing Planck's constant.

1.1.3. Power Flow of Electromagnetic Radiation

The Maxwell's equations in material media in MKS units are expressed as

$$\nabla \times \mathbf{H} = \mathbf{j} + \partial \mathbf{D}/\partial t, \quad \nabla \cdot \mathbf{D} = \rho$$

$$\nabla \times \mathbf{E} = -\partial \mathbf{B}/\partial t, \quad \nabla \cdot \mathbf{j} = -\partial \rho/\partial t$$

$$\mathbf{D} = \varepsilon_0 \mathbf{E} + \mathbf{P} \quad \nabla \cdot \mathbf{B} = 0$$

$$\mathbf{B} = \mu_0 (\mathbf{H} + \mathbf{M}), \qquad (1.31)$$

where $\mathbf{j}$ is current density (amp/m^2), $\mathbf{P}$ and $\mathbf{M}$ are electric and magnetic polarizations (dipole moment per unit volume), ε_0, μ_0 are electric and magnetic permeabilities of vacuum. $\mathbf{E}$ and $\mathbf{H}$ are field vectors, and $\mathbf{D}$ and $\mathbf{B}$ are displacement vectors.
Using Eq. 1.31 and noting that

$$(1/2)\, \partial/\partial t\, (\mathbf{E} \cdot \mathbf{E}) = \mathbf{E} \cdot \partial \mathbf{E}/\partial t$$

$$\nabla \cdot (\mathbf{A} \times \mathbf{B}) = \mathbf{B} \cdot (\nabla \times \mathbf{A}) - \mathbf{A} \cdot (\nabla \times \mathbf{B})$$

and applying Gauss theorem

$$\int_V (\nabla \cdot \mathbf{A})\, dV = \int_S (\mathbf{A} \cdot d\mathbf{n})\, dS,$$

where $\mathbf{A}$ is any vector function, $\mathbf{n}$ is unit vector normal to surface S enclosing volume V, dV and dS are differential volume and surface elements, it can be shown that

$$-\int_V \nabla \cdot (E \times H) dV = -\int_S (E \times H) \cdot n \, dS$$

$$= \int_V [(E \cdot j) + \partial/\partial t (\varepsilon_o/2) E \cdot E + \partial/\partial t (\mu_o/2) H \cdot H$$

$$+ E \cdot \partial P/\partial t + \mu_o H \cdot \partial H/\partial t] dV. \quad (1.32)$$

The expression on left hand side of Eq. 1.32 represents total power flowing into the volume bounded by surface S. On RHS, the product $E \cdot j$ represents power expended by field on moving charges, whereas the expression $[\partial/\partial t (\varepsilon_o/2) E \cdot E + \partial/\partial t (\mu_o/2) H \cdot H]$ gives the rate of increase of vacuum electromagnetic energy. The terms $E \cdot \partial P/\partial t$ on RHS represent power per unit volume expended on the electric dipoles while the last term corresponds to magnetic dipole moment. The power corresponding to the last two terms goes into an increase in the potential energy stored by dipoles as well as to supply the dissipation that may accompany the change in **P** (and **M**).

1.2. Interaction of Radiation with Matter - Emission and Absorption of Radiation

The interaction of electromagnetic field with atomic system is described either semiclassically or by quantum mechanical perturbation. The effects of this interaction are absorption and emission of electromagnetic radiation by atoms. The interacting radiation excites transitions between atomic states (Fig. 1.1). The field may be quantized and the quantized field perturbations may induce transitions between states of the combined atom plus field system. Further, the atomic energy levels may be perfectly sharp or have finite energy width and the transition to an excited state may be to a single level among a large number of accessible final states.

It is seen from quantum theory that emission from excited state persists even when the stimulating field is absent. This (spontaneous) emission is a purely quantum effect and if this were absent or unimportant, there would be no difference in the semiclassical and fully quantized treatment of radiative interactions. Differences between the two approaches occur whenever the operator nature of the field is important and radiative corrections due to vacuum fluctuations and spontaneous decay need to be taken into account. Otherwise there is no need to invoke the photon

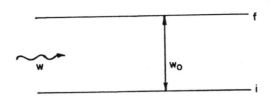

Fig. 1.1. Two level atom excited by radiation field. The radiation field can either be monochromatic of frequency ω or a broadband. ω_0 is the atomic transition frequency. At resonance, $\omega=\omega_0$ and probability of transition increases.

concept and semiclassical theory suffices for induced absorption and emission of radiation.

1.2.1. Transition Probability

A typical problem relates to finding the probability that a system known to be in state $|\psi_i(0)\rangle$ at $t=0$ be found in an eigen state $|E_f\rangle$ at time t. This probability is given by

$$P_f = |\langle E_f|\psi_i(t)\rangle|^2. \tag{1.33}$$

We now introduce an operator S, called the scattering operator (its matrix elements form the scattering matrix) and write

$$|\psi_i(t)\rangle = S|\psi_i(0)\rangle. \tag{1.34}$$

The scattering operator transforms the state vector at t=0 to the appropriate state vector at time t. The interaction starts at t=0 at which time the system is described by the state vector $|\psi_i(0)\rangle$. The influence of the interaction changes $|\Psi_i(0)\rangle$ to $|\Psi_i(t)\rangle$ which is shown by Eq.(1.34). We can thus write

$$\langle E_f|\Psi_i(t)\rangle = \langle E_f|S|\Psi_i(0)\rangle \tag{1.35}$$

$$|\Psi_i(0)\rangle = |E_i\rangle$$

so that the elements of the scattering matrix

$$S_{fi} = \langle E_f | S | E_i \rangle \qquad (1.36)$$

are the probability amplitudes for transition from an initial state $|E_i\rangle$ to a final state $|E_f\rangle$.

Considering interaction Hamiltonian $\mathbb{H}_{int}$ as time independent and $|E_f\rangle$ and $|E_i\rangle$ as energy eigen vectors of the unperturbed Hamiltonian $\mathbb{H}_0$, the matrix element S_{fi} can be written to first approximation as [1.1]

$$S_{fi} = (1/j\hbar)\langle E_f | \mathbb{H}_{int} | E_i \rangle \int_0^t \exp[j(E_f-E_i)\tau/\hbar]d\tau. \qquad (1.37)$$

Eq. 1.37, on performing integration and rewriting, becomes

$$S_{fi} = -2j \ \exp[j(E_f-E_i)t/2\hbar]\langle E_f | \mathbb{H}_{int} | E_i \rangle$$
$$\cdot [\sin(E_f-E_i)(t/2\hbar)/(E_f-E_i)] \qquad (1.38)$$

and gives

$$P_f = |S_{fi}|^2$$
$$= 4\pi|\langle E_f | \mathbb{H}_{int} | E_i \rangle|^2 [\sin(E_f-E_i)(t/2\hbar)/(E_f-E_i)]\delta(E_f-E_i) \qquad (1.39)$$

provided t is large. The δ function is Dirac δ defined by

$$\delta(x) = \lim_{K\to\infty} (1/\pi) \cdot \sin Kx/x$$

and has the property that all functions multiplied with it can be taken constant at the value $E_f = E_i$ so that we can write

$$P_f = (2\pi t/\hbar)|\langle E_f | \mathbb{H}_{int} | E_i \rangle|^2 \delta(E_f-E_i). \qquad (1.40)$$

Eq. 1.40 shows that only transitions with $E_f = E_i$ are physically possible if enough time t is available for the transition to take place.

In most practical cases we are not interested in the transition from the state $|E_i\rangle$ to one exact state $|E_f\rangle$ but rather in the transition to all states infinitesimally close to $|E_f\rangle$. The number of states available in the immediate vicinity of any given state $|E_f\rangle$ is usually very large so it is impossible to distinguish which of the many possible closely spaced states is being assumed for the system. The probability of finding the system in any of a number of possible states is obtained by summing over the probabilities P_f for each of these states. For the density of states to be sufficiently high,

$$\sum_{E_f} P_f = \int P_f \, \rho_f \, dE_f, \qquad (1.41)$$

where ρ_f is the density of states function. The transition probability per unit time, W, is thus

$$W = (1/t) \sum_{E_f} P_f$$

$$= (2\pi/\hbar) |\langle E_f|H_{int}|E_i\rangle|^2 \, \rho_f. \qquad (1.42)$$

Another situation of interest is when the incident radiation is a strong monochromatic field, near resonant with a particular excited state and would cause large population changes in that state but not in others. For this analysis, the perturbation theory is not adequate, because the perturbation approach assumes that the initial state population is hardly changed by the excitation and that the probability amplitude for being in any other state remains small. The resonant strong field situation, therefore, needs the solution to be more exact and explicit. The probability of excitation in this case can be unity and thereby lead to complete population inversion.

It is to be noted that atomic energy levels are never perfectly sharp, because excited states decay by spontaneous emission and collisions de-excite levels. The uncertainty principle (Eq.1.11) associates a finite energy width with finite lifetime of the state. A useful way to model this situation is to assume that the excited state probability for any state j decays at a rate $r_j = 1/\tau_j$ where τ_j is the life time of the state. The steady state excitation probability P_f for this case, for an excitation turned on at t=0, is seen to be (Fig.1.2)

$$P_f \propto 1/[(\omega-\omega_0)^2 + r_f^2/4]. \qquad (1.43)$$
$$t \gg \tau_f$$

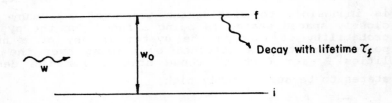

Fig. 1.2. Excitation of a decaying state by a monochromatic field.

Here energy of excited state E_f is considered as equal to

$\hbar(\omega_o - jr_f/2)$.

1.2.2. Emission and Absorption of Radiation by Bound Electrons

Interaction of bound electrons with radiation field is important for the understanding of lasers. Most lasers use bound electrons as their gain medium for amplifying electromagnetic radiation. In the semiconductor laser, however, electron exists initially in a free state in the conduction band from where it decays to a bound state in the valence band (or an impurity state), radiating the energy difference in the form of a photon. In most applications the interaction is between the electromagnetic field and one or more electrons. This is analyzed by first considering the two parts of the system (electron and field) as independent and then couple them together by an interaction Hamiltonian. This ensures that the interaction between the two systems works both ways. The field acts on the electron changing its behaviour and conversely the electron acts on the field changing its value.

An electron is considered bound if the probability of finding it at infinity is zero. For a bound electron, coupled to the radiation field with vector potential **A**, the interaction Hamiltonian is given by

$$\mathbb{H}_{int} = (-e/m) \; \mathbf{p}\cdot\mathbf{A} + (e^2/2m) \; \mathbf{A}^2,$$

which is approximated for most cases to

$$\mathbb{H}_{int} \cong - (e/m) \; \mathbf{p}\cdot\mathbf{A}, \qquad (1.44)$$

since e^2 term is very small. An alternative way of looking at Eq. 1.44 is by noting

$\mathbf{A} \cong (1/\omega)\, \mathbf{E}, \quad \mathbf{p} \cong m\mathbf{v} = m\, d\mathbf{r}/dt \cong m\,\omega\,\mathbf{r}$,

so that

$$\mathbb{H}_{int} \cong -e\mathbf{E}\cdot\mathbf{r}. \tag{1.45}$$

The radiation field causes the electron to undergo transitions from one eigen state to another. An excited electron in quantum electrodynamics decays to lower states and eventually to the ground state even if the radiated field is empty i.e. it contains no photons at all. The fact that electron is coupled to the radiation field, even if it is empty, prevents it from remaining in an excited state. The quantized electron decays to its ground state and persists there without any further radiation even though it has not come to rest at the location of the nucleus.

For a bound electron, the transition probability per unit time for a transition of the electron from a state $|\psi_i\rangle$ to a state $|\psi_f\rangle$ with a simultaneous change of a photon state from $|n_l\rangle$ to $|n_l \pm 1\rangle$ is given by

$$W = (2\pi/\hbar)\,|\langle\phi_f|\,\mathbb{H}_{int}\,|\phi_i\rangle|^2 \rho_f, \tag{1.46}$$

with $\mathbb{H}_{int}$ given by Eq. 1.44 and the initial and final states described by

$$|\phi_i\rangle = |\psi_i\rangle|n_1,\, n_2\, \ldots n_l \ldots\rangle$$

$$|\phi_f\rangle = |\psi_f\rangle|n_1,\, n_2\, \ldots n_l \pm 1 \ldots\rangle. \tag{1.47}$$

The initial and final states of the system are products of an electron state and the photon state described by

$$|n_1\rangle|n_2\rangle|n_3\rangle\,\ldots = |n_1,\, n_2\ldots n_l\ldots\rangle \tag{1.48}$$

with $E_l = \hbar\omega_l(n_l + 1/2)$,

where the index l indicates the eigen vector $|n_l\rangle$ as well as the eigen value E_l belonging to the lth radiation mode (Fig. 1.3). The photon state describes the photon occupation number of all the radiation modes. The energy of a system, where eigen vector can be described as a product of the individual eigen states of the noninteracting components of

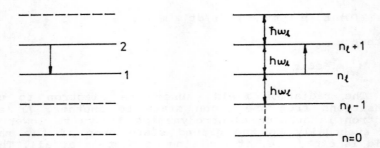

Fig. 1.3. Atomic levels and radiation modes involved in emission process. The initial state of the atom is in level 2 and mode ı has n_1 quanta. The final state of the atom is in the lower state 1 and the field has gained a photon.

the system, is a sum of the energies of its components. The initial and final energy of the system (E_i and E_f, respectively) are thus

$$E_i = W_i + \hbar\omega_1(n_1+1/2) + \hbar\omega_2(n_2+1/2) + .. + \hbar\omega_1(n_1+1/2) + ..$$

$$E_f = W_f + \hbar\omega_1(n_1+1/2) + \hbar\omega_2(n_2+1/2) + .. + \hbar\omega_1(n_1\pm 1+1/2) + ..$$

(1.49)

where W_i and W_f are the initial and final energies of the bound electron. We use plane wave representation of vector potential

$$\mathbf{A} = \sum_{k=1}^{2} \sum_{1} (\hbar/2\epsilon_1 V)^{1/2} \mathbf{e}_{1k}[\bar{a}^+_{1k}\exp(-j\beta_1 \cdot \mathbf{r}) + \bar{a}_{1k}\exp(j\beta_1 \cdot \mathbf{r})],$$

(1.50)

where $\mathbf{e}_{1k}$ are unit vectors describing the direction of $\mathbf{A}$, k determines the two polarization of $\mathbf{A}$, V is volume and β_1 is propagation constant of ıth mode. Further, postulating that for any arbitrary vector $|\psi\rangle$

$$\langle q|p|\Psi\rangle = -j\hbar(\partial/\partial q)\langle q|\psi\rangle = -j\hbar(\partial/\partial q)\psi(q) \tag{1.51}$$

and creation and annihilation operators $\bar{a}^+$ and $\bar{a}$ obey

$$\bar{a}|E_n\rangle = n^{1/2}|E_{n-1}\rangle$$

$$\bar{a}^+|E_n\rangle = (n+1)^{1/2}|E_{n+1}\rangle, \quad (1.52)$$

then with the help of Eqs. 1.46 and 1.44 it can be shown that the emission probability W_e, or absorption probability W_a, is given by [1.1]

$$W_e = (\pi e^2 \hbar^2/\epsilon m^2 \omega_1 V)(n_1+1)\rho_f |\int \psi_f^*(e_{1k}\cdot\nabla\Psi_i)\exp(\pm j\beta_1\cdot\mathbf{r})dV|^2$$

$$W_a = (\pi e^2 \hbar^2/\epsilon m^2 \omega_1 V)n_1\rho_f |\int \psi_f^*(e_{1k}\cdot\nabla\Psi_i)\exp(\pm j\beta_1\cdot\mathbf{r})dV|. \quad (1.53)$$

First, we consider spontaneous emission and take $n_1=0$. Spontaneous emission is described by the 1 in (n_1+1) and the spontaneous emission probability W_{sp}, contributed by this term, exists regardless of whether $n_1=0$ or not. This situation changes if n_1 is large enough so that first order approximation is no longer valid. In higher orders of approximation, there is a coupling between spontaneous and stimulated emission probabilities.

To obtain the density of states function ρ_f, we allow the photon to be emitted into a range of directions described by the solid angle $d\Omega$ centered around the direction of β and also to have a spread of values of β between β and $\beta+d\beta$. The choice of admissible photon states means that the final electron states are also forced to lie within a range of values. We obtain the number of final states from the volume element in spherical polar coordinates in number space

$$\rho_f dE_f = n^2 dn\, d\Omega = [V/(2\pi)^3]\beta^2 d\beta\, d\Omega$$

Thus, for spontaneous emission into a certain element of solid angle $d\Omega$ centered around a certain direction β_1 in space, the density of states function ρ_f is given by

$$\rho_f dE_f = [2V/(2\pi)^3]\beta_1^2\, d\beta_1\, d\Omega.$$

The factor of 2 is added to account for both directions of

polarization. Since electronic level W_f is considered well defined, $dE_f = \hbar\omega_1$ and with $\beta_1 = (\epsilon/\epsilon_0)^{1/2}\omega_1/c$, we obtain

$$\rho_f = [2V/(2\pi^3)] \, (\epsilon/\epsilon_0)^{3/2} (\omega_1^2/\hbar c^3) \, d\Omega. \qquad (1.54)$$

If we limit ourselves to dipole radiation by considering the zero order approximation, the term $\exp(\pm j\beta_1 \cdot r)$ is omitted from the integral in Eq. 1.53. Then, using Eq. 1.54, we obtain from Eq. 1.53 with $n_1 = 0$, the probability for spontaneous emission of one photon per unit time into the element of solid angle $d\Omega$ centred around the direction β_1.

$$W_{sp} = [e^2 \hbar (\epsilon/\epsilon_0)^{1/2} \omega_1 / 4\pi^2 m^2 c^3 \epsilon_0] |\int \psi_f^* e_{1k} \cdot \nabla \psi_i dV|^2 d\Omega. \qquad (1.55a)$$

If we use $\mathbb{H}_{int}$ written in the form following Eq. 1.45,

$$\mathbb{H}_{int} = e\omega_1 A \cdot r,$$

we would obtain

$$W_{sp} = [(\epsilon/\epsilon_0)^{1/2} \omega_1^3 / 4\pi^2 \hbar c^3 \epsilon_0] |\mu|^2 d\Omega \qquad (1.55b)$$

with dipole moment matrix element

$$\mu = e\int \psi_f^* (e_{1k} \cdot r) \psi_i dV.$$

Eqs. 1.55a and 1.55b are identical as far as dipole radiation is concerned.

In many atoms and molecules, the initial and final levels of the atom are degenerate. We must, therefore, consider the possibility that there are several states of the atom for each energy level. If final state with energy W_f is considered g_1 fold degenerate, the transition probability for transition from state i to any of the final state f, is obtained by adding the transition probabilities of the g_1 possible transitions. Thus

$$\overline{W}_{sp} = [(\epsilon/\epsilon_0)^{1/2} \omega_1^3 g_1 / 4\pi^2 \hbar c^3 \epsilon_0] \int |\mu|^2 d\Omega, \qquad (1.56)$$

where

$$\mu = e\int \psi_f^* (e_{1k} \cdot r) \psi_i dV$$

is the dipole moment matrix element.

If $|\mu|^2$ is not the same for all possible g_1 transitions, its value in Eq. 1.56 must be considered as a suitable average.
The frequency of the emitted photon is given by

$$W_i - W_f = \hbar\omega_1. \tag{1.57}$$

This is seen by applying condition of conservation of energy $E_i = E_f$ to the Eq. 1.49

Absorption of radiation by an atom can be obtained in an analogous manner, except that now it is the initial state that is arbitrary to a certain extent while the final state is definitely determined. The atom is brought in an excited state by absorption of a photon. The transition probability per unit time is again given by Eq. 1.46 with the density of the final state ρ_f replaced by the density of initial state ρ_i. However, ρ_i is of the same form as ρ_f of Eq. 1.54. The $d\Omega$ is the element of solid angle into which the radiation modes, which lose a photon by absorption, spread. The absorption probability per unit time is then

$$W_a = [(\epsilon/\epsilon_0)^{1/2}/4\pi^2\hbar c^3 \epsilon_0]\omega_1^3 g_2 \bar{n}_1 |\mu|^2 d\Omega, \tag{1.58}$$

where g_2 is the degeneracy of the state with energy W_1. The photon number $\bar{n}_1$ is the average number of photons in each radiation mode. To estimate $\bar{n}_1$, it is noted that $\bar{n}_1$ is related to power density ds flowing into the solid angle $d\Omega$ in the unit radian frequency range such that

$$ds \quad dw = \hbar\omega_1[\bar{n}_1/V](\epsilon_0/\epsilon)^{1/2} \quad c\rho_i \hbar d\omega$$

which gives, by putting $\rho_f = \rho_i$ in Eq. 1.54,

$$ds = [2\hbar\omega_1^3 \bar{n}_1 (\epsilon/\epsilon_0)/(2\pi)^3 c^2] d\Omega. \tag{1.59}$$

Eq. 1.59 thus enables the estimation of $\bar{n}_1$ in terms of $ds/d\Omega$.

The power of spontaneous radiation that the bound electron radiates into the solid angle $d\Omega$ is given by

$$dP_{sp} = \hbar \omega_1 \overline{W}_{sp}$$
$$= [(\epsilon/\epsilon_0)^{1/2}\omega_1^4/4\pi^2c^3\epsilon_0]g_1|\mu|^2 \, d\Omega \qquad (1.60)$$

so that total power radiated in all directions is

$$P_{sp} = [(\epsilon/\epsilon_0)^{1/2}\omega_1^4/4\pi^2c^3\epsilon_0] \, g_1 \int |\mu|^2 \, d\Omega. \qquad (1.61)$$

An atom that has been excited to a state, with an energy larger than the ground state energy, returns spontaneously to the ground state emitting photons. The radiative lifetime of an atom is related to the total spontaneous emission probability. Since $\overline{W}_{sp}$ is the probability per second for spontaneous emission of a photon into any of the possible radiation modes, the average lifetime of the atom due to this transition is given by

$$1/\tau_{sp} = \overline{W}_{sp}. \qquad (1.62)$$

If the atom can decay to more than one energy level, a sum over the spontaneous emission probabilities to all these levels must be taken.

In Eq. 1.56, if $|\mu|^2$ is not the same for all possible g_1 transitions, its value must be considered as a suitable average. This average of the square of dipole moment matrix element can no longer depend on the direction of solid angle and can be taken out of the integral. The integral over all elements of solid angle results in 4π. Thus, using Eqs. 1.56 and 1.62 we get,

$$1/\tau_{sp} = [(\epsilon/\epsilon_0)^{1/2} \omega_1^3 g_1/\pi\hbar c^3\epsilon_0]|\mu|^2$$
$$= 2n^3 \omega_1^3 g_1/hc^3\epsilon|\mu|^2, \qquad (1.62a)$$

where the refractive index $n=(\epsilon_r)^{1/2}=(\epsilon/\epsilon_0)^{1/2}$.

1.3. Spontaneous and Stimulated Radiation

It has been stated in section 1.2 that a bound electron can emit as well as absorb radiation at a frequency given by Eq. 1.57. The emission has two parts (i) spontaneously emitted power appearing in all radiation modes which may or may not be already occupied by photons and (ii) stimulated emission power which adds onto the energy of the already excited mode (i.e. to the mode already occupied by photon).

The increase of power by stimulated emission is a coherent process since it increases the existing excitation of the radiation mode at a rate proportional to the energy already stored in that mode. Spontaneous emission, on the other hand, is not related to the existing excitation of a mode and occurs at a rate independent of the energy stored in that mode. Spontaneous emission is incoherent and appears as noise in optical amplifier or laser. Atoms radiating spontaneously do so in much the same way as classical radiating dipoles. Stimulated emission goes into radiation modes already occupied by photons. If only one radiation mode is occupied by photons with all others being empty, then only this mode receives power by stimulated emission. This is of course in addition to the power it gains by spontaneous emission.

One may ask why in the presence of stimulating radiation there is stimulated emission and not absorption. The quantum mechanical reason for this is the existence of a lower energy level, distant $\hbar\omega$ below the excited level occupied by the electron and the absence of a corresponding higher level to receive the energy.

A semiclassical first order perturbation theory can be applied to the absorption and emission resonance in a two level atom with weak field approximation [1.5]. This approach assumes linear response of the atom to the field where the atom is quantized but not the field. Further, the ground state population is not significantly disturbed by the optical perturbation. A phenomenological description of the decay state is adopted whereby the probability of finding the atom in state $|f>$ at time t is $\exp(-r_f t)$ where r_f is decay constant of the state.

Considering the interaction Hamiltonian, $\mathbb{H}_{int}$, given by

$$\mathbb{H}_{int} = -\mathbf{E}_i(\mathbf{r'},t) \cdot \hat{\mathbf{D}}, \qquad (1.63)$$

where $\hat{\mathbf{D}}$ is electric dipole moment operator for atom situated at position $\mathbf{r'}$

$$\mathbf{E}_i(\mathbf{r},t) = \mathbf{e}\, E_o \cos(\omega t - \mathbf{K}\cdot\mathbf{r}). \qquad (1.64)$$

In Eq. 1.64, $\mathbf{e}$ is the unit polarization vector, $\mathbf{K}$ the propagation vector, so that $\mathbf{e}$, $\mathbf{K}$ define direction of incident field. The steady state of atom at time t is given by

$$|t> = |i> \pm \frac{E_o <f|\mathbf{e}\cdot\hat{\mathbf{D}}|i>}{2\hbar} |f> \frac{\exp[\mp j(\omega t - \mathbf{K}\cdot\mathbf{r'})]}{\omega_f - \omega \mp j r_f/2} \qquad (1.65)$$

The upper sign refers to absorption resonance from a ground state |i> and lower sign for an emission resonance from a pumped excited state |i> (Fig. 1.4).

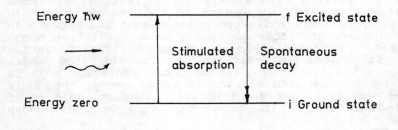

(a)

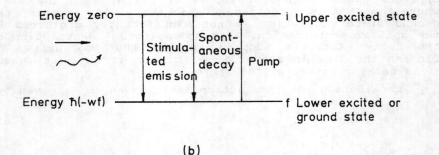

(b)

Fig. 1.4. (a) Two level resonant absorption. Incident field stimulates absorption transition from ground state |i> to excited state |f>. In steady state stimulated absorption is balanced by spontaneous decay. For weak field, stimulated decay is negligible compared to spontaneous decay.

(b) Two level resonant emission. Incident field stimulates emission transition from upper excited state |i> to a lower state |f>. In steady state population of state |i> is maintained against the combined effects of spontaneous and stimulated decay by a strong externally applied pumping.

The radiation field (induced field) is then given by

$$E(r,t) = \pm(\omega^2/4\pi\epsilon_0 c^2 R) E_0 [e - (n.e)n]\cos(\omega t - K.r' - \alpha - KR), \quad (1.66)$$

where
$$K = |K| = \omega/c, \quad (1.66a)$$

$$n = (r-r')/|(r-r')|, \quad (1.66b)$$

$$R = |r - r'|, \quad (1.66c)$$

ω = angular frequency of dipole,

$$X = \frac{|<f|e.\hat{D}|i>|^2/\hbar}{[(\omega_f - \omega)^2 + r_f^2/4]^{1/2}}, \quad (1.66d)$$

$$\tan\alpha = (r_f/2)/(\omega_f - \omega). \quad (1.66e)$$

The induced field propagates outwards from the atom with an angular distribution given by dipole factor in square brackets in Eq. 1.66. For forward direction i.e. $(r-r')$ in the direction of K, we can put $n.e=0$, $K.r'+KR = K.r$ and we get the field in forward direction

$$E_F(r,t) = \pm(\omega^2/4\pi\epsilon_0 c^2 R) E_0 e.\cos(\omega t - K.r - \alpha). \quad (1.67)$$

At peak of emission resonance ($\omega = \omega_f$) Eq. 1.66e gives $\alpha = \pi/2$ so that forward emission is not in phase with incident light but leads by $\pi/2$. However, this fixed phase relationship makes the forward emission coherent. For emission direction n outside the forward direction, the phase (Eq.1.66) is not independent of the random position r' of the emitting atom and gives an incoherent component to the coherent forward emission.

Considering an optically thin (thickness d) infinite plane slab of noninteracting atoms placed normal to the incident light and assuming N atoms per unit volume, the induced field at a point P is found by summing contributions from all atoms (each atom's contribution given by Eq. 1.67). Thus

$$E_p(r,t) = \pm e\, E_0 (N\pi d/\epsilon_0 \lambda).\cos(\omega t - K.r - \alpha - \pi/2). \quad (1.68)$$

Eq. 1.68 has the form of a plane wave propagating in the same mode as the incident wave. At peak of resonance $\alpha =$

$\pi/2$, the stimulated emission is thus in phase with the incident light and enables amplification. Contrary to this, in absorption case, a forward propagating wave of spontaneous emission at resonance is in antiphase and gives rise to attenuation.

In the assumption of two state model, ground state $|i\rangle$ is the only lower state accessible from $|f\rangle$ by spontaneous decay (Fig. 1.4a). If we remove this restriction and allow other states $|c\rangle$ etc. which may or may not be degenerate with $|i\rangle$, then emitted radiation field will contain additional terms representing the spontaneous emission from the upper state $|f\rangle$ to lower states $|c\rangle$ etc. These additional terms will contain phase factors $\exp(-j\delta_c)$ etc. Their origin is in the detailed mechanisms of the relaxation processes responsible for maintaining the steady state population and must therefore be regarded as unknown random variables. The phase of the additional spontaneous emissions are unrelated to the phase of the incident optical field, i.e. these spontaneous emissions are incoherent.

To summarize, we note that stimulated emission from a single atom is coherent with the incident stimulating beam. The spontaneous emission from an atom in resonant absorption with an incident field is also coherent with the incident field provided the atom returns to its original quantum state. Other spontaneous emissions are incoherent.

For electric dipole transition, the coherent emission from a single atom, stimulated or spontaneous, has a dipole like angular distribution. For macroscopic sample, this dipole angular distribution is drastically modified by interference effects. Thermal density fluctuations in the sample degrade the coherence of the emitted light and give rise to a lateral dipole like field which is incoherent with incident beam.

1.4. Einstein Coefficients

Einstein, while discussing absorption and emission of radiation [1.6], assumed that momentum is transferred to the molecules in the interaction between radiation and matter, and that this momentum transfer provides the molecules a velocity distribution which must be the same as the one the molecules attain through the action of their mutual collisions alone, that is it must be the same as the Maxwell distribution. The process of momentum transfer was considered as completely directional, namely, if a body emits the energy E it receives a recoil E/c if all the radiation is emitted in the same direction and no recoil if the emission is isotropic.

In the energy exchange process, if two theoretically possible quantum states exist with energies E_m and E_n with inequality $E_m > E_n$, and the molecule is able to make a transition from state n to m by absorbing radiative energy (E_m-E_n) and a transition from m to n with radiative emission of same energy, then the probability for the molecule to make the transition m to n without external stimulation and emitting the radiation energy (E_m-E_n) at frequency ν in time dt is

$$dw = A_m^n \, dt, \qquad (1.69)$$

where A_m^n denotes a constant, characteristic for the combination of indices m,n.

Similarly for induced radiation process, under the influence of a radiation density ρ of frequency ν, a molecule could make a transition from the state n to m by absorbing the radiative energy (E_m-E_n) and the probability for the process is

$$dw = B_n^m \, \rho \, dt. \qquad (1.70)$$

Also, a transition m to n may also be possible under the influence of the radiation in which radiative energy (E_m-E_n) is released according to the probability

$$dw = B_m^n \, \rho \, dt. \qquad (1.71)$$

In Eqs 1.70 and 1.71, B_n^m and B_m^n are constants.

Regarding the direction of momentum transfer, it was assumed that in each process momentum of magnitude $(E_m-E_n)/c$ is transferred to the molecule in the direction of propagation of the beam. If the molecule is simultaneously subjected to several radiation beams, the total energy (E_m-E_n) of an elementary process is absorbed or added to one of these beams and the momentum $(E_m-E_n)/c$ is transferred to the molecule.

The process of Eq. 1.69 is identified with spontaneous emission, while Eqs. 1.70 and 1.71 are identified with induced absorption and emission, respectively, in the presence of radiation of energy density ρ. If we use the Maxwell Boltzmann distribution for the occupancy of energy state n of a molecule in a gas at temperature T we can write

$$W_n = p_n \exp(-E_n/kT), \qquad (1.72)$$

$$p_n \exp(-E_n/kT) B_n^m \rho = p_m \exp(-E_m/kT) (B_m^n \rho + A_m^n), \quad (1.73)$$

where k is the Boltzmann constant and p_n is the statistical weight of the state, characteristic of the molecule for the nth quantum state and independent of T.

Eq. 1.73 expresses exchange of energy between radiation and molecules under the condition that, on the average, as many elementary processes represented by Eq. 1.70 take place as of Eqs. 1.69 and 1.71, per unit time. Eq. 1.73 can be rearranged as

$$\rho = (A_m^n/B_m^n) / [(p_n B_n^m / p_m B_m^n) \exp(E_m - E_n)/kT - 1]. \quad (1.74)$$

Comparing Eq. 1.74 with the Planck's law (Eq. 1.5), we find

$$A_m^n/B_m^n = 8\pi^2 h\nu^3/c^3 \quad (1.75)$$

$$p_n B_n^m = p_m B_m^n \quad (1.76)$$

and

$$E_m - E_n = h\nu. \quad (1.77)$$

The A and B coefficients defined by Eqs. 1.75 and 1.76 are known as Einstein coefficients and relate to spontaneous and stimulated processes. The spontaneous and induced transition rates are expressed in terms of A and B coefficients as per Eqs. 1.69 and 1.71.

1.5. Optical gain

To develop an expression for optical gain we assume that an active laser medium fills the volume V and a plane wave travels in single transverse mode in the medium. (The transverse and longitudinal modes are discussed in section 1.7). Each atom contributing to the gain mechanism is further assumed to have two energy levels W_1 and W_2. (In Fig. 1.1, for emission $W_2 = W_i$ and $W_1 = W_f$ while for absorption $W_2 = W_f$ and $W_1 = W_i$). We consider emission (and absorption) of radiation into well defined radiation mode whose initial photon number is n_{lk}.

Since the initial and final states (i and f in Fig.1.1) of the system of radiation mode and any given atom are very definitely determined, we consider transition probability P (Eq.1.40), instead of transition probability per unit time W

(Eq.1.42), so that

$$P_e = (2\pi t/\hbar)|<\phi_f|\mathbb{H}_{int}|\phi_i>|^2 \delta(E_f-E_i). \quad (1.40)$$

Using

$$\mathbb{H}_{int} = (-e/m)(\mathbf{p}\cdot\mathbf{A}) \quad (1.44)$$

and Eq. 1.55, we get

$$P_e = (\pi t\omega_1/\epsilon V)(n_{1k}+1)|\mu_{1k}|^2 g_1 \delta(E_f-E_i). \quad (1.78)$$

Here g_1 allows g_1-fold degeneracy of lower level. The dipole moment matrix element μ must be considered as an average.

It is unrealistic to assume that all the atoms have the same energy difference W_2-W_1 since different atoms are subjected to slightly different external forces resulting in slight energy shifts. The fractional number of atoms with energy level at W_2-W_1 in the interval $d(W_2-W_1)$ is given by

$$dN = N_2\rho(W_2-W_1)\,d(W_2-W_1) \quad (1.79)$$

with N_2 being the total number of atoms with initial energy W_2 and

$$\int_0^\infty \rho(W_2-W_1)\,d(W_2-W_1) = 1 \quad (1.80a)$$

$$d(W_2-W_1) = d(E_f-E_i). \quad (1.80b)$$

We use Eqs. 1.41, 1.42, and 1.57 and then integrate Eq. 1.78 to get

$$W_e = (\pi\omega_1/\epsilon V)(n_{1k}+1)\,N_2|\mu|^2 g_1^2 \rho(\hbar\omega_1). \quad (1.81)$$

Again, if μ is different for different atoms, its value in Eq. 1.81 has to be considered as a suitable average. In particular, we must replace $|\mu|^2$ by its average value over all possible orientations if the atoms are randomly oriented.

Absorption probability per unit time W_a is obtained by

replacing $(n_{lk}+1)$ by n_{lk} and N_2 by the number N_1 of atoms in the lower energy state. The function $\rho(W_2-W_1)$ is the same in either case, since it describes the distributions of energy difference (W_2-W_1) and not the number of atoms in any given initial state. The absorption probability per unit time W_a is therefore

$$W_a = (\pi\omega_1/\epsilon V) \; n_{lk} \; N_1 |\mu|^2 \; g_2 \; \rho(\hbar\omega_1), \qquad (1.82)$$

factor g_2 is degeneracy of the upper level with energy W_2. The emission probability (Eq.1.81) contains the combined effect of stimulated and spontaneous emission. Concentrating on stimulated emission alone, we omit 1 in $n_{lk}+1$ so that stimulated emission probability,

$$W_{ST} = (\pi\omega_1/\epsilon V) \; n_{lk} \; N_2 \; |\mu|^2 \; g_1 \; \rho(\hbar\omega_1). \qquad (1.83)$$

Thus, the stimulated emission power is given by

$$\Delta P_{ST} = (W_{ST} - W_a)\hbar\omega_1. \qquad (1.84)$$

Substituting for W_a and W_{ST} by using Eqs. 1.82 and 1.83,

$$\Delta P_{ST} = (\pi w_1/\epsilon V) \; n_{lk} |\mu|^2 \rho(\hbar\omega_1)[N_2 g_1 - N_1 g_2], \qquad (1.85)$$

where $|\mu|^2$ is the average value over all the transitions between all possible degenerate initial and final states. If we now express power carried in mode l,k by P,

$$P = (n_{lk}/L) \; \hbar\omega_1 \; (\epsilon_o/\epsilon)^{1/2} c, \qquad (1.86)$$

where energy $n_{lk}\hbar\omega_1$ in a volume of length L and unit cross-section flows in a time $\tau = L/c \; (\epsilon_o/\epsilon)^{1/2}$.

We can rewrite Eq. 1.85, by using Eq. 1.86 and expressing

$N_i/V = \bar{N}_i$, $i = 1,2$, as

$\Delta P_{ST} = \alpha \; P \; L$, where

$$\alpha = [\pi\omega_1/(\epsilon/\epsilon_o)^{1/2} c] \; |\mu|^2 \; \rho(\hbar\omega_1) \; (\bar{N}_2 g_1 - \bar{N}_1 g_2) \qquad (1.87)$$

OPTICAL RADIATION AND PHOTONS   29

The factor α has a physical meaning signifying a power gain per unit length by stimulated emission. L denotes the interaction region, i.e. the field gains power ΔP_{ST} over this length. The power gain depends on the difference of the density of atoms (number of atoms per unit volume) between the upper and lower laser levels of the amplifying medium. If both levels are equally populated there is no power gain. If $g_1 \bar{N}_2 > \bar{N}_1 g_2$ the field gains power by stimulated emission of radiation and if $\bar{N}_2 g_1 < \bar{N}_1 g_2$, it loses power by absorption. Eq. 1.87 considers atom interacting with a single mode. However, one would like to consider an interaction with a travelling monochromatic wave at frequency ν having a finite width of line shape function $g(\nu)$. Also one could replace $|\mu|^2$ in terms of radiation lifetime τ_{sp} by using Eq. 1.62a. The energy distribution function $\rho(\hbar\omega_1)$ is replaced in terms of the frequency distribution function $\rho(\nu)$. The relation between $\rho(\hbar\omega_1)$ and $g(\nu)$ is obtained by comparing their normalization condition

$$\int_0^\infty \rho(\hbar\omega_1) \hbar d\omega = \int_0^\infty g(\nu) d\nu = 1$$

so that $g(\nu) = 2\pi\hbar \, \rho(\hbar\omega)$. \hfill (1.88)

Replacing the dipole moment matrix element μ and the energy distribution function $\rho(\hbar\omega_1)$ in terms of τ_{sp} and $g(\nu)$, we get

$$\alpha = [\lambda^2/8\pi(\epsilon/\epsilon_0)](\bar{N}_2 - \bar{N}_1 \, g_2/g_1) \, g(\nu)/\tau_{sp}, \hfill (1.89)$$

where $\lambda = c/\nu$ and $\omega = 2\pi\nu$.

1.6. Gain saturation

The gain expression (Eq. 1.89) shows that gain coefficient α depends on the population inversion ΔN

$$\Delta N = (\bar{N}_2 - \bar{N}_1 g_2/g_1) \hfill (1.90)$$

and so long as ΔN is constant, the gain α per unit length due to stimulated emission also remains constant. Eq. 1.87 further shows that the intensity of radiation field I (power P per unit area) builds up due to stimulated emission. Eq. 1.87 gives

$$\Delta P_{ST} = \alpha \, L \, P$$

or $dI/dZ = \alpha I$

or $I = I_o \exp \alpha Z.$ (1.91)

At threshold of laser oscillation, the gain per transit of the active medium must equal the losses in the cavity (we shall describe the resonant cavity in section 1.7). If however the population inversion is increased beyond this threshold value, the gain of the active medium will exceed the cavity loss for a particular mode and the radiation field will continue to build up. This practice however, for sustained (steady state) oscillation, is not desirable because when a certain field intensity is reached the gain must decrease, or saturate, so as to equal the cavity loss. Here again, as in all oscillators, the gain readjusts itself through nonlinear effects. The linear relation (Eq.1.91) showing growth of field intensity,

$dI = \alpha I dZ$

is based on small signal approximation. A more general description however involves nonlinear terms in the field intensity described by

$dI = (\alpha I - \beta I^2) dz$ (1.92)

which takes into account the dependence of population inversion ΔN on the field intensity. This dependence is seen by writing the rate equations. On substituting field intensity dependent ΔN in the gain expression one can realize the presence of nonlinearity in the process. The field intensity at saturation (steady state oscillation) can thus be derived by equating the saturated gain to the cavity loss.

The gain profile may also vary with the mechanism of line broadening i.e. whether the emission line is broadened homogeneously or inhomogeneously. In the homogeneous broadening the atoms are indistinguishable and have the same transition energy ($E_2 - E_1$), the broadening is mainly due to collisional processes (elastic and inelastic) and the emission line follows a Lorentzian response curve. In the inhomogeneous broadening the atoms are distinguishable and the broadening gives a spread in the individual resonant (transition) energies of the atom. Examples of this type of broadening are impurity ions in a host crystal and molecules in low pressure gases. Crystal imperfections and random strain cause the crystal surroundings to vary from one ion to the next, thereby causing a line spread. The transition frequency of a gaseous atom (or molecule) is Doppler shifted due to the finite velocity of the atom.

When systems with homogeneous and inhomogeneous broadening are used as laser medium, their gain decreases with increasing field intensity and the amount of decrease and its spectral dependence differs in the two cases. In the case of homogeneous broadening

$$\alpha = \alpha_o / (1+I/I_s), \tag{1.93}$$

where α_o is the unsaturated gain (at zero field), I is field intensity and I_s the saturation intensity at which unsaturated gain is reduced by 1/2.
In the inhomogeneous broadening,

$$\alpha = \alpha_o / (1+I/I'_s)^{1/2}, \tag{1.94}$$

where I'_s is the saturation intensity of the inhomogeneously broadened line and is different from saturation intensity (I_s) of the homogeneously broadened system.

An important consequence of the saturation behaviour of homogeneous and inhomogeneous media is the phenomenon of hole burning. In the presence of a strong saturating field at frequency ν', the gain experienced by a weak probing signal at frequency ν is shown in Fig. 1.5. The gain curve for homogeneously broadened medium shows a depression over a frequency interval where the gain is reduced. The depressed region is usually referred to a 'hole' and the phenomenon is described as 'hole burning'. This is discussed again in section 1.15.

1.7. Optical Resonators

Optical resonators form the resonant cavity of a laser. Like the microwave counterparts, these are needed to build up large field intensities at specified frequencies and act as spatial and frequency filters. The radiation travels back and forth between the mirrors of the resonator thereby increasing the path length through the active medium. This large folded path increases the total gain (gain per unit length times the path length) of the system. The condition of oscillation requires the total gain of the system to exceed the total losses in the system for build up of oscillation and to equal the losses for sustained oscillation. Like any other resonant cavity, the figure of merit of an optical resonator is also quantified by its quality factor Q.

Depending upon the geometry of the resonator mirrors, standing waves or travelling waves may result in the cavity. The reflecting mirrors may be plane parallel or spherical to give standing waves or retroreflector prisms or corner cubes to result in travelling waves.

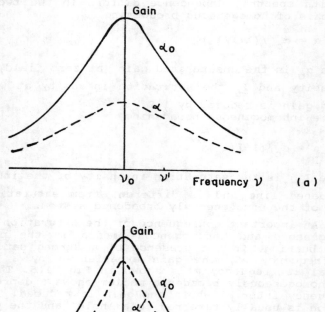

Fig. 1.5. Gain constant α experienced by a weak probing signal at ν in presence of a strong saturating field at ν'. (a) Homogeneously broadened gain medium. (b) Inhomogeneously broadened gain medium.

A cavity mode is defined as a field distribution that reproduces itself in spatial distribution and phase as the wave travels between the cavity reflectors. Amplitude is not reproduced because of diffraction and reflection losses. Also, because of the loss, a given mode has a specific lifetime defined as the time for the amplitude to decay to $1/e$

of its initial value.

Modes are classified as longitudinal and transverse. Longitudinal modes are designated according to number of nodes along the axis of cavity between the mirrors. Transverse modes are defined by number of nodes in the plane of the mirror i.e. the plane normal to the laser axis.

1.7.1. Resonant Modes of Optical Cavity

Because of the small value of wavelength in the optical range compared to microwaves, the optical cavity is necessarily multimode. But the number of oscillating modes is kept low by retaining high Q in one direction only. This is done by having two parallel mirrors separated by an arbitrary distance to form a Fabry Perot resonator. This is called an open cavity or an open resonator.

In a closed cavity resonator of volume V, the number of modes within a frequency interval $d\nu$ are given by

$$N \simeq (8\pi\nu^2/c^3) \, V \, d\nu. \tag{1.95}$$

The Eq. 1.95 implies that for $\nu=3\times10^{14}$ Hz ($\lambda \approx 1$ μm) and $d\nu=3\times10^{10}$ Hz, nearly 2×10^9 modes per cubic cm volume would have a comparable high Q value making the power to go into all these modes with different frequencies and different spatial characteristics. The open resonator, of the Fabry Perot type, however, does not allow the travel of energy at right angles to the mirror axis (i.e. the line joining the mirrors AB in the longitudinal direction, Fig. 1.6a). This is because it offers high loss in single traversal normal to mirror axis and consequent low Q in that direction.

Two of the earliest proposals for open resonator were almost simultaneously given by Schawlow and Townes [1.8] and Prokhorov [1.9]. It was suggested that an open resonator will discriminate heavily against modes whose energy propagates along directions other than that normal to the reflectors. This discrimination will reduce the number of oscillating modes in a laser cavity, thereby improving spectral purity and coherence of radiation.

Let us consider the cavity of Fig.1.6a with no side walls and plane parallel perfect reflectors. Let us also assume that there are no diffraction and other losses in the cavity. In this cavity, the travelling of radiation back and forth between the mirrors sets up a standing wave pattern provided the radiation travels from A to B and back to A in correct phase. This requirement of correct phase is satisfied only at certain frequencies. Only radiation at these frequencies can exist in the cavity and each of these frequencies is a possible longitudinal mode of the cavity oscillation. For a wavelength of 1 μm, there would be about a million of these modes in a cavity length of one meter. If

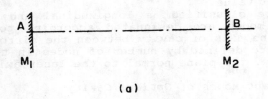

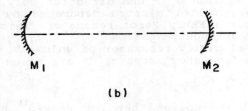

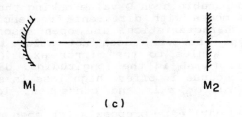

Fig. 1.6. Open resonator configuration of an optical cavity with (a) plane parallel mirrors, (b) spherical mirrors and (c) plane and spherical mirrors.

c is the velocity of radiation in the medium of the cavity, then the frequency interval by which longitudinal modes are separated is given by

$$\nu_q - \nu_{q-1} = \delta\nu_l = [q-(q-1)]c/2d = c/2d, \tag{1.96}$$

where d is the cavity length. For the case of transverse modes (Section 1.72) the resonance condition is

$$4d/\lambda = 2q + (m+n+1), \tag{1.97}$$

where q refers to the longitudinal mode and m and n to the transverse mode. From Eq. 1.97 the separation $\delta\nu$ between modes of index m and m-1 is c/4d since

$$\delta\nu_m = \nu_{m,n,q} - \nu_{m-1,n,q} = (c/4d)4d\,[(1/\lambda_{m,n,q})-(1/\lambda_{m-1,n,q})]$$

$$= (c/4d)\,[(2q+m+n+1)-(2q+m-1+n+1)] = c/4d. \qquad (1.98)$$

The indices m and n refer to the number of nodes in the two orthogonal directions in the plane normal to the resonator axis, much as the index q refers to the number of nodes in the direction of resonator axis (i.e. the number of zeros of intensity in the standing wave system along the laser axis) (Fig. 1.7). The transverse mode determines the spatial distribution of the mode amplitude (or intensity) in the plane normal to the resonator axis. The transverse modes are transverse electromagnetic and are represented as TEM_{mn}. The intensity distribution of some of the low order transverse modes are shown in Fig. 1.8. For each transverse mode there are a number of longitudinal modes characterized by the value of q (q is generally very large and hence not written). The lowest transverse mode is TEM_{ooq} which has the lowest (diffraction limited) divergence (see section 1.73 on Gaussian beams). The TEM_{mnq} modes are 'off axis' modes.

Longitudinal modes differ from one another only in their oscillation frequency whereas transverse modes differ from one another both in oscillation frequency and field distribution in a plane normal to laser axis. Several transverse modes may also oscillate simultaneously if conditions so permit. Modes with axial symmetry may also oscillate. These

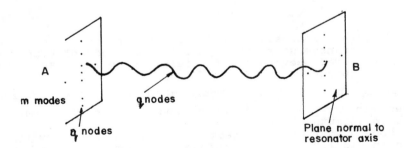

Fig. 1.7. Indices q and m,n for the longitudinal and transverse modes. Transverse mode refers to the spatial distribution in the plane normal to the resonator axis.

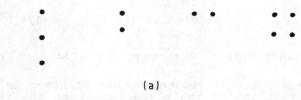

(a)

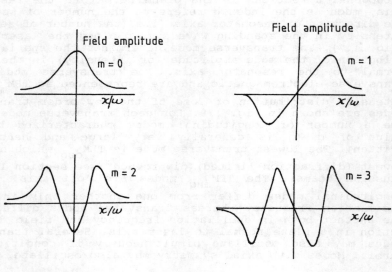

(b)

Fig.1.8. (a) Intensity distribution of some low order transverse modes. (b) Field amplitude of transverse mode. Reversal of field indicates 180° phase change.

are designated as TEM_{plq} where p gives number of zeros (nodes) in the radial direction and l gives the number of nodes in azimuthal direction.

Some commonly used geometries of optical cavity are shown in Fig 1.9. In the confocal geometry distance between the mirrors is equal to the radius of curvature of each mirror while for concentric resonator the distance between mirrors is twice the radius of curvature. If one of the mirrors of a confocal resonator is replaced by a plane mirror (i.e. mirror with infinite radius of curvature R=∞) we get the hemispherical resonator.

OPTICAL RADIATION AND PHOTONS   37

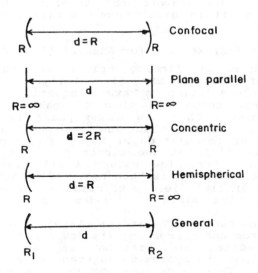

Fig. 1.9. Some commonly used resonator geometries for optical cavity.

The two essential requirements for a laser are (i) a cavity and (ii) an active medium within the cavity. The active medium provides amplification necessary to establish oscillation in the cavity. A passive cavity does not contain active medium. The resonator modes for an active cavity are likely to be different from that of a passive cavity because the optical length between the resonator mirrors will be affected by the refractive index of the active medium. Also, due to the thermal effects, the active medium may lead to thermal lensing thus affecting spatial distribution of the field. The effective aperture of the active medium (e.g. thin laser rods) may also have an effect on the low order transverse modes. Any refractive index gradients in the active medium may also lead to variation of transverse distribution of the beam and consequent change in mode structure. The active medium provides radiation and gain over a limited frequency range $\Delta\nu$. The frequency range is determined by the width of spectral line $\delta\nu$. The number of longitudinal modes possible within the range is thus

$$N = \Delta\nu/\delta\nu. \tag{1.99}$$

N is the number of longitudinal modes which will oscillate in the cavity if the gain condition is satisfied. The laser gain should be well above the threshold necessary for oscillation to obtain simultaneous oscillation of several modes.

1.7.2. Theoretical Methods for Analyzing the Modes

The problem of finding transverse modes has been tackled in more than one way [1.10-1.12]. In one approach one seeks simple solution to Maxwell equations that take the form of narrow beams and then to make the reflection surfaces intersect the beam along phase fronts ensuring reflection of the wave back on itself. In the other approach, scalar formulation of Huygen's principle is used to compute the field at one mirror caused by the illumination of the other. The return field configuration is similarly calculated and is then required to match, within a constant, the initial field configuration. Solutions of the resulting equation yield the modes and the diffraction losses.

In the approach of Fox and Li [1.10], a plane wavefront was set off from one mirror and its development was followed as it travelled back and forth between the mirrors. It was demonstrated that the system would eventually settle down to a stable mode. This is because the dominant mode is characterized by fields which are considerably smaller at the edges of the mirrors than at the centres (see section 1.73 on Gaussian beams), thus making diffraction loss smaller than that predicted by consideration of plane waves.

The cross-sectional amplitude distribution of transverse modes, for curved mirror resonators with rectangular symmetry, is given by

$$A(x,y) = A_{mn} [H_m(2^{1/2}x/w) \; H_n(2^{1/2}y/w)] \exp[-(x^2+y^2)/w^2],$$

where x and y are the transverse coordinates and A_{mn} is a constant whose value depends on the field strength of the mode, w is the radius of the fundamental mode (m=0, n=0) at 1/e maximum amplitude, $H_l(b)$ is the lth order Hermite polynomial with argument (b) and m and n are the transverse mode numbers. This distribution is independent of the longitudinal mode number. Fig 1.8b is a plot of A along the axis for different m values. It is seen that as transverse mode number increases, the energy is spread further and further from the axis of resonator.

For a resonator to be able to support low loss (high Q) modes, it must satisfy two criteria. First there must be a family of rays that, on suffering sequential specular reflections from the two reflectors, do not miss either reflector before making a reasonable number (10-100) of traversals. Second, the dimensions of the reflectors must satisfy the relation

$$d_1 d_2 / \lambda d \geq 1, \qquad (1.100)$$

where $d_{1,2}$ are the half widths of the two reflectors respectively in any arbitrary direction perpendicular to the resonator axis and d is the distance between reflectors. λ is the wavelength of radiation undergoing reflection between the mirrors. The first criterion follows from geometric optics, since reflecting areas are large compared to wavelength. The second criterion follows from the requirement that half the angle subtended by one reflector at the second (i.e. d_1/d) be somewhat greater than half the angle of the far field diffraction pattern of a nearly plane wave originating at and restricted to the dimension of the second (i.e. $\lambda/2d_2$).

Consideration of a plane wave suffering repeated reflection between the two mirrors of the cavity leads to a stable mode configuration of a Gaussian beam which is characterized by a high field along the axis and tapering towards the edges thereby signifying reduced diffraction losses at the edges of mirror geometry.

1.7.3. Gaussian Beams

Following Fox and Li [1.10] it is observed that the beam travelling back and forth between the mirrors of an optical resonator eventually settles down to a Gaussian distribution. It is to be remembered that each reflection produces a Fourier transformation of the distribution and the Fourier transform of a Gaussian is a Gaussian.

Propagation characteristics of a Gaussian beam differ from that of a plane wave or a spherical wave. In a plane wave, the amplitude of the electric field is independent of the location of the point considered in the xy plane and is independent of z, the location of the plane considered (Fig. 1.10a). In a spherical wave (Fig. 1.10b) the amplitude is uniform over any spherical surface which can be drawn about the centre of propagation but depends upon the distance of the surface from the centre. A Gaussian beam (Fig. 1.10c) is characterized by waves which are approximately spherical near the z axis (the longitudinal axis of laser resonator) but the approximation becomes worse as the distance from the axis increases. Further, the centre from which the spherical wavefront appears to originate is not fixed. The amplitude of electric field is a Gaussian function of distance from the z axis.

At z=0, the beam has a minimum width of radius w_o known as the beam waist. At distance z along the axis of propagation, the beam has a width given by

$$w^2(z) = w_o^2[1+(2z/kw_o^2)^2]. \tag{1.101}$$

$w(z)$ is the radial distance r at which the field amplitude is down by a factor $1/e$ compared to its value on the axis (Fig. 1.10c ii). The parameter w is known as the beam radius, or the spot size, so that w_o specifies the minimum spot size at z=0. At the point of intersection of the wavefront with z axis, the radius of curvature of Gaussian beam is

$$R(z) = z [1+(w_o^2 k/2z)^2]. \tag{1.102}$$

At the beam waist (z=0), the wavefront is that of a plane wave, while for large z it approaches that from a point source located at the origin.

The form of the fundamental Gaussian beam is uniquely determined once its minimum spot size w_o and its location (i.e. plane z=0) are specified. Its spot size w and radius of curvature R at any plane z are then found from Eqs.(1.101 and 1.102). The beam parameters $R(z)$ and $w(z)$ are the same for modes of all orders.

The angle of divergence θ of the Gaussian beam, as z becomes very large, is obtained from $\tan\theta = w/R \approx \theta$ which gives

$$\theta = 2/kw_o = \lambda/\pi\omega_o. \tag{1.103}$$

This result is a manifestation of wave diffraction according to which a wave confined in the transverse direction to an aperture of radius w_o will spread (diffract) in the far field ($z >> \pi\omega_o^2/\lambda$) according to Eq. 1.103. The complete expression for the field distribution $E(x,y,z)$ of a Gaussian (TEM_{oo} mode) is

$$E(x,y,z) = E_o(w_o/w) \exp[-j(kz+\phi)-(x^2+y^2)(jk/2)\{(1/R)-(2j/kw^2)\}] \tag{1.104}$$

where x and y are the orthogonal coordinates in the xy plane normal to the direction of propagation along z, w_o is the waist radius, $k=2\pi/\lambda$, E_o is the field along the z direction, R is radius of curvature and w, the beam radius at distance z. ϕ is given by

$$\tan\phi = kw_o^2/2z. \tag{1.105}$$

Eq. 1.104 can also be expressed as

OPTICAL RADIATION AND PHOTONS

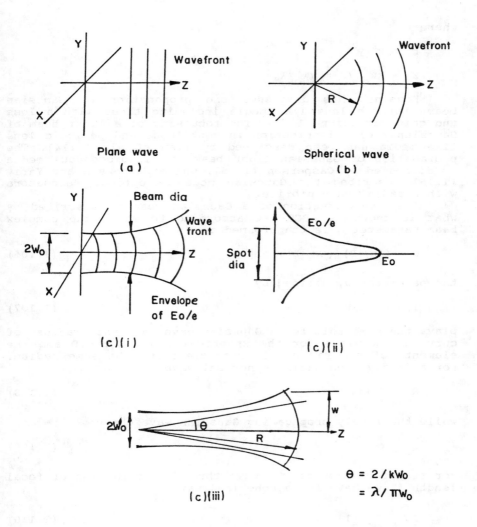

Fig. 1.10. Representation of a plane wave, spherical wave and Gaussian wave.

$$E(x,y,z)/E_o = [w_o/w(z)] \exp[-r^2/w^2(z)] \qquad \text{amplitude factor}$$
$$\cdot \exp[-j\{kz-\tan^{-1}(z/z_o)\}] \qquad \text{longitudinal phase}$$
$$\cdot \exp[-jkr^2/2R(z)], \qquad \text{radial phase}$$

where

$$r^2 = x^2+y^2$$

$$z_o = kw_o^2/2 = \pi w_o^2/\lambda.$$

Kogelnik has discussed the propagation of Gaussian beams through lens like media including those with a loss and gain variation [1.13]. The focussing of a light beam of Gaussian field distribution in continuous and periodic lens like media has been described by Tien *et al.* [1.14]. The propagation of Gaussian light beams in inhomogeneous media is discussed by Casperson [1.15], while Casperson and Yariv [1.16] described the Gaussian mode in optical resonators with a radial gain profile.

The transformation of a Gaussian beam is described by what is known as ABCD law according to which the complex beam parameter q is transformed by

$$q_2 = (Aq_1+B)/(Cq_1+D), \tag{1.106}$$

the parameter q, defined by

$$1/q = (1/R) - 2j/kw^2 \tag{1.107}$$

plays the same role for a Gaussian wave as the radius of curvature R plays for the spherical wave. A,B,C,D are the elements of the ray matrix characterizing the same medium. For a freely propagating spherical wave

$$R_2 = R_1+z, \tag{1.108}$$

while the freely propagating Gaussian wave obeys

$$q_2 = q_1+z. \tag{1.109}$$

For transformation of the wave through a thin lens of focal length f, we have for a spherical wave

$$\frac{1}{R_2} = \frac{1}{R_1} - \frac{1}{f} \tag{1.110}$$

while for Gaussian wave,

$$\frac{1}{q_2} = \frac{1}{q_1} - \frac{1}{f} \tag{1.111}$$

The comparison of the effects of a lens on a spherical wave, and on a Gaussian wave are shown schematically in Fig. 1.11.

If a ray cuts the input and output planes normal to the optic axis at distance x_1, x_2 from the axis and at angle θ_1, θ_2 to the axis (Fig. 1.11), then

$$\begin{bmatrix} x_2 \\ \theta_2 \end{bmatrix} = \begin{bmatrix} A & B \\ C & D \end{bmatrix} \begin{bmatrix} x_1 \\ \theta_1 \end{bmatrix}, \qquad (1.112)$$

where the matrix ABCD is the ray transfer matrix and A,B,C,D are functions characteristic of the particular optical system under consideration. It is easy to see that for a ray travelling an optical distance d between two planes in a

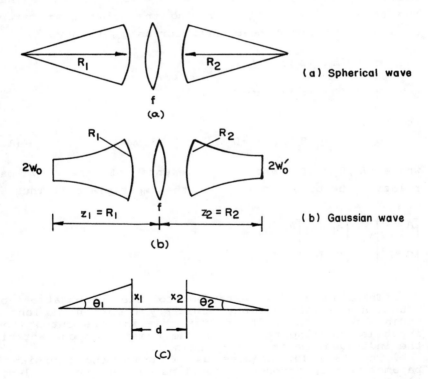

Fig. 1.11. Effect of a thin lens on (a) spherical wave and (b) Gaussian wave.

homogeneous medium, the ray transfer matrix is

$$\begin{bmatrix} 1 & d \\ 0 & 1 \end{bmatrix}.$$

For a thin lens, the input and output planes coincide (Fig. 1.11) so that the ray transfer matrix becomes

$$\begin{bmatrix} 1 & 0 \\ -1/f & 1 \end{bmatrix}.$$

Following the ABCD law for transformation of a Gaussian wave, we consider the propagation of a Gaussian beam through two lens like media that are adjacent to each other, the ray matrix describing the first one is A_1, B_1, C_1, D_1 while that of the second is A_2, B_2, C_2, D_2. Taking the input beam parameter as q_1 and the output beam parameter as q_3, we have

$$q_2 = (A_1 q_1 + B_1)/(C_1 q_1 + D_1)$$

$$q_3 = (A_2 q_2 + B_2)/(C_2 q_2 + D_2)$$

so that,

$$q_3 = (A_3 q_1 + B_3)/(C_3 q_1 + D_3), \qquad (1.113)$$

where A_3, B_3, C_3, D_3 are the elements of the ray matrix relating the output plane 3 to the input plane 1. Thus

$$\begin{bmatrix} A_3 & B_3 \\ C_3 & D_3 \end{bmatrix} = \begin{bmatrix} A_2 & B_2 \\ C_2 & D_2 \end{bmatrix} \begin{bmatrix} A_1 & B_1 \\ C_1 & D_1 \end{bmatrix}. \qquad (1.114)$$

This is easily extended to the propagation of a Gaussian beam through any arbitrary number n of lens like media and elements. The matrix relating the output to the input is then the product of the n matrices characterizing the individual parts of the chain.

The ABCD law enables us to trace the Gaussian beam parameter $q(z)$ through a complicated sequence of lenslike elements. The beam radius $R(z)$ and spot size $w(z)$ at any plane z can be obtained through the use of Eq. 1.107.

Remembering that q plays the same role for Gaussian

waves as R does for the spherical waves, the transformation law for spherical waves takes the form

$$R_2 = (AR_1+B)/(CR_1+D) \tag{1.115}$$

1.7.4 Design of Open Resonator Cavity

It is often desired to know the beam parameters, such as waist size and spot size at the mirrors of a laser cavity. Consider the cavity to be formed with mirrors of radii of curvature R_1, R_2 placed at distance d apart. Let us also assume that the Gaussian mode in the cavity has a self consistent field configuration and that beam parameters remain unchanged after a round trip between the cavity mirrors.

Alternatively, for given beam parameters, we may like to choose the mirrors (their radii of curvature) and the spacing between them.

Both the problems may be analyzed by considering the resonator cavity structure as a series of lenses and assuming that the cavity is of infinite aperture so that diffraction effects at the mirror surface are negligible. In this representation, a lens produces the same effect as a spherical mirror and we can replace the mirrors with a periodic sequence of lenses in which there is one lens for each reflection. The cavity is thus unfolded into a sequence of lenses which forms an optical transmission line. The focal lengths of the lenses are determined by radii of curvature of the mirrors that they replace and their spacing by the separation of the mirrors. We can then treat the problem as one of a beam propagating through a periodic lens sequence.

Consider a resonant cavity with spherical mirrors A and B (radius of curvature R_1, R_2) placed at distance d apart. Suppose that the Gaussian beam has a complex beam parameter q_1, q_2 immediately after leaving mirror A (Fig. 1.12). When it reaches mirror B the complex beam parameter is (from Eq. 1.109)

$$q_2 = q_1+d.$$

After reflection by mirror B we get from Eq. 1.111,

$$\frac{1}{q_3} = \frac{1}{q_1+d} - \frac{2}{R_2}$$

On reaching mirror A again, (this time travelling in opposite direction),

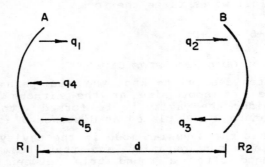

Fig. 1.12. Resonant cavity with mirrors A and B of radii R_1 and R_2 and distance d apart. The complex parameter q_5 obtained after one full round trip is equated with q_1 to obtain stability condition.

$q_4 = q_3 + d$

so that after reflection by mirror A,

$$q_5 = \frac{R_2 R_1 (2d+q_1) - 2R_1(d+q_1)}{R_2(R_1-2d) - 2(R_1+R_2)(q_1+d) + 4d(q_1+d)}. \quad (1.116)$$

Since the beam has now undergone a complete round trip inside the cavity, the condition for a self consistent field configuration (and hence for the Gaussian wave to be a mode) is

$$q_5 = q_1 \quad (1.117)$$

From Eqs. 1.116 and 1.117, we obtain a quadratic equation for the beam parameter

$$dR_1(R_2-d)(1/q_1)^2 + 4d(R_2-d)(1/q_1) + 2(R_1+R_2-2d) = 0$$

which has the solutions

$$1/q_1 = -(1/R_1) \pm (j/R_1)[\{R_1(R_1+R_2-2d)/(R_2-d)d\}-1]^{1/2}. \quad (1.118)$$

Since all real Gaussian beams must have a non-zero width,

the complex beam parameter must have a non-zero imaginary part (which by definition contains information about the beam width (see Eq. 1.107). Therefore, we get

$$[(R_1+R_2-2d)R_1/(R_2-d)d] > 1. \tag{1.119}$$

Similarly we could also write

$$[(R_1+R_2-2d)R_2/(R_1-d)d] > 1 \tag{1.120}$$

since our choice of mirrors for launching the beam is arbitrary.

With some algebraic manipulation, the Eqs. 1.119 and 1.120 can be combined into

$$0 < (1-d/R_1)(1-d/R_2) < 1 \tag{1.121}$$

The relation 1.121 is known as the stability criterion and puts a limit on the cavity parameters which enable the supporting of a stable mode in the cavity. The choice of R_1, R_2 and d should be such that they satisfy the stability criterion for supporting a stable mode. If we define

$$g_1 = 1 - d/R_1 \tag{1.122}$$

$$g_2 = 1 - d/R_2 \tag{1.123}$$

and plot g_2 against g_1, we get Fig. 1.13 in which each point in the g_1 g_2 plane defines a particular cavity configuration. The right hand inequality in the Eq. 1.121 defines a hyperbola and is satisfied for all points lying below the upper curve A and above the lower curve B. The left hand inequality is satisfied so long as g_1 and g_2 are of the same sign. Therefore, the area of the plane for a stable cavity configuration is as shown shaded in the stability diagram.

The confocal resonator ($R_1=R_2=d$) with identical mirrors lies at the origin ($g_1,g_2=0$). It is worth noting that slight deviations (say due to manufacturing tolerances) makes this confocal system unstable and may result in serious losses. It is therefore better to avoid this region. The cavities with $R_1=\infty$ (planar) and $R_1=R_2=2d$ (concentric) are also critically stable since they lie on the curves (A and B respectively) which form the boundary between the stable and

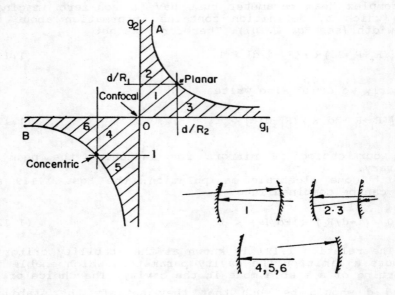

Fig. 1.13 Stability diagram for cavities with spherical mirrors. Points in shaded region correspond to stable cavities.

unstable regions. Each point on the diagram represents a particular geometry. Points within the shaded area represent stable systems. Stability implies that the system is not subject to the type of loss which can be predicted by geometric optics. If the mirrors are regarded as a periodic focussing system, a ray within a stable resonator remains within the resonator after any number of round trips but a ray in an unstable resonator walks out. The boundaries between the stable and unstable regions are the axis and the hyperbola (curve A and B) defined by $g_1 \cdot g_2 = 1$.

In addition to geometric losses, both stable and unstable resonators are subject to diffraction losses. These losses are much greater for unstable geometries than for stable resonators. This is because, the fields of the modes in stable resonators remain concentrated near the axis of the resonator whereas the fields in unstable resonators move progressively towards the periphery of the mirrors where diffraction losses are greater. It is, therefore, more appropriate to describe the regions of the stability diagram as low loss and high loss regions.

OPTICAL RADIATION AND PHOTONS 49

To find the beam parameters of the Gaussian mode (waist radius w_o and spot radius at mirrors w_1, w_2) in terms of the radii of curvature R_1, R_2 of the mirrors and the separation d between them, we note that the radii of curvature of the Gaussian beam at the mirrors must match the respective mirror radii of curvature. Further, the required width of the Gaussian at the mirrors is given by imaginary part of Eq. 1.107. Thus, using Eq. 1.118 along with Eq. 1.107,

$$2/kw_1^2 = [\{(R_1+R_2-2d)R_1/(R_2-d)d\}-1]^{1/2}/R_1,$$

which gives,

$$w_1^4 = \frac{4 R_1^2 (R_2-d)d}{k^2(R_1-d)(R_1+R_2-d)}. \qquad (1.124)$$

To find the waist size and its position, it is noted from Eq. 1.102 that at the waist (z=0) the wavefront is that of a plane wave. From Eq. 1.107 the complex beam parameter q is seen to be purely imaginary (i.e. $2j/kw_o^2$). If z_1 is the distance of the waist from mirror A (Fig 1.14), then from Eq. 1.102

$$R_1 = z_1 + (1/z_1)(w_o^2 k/2)^2, \qquad (1.125)$$

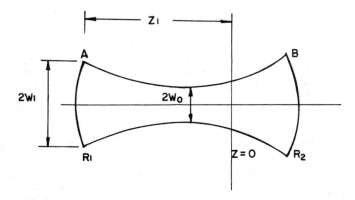

Fig. 1.14. Location of waist in an optical cavity with mirrors of radii R_1 and R_2 and distance d apart.

$$R_2 = (d-z_1)+[1/(d-z_1)](w_o^2 k/2)^2. \tag{1.126}$$

Eliminating w_o from Eqs. 1.125 and 1.126,

$$z_1 = d(R_2-d)/(R_1+R_2-2d) \tag{1.127}$$

and substituting Eq. 1.127 in Eq. 1.125

$$w_o^2 = \frac{(2/k)^2 d(R_1-d)(R_2-d)(R_1 R_2-d)}{(R_1+R_2-2d)^2}. \tag{1.128}$$

Knowing beam parameters from Eqs. 1.127 and 1.128 therefore, enables us to follow the propagation of a Gaussian beam both inside and outside a cavity.

For the cavity to resonate (i.e. for the field to be setup inside the cavity with a well defined phase structure in order to form a standing wave) the phase shift per transit along the Z axis must be an integral multiple of π. The phase term from Eq. 1.104 is

$$(kz+\phi)+(x^2+y^2)k/2,$$

where ϕ is given by Eq. 1.105. The condition of resonance, therefore, becomes

$$kd + \tan^{-1}(kw_o^2/2z_1) + \tan^{-1}[kw_o^2/2(d-z_1)] = \pi(q+1) \tag{1.129}$$

where q is an integer which is equal to the number of nodes of the standing wave (section 1.7.1, Fig. 1.7). The values of k which satisfy Eq. 1.129 determine the frequencies of the longitudinal modes of the cavity designated by their different q values.

In many situations we need to match the Gaussian mode of one cavity to that of another by the use of a component such as a lens. This matching is necessary when beam corresponding to a particular mode (say the lowest order mode TEM_{oo}) in one system is injected into another. If not matched, a single mode from one system will couple to several modes of other system and mode conversion occurs. To do mode matching one uses the fact that a lens transforms the position and magnitude of the beam waist of one cavity into that of the other (Fig. 1.15).

Let us suppose that the position and diameters of the beam waists of two cavities are known and that we wish to match the two cavities. Let w_{o1} and w_{o2} be the waist radii

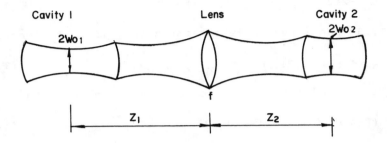

Fig. 1.15. Matching of Gaussian beam from one resonant cavity to another.

and z_1 and z_2 the distances of the waist of the two cavities from the lens of focal length f.

From Eq. 1.107, at the beam waist,

$$\frac{1}{q_o} = \frac{1}{\infty} - \frac{2j}{kw_o^2}$$

so that at the waist of the two cavities

$$q_{o1} = (jk/2)w_{o1}^2$$

$$q_{o2} = (jk/2)w_{o2}^2$$

Using Eqs. 1.109 and 1.111

$$q_{o2} = [\{1/(q_{o1}+z_1)\} - 1/f]^{-1} + z_2. \qquad (1.130)$$

Since both q_{o1} and q_{o2} are imaginary, we can substitute for q_{o1} and q_{o2} and then split Eq. 1.130 into real and imaginary parts to give

$$(z_1-f)(z_2-f) = f^2 + q_{o1}q_{o2} \qquad (1.131)$$

$$q_{o2}(f-z_1) - q_{o1}(f-z_2) = 0. \qquad (1.132)$$

For a lens of given focal length, we can find z_2 (say) by eliminating z_1 so that

$$z_o^2 - 2fz_2 - q_{o2}^2 - f^2[(q_{o2}/q_{o1}) - 1] = 0. \qquad (1.133)$$

Eq. 1.133 has a real solution for z only if $f > f_o$, where

$$f_o = [|q_{o1}||q_{o2}|]^{1/2} = w_{o1} w_{o2} k/2. \qquad (1.134)$$

Eqs. 1.131 and 1.132 can also be written in the form

$$(f-z_1)/(f-z_2) = (w_{o1}/w_{o2})^2 \qquad (1.135)$$

$$(z_1-f)(z_2-f) = f^2 - f_o^2. \qquad (1.136)$$

Eqs. 1.135 and 1.136 enable the estimation of z_2 at w_{o2} for given known values of w_{o1}, z_1 and f. For this we solve Eqs. 1.135 and 1.136 simultaneously and get

$$(z_2 - f) = (z_1-f) f^2/[(z_1-f)^2 + (\pi w_{o1}^2/\lambda)^2] \qquad (1.137)$$

$$1/w_{o2}^2 = (1/w_{o1}^2)[1-(z_1/f)]^2 + (1/f^2)(\pi w_{o1}/\lambda)^2. \qquad (1.138)$$

Eqs. 1.137 and 1.138 provide a more practical tool for the matching of cavities.

1.7.5. 'Q' of Optical Resonator

In a laser oscillator, the cavity mode is receiving coherent energy as well as losing energy through the various dissipative processes. In this case, the definition for the Q of a mode becomes

$$Q = \frac{2\pi\nu \text{ (stored energy)}}{\text{net loss of coherent energy per second}}$$

$$= \frac{2\pi\nu(h\nu N)}{P}, \qquad (1.139a)$$

where N is the number of photons of energy $h\nu$ stored in the cavity and P is the rate of power dissipation.

If the fractional power lost by transmission and diffraction at each mirror per pass is α and the mirrors are separated by a distance d,

$$Q = 2\pi d/\alpha\lambda. \qquad (1.139b)$$

Eq. 1.139b is obtained by considering

$$P = \alpha(h\nu N)/t, \quad \nu = c/\lambda = d/t,$$

where λ is the wavelength of laser radiation and t is the time of travel of radiation through distance d. A smaller value of the angle λ/d increases the possibility of the energy to remain in the cavity for a larger number of multi-passes between the mirrors and thereby increases the Q. The Q may thus be increased by increasing d, but with increase of d the diffraction losses contributing to α also increase. An upper limit of Q is thus reached when diffraction and reflection losses (due to transmission through mirrors) are about equal. At values of d greater than this, the value of Q decrease on account of greater diffraction loss. The greater diffraction loss (compared to reflection loss) enables mode selection by making Q lower for higher modes in relation to Q for lower modes. However, this is not the only method of mode selection.

The full frequency width of the cavity at half maximum response is obtained by using the definition of Q as

$$Q = \nu/\Delta\nu \tag{1.140}$$

so that, using Eq. 1.139

$$\Delta\nu = c\alpha/2\pi d, \tag{1.141}$$

where c is the velocity of light in the resonator medium.

As radiation travels back and forth within a cavity, losses occur by diffraction and transmission at mirror, absorption and scattering in the components etc and the radiation field in the cavity decays. The Q of the cavity therefore is also defined in terms of the photon lifetime in the cavity. The time required for the energy in the cavity to decrease to the fraction 1/e of its initial value is the lifetime τ_c of radiation in the cavity and is given by

$$\tau_c P = Nh\nu.$$

Using Eqs. 1.139 and 1.140

$$\tau_c = \frac{Nh\nu}{P} = \frac{Q}{2\pi\nu} = \frac{1}{2\pi\Delta\nu}.$$

On substitution for $\Delta\nu$ from Eq. 1.141

$$\tau_c = d/c\alpha. \tag{1.142}$$

The Q is thus given by

$$Q = 2\pi/[1-\exp(-\tau/\tau_c)] \approx \frac{2\pi\tau_c}{\tau} = 2\pi\nu\tau_c, \qquad (1.143)$$

where τ is the period of oscillation and ν is the frequency of oscillation.

1.7.6. Unstable Resonators

Referring to the stability diagram (Fig. 1.13), it is seen that stability of a resonator, with mirrors of radii R_1 and R_2 separated by distance d, is described by a single point in the g_1, g_2 plane, where

$$g_1 = (1-d/R_1)$$

and $\quad g_2 = (1-d/R_2).$

The stability is determined by the location of point on the diagram. Stable region is governed by the condition $0 < g_1 g_2 < 1$ and is characterized by small diffraction losses. Near the boundaries of the stable region the diffraction losses increase.

Unstable cavities have been described by Siegman [1.17]. The rays which miss the mirrors are designated as geometric losses. The loss of radiation due to finite aperture mirrors in stable cavities is described as diffraction losses.

It has been shown that the fraction of power that survives a round trip in an unstable resonator cavity (Fig. 1.16) is given by

$$\Gamma^2 = \Gamma_1 \Gamma_2 = [r_1 r_2 / (r_1+1)(r_2+1)]^2,$$

where

$$\Gamma_{1,2} = \frac{\pi a_{2,1}^2 / 4\pi (r_{1,2}+1)^2 d^2}{\pi a_{1,2}^2 / 4\pi r_{1,2}^2 \, d^2},$$

$$r_{1,2} = \frac{[\{1-(g_1 g_2)^{-1}\}^{1/2} - 1 + g_{1,2}^{-1}]}{(2 - g_{1,2}^{-1} - g_{2,1}^{-1})},$$

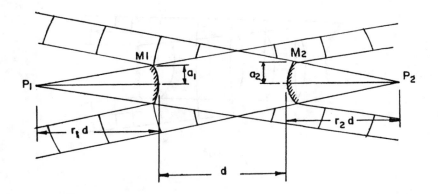

Fig. 1.16. Unstable cavity configuration.

$g_{1,2} = 1 - d/R_{1,2}$

and a_1, a_2 are aperture radii of the mirrors M_1 and M_2 respectively. Γ_1 signifies the fraction of power reflected from M_1 and Γ_2 the fraction reflected from M_2 which originates at M_1. An example of unstable cavity is shown in Fig. 1.17.

The energy loss due to diffraction effects increase rapidly as one passes the boundary from the stable to the unstable region. Though resonant modes of unstable resonators have high losses, these resonators still have practical utility. One of the configurations used is shown in Fig. 1.17a. It is based on the cassegrainian mirror configuration.

Using unstable resonators efficient output coupling from laser can be obtained via diffraction by employing diffracted energy output from resonator as the useful output rather than as undesired loss. In this diffraction coupling the laser output emerges past the edges of the mirrors which can be opaque rather than partially transparent. The cassegrainian mirror configuration of Fig. 1.17 shows diffraction coupling of unstable resonator. Such configurations are particularly suitable for high gain systems using large output coupling.

The laser emission from unstable resonator configuration consists of a diffraction coupled annular ring or halo of laser light which is clearly observed in an arrangement

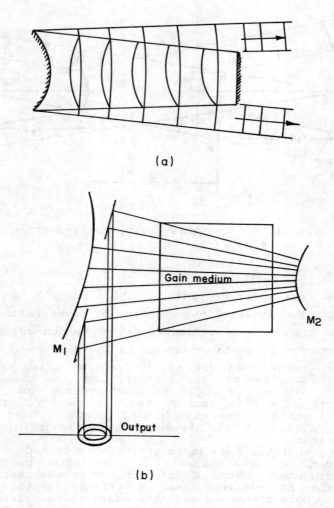

Fig. 1.17. (a) Cassegrainian unstable resonator configuration having divergent spherical surface as one end of resonator (b) Laser arrangement using stable resonator.

shown in Fig. 1.17b and predicted by the theory of unstable resonator [1.17].

A significant advantage of cassegrainian unstable resonator configuration is large mode volume which forms a large fraction of the useful volume of the active medium (e.g. a laser rod). At the same time, the size of the smaller mirror can be adjusted so as to just avoid any contact of the mode pattern with the side walls of the laser rod. The entire diffraction coupled output is available as useful output from one end of the rod. Although the near field output is an annular ring, the far field pattern after collimation still contains a single major on-axis lobe with no hole on the axis in the far field. Since the output wave is a divergent spherical wave, collimation of this wave by a lens is required to obtain a beam of minimum divergence.

1.8. Threshold Condition for Laser Oscillation

The threshold condition for establishing laser oscillation in any cavity mode requires the gain of the travelling wave passing through the amplifier to just balance the loss at each mirror. This loss, including all forms of optical losses, is expressed in terms of an equivalent reflectivity R, such that

$$R \exp \alpha d = 1, \tag{1.144}$$

where d is the active length of the amplifier section. If unequal losses are contributed by the two mirrors, R is replaced by its geometrical mean $(R_1 R_2)^{1/2}$. Eq. 1.144 holds above threshold for oscillation if α is taken to be the saturated gain α_s of the amplifier at the operating level of oscillation.

For small losses, $R \simeq 1$, so that Eq. 1.144 takes the form

$$\alpha d = \ln(1/R) \simeq 1-R.$$

Substituting for the gain factor α from Eq. 1.89

$$[\lambda^2/8\pi(\epsilon/\epsilon_0)] \, (\bar{N}_2 - \bar{N}_1 g_2/g_1) \, g(\nu) d/\tau_{sp} = 1-R$$

or

$$(\bar{N}_2 - \bar{N}_1 g_2/g_1) = \frac{8\pi(1-R)\tau_{sp}\epsilon/\epsilon_0}{g(\nu)\lambda^2 d}. \tag{1.145}$$

An alternative way to express Eq. 1.145 is to consider the loss term $(1-R)/d$ in terms of the cavity decay time τ_c.

Equating the cavity loss rate

$$1/\tau_c = (1-R)/d,$$

we can rewrite Eq. 1.145 as

$$(\overline{N}_2 - \overline{N}_1 g_2/g_1)_{th} = \frac{8\pi\nu^2 \tau_{sp}}{c^3 \tau_c g(\nu)} \simeq \frac{8\pi\nu^2 \tau_{sp} \Delta\nu}{c^3 \tau_c}, \qquad (1.145a)$$

where $1/g(\nu) \simeq \Delta\nu$ is the transition line width.

Expressing the population inversion at threshold

$$(\overline{N}_2 - \overline{N}_1 \, g_2/g_1)_{th} = (\Delta N)_{th} \qquad (1.146)$$

the minimum pump power P_{min} required to sustain the inversion Δn_i is given by

$$P_{min} = \eta(\Delta n)_{th} \, Vh\nu/\tau_{sp}. \qquad (1.147)$$

where η represents the efficiency of conversion of pump photon to laser photon and V is the volume of the amplifying medium contributing to the gain. The atoms filling the volume V radiate simultaneously. It is worth noting that since $(\Delta n)_{th}$ is proportional to $1/\lambda^2$ or ν^2, the required pump power P_{min} from Eq. 1.147 becomes proportional to ν^3. This proportionality of required pump power to the cube of frequency makes laser action at higher frequencies increasingly harder to obtain. It would therefore be much more difficult to get laser action at X-rays compared to optical frequencies.

A laser oscillator is basically obtained by enclosing an optical amplifier in an optical resonator. The role of the resonator is to maintain some chosen electromagnetic field configuration with small energy loss so that energy fed into the field by the amplifier is enough to maintain it against the loss. In the case of two mirror resonator, the optical field is in the form of a standing wave or a travelling wave bouncing back and forth at velocity of light between the two mirrors. The loss of energy from the cavity is by way of scattering, absorption and the desired useful transmission from output mirror. In order to maintain oscillation the net round trip gain must equal unity and for oscillation to build up from spontaneous emission, the net round trip gain must be greater than unity.

The condition of oscillation essentially requires that

OPTICAL RADIATION AND PHOTONS   59

round trip (amplitude) gain be equal to one and round trip phase be a multiple of 2π. The first condition provides the expression for $(\Delta N)_{th}$ (Eq. 1.145) whereas the phase condition provides the frequency of oscillation.

1.9. Cavity Coupling

The optical gain in a laser depends upon stimulated emission in the active medium of the cavity. Since stimulated emission, in turn, depends upon the energy density (or intensity) of the radiation field in the cavity, the gain too is a function of the radiation energy density. Thus power extracted from the cavity affects the gain conditions and the need for optimization arises.

For a homogeneously broadened line, the gain α is given by

$$\alpha = \alpha_o/(1+I/I_s), \tag{1.93}$$

where α_o is the unsaturated gain coefficient and I_s is the saturation parameter equal to the power density at which gain falls to $\alpha_o/2$. Consider now a laser of active length d, output mirror transmittance t and other cavity losses per unit length represented by β. In the steady state, the intensity of the radiation in the cavity is unchanged after one complete round trip in the cavity. If the intensity at the start is I, then after one round trip we have

$$I[\exp(\alpha-\beta)d]\,[\exp(\alpha-\beta)](1-t) = I. \tag{1.148}$$

On the assumption that β is small, we can write Eq. 1.148 as

$$(1-t-2\beta d)\exp[2\alpha d] = 1. \tag{1.149}$$

The power output P is related to the energy density in the cavity and the transmittance t so that

$$P = KIt, \tag{1.150}$$

where K is a constant. From Eqs. 1.193 and 1.149

$$I = -I_s[1+2\alpha_o d/\ln(1-t-2\beta d)] \tag{1.151}$$

and substituting Eq. 1.151 in Eq. 1.150

$$P = -KI_s t\,[1+2\alpha_o d/\ln(1-t-2\beta d)]. \tag{1.152}$$

For a representative case of a cavity of 100 cm long an

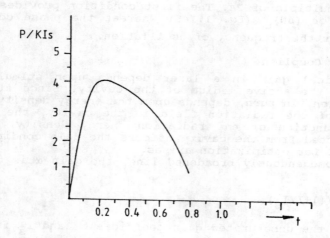

Fig. 1.18. Power output vs. mirror transmission in a laser cavity. Eq. 1.152 with $2\alpha d=10$ and $2\beta d=0.2$ is used. The constant value kI_s is taken as unity.

unsaturated gain of 5% and loss per unit length as 0.1% ($2\alpha_o d=10$, $2\beta d=0.2$), the general shape of the variation of power output P' with mirror transmission t is seen to be as shown in Fig. 1.18. It is seen to have an optimum value for achieving maximum power output for given excitation and cavity conditions.

For given conditions of cavity therefore, there is an optimum value of mirror transmission for which the output is dependent on the excitation of laser material and losses in the optical components. Pumping level determines α_o. Scattering and absorption losses in the active and passive components in the cavity determine β. The round trip losses are lumped in the product $2\beta d$.

1.10. Frequency of Resonant Modes

Since resonators in lasers have dimensions large compared to wavelength, they have large number of closely spaced modes. If gain medium in the resonator provides gain at several of these mode frequencies, laser output will consist of output at a number of closely spaced frequencies.

Mode has been defined as a self consistent field configuration. Optical field distribution reproduces itself after one round trip in the resonator. Longitudinal modes are sets of modes having the same spatial energy distribution transverse to the resonator axis, but having different number of half wavelengths of light along the axis of resonator. Corresponding to each longitudinal mode number (q), i.e. to a given number of half wavelengths of light along the resonator axis, there exists a set of modes which have different distributions of energy in the plane transverse to the resonator axis. These are the transverse modes of the resonator determined by the indices m,n. For a given laser resonator, the resonant frequency of a given mode is

$$\nu_{m,n,q} = (c/2d)[(q+1) + \{(1+m+n)/\pi\}\cos^{-1}(g_1 g_2)^{1/2}], \qquad (1.153)$$

where q, the longitudinal mode order number, is the number of nodes in the axial standing wave pattern (the number of half wavelengths is q+1), m and n are the transverse mode order numbers and $g_i = (1-d/R_i)$ where R_i is the radius of curvature of mirror i (i=1,2). The adjacent longitudinal modes are spaced in frequency c/2d and different transverse modes have, in general, different frequencies.

The longitudinal modes are designated by q and transverse modes by m and n. When the value of (m+n) is fixed the spectral separation between longitudinal modes corresponding to Δq=1 is given by

$$\delta\nu = c/2d. \qquad (1.96)$$

When q is fixed, the frequency separation $\Delta\nu$ between transverse modes corresponding to Δ(m+n) is

$$\Delta\nu = \Delta(m+n)(c/2d)\{(1/\pi)\cos^{-1}[(1-d/R_1)(1-d/R_2)]^{1/2}\}. \qquad (1.154)$$

For stable systems the factor within curly brackets { } varies between 0 and 1/2. For each value of (m+n) there is a longitudinal set of frequencies and for each value of q, there is a transverse set (compare Eqs. 1.97-1.98 of section 1.7.1 with Eqs. 1.153-1.154).

To observe beat frequencies between the various modes which may oscillate simultaneously within the gain profile of a given laser, one may arrange the laser beam to fall on a photodetector whose output is fed to a spectrum analyzer.

1.11. Frequency Selection of Laser Oscillation

When gain is available from the inverted medium over a band of frequencies, or at a number of discrete frequencies, it is desired to make the cavity oscillate at a given frequency. To remove the competition for oscillation between different frequencies it is desirable for most situations that the dominant transition is usually the one at longer wavelength since gain and pump power requirements are dependent on square and cube powers of frequency ν (or $1/\lambda$) respectively. Further, the diffraction losses also increase with wavelength, but at much lower rate than increase in gain. The use of dielectric layers on mirror to achieve high reflectivity at desired wavelength and low reflectivity at undesired wavelengths helps in frequency selection. The use of dispersive element such as a prism or grating in the cavity also enables the mode of one transition to be reflected perpendicular to the mirror while that of the other is reflected at an angle to the mirrors and hence suffers large diffraction losses (Fig. 1.19).

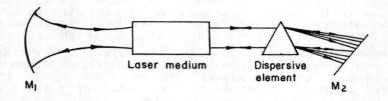

Fig. 1.19. Resonant cavity with dispersive element. Only the radiation which falls normal to mirror M_2 is reflected back with minimum loss. The tilt of mirror enables the tuning of frequency of oscillation.

1.12. Transverse Mode Selection

Once frequency selection is adopted, it is desired to restrict the oscillating mode to a given transverse mode, usually the TEM_{00} mode. This is done by putting aperture stops in the cavity at appropriate position. Higher and higher order transverse modes have their energy less and less concentrated along the axis of the laser resonator. A circular aperture in the laser resonator gives progressively higher diffraction losses as one goes to higher and higher order transverse modes. This fact enables the selection of lowest order (fundamental) transverse mode. As an example, one could put a circular aperture with opening of the size of $2w_o$ at the location of waist such that the beam is confined to the lowest order Gaussian mode in the cavity. The advantage is that the parameters of the beam (e.g. spatial variation of intensity, its wavefront and angle of divergence) are then well defined and the beam has the lowest spot size and minimum spread due to divergence. This however has the disadvantage that the mode volume in the active medium is reduced and effective use of the active medium is rather restricted. The mode volume ($\approx \pi w_o^2 d$) determines the number of effective excited atoms subjected to a high radiation density and it increases for the higher order modes. The confocal mirror configuration is particularly undesirable for reason of small mode volume. The resonator with mirrors of large radii of curvature is attractive for reason of large mode volume. The unstable resonator also provides the advantage of large mode volume.

An arrangement for having large mode volume and at the same time operation in TEM_{oo} mode with aperture control is shown in Fig. 1.20. This is the 'cat's eye' resonator and provides larger output in the lowest order mode.

Higher order transverse modes can be selected by the use of complex apertures or reflectors with certain regions of low reflectivity which discriminate against all but the desired mode. For some applications it may be desirable to have the ability to switch from one transverse mode to another.

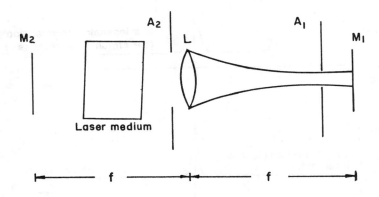

Fig. 1.20. The cat's eye resonator. M_1, M_2 are mirrors, L is a lens of focal length f and A_1, A_2 are aper- aperture stops.

1.13. Longitudinal Mode Selection

As stated earlier, the longitudinal modes of a simple resonator are spaced in frequency by $c/2d$ where c is velocity of light and d is the mirror spacing. It is always possible to increase the resonator losses (Fig. 1.21) to the

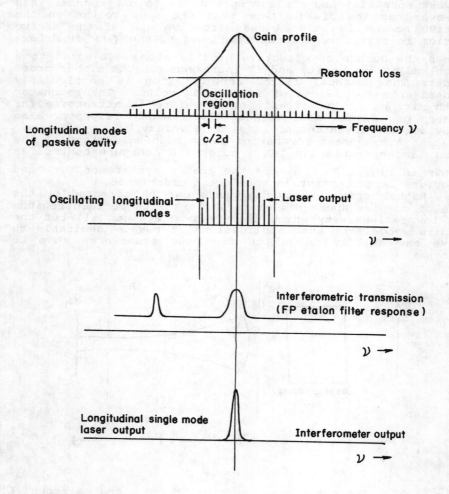

Fig. 1.21. Longitudinal mode selection. Use of interferometer filter for obtaining single mode output.

point that only one or a few longitudinal modes are above threshold for oscillation. In this way we get narrow bandwidth laser operation but at the expense of output power. A somewhat more practical approach is to reduce the length of the laser resonator and thus increase the frequency spacing between longitudinal modes to the point that only one mode is above threshold for oscillation. Although this also sacrifices output power to some extent, nevertheless this is a useful technique for constructing a practical single frequency laser. The use of Fabry Perot etalon as mode filter is based on interferometric technique (Fig. 1.21). Tilted FP etalon is also used for mode selection whereby the angle of tilt enables frequency tuning.

1.14. Mode Competition

When laser gain medium is placed in the resonator, optical energy begins to build up in those resonator modes for which the net gain exceeds the losses (Fig. 1.21). The oscillation bandwidth, with no mode control, is governed by the width of the gain curve. The total output of the laser as a function of time will depend upon the amplitudes, frequencies and phases of the oscillating modes. Due to random phase fluctuations the output will fluctuate in a random way as a function of time. One way to control this is to fix the amplitudes and phases of the modes. This is called mode locking of a laser (It will be described later in detail). This technique is useful when an output pulse train is desired. Alternatively, mode selection techniques may be used to obtain single frequency CW laser output.

The presence of the laser medium will affect the transverse mode distributions in the resonator when very high gain system with possibility of thermal gradients in the cavity are present. The dispersive effects of the laser medium will modify the mode frequencies of the passive cavity; again the effect is significant only for the high gain systems.

For most lasers, the major effect of the gain medium on the oscillating laser modes is that of mode competition. Fig. 1.5 shows the gain saturation of homogeneously and inhomogeneously broadened resonance lines under the influence of a travelling wave of intense monochromatic radiation at frequency ν'. For an inhomogeneous line it is possible to 'burn a hole' i.e. saturate only those atoms with resonant frequencies close to that of the incident radiation. The half width of the hole $\Delta\nu_h$ in the population inversion versus frequency curve is

$$\Delta\nu_h = \Delta\nu \ (1+I/I_s)^{1/2}, \qquad (1.155)$$

where $\Delta\nu$ is the homogeneous line width (half width of the atomic response of an individual atom), I is the incident light intensity and I_s a saturation parameter involving atomic constants. Radiation in a single resonator mode can interact with atoms over a frequency range of $\sim 2\Delta\nu_h$. If resonator mode spacing is less than $2\Delta\nu_h$ then several resonator modes will be interacting with the same group of atoms. This is called mode competition since a given atom can only contribute its energy to one of the modes and this energy is then unavailable for the others. If mode suppression technique is used to favour one mode within $2\Delta\nu_h$ frequency region, this mode will compete more favourably for the available energy and grow at the expense of other less favoured modes. Mode competition can thus help oscillation on a single resonator mode.

1.15. Hole Burning

The term hole burning is used to describe the selective depletion of the population. Hole burning occurs in predominantly inhomogeneously broadened lines.

Hole burning occurs at frequencies near each oscillating mode and Eq. 1.155 shows the dependence of the width of hole on the power in the mode and the natural line width. At oscillation threshold, the hole width is simply the natural line width. As the laser field intensity increases, the hole width increases.

In a gas discharge laser, broadening is by Doppler effect and is inhomogeneous. In the two mirror resonator, the radiation consists of approximately equal intensities of light travelling in each direction and standing waves are formed. In a inhomogeneously broadened medium with Doppler broadening due to distribution of thermal velocities, a single frequency of radiation will in general, interact with two groups of atoms, one in resonance with light in one direction and the other with light travelling in opposite direction. Since the oscillation consists of waves travelling in both directions along the laser axis, atoms with both positive and negative velocity components can contribute to radiation at a given frequency. Thus each mode frequency depletes the population of the groups of atoms, one group with velocity ($+\nu$) and the other with velocity ($-\nu$) and, provided the mirrors are of equal reflectivity, the holes burned in the gain curve are symmetrical about the centre (Fig. 1.22). There are two holes for one resonance at all resonant frequencies above threshold, unless the frequency coincides with the line centre, in which case of course, there is only one.

Fig. 1.23a shows two holes burned in the gain profile radiation in single mode at frequency ν'. When ν' coincides with the atomic resonance frequency, both travelling waves will interact with the same group of atoms (those with zero axial velocity, Fig. 1.23b). Since at line centre fewer

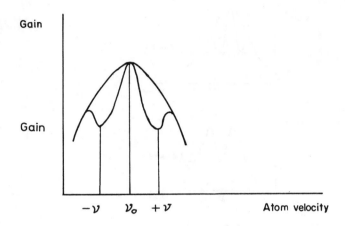

Fig. 1.22. Symmetrical hole burning in gain curve.

atoms give energy to the laser mode, the output power is reduced (Fig. 1.23c). This provides the so called 'Lamb dip' at the centre of the output power versus frequency curve. In the Lamb dip [1.18] the two holes overlap at the centre of the gain curve and there is a dip in the power output of the laser. This fact is utilized for mode stabilization in a laser [1.19]. Also, when I is small (Eq. 1.155), $\Delta\nu_h \to \Delta\nu$ and the Lamb dip gives a measure of the homogeneous line width.

1.16. Mode Pulling

In a passive cavity (evacuated resonator cavity with refractive index n=1) the oscillation frequency is at frequency ν_o say. However, when the amplifying medium is introduced, it changes the refractive index and hence the optical path length so that the system oscillates at a different frequency ν_a. Part of the change in refractive index is due to the unexcited particles of the medium and part is due to the excited particles. This change in refractive index leads to mode pulling [1.20].

The phase shift ϕ resulting from a wave travelling once through an interferometer of length d at a phase velocity c/n is

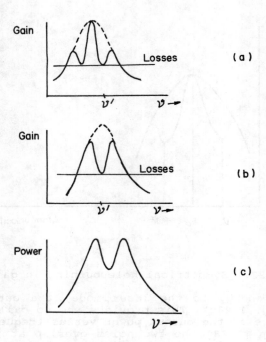

Fig. 1.23. Plots of laser gain as function of frequency. (a) single mode at ν' away from atomic line centre, (b) single mode at ν' at line centre and (c) single mode output power as function of frequency.

$$\phi = 2\pi\nu d\, n/c. \qquad (1.156)$$

For standing wave to build up in the cavity, the single pass phase shift must be an integral multiple of π. Hence from Eq. 1.156 for the passive cavity (n=1) the resonant frequencies are seen to be separated by

$$\Delta\nu_o = (\nu_{m+1} - \nu_m) = (\phi_{m+1} - \phi_m)c/2\pi d = \pi c/2\pi d = c/2d$$

and

OPTICAL RADIATION AND PHOTONS 69

$$(\partial\phi/\partial\nu)_o = 2\pi d/c. \tag{1.157}$$

If energy U is placed in the evacuated cavity in the mode of interest, this energy decays exponentially with time at the rate (cf/d) where f is the fractional energy loss per pass. Using Eqs. 1.139a and 1.140, the Q of evacuated cavity is then

$$Q = 2\pi\nu_o U/(cf/d)U = \nu_o/\Delta\nu_o \tag{1.158}$$

so that from Eqs. 1.157 and 1.158

$$\partial\phi/\partial\nu = 2\pi d/c = f/\Delta\nu_o. \tag{1.159}$$

The introduction of the amplifying medium changes the refractive index in the system thereby altering the single pass phase shift from that obtained in the evacuated case. Oscillation therefore occurs at another frequency ν_a, differing from the cavity resonance ν_o such that the single pass phase shift is still an integral multiple of π. Since the cavity dispersion $\partial\phi/\partial\nu$ is large compared to that for the amplifying medium, oscillation occurs close to ν_o and pulling is small. It is convenient to consider the problem in terms of the difference in frequency from the cavity resonance so that

$$(\partial\phi/\partial\nu)(\nu_a - \nu_o) + \Delta\Phi_m(\nu_a) = 0, \tag{1.160}$$

where $\Delta\Phi_m(\nu_a)$ is the total change in single pass phase shift at the actual frequency of oscillation which is caused by the insertion of the medium. $\Delta\Phi_m(\nu_a)$ is composed of two parts and may be expressed (using Eq. 1.156) as

$$\Delta\Phi_m(\nu_a) = (2\pi d/c)[(n_o-1) + \delta n]\nu_a$$
$$= (f/\Delta\nu_o)(n_o-1)\nu_a + \Delta\phi_m(\nu_a). \tag{1.161}$$

The first term in Eq. 1.161 arises from a refractive index which is essentially independent of frequency over the range of interest whereas the second term is frequency dependent. From Eqs. 1.159 and 1.161, Eq. 1.160 can be expressed as

$$\nu_a = (\nu_o/n_o) - (\Delta\nu_o/n_o f)\Delta\phi_m(\nu_a).$$

Writing

$$\nu_c = \nu_o/n_o, \quad \Delta\nu_c = \Delta\nu_o/n_o \tag{1.162}$$

oscillating frequency is expressed as

$$\nu_a = \nu_c - (\Delta\nu_c/f)\,\Delta\phi_m(\nu_a). \tag{1.163}$$

The term $\Delta\phi_m(\nu_a)$ is a function of the fractional energy gain per pass. Since the latter varies with frequency over the transition responsible for the amplification, $\Delta\phi_m$ does also. Generally, $\Delta\phi_m$ goes through zero at the line center (ν_m), becomes a negative (anomalous dispersion) for frequencies less than ν_m and is positive for frequencies greater than ν_m. Eq. 1.163 predicts a shift in the direction of the line center (Fig.1.24)

For a homogeneously broadened Lorentzian line, it has been shown [1.20] that

$$\nu_a = (\nu_c \Delta\nu_m + \nu_m \Delta\nu_c)/(\Delta\nu_m + \Delta\nu_c), \tag{1.164}$$

where ν_a is the oscillation frequency and $\Delta\nu_m$ is the full

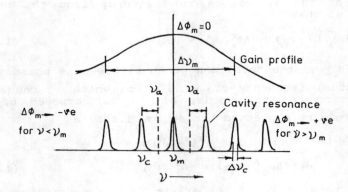

Fig. 1.24. Frequency pulling effect. ν_a is actual lasing frequency. ν_c is frequency of cavity resonance. ν_m is the centre frequency of gain profile. When $\nu < \nu_m$, the lasing frequency ν_a shifts towards ν_m.

width of the line at half maximum energy gain.
In the limit $\Delta\nu_c < \Delta\nu_m$, Eq. 1.164 can be rewritten as

$$\nu_a = \nu_c + (\nu_m - \nu_c)(\Delta\nu_c/\Delta\nu_m)[1-(\Delta\nu_c/\Delta\nu_m) + ..] \qquad (1.165)$$

showing thereby that in the Lorentzian case, the pulling is linearly dependent on the frequencies. This is of course not true in the case of inhomogeneously broadened Lorentzian line above threshold.

For the Gaussian line, the pulling is nonlinear even in the homogeneously broadened case. For frequencies very near the line centre in the limit $\Delta\nu_c < \Delta\nu_m$,

$$\nu_a = \nu_c + (\nu_m - \nu_c)(0.94\Delta\nu_c/\Delta\nu_m)[1-0.9(\Delta\nu_c/\Delta\nu_m) + --]. \qquad (1.166)$$

The main difference between the Gaussian and the Lorentzian is a 6% reduction in the pulling factor. Generally for lasers, $\Delta\nu_m \gg \Delta\nu_c$ and $\Delta\phi_m(\nu)$ will be a slowly varying function of the frequency over the cavity resonance. We may therefore expand $\Delta\phi_m$ about the cavity resonance frequency and get

$$\Delta\phi_m = (\Delta\phi_m)_{\nu_c} + (\partial\Delta\phi_m/\partial\nu)_{\nu_c}(\nu-\nu_c) + --- . \qquad (1.167)$$

Substituting Eq. 1.167 in Eq. 1.163 and noting that

$[\Delta\phi_m/(\nu-\nu_c)]_{\nu_c} \simeq [\partial\phi_m/\partial\nu]_{\nu_c} = f/\Delta\nu_c$ from Eq 1.159, we get

$$\nu_a = \nu_c - (\Delta\nu_c/f)\Delta\phi_m(\nu_c)[1-(\Delta\nu_c/f)(\partial\Delta\phi_m/\partial\nu)_{\nu_c} + --]. \qquad (1.168)$$

Comparing Eq. 1.165 and Eq. 1.166, the last approximation is equivalent to an expansion of the pulling terms in powers of $(\Delta\nu_c/\Delta\nu_m)$ which still retains the nonlinear properties of the phase characteristics. Typically $\Delta\nu_c/\Delta\nu_m \sim 10^{-3}$. The oscillator frequency may therefore be approximated from Eq. 1.168 by

$$\nu_a \simeq \nu_c - (\Delta\nu_c/f)\Delta\phi_m(\nu_c), \qquad (1.169)$$

where errors in the second term of about 1 part in 10^3 may be expected. $\Delta\phi_m(\nu_c)$ represents the actual phase shift introduced by the amplifying transition at the cavity resonance in the presence of oscillation. Eq. 1.169 may be solved numerically in the general case for both homogeneous and inhomogeneous broadening.

1.17. Frequency Stability of Laser Output

Frequency stability of a laser is dependent on control of environment changes. Small changes in ambient temperature and pressure can cause significant changes in the frequency of oscillation. The theoretical limit of stability is about 1 part in 10^{14}. However, there are several factors, including environment changes, which cause the stability in practice to fall short of this limit.

a) Many longitudinal and transverse modes may oscillate simultaneously within the transition line profile since cavity resonances are much narrower than the atomic resonance. The multimode output thus covers a range of frequencies.

b) The stability of cavity resonance depends on stability of cavity dimensions which in turn depends on mechanical vibrations and thermal fluctuations. The longer the cavity, the greater this problem. A change of optical length by $\lambda/2$ will change the laser frequency by longitudinal mode spacing $c/2d$. This, for one meter resonator, is 150 MHz.

c) The refractive index of the active medium, n, determines the optical length nd of the cavity. A change in optical length of $\lambda/2$ is caused by a variation of refractive index of $\lambda/2d$ which in turn causes a frequency shift of $c/2d$. Such changes in n are possible even in low pressure gases.

d) Long term drifts result from changes in the ambient temperature and pressure. The frequency shift $\Delta \nu$ due to pressure and temperature changes is approximately given by [1.7]

$$|\Delta \nu| = \nu_o (\alpha \ \Delta T + 3.63 \cdot 10^{-7} \Delta p \ x), \qquad (1.170)$$

where α is the linear expansion coefficient of the cavity spacers, ΔT is the temperature change (°C), Δp is the pressure change (in torr) and ν_o is the centre frequency of resonance. Typically, uncompensated invar cavities, with x=0.1, have a drift of 500 MHz/°C, and 20 MHz/torr, for the 632.8 nm wavelength.

e) Many lasers operate in significant magnetic fields caused by electric discharges which cause Zeeman splitting. Lasers with high current flash tubes may cause this effect.

f) Pulsed lasers have a bandwidth of at least $1/\Delta \tau$, where

$\Delta\tau$ is the pulse width.

g) Variation of electron density in the electric discharge of a gas laser results in variation of refractive index and hence variation in optical path length. The wavelength shift can be calculated from the relation

$$n = [1-(4\pi e^2 \mu / m\omega^2)]^{1/2} \tag{1.171}$$

and $\delta\nu = c/2nd$,

where n is refractive index, ω the angular frequency, μ is the mean electron density and e and m are the electronic charge and mass respectively.

h) Frequency may also vary due to mode jumping. Usually several modes oscillate simultaneously within the profile of the particular transition. Oscillation of a mode depletes the population and since pumping process is usually quite random in its selection of atoms, the excited atoms needed to maintain a given mode may not be supplied and the mode may go out of oscillation. Thus modes may get established or go out of oscillation quite randomly.

The frequency stability of a laser can be improved by reducing environment related effects by building mechanically stable resonators shielded from air pressure fluctuations and providing temperature control. It has often been more practical however to build electronic feedback systems to stabilize the laser.

1.18. Single Mode Operation of a Laser

For many applications, it is desired to have a stabilized single mode laser for high spectral purity and good coherence properties. Such a laser is first confined to lowest order transverse mode (TEM$_{00}$) and then made to oscillate on a single longitudinal mode under controlled environment.

Operation on a single axial mode can be realized in a number of different ways utilizing a variety of multiple mirror resonator configurations to achieve the appropriate limitations of resonances. One representative method utilizes an intra-cavity Fabry Perot etalon as shown in Fig. 1.25. The etalon is inserted at the position of the beam waist in the cavity with a tilt angle that provides sufficient reflection loss to prevent oscillation of all the unwanted axial modes of the cavity. Oscillation occurs at a transmission maximum of the etalon. This maximum is made to coincide with a laser cavity mode near the centre of the gain profile. The optimum design of the scheme and its use for stabilizing the output is described in reference [1.21].

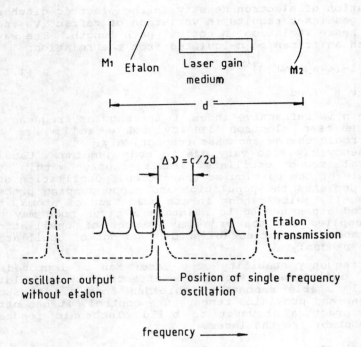

Fig. 1.25. Single frequency operation of a laser using tilted etalon in the cavity.

1.19. Coherence of Laser Radiation

Laser radiation is characterized by high intensity, narrow spectral width, directionality and high degree of coherence compared to conventional light sources. The laser radiation has a different statistical nature to that from thermal sources. The distinctive properties of laser essentially arise on account of high intensity and a reasonably high degree of coherence associated with it.

Coherence implies a constant phase difference between two points on a series of equal amplitude wavefronts and a correlation in time between the same points on different wavefronts. If at a particular time the radiation has the same phase every where across the uniphase wavefront, it is said to be completely spatially coherent. It is also possible to compare the phase of radiation on one travelling uniphase wavefront at a particular time with its phase on the displaced wavefront after it has travelled a distance s in a time interval s/c where c is the velocity of light. If these phases agree for all time intervals, the radiation is said to be completely time coherent. A perfectly monochromatic spherical wave or plane wave of infinite extent is completely coherent in both space and time. It is however not necessary to restrict the definition to uniphase modes. Any pure lossless mode of a cavity is completely coherent in space and time. The phase pattern in both space and time is fixed for all other points in space and time by its specification over one surface at one particular time. Further, a superposition of lossless modes of a cavity may also be considered coherent provided coherence of radiation field is identified with a unique determination of its value over space and time.

In principle coherence does not imply either plane waves or monochromaticity. However, single mode lasers are essentially monochromatic and most useful modes of oscillation are uniphase ones, which are either essentially plane waves or can be converted to them. For most purposes therefore, criteria of coherence based on quasi monochromatic quasi plane wave are quite adequate for lasers.

In measurement of coherence, the time (or space) resolution of the detector limits the precision with which the actual coherence of the radiation can be measured. Normally observation of fringes in interference effect is used for coherence measurement. Optical coherence can be studied by classical or quantum mechanical methods.

1.19.1 Time Coherence

To obtain a measure of time coherence, an arrangement like the Michelson interferometer may be adopted in which the phase of light at two different times, corresponding to the path difference, is compared at the same point in space (Fig. 1.26). In the limit of long observation time 2τ the intensity I of the recombined beams is given by

$$I = \lim_{\tau \to \infty} (K/2\tau) \int^{\tau} [E(t+T)+E(t)][E^*(t+T)+E^*(t)]dt,$$

where K is a constant of proportionality and T is the time difference arising due to path difference between beams of amplitude E.

With I_0 as the intensity of each component of the two combining beams we can express Eq. 1.171 as

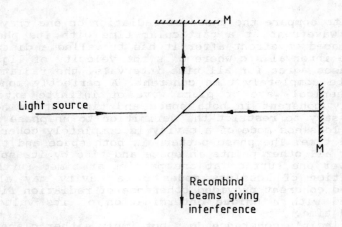

Fig. 1.26. Michelson Interferometer arrangement for providing a measure of coherence.

$$I = 2I_o + K[\overline{\Gamma}(T) + \Gamma(T)] \qquad (1.172)$$

with

$$\Gamma(T) = \lim_{\tau \to \infty} (1/2\tau) \int_{-\tau}^{\tau} E(t+T) E^*(t) dt$$

$$\overline{\Gamma}(T) = \lim_{\tau \to \infty} (1/2\tau) \int_{-\tau}^{\tau} E^*(t+T) E(t) dt \qquad (1.173)$$

and

$$K\Gamma(o) = I_o.$$

$\Gamma(T)$ is known as the autocorrelation function of the field and, in the normalized form, is given by

$$\gamma(T) = \frac{|\Gamma(T)|}{\Gamma(o)} = \frac{K|\Gamma(T)|}{I_o}. \qquad (1.174)$$

γ lies between 0 and 1. To associate interference fringes with correlation function, we note that

$$I_{max} = 2I_o + 2K|\Gamma(T)|$$

$$I_{min} = 2I_o - 2k|\Gamma(T)|$$

and fringe visibility,

$$V = \frac{(I_{max} - I_{min})}{(I_{max} + I_{min})}$$

$$= \gamma(T). \tag{1.175}$$

For the determination of temporal coherence of a single mode laser, we note that for the emission

$$E = E_o \cos(\omega t + \Psi)$$

only the phase perturbations $(\overline{\Delta \Psi})^2$ contribute to incoherence. It can be shown [1.22] that for one photon added by spontaneous emission

$$(\overline{\Delta \Psi})^2 = \frac{1}{N}, \tag{1.176}$$

where N is the number of photons already present in the mode. If there are n spontaneous emissions per unit time, the mean squared deviation in oscillator phase after a time T will be

$$(\Delta \Psi)^2 = nT(\overline{\Delta \psi})^2 = \frac{nT}{N}, \tag{1.177}$$

where $\Delta \Psi$ is $\sum \Delta \psi$ summed over all the emissions occurring in time T. For laser oscillator, the rate of spontaneous emission can be expressed in terms of τ_c, the cavity decay time by

$$n = 1/\tau_c \tag{1.178}$$

so that

$$(\overline{\Delta \Psi})^2 = T/N \tau_c. \tag{1.179}$$

The power spectral density of the emitted radiation has the form

$$I(\omega) \simeq \frac{1}{(\Delta \omega)^2 + (1/2N\tau_c)^2}. \tag{1.180}$$

At full width half maximum (FWHM) power $I(\omega) = 1/2$ giving

$$2\Delta\nu = \frac{1}{2\pi\tau_c N} \quad . \tag{1.181}$$

Since $1/2\pi\tau_c = 2\Delta\nu_c$, where $2\Delta\nu_c$ is the FWHM line width of passive cavity resonance, we get

$$\Delta\nu = \Delta\nu_c/N \tag{1.182}$$

Further, for the oscillation function of Eq. 1.180 we define the coherence time T_c given by

$$T_c = 2^{1/2} N \tau_c. \tag{1.183}$$

Comparison of Eq. 1.183 with Eq. 1.181 relates T_c with $\Delta\nu$

$$T_c = [1/2(2)^{1/2}\pi](1/\Delta\nu). \tag{1.184}$$

Temporal coherence time is thus proportional to $1/\Delta\nu$.

1.19.2. Spatial Coherence

The spatial coherence of a laser oscillator is determined primarily by its mode selection properties. A source may have very little time coherence and still have essentially perfect space coherence. The space coherence of a quasi monochromatic nearly plane wave may be given a quantitative interpretation in terms of the visibility of the interference fringes observed from two pinholes in a mask that is placed normal to the wavefront (Fig. 1.27).

The space coherence defined by

$$\Gamma(r_1, r_2) = \lim_{\tau \to \infty} (1/2\tau) \int_{-\tau}^{\tau} E(r_1, t) E^*(r_2, t) dt \tag{1.185}$$

and its normalized form as

$$\gamma(r_1, r_2) = |\Gamma(r_1, r_2)|/(\Gamma_{11}\Gamma_{22})^{1/2} \tag{1.186}$$

are related to the fringe visibility

$$V = 2\gamma(r_1, r_2)/[(I_1/I_2)^{1/2} + (I_2/I_1)^{1/2}], \tag{1.187}$$

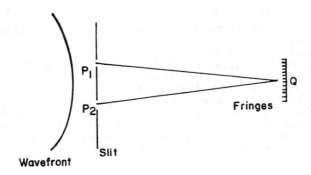

Fig. 1.27. Two slit arrangement for observing spatial coherence by interference fringes.

where I_1 is the intensity at the point of observation of the fringes arriving from P_1 alone and I_2 that arising from P_2 alone. This is seen by considering

$$I = [E(r_1,t)+E(r_2,t)][E^*(r_1,t)+E_2^*(r_2,t)]$$
$$= (I_1+I_2)+k[\Gamma(r_1,r_2)+\overline{\Gamma}(r_1,r_2)]$$

and other steps as in section 1.19.1.

With reference to Eq. 1.187, the distance between r_1 and r_2 at which the fringes begin to be washed out is a measure of the (transverse) space coherence. By contrast the longitudinal coherence length cT_c is a measure of the coherence time.

1.19.3. Time and Space Coherence

Though we have considered time and space coherence separately, it is possible to combine the definition of $\Gamma(T)$ and $\Gamma(r_1,r_2)$ into a single measure of both time and space coherence in the mutual coherence function $\gamma(r_1,r_2,T)$. It is obtained by replacing $E^*(r_2,t)$ by $E^*(r_2,t+T)$ in Eqs. 1.185

to 1.187 so that $\gamma(r_1,r_2)$ becomes $\gamma(r_1,r_2,T)$ and it measures the coherence in the vicinity of a point Q such that the time difference for light signal from r_1 and r_2 to Q is cT. This function allows us to visualize coherence properties at any point in space.

Traditionally, coherence is related to the ability of two light beams to give rise to interference fringes on superposition. The two light beams might originate from a single space point at two different times (Michelson interferometer), from two space points at the same time (symmetrical position in Young's slits) or different space and time points together (general position in Young's slit experiment). Both spatial and temporal coherence effects are characterized by a single mutual coherence function $\Gamma_{12}(T)$ given by

$$\Gamma_{12}(T) = <E_1(t+T)E_2^*(t)>_{time\ avg} \qquad (1.188)$$

$$= \lim_{\tau \to \infty} (1/2\tau) \int_{-\tau}^{\tau} E_1(t+T)E_2^*(t)dt,$$

where light disturbance at point r_1 at time t+T is $E_1(t+T)$ and at point r_2 and time t is $E_2(t)$. The normalized function is

$$\gamma_{12}(T) = \frac{\Gamma_{12}(T)}{[\Gamma_{11}(0)\Gamma_{22}(0)]^{1/2}}. \qquad (1.189)$$

Interference due to superposition of beams is obtained only when $|\Gamma_{12}(T)|$ is nonzero. Modulus of $\gamma_{12}(T)$ lies between 0 and 1, zero for incoherene and unit for complete coherence. Intermediate values designate conditions of partial coherence. Temporal and spatial coherence are described by $|\gamma_{11}(T)|$ and $|\gamma_{12}(0)|$ respectively. $\Gamma_{11}(T)$ is the autocorrelation function of $E_1(t)$, $\Gamma_{12}(T)$ is the cross-correlation function of $E_1(t)$ and $E_2(t)$. Their Fourier transforms are given by

$$G(\nu) = \int_{-\infty}^{\infty} \Gamma_{11}(T) \exp(2\pi j\nu T) dT, \qquad (1.190)$$

$$G_{12}(\nu) = \int_{-\infty}^{\infty} \Gamma_{12}(T) \exp(2\pi j\nu T)\,dT. \qquad (1.191)$$

Eq. 1.190 gives power spectrum (spectral density) of $E_1(t)$ while Eq. 1.191 gives cross (mutual) spectral density of $E_1(t)$ and $E_2(t)$. The fringe visibility is given by

$$V = \frac{2|\gamma_{12}(T)|(I_1 I_2)^{1/2}}{(I_1+I_2)}. \qquad (1.192)$$

When intensities at observation point, due to each source separately, are equal, the visibility is just $|\gamma_{12}(T)|$. For completely coherent light, the maximum value of $|\gamma_{12}(T)|$ with respect to T, is unity. The argument of $\gamma_{12}(T)$ signifies location of the intensity maxima in the fringe pattern.

Generally it is not possible to treat temporal and spatial coherence independently, since one type of coherence can be transformed into the other through the propagation of the field. Also spatially incoherent light may become partially coherent through the process of propagation as in the case of starlight. The passive laser cavity, through the process of propagation and diffraction, imparts spatial coherence to the spatially incoherent quasi monochromatic light. The gain medium improves temporal coherence by spectral narrowing.

1.19.4. Transient Coherence

Interference fringes are observed between two independent oscillators of frequency ν provided the time constant of observation is sufficiently small. If we consider the mutual coherence function Γ_{12} for two monochromatic sources of different frequencies ν_1, ν_2 then for long observation time the value tends to zero. However, if the expression is integrated over a finite interval of time τ, then it can be shown that for observation time τ less than the reciprocal of the frequency separation $\nu_2-\nu_1$, Γ_{12} is finite and transient interference effects may be observed. If instead of monochromatic, we consider quasi monochromatic sources, there is additional requirement that observation time be less than the reciprocal of the bandwidth. With the advent of gas lasers such effects have become observable, since it is possible to arrange two separate sources with bandwidths of the order of a couple of hundred Hertz and mean frequencies about a kHz apart. Transient interference effects can then be observed for times of the order of a millisecond. With thermal sources, the bandwidth limitation by itself usually makes it difficult to observe transient interference effects, since very fast detectors are required.

1.19.5. Higher Order Coherence Functions

Eqs. 1.173 and 1.185 are not entirely sufficient for describing coherence properties of light though, as first order expressions, they are convenient to use. They are intended to be used in situation where measuring instruments record average values of E^2 i.e. intensity of light. These measurements, however, can be extended to (i) observe correlations in fluctuations of intensity at two or more points in space and time or (ii) measure at one point quantities proportional to higher powers of E^2. Expressions similar to Eqs. 1.173 and 1.185 involving higher powers of E have been described classically [1.23] as well as quantum mechanically [1.24].

In the photon correlation experiments where correlation between counts of two photon detectors placed at different parts of a beam are measured, it has been observed that though a correlation would exist for a coherent beam to first order, for the second order coherence there would be no measurable photon correlation. Thus light which is coherent to first order may be much less coherent to second order. Alternatively there may be laser beams which are coherent to a higher order.

It can be shown that correlation in the intensity fluctuation is

$$<\Delta I_1(t+T) \Delta I_2(t)> = |\Gamma_{12}(T)|^2. \qquad (1.193)$$

By measuring correlation in intensities or intensity fluctuations, it is possible to evaluate the mutual coherence function without recourse to more conventional second order interference experiments based on observation of fringes.

Measurement of correlation in intensity fluctuations is the basis of 'Hanbury-Brown and Twiss Interferometer' first used for radio frequency radiation and later extended to optical region [1.25].

1.19.6. Factors Responsible for Imparting Coherence to Laser Radiation

In considering radiation from a laser, we are particularly concerned with light that has a high degree of spatial and temporal coherence. The completely coherent light requires that

$$\max |\gamma_{12}(T)| = 1 \quad (1.194)$$

for all pairs of points r_1 and r_2 within the light beam. We take maximum value of the modulus with respect to T since, in general, it is a function of T and it is this maximum value that determines the visibility of the fringes. Expressed in another way, the mutual coherence function $\Gamma_{12}(T)$ for completely coherent light is of the form

$$\Gamma_{12}(T) = U_1 U_2^* \exp(-2\pi j \nu T) \quad (1.195)$$

for a range of values of T short compared to the coherence time T_c. Here U_1 and U_2 are functions only of the coordinates of the points r_1 and r_2 respectively and are independent of T.

The frequency filtering effect of the cavity resonance of laser resonator helps in improving temporal coherence only slightly. A significant contribution to the improvement of temporal coherence comes from the properties of the active medium (see section 1.20 Eqs. 1.207 and 1.210 for spectral narrowing due to gain of medium).

For the passive cavity, the coherence time is simply $2\pi\tau_c$ ($T_c = 1/\Delta\nu_c = 2\pi\tau_c$), whereas the active medium in the cavity improves it to $N\tau_c$ (for the Lorentzian profile, Tc= $2^{1/2}N\tau_c$ from Eq. 1.183). For τ_c of the order of 1 μs, T_c for passive cavity would be of the same order while for the active cavity with N ~ 10^{10} it would become ~10^4 s or 3 hr. This gives a line width $\Delta\nu$ ~ $1/N\tau_c$ ~ 10^{-4} Hz which is much less than what is practically obtained. The reason for the practical limitation is the other fluctuations in laser properties (not related to line narrowing by radiation density) such as thermal and acoustic fluctuations in the optical resonator length which cause the resonance frequencies to shift about rapidly and hence the need for stabilization. The practical limitations restrict the spectral width to a few Hz.

The properties of, and diffraction of radiation through, the passive cavity imparts spatial coherence to the field. The high degree of spatial coherence found in laser radiation is a consequence of the cavity properties alone. The active medium, through the process of stimulated emission, maintains the radiation field against the losses, as the radiation propagates from one mirror to the other in the cavity, but does so without altering the spatial coherence of the field.

1.20. Laser Noise

There are essentially two related types of noise arising from spontaneous emission always present in lasers. One is random noise accompanying the coherent emission of a laser which gives rise to low frequency fluctuations. The other is the added optical noise generated in the laser amplifier referred to the input end of the amplifier as compared to the signal entering the amplifier. The two types have been referred some times as the laser oscillator noise and the laser amplifier noise.

The low frequency contribution of noise arises from the beating of noise components with a frequency difference within the bandwidth of the detector. This component however becomes less and less significant as power of the laser output increases,i.e. there is a sharp decrease in the mean squared fluctuations as the laser oscillates more and more strongly.

A wave propagating through the inverted medium grows coherently due to stimulated emission. However, this radiation gets contaminated by noise due to spontaneous emission. This results in laser output to have finite spectral width and makes the signal-to-noise ratio (S/N) limited.

The signal-to-noise ratio (SNR) in the presence of spontaneous emission noise for an optical amplifier can be estimated if we imagine a detector placed on the optic axis very far away from the amplifier and the detector is allowed to accept only radiation emitted into a solid angle $\Omega=\lambda^2/A$ where A is the area of aperture stop in the output plane which provides the passage of signal. If we insert a polarizer in front of the detector, the noise power received is [1.26]

$$N_o = \frac{h\nu \, d\nu \, N_2}{N_2 - N_1 (g_2/g_1)} (G-1), \qquad (1.196)$$

where N_2 and N_1 are populations in upper and lower level of the transition and G is the gain of the amplifier. Assuming that the signal arrives at the output plane of the amplifier with power S_o, the SNR at the detector is S_o/N_o, neglecting the noise contribution from all other sources except the amplifier. This is the ideal SNR value which is always higher than the actual obtained, since the far field conditions are not always obtained in the laboratory. Kogelnik and Yariv [1.26] have described some arrangements by which SNR can be improved under practical conditions. One simple arrangement is shown in Fig 1.28 where the apertures A_1 and

A_2 have to satisfy the condition

$$A_1 A_2 / f^2 \lambda^2 = 1. \tag{1.197}$$

Effect of spontaneous emission noise on the spectral distribution of laser output is viewed by representing a laser oscillator by an LCR circuit in which presence of laser medium with gain (negative loss) is accounted for by including a negative conductance ($-G_a$) while the ordinary loss mechanisms are represented by the positive conductance G_1.

In analogy to the Johnson noise for positive conductance, the noise generator associated with losses in a bandwidth $\Delta\nu$ is

$$\overline{(i_N^2)}_1 = 4h\nu G_1 \Delta\nu / [\exp(h\nu/kT_1) - 1], \tag{1.198}$$

while for the spontaneous emission in gain medium

$$\overline{(i_N^2)}_a = -4h\nu G_a \Delta\nu / [\exp(h\nu/kT_a) - 1], \tag{1.199}$$

where T_1 is actual temperature and T_a is the temperature determined by the population ratio according to

$$(g_1/g_2)(N_2/N_1) = \exp(-h\nu/kT_a). \tag{1.200}$$

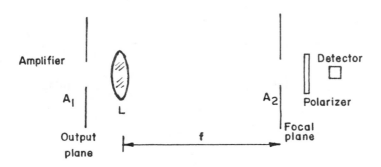

Fig. 1.28. Noise reduction by an iris in the focal plane.

It is easy to see the relevance of Eq. 1.196 in view of Eqs. 1.199 and 1.200.

We note that for the inverted medium, $N_2 > (g_2/g_1)N_1$, so that, when T_a is negative, $\overline{(i_N^2)}_a$ in Eq. 1.199 becomes positive. We can thus represent an equivalent circuit of the laser oscillator (Fig. 1.29) and write

$$Q = 2\pi\nu_o C/(G_1 - G_a) \qquad (1.201)$$

$$Z(\nu) = \frac{1}{[(G_1 - G_a) + (-j/2\pi\nu L) + j2\pi\nu C]}, \qquad (1.202)$$

where $\nu_o^2 = 1/4\pi^2 LC$, Q is quality factor, and Z is the impedance.

For a current source with a complex amplitude $I(\nu)$, the voltage across impedance Z, near the fre/quency $\nu = \nu_o$ becomes

$$|\overline{V(\nu)}|^2 = \frac{|\overline{I(\nu)}|^2}{16\pi^2 C^2 [(\nu_o - \nu)^2 + (\nu_o^2/4Q^2)]}. \qquad (1.203)$$

Since the driving current sources $\overline{(i_N^2)}_1$ and $\overline{(i_N^2)}_a$ are not correlated, $|I(\nu)|^2$ can be considered as the sum of their mean square values so that, using Eqs. 1.198 to 1.200,

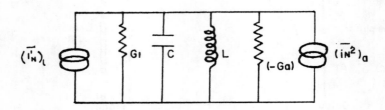

Fig. 1.29. Equivalent circuit of a laser oscillator. LC represents the resonant structure, G_1 the various loss mechanisms and G_a the gain mechanism.

$$|\overline{I(\nu)}|^2 = 4h\nu \left[\frac{G_a N_2}{N_2-(g_2/g_1)N_1} + \frac{G_1}{[\exp h\nu/kT_1 - 1]} \right] d\nu. \qquad (1.204)$$

At room temperature T=300°K, for λ=1 μm, $h\nu/kT \approx 50$. Also near oscillation, $G_a \approx G_1$. Therefore second term in bracket (Eq. 1.204) is neglected giving thereby

$$|\overline{V(\nu)}|^2_{\nu=\nu_o} = \frac{hG_a}{4\pi^2 c^2} \left[\frac{N_2}{N_2-(g_2/g_1)N_1} \right] \left[\frac{\nu d\nu}{(\nu_o-\nu)^2+(\nu_o^2/4Q^2)} \right] \qquad (1.205)$$

Eq. 1.205 represents the spectral distribution of the power output and applies to both situations of below threshold as well above threshold of oscillation. The spectral width of the mode field, $\Delta\nu$, is given by

$$\Delta\nu = \nu_o/Q$$
$$= (1/2\pi C)(G_1-G_a). \qquad (1.206)$$

Expressing $G_1/2\pi C = \Delta\nu_c$ for the passive resonator (where only losses are present with no gain medium)

$$\Delta\nu = \Delta\nu_c (1-G_a/G_1) \qquad (1.207)$$

This shows spectral narrowing with gain $\Delta\nu=\Delta\nu_c$ for passive cavity ($G_a=0$). As threshold ($G_a=G_1$) is approached, $\Delta\nu$ becomes smaller than $\Delta\nu_c$.

Again, we could link up the spectral narrowing of laser line width in respect of emitted power P by noting that

$$P = G_1 \int_0^\infty [|V(\nu)|^2/d\nu] d\nu.$$

Using Eq. 1.205, P becomes

$$P = (h/4\pi^2 c^2) G_a G_1 [N_2/(N_2-N_1 g_2/g_1)]$$
$$\cdot \int_0^\infty [\nu/\{(\nu_o-\nu)^2+(\nu_o/2Q)^2\}] d\nu. \qquad (1.208)$$

Since the integrand peaks sharply near $\nu=\nu_o$, we may replace ν by ν_o and further consider $G_a \simeq G_1$ at threshold so that

$$P = [hG_1^2 Q/2\pi c^2] \, [N_2/\{N_2-N_1 g_2/g_1\}]_{threshold}. \qquad (1.209)$$

Substituting Eq. 1.209 in Eq. 1.206

$$\Delta\nu = [\nu_o h G_1^2/2\pi c^2 P][N_2/\{N_2-N_1 g_2/g_1\}]_{threshold}.$$

Since $\Delta\nu_c = G_1/2\pi c$

$$\Delta\nu = [2\pi h\nu_o (\Delta\nu_c)^2/P][N_2/\{N_2-N_1 g_2/g_1\}]_{threshold}. \qquad (1.210)$$

This shows spectral narrowing of laser line width as P increases. It is to be noted that although gain is proportional to population inversion N_2-N_1, the spontaneous emission noise is proportional to the upper laser level population N_2.

Both Eqs. 1.207 and 1.210 show that in the absence of amplifier gain, or the population inversion, the line width obtained is given by the line width $\Delta\nu_c$ of the passive cavity resonance. The laser line width becomes narrower as gain or population inversion increases on account of pumping the laser medium. When the laser is not oscillating (below threshold), the Eq. 1.207 applies; while above threshold, when laser is oscillating, the Eq. 1.210 applies.

The equivalent circuit approach helps in analyzing the laser on the basis of a noise driven resonant circuit. Besides providing simplicity in approach, it relates to the build up of oscillation from spontaneous emission noise. One may compare this with building up of electrical oscillation initiated by electrical noise. The similarity between thermal Johnson noise and the spontaneous emission noise is relevant in this context.

To obtain a more realistic model, however, one has to take into account the dynamic aspects of the saturation behaviour of the laser medium. This saturation behaviour is important since it relates to the nonlinearity which brings the growth of oscillation to a steady state condition for sustained oscillation.

If one considers the coupling between intensity and phase fluctuations which is not important for gas and solid state lasers but significant to semiconductor lasers, then the Eq. 1.210 is modified to

$$\Delta\nu = [2\pi h\nu_o (\Delta\nu_c)^2/P][N_2/\{N_2-N_1 g_2/g_1\}]_{th} (1+\alpha^2), \qquad (1.211)$$

where α the phase amplitude coupling factor, accounts for the effect of the modulation of the index of refraction of the gain medium by the inversion density fluctuations caused by spontaneous transitions. The increase in laser noise represented by α is due to random modulation of the carrier density by spontaneous emission which in turn modulates the index of refraction and consequently the frequency. In most gas and solid state lasers α=0, whereas in semiconductor laser α≠0. Any laser in which α≠0 should have a wavelength (or frequency) modulation attendant on any population modulation. This causes undesirable wavelength chirping. Experimental observations of line width indicate a value of α around 3 to 4.

1.21. General Treatment of Laser Oscillation

In the discussion of condition of oscillation in section 1.8, we have considered, for the geometry of an optical resonator with mirrors of reflectivity R, the condition that the round trip gain equals the round trip losses. This provided expressions for the population inversion at threshold and the frequency of oscillation. A more general situation would be when condition of oscillation is obtained independent of the geometry of resonator. Such a generalized model considers a resonator that contains an inverted medium with no specific geometry. In this medium a mode l of the resonator is excited and is oscillating. This mode, with an electric field $E_l(r,t)$ induces a coherent polarization $P_l(r,t) = \epsilon_0 \chi E_l(r,t)$ where χ is the susceptibility. It is then required that $P_l(r,t)$ acting as a driving source should give rise to an oscillation field $E(r,t)$. This approach has been used in Lamb's analysis of an inhomogeneous laser [1.18] with the advantage that calculations include non-linear effects so that phenomenon of frequency pulling and pushing, mode competition, frequency locking etc. can be described. This semiclassical theory treats the atoms quantum mechanically while considering the radiation as a classical electromagnetic field. Extension of the theory to allow for the presence of a magnetic field or cavity anisotropy and various forms of modulation have been considered by several authors. Further generalization of the theory has been brought out by Scully and Lamb [1.27] by using quantized electromagnetic field. It has been pointed out that Lamb's theory assumes a monochromatic radiation in an ideal steady state while an actual laser has mechanical and statistical disturbances and these give rise to finite radiation bandwidth. Also, oscillations will not grow spontaneously, but require an initial optical frequency field for initial starting. It is desirable to know how oscillation can develop from a state with no radiation initially present. In view of the relevance of spontaneous radiation, the generalized version would require quantum theory of radiation. The fully quantum mechanical theory determines the photon statistics as well.

Another approach that is generally adopted for analyzing a specific laser system (three level or four level solid state laser, to be described later in chapter 2) involves the writing of rate equations for the specific energy levels involved in the laser transition and describe dynamically the population of the relevant laser levels. The equations describe the change in level population due to the combined effect of pumping, spontaneous and induced radiative transitions and relaxation processes. The solution of rate equation provides expressions relating the laser power output to atomic, optical and pumping parameters.

References

1.1   D. Marcuse, *Principles of Quantum Electronics* (Academic Press, 1980).

1.2   A. Yariv, *Quantum Electronics,* (Wiley, 1975).

1.3   M. Sargent, M.O. Scully and W.E. Lamb Jr., *Laser Physics* (Addison Wesley, 1974).

1.4   P.L. Knight and L. Allen, *Concepts of Quantum Optics* (Pergamon Press, 1983).

1.5   A.V. Durrant, *Am. J. Phys* **44** (1976) 630.

1.6   A. Einstein, *Phys Z.* **18** (1921) **121**; English Translation in *The Old Quantum Theory* (Pergamon, 1967) 167.

1.7   A. Maitland and M.H. Dunn, *Laser Physics* (North Holland, 1969).

1.8   A.L. Schawlow and C.H. Townes, *Phys.Rev.* **112** (1958) 1940.

1.9   A.M. Prokhorov, *Sov. Phys.* JETP **7** (1958) 1140.

1.10 A.G. Fox and T. Li, *BSTJ.* **40** (1961) 453.

1.11 G.D. Boyd and J.P. Gordon. *BSTJ.* **40** (1961) 489.

1.12 G.D. Boyd and H. Kogelnik, *BSTJ.* **41** (1962) 1347.

1.13 H. Kogelnik, *Appl. Opt.* **4** (1965) 1562.

1.14 P.K. Tien, J.P. Gordon and J.R. Whinnery, *Proc. IEEE* **53** (1965) 129.

1.15 L. Casperson, *Appl. Opt.* **12** (1973) 2434.

1.16 L. Casperson and A. Yariv, *Appl.Phys. Lett.* **12** (1968) 355.

1.17 A.E. Siegman, *Proc. IEEE* **53** (1965) 277.

1.18 W.E. Lamb Jr., *Phys. Rev.* **134 (A1429)** (1964) 219.
1.19 A.L. Bloom and D.L. Wright, *Proc. IEEE* **54** (1966) 1290.
1.20 W.R Bennett Jr, *Phys. Rev.* **126** (1962) 580.
1.21 H.G. Danielmeyer, *IEEE J. Quantum. Electron.* **QE6** (1970) 101.
1.22 M.V. Smith and P.P. Sorokin, *The Laser* (Mc-Graw Hill, 1966).
1.23 L. Mandel and E.Wolf, *Phys. Rev.* **124** (1961) 1696.
1.24 R.J. Glauber, *Phys.Rev.* **131** (1963) 2766 ; **130** (1963) 2529.
1.25 R. Hanbury-Brown and R.Q. Twiss, *Proc. Roy. Soc.* **A243** (1957) 291.
1.26 H. Kogelnik and A. Yariv, *Proc. IEEE* **52** (1964) 165.
1.27 M.O. Scully and W.E. Lamb Jr., *Phys. Rev.* **159** (1967) 208.

CHAPTER 2

SPATIAL, TEMPORAL AND SPECTRAL CHARACTERISTICS OF LASER

2.1. Introduction

Typically, a laser consists of an optical resonator formed by two coaxial mirrors and a gain medium in it. The frequency over which laser oscillation occurs is determined by the frequency region over which the gain exceeds the resonator losses. There are many modes of the resonator which may fall within the frequency region of oscillation and the output consists of radiation at a number of closely spaced frequencies. The total output of such a laser as function of time will depend on the amplitudes, frequencies and relative phases of all these oscillating modes. If there is nothing which fixes these parameters, random fluctuations and nonlinear effects will cause them to change with time, and the output will vary in an uncontrolled way. If the oscillating modes are forced to maintain equal frequency spacings with a fixed relationship to each other, the output as a function of time will vary in a well defined manner. The form of this output will depend on which laser modes are oscillating and what phase relationship is maintained.

Lasers operate in continuous wave (CW) or pulsed mode. In the latter operation, the pump energy can be concentrated into extremely short durations, thereby increasing the peak power. This is of importance in numerous applications such as ranging, material processing and probing short lived transient phenomena. The techniques of Q-switching and mode locking are used to generate short laser pulses.

The resonators that are used for typical lasers have dimensions which are large compared to an optical wavelength, and in general they have a large number of closely spaced modes. However, there is a need for narrow bandwidth laser sources for many applications. Spectral purity is an important and desirable feature. Many techniques for reducing the number of oscillating modes have been developed [2.1]. We have already mentioned the relevance of mode selection in sections 1.11-1.13.

A mode of a resonator is a self consistent field configuration. The optical field distribution reproduces itself after one round trip in the resonator. A set of longitudinal (or axial) modes can be identified which have all the same form of spatial energy distribution in a transverse plane, but have different axial distributions corresponding to different number of half wavelengths along the axis of the resonator. These longitudinal modes are spaced in frequency by $c/2d$ where c is the velocity of light and d is the optical path length between the mirrors. To obtain laser operation at a single longitudinal mode, it is required to design a resonator sufficiently short so that $c/2d$ is greater than the bandwidth of the gain of laser material, or to design a complex resonator which has high loss for all modes within the oscillation bandwidth except the favoured one.

For each longitudinal mode number q, there exist a set of solutions called the transverse modes with mode number m, n which correspond to different energy distributions in a plane transverse to the resonator axis. As the mode number m,n increases, the energy is spread further and further from the axis of the resonator. In order to obtain oscillation in a single transverse mode it is necessary to use some device which gives high losses to all transverse modes except the desired one. Since higher order modes spread further from the resonator axis, the easiest way to accomplish single transverse mode operation is to insert into the cavity a circular aperture whose size is such that the fundamental mode TEM_{oo} experiences little diffraction loss while higher order modes suffer appreciable attenuation.

If we consider the output of a laser operating in a number of longitudinal and transverse modes, the total field will be the sum of the individual fields of each of the modes. The optical length d will not be the same for all the modes due to the dispersion of the laser material. In general, both the amplitude and phase of these modes will vary with time due to random mechanical fluctuations of the resonator length and the nonlinear interaction of those modes in the laser medium. The total field will thus vary with time in some uncontrolled way with a characteristic time which is of the order of the inverse of the bandwidth of the oscillating mode frequency spectrum. The technique which fixes the frequency spacings and phases of the oscillating modes and obtains a regular train of narrow well defined pulses, is called mode locking.

2.2. Mode Locking

Mode locking is one of the methods that utilize intra-cavity modulation to generate intense laser pulses with durations as short as a few femtoseconds. As stated earlier, the technique involves fixing the frequency spacings and phases of the oscillating modes. We have also seen that oscillating modes are longitudinal as well as transverse and as such we would have mode locking of longitudinal as well as transverse modes [2.2]. We shall first discuss the longitudinal mode locking.

In a free running laser, both longitudinal and transverse modes oscillate simultaneously without fixed mode to mode amplitude and phase relationships. The resulting laser output is a sort of time arranged statistical mean value. If we restrict the transverse modes to the lowest order TEM_{oo} mode, then there are a large number (a few hundred may be) of axial modes which fall within the frequency region where the gain of the laser medium exceeds the losses of the resonator. In the frequency domain, the radiation consists of a large number of discrete spectral lines spaced by the axial mode interval $c/2d$. Each mode oscillates independent of the others and the phases are randomly distributed in the range $-\pi$ to $+\pi$. In the time domain, the field consists of an intensity distribution which has the characteristic of thermal noise.

If the oscillating modes are forced to maintain a fixed relationship to each other, the output as a function of time will vary in a well defined manner. The laser is then said to be mode locked or phase locked. Mode locking corresponds to correlating the spectral amplitudes and phases. When all the initial randomness has been removed, the correlation of the modes is complete and the radiation is localized in space in the form of a single pulse.

Consider a set of oscillating longitudinal modes as shown in Fig. 2.1. If the frequency spacing ($c/2d$) and relative phases and amplitudes of these modes are fixed, then laser output is a well defined function of time. Let the nth mode have amplitude E_n, angular frequency ω_n and phase ϕ_n. Then the total laser output field E_T is

$$E_T = \sum_n E_n \exp j[\omega_n(t-z/c) + \phi_n] + c.c. \qquad (2.1)$$

The radiation is assumed to travel in z direction and c.c represents complex conjugate. If the mode spacing is equal between the modes

$$\nu_n = \nu_o + n\, \delta\nu, \qquad (2.2)$$

where $\delta\nu = c/2d$ and ν_o is the optical frequency of the laser output. Noting that $\omega_n = 2\pi\nu_n$, we can rewrite Eq. 2.1 as

$$E_T = [\sum_n E_n \exp j\{2\pi n\, \delta\nu(t-z/c) + \phi_n\}] \exp j 2\pi\nu_o(t-z/c) + c.c \qquad (2.3)$$

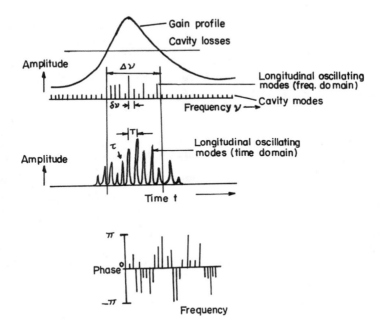

Fig. 2.1. Multimode operation of laser on several longitudinal modes. Laser output contains N oscillating modes covering a frequency range $\Delta\nu$. The ratio of pulse separation (T) to the pulse width (τ) is N, the number of oscillating modes.

Comparing Eq. 2.3 with the expression of a carrier wave of frequency ν travelling in +z direction

$$E = \exp j(t-z/c)2\pi\nu + c.c.,$$

we find that Eq. 2.3 corresponds to a carrier wave of frequency ν_o whose envelope depends on the values of E_n and ϕ_n. This envelope is seen to travel with the velocity of light and is periodic with a period

$$T \sim 1/\delta\nu = 2d/c. \tag{2.4}$$

The periodic property of E_T depends on the fact that the phases ϕ_n are fixed. In typical lasers, however, the phases ϕ_n are likely to vary randomly with time. This causes the intensity of the laser output to fluctuate randomly and thereby reduces its coherence. To make the situation better, one can either make the laser operate at single frequency so as to eliminate mode interference or force the phases ϕ_n to maintain their relative values. The former approach (single frequency operation) includes, among other ways, shortening of the resonator length d in order to increase the mode spacing $\delta\nu=c/2d$ to a point where only one mode falls within the gain line width. The latter approach (mode locking) causes the oscillation intensity to consist of a periodic train with a period $T=2d/c$.

For ϕ_n=constant and independent of n, the envelope (i.e. the term within [ ] in Eq. 2.3) consists of a single pulse in the period T whose width τ is approximately the reciprocal of the frequency range over which the E_n's have an appreciable value. In other words, the ratio of the pulse space (T) to the pulse width (τ) is approximately the number of oscillating modes. Expressed in symbols, the relations

$$N(\delta\nu) = \Delta\nu,$$

$$\tau = 1/\Delta\nu$$

along with Eq. 2.4 give

$$T/\tau = N \tag{2.5a}$$

so that

$$\tau \simeq 1/\Delta\nu \simeq T/N. \tag{2.5b}$$

Eq. 2.5b expresses the well known fact that the narrower the pulse width τ the larger the bandwidth $\Delta\nu$ required to generate the pulse. Also, the larger the number N of the oscillating modes that are locked, the narrower the pulse width.

A simple but useful analysis can be made by considering the situation where the phases ϕ_n are made equal to zero. We further assume that there are N oscillating modes with equal amplitude. Thus taking $E_n = 1$ and $\phi_n = 0$, Eq. 2.3 gives

$$E_T = \exp(j2\pi\nu_0 t) \sum_{-(N-1)/2}^{(N-1)/2} \exp(j2\pi n\, \delta\nu\, t)$$

$$= \exp(j2\pi\nu_0 t)[\sin N\pi(\delta\nu)t]/[\sin \pi(\delta\nu)t]. \qquad (2.6)$$

The laser power P(t) is then

$$P(t) = E_T(t)E_T^*(t)$$
$$= \sin^2[N\pi(\delta\nu)t]/\sin^2[\pi(\delta\nu)t]. \qquad (2.7)$$

The mode locked laser power therefore has the time variation as shown in Fig. 2.2. This is the time averaged statistical mean output of the N oscillating modes of equal amplitude ($E_n=1$) and zero fixed phase ($\phi_n=0$). The averaging is performed over a time T long compared to $1/\nu_0$ but short compared to $1/\delta\nu$.

The peak field amplitude of E_T is obtained by substituting t=0 in Eq. 2.6. Since

$$[\sin N\pi(\delta\nu)t]/[\sin\pi(\delta\nu)t]_{t=0} \longrightarrow N ,$$

the peak field amplitude $E_T(0)$ of the mode locked pulse is seen to be equal to N times the amplitude of a single mode. Further, if P_0 and P_{av} denote the peak power and average power of the mode locked pulse [P(t=0) and P(t) averaged over T] respectively and $\Delta\tau$ denotes the pulse width (Fig. 2.2), then

$$P_0 \Delta\tau = P_{av} T$$

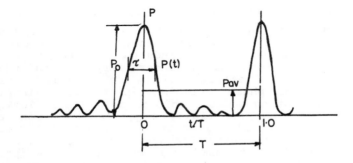

Fig. 2.2. Time variation of mode locked pulse power.

or, using Eq. 2.5b,

$$P_o = P_{av}(T/\Delta\tau) \simeq P_{av} N$$

Thus peak power of the mode locked pulse is equal to N times the average power where N is the number of oscillating modes which are locked. P_{av} represents the power resulting from incoherent phasing of the axial modes and averaged over the coherence time T. The mode locked power circulates around inside the cavity with a repetition rate determined by the round trip transit time. These signals repeat themselves and appear in the laser output at a rate of c/2d. Therefore, mode locking results in a train of pulses whose repetition period is twice the cavity transit time (Eq. 2.4).

Mode locked lasers can thus have high peak powers because the power contained in the entire output of the uncoupled laser is now contained within the more intense ultra fast mode locked pulse.

The structure of an optical pulse is completely defined by phase and intensity profiles which may refer to either time or frequency. If the description in one domain is complete the profiles in the other are obtained from Fourier transformation. However, there is no one to one correspondence between the two intensity profiles I(t) and I(ν) since each depends not only on the other but also on the associated phase function. The FWHM widths of the two profiles are however related by

$$\Delta t \, \Delta\nu = \text{constant}.$$

For mode locking, $\Delta\nu$ corresponds to the gain bandwidth of the laser.

2.3. Methods of Mode Locking

Mode locking can be achieved by modulating the losses (or gain) of the laser at the frequency of mode spacing $\delta\nu$ =c/2d. Mode locking can also be induced by internal phase, rather than loss, modulation. Under certain conditions, however, the nonlinear effects of the laser medium itself may cause a fixed phase relationship to be maintained between the oscillating modes. This is called self locking.

Loss modulation essentially requires the condition that cavity loss be modulated at the frequency c/2d whereby one cycle of the modulation frequency corresponds to the time it takes the light to make a round trip in the optical resonator. This way, light which sees a loss at one time will again see a loss after one round in the resonator. This means all the light in the resonator will experience loss except that which passes through the modulator when modulator loss is zero. Light will thus tend to build up in narrow pulses in those low loss time positions. These pulses will then have a width given by the reciprocal of the gain bandwidth product. Pulses wider than this will experience more loss in the modulator while pulses narrower in time will experience less gain because of their frequency spectrum being wider than the gain bandwidth product.

Mode locking by phase modulation is usually done by using an electro-optic crystal inside the resonator such that the travelling wave undergoes a phase delay proportional to the instantaneous electric field across the crystal. The frequency of the modulating signal is, as in loss modulation case, equal to the inverse of the round trip delay time or the intermode frequency separation $\delta\nu = c/2d$.

2.3.1. Mode Locking with Saturable Absorber- Passive Mode Locking

The technique of mode locking with a saturable absorber has proved to be very useful for lasers with light intensity in the cavity high enough to saturate the absorber material. A saturable absorber is a material whose absorption coefficient for laser light decreases as the light intensity is increased. A typical arrangement of a passive mode locked laser with saturable absorber is shown in Fig. 2.3.

The operation of a saturable absorber to produce mode locking can be understood in terms of an electronic pulse generator described by Cutler [2.3] and shown in Fig. 2.4a. The feedback loop of this arrangement which includes an amplifier, a filter, a delay line and a nonlinear element that provides less attenuation for a high level signal than for a low level signal, behaves as a regenerative pulse generator. When the loop gain exceeds unity, a pulse recirculates indefinitely around the loop and each traversal gives rise to a pulse at the output terminal. Such a pulse would however be degraded unless the effects of noise and distortion can be countered. The nonlinear element called the expander, has threefold effect. First, it emphasis the

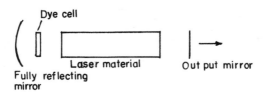

Fig. 2.3. A typical arrangement of a passively mode locked laser using saturable absorber.

peak region of the recirculating pulse while reducing the lower amplitude region, secondly, it discriminates against noise and reflections and thirdly, it acts to shorten the pulse until the width is limited by the frequency response of the circuit. The output of the regenerative oscillator has a pulse rate equal to the reciprocal of the loop delay, pulse widths equal to the reciprocal of the overall system bandwidth and a centre frequency determined by the filter frequency.

Optical analogue of the electronic regenerative pulse generator is provided by the laser with the exception of the expander element. The laser medium serves as the amplifier, the combination of the Fabry Perot resonances and the line width of the laser transition serve as the filter, and the time required for an optical pulse to traverse twice the distance between the reflectors serves as the loop time delay (Fig. 2.4b). The optical analogue of the electronic expander is a saturable absorber, such as the reversible bleachable dye solution. The fundamental requirements of the saturable absorber are (i) it should have an absorption line

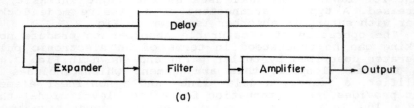

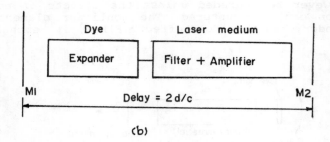

Fig. 2.4. Regenerative pulse generator, (a) electronic and (b) optical analogue (laser with dye modulator).

at the laser wavelength, (ii) it should have a line width equal to or greater than the laser line width and (iii) the recovery time should be shorter than the loop time delay of the laser.

The system shown in Fig. 2.4a can be analyzed by repeated application of Fourier transforms, applying the frequency or amplitude characteristics of each element and finally equating the characteristics of the returning signal to the characteristics of the assumed initial signal [2.4]. Consider a signal entering the expander as $P_1(t) \exp j\phi t$. The signal leaving the expander is then

$$P_2(t) = K [P_1(t)]^q \exp j\phi t,$$

where K is a constant indicating an amplitude change of the signal and q indicates a nonlinear operation on the envelope portion of the signal. If this signal is now passed through a filter having a frequency function $F(w)$, the output from the filter has the form $F_3(w) = F(w) F_2(w)$, where

$$F_2(w) = \int_{-\infty}^{\infty} P_2(t) \exp(-j\omega t) \, dt$$

is the transform of $P_2(t)$ in the frequency domain. The transformation of the signal $F_3(w)$ into the time domain gives

$$P_3(t) = (1/2\pi) \int_{-\infty}^{\infty} F_3(\omega) \exp j\omega t \, d\omega.$$

The amplifier and circuit losses give a net gain (say g) so that the output of the amplifier is $P_4(t) = g \, P_3(t)$. The signal $P_5(t)$ leaving the delay element is then delayed by τ and brings us back to the expander input. $P_5(t)$ is now required to equal $P_1(t)$ so that

$$P_1(t) \exp j\phi t = (gk/2\pi) \exp(-j\theta) \int_{-\infty}^{\infty} F(\omega) \exp j\omega (t-\tau) d\omega$$

$$\cdot \int_{-\infty}^{\infty} [P_1(t-\tau)]^q \exp(-j\omega)(t-\tau) \exp j(\phi t-\tau) dt,$$

(2.8)

where θ is the phase shift of the optical wave relative to the pulse time. Given a filter characteristic $F(w)$, an expander nonlinear law q and time delay t, Eq. 2.8 specifies the time function $P_1(t)\exp j\phi t$. Solutions of Eq. 2.8 have been found by assuming Gaussian functions and a power law for q, which gives pulse rate equal to the reciprocal of the group delay $1/\tau = d\omega/d\phi = 2\pi\beta$ around the loop and the pulse width $\Delta\tau$ given by

$$\Delta\tau = [4/\pi\Delta\omega(1-1/q)^{1/2}] \left[\frac{2(1+r^2\Delta\omega^4/16)}{1+\{1+qr^2\Delta\omega^4/4(q^2+2q+1)\}^{1/2}} \right]^{1/2},$$
(2.9)

where $\Delta\omega$ is the bandwidth at the 1 neper point of a Gaussian filter. β and r are the coefficients of linear and quadratic terms in the power series expansion of the loop's phase shift

$$\phi = \alpha + \beta(\omega-\omega_0) + r(\omega-w_0)^2 + \ldots \quad (2.10)$$

This power series expansion of the phase in the feedback loop can be obtained by taking a Taylor series expansion of the refractive index of the medium within the feedback loop,

$$n = n_0 + (dn_0/d\nu)(\nu-\nu_0) + (d^2n_0/d\nu^2)(\nu-\nu_0)^2 + \ldots \quad (2.11)$$

and substituting into the phase $\phi = 2\pi nL/c$ where L is the length of the media within the feedback loop. For $r=0$, the pulse width is a minimum,

$$\Delta\tau_{min} \simeq 1/\Delta\nu$$

2.3.2. Active Mode Locking

Laser oscillator consists of a resonant system with dimensions large compared to lasing wavelengths. Mode density is, therefore, high and there are $[2d\,\Delta\lambda/\lambda_0^2]$ axial resonances within the line width $\Delta\lambda$ of a laser transition having centre wavelength λ_0 and mirrors separated by optical length d. In the normal operation of laser, these modes are largely uncoupled and have no fixed phase relationship between the discrete oscillating frequencies. In the passive (dye) mode locking, the required coupling for locking is supplied by the passive modulator (saturable absorber). The required mode coupling for locking the phase of axial modes can also be supplied by an active modulator, such as the intracavity acousto-optic or electro-optic device, which provides time varying loss or phase modulation in the optical path. KDP Pockels cells and acousto-optic modulators have been used for this purpose [2.5a].

Active mode locking of a laser is achieved by modulating the mode locking device with a tunable (electronic) oscillator and adjusting the modulator frequency to agree within the cavity length or, alternatively, selecting a fixed frequency and adjusting the mirror spacing accordingly. Slow shifts in the modulator frequency, or more commonly the changes in the cavity lengths due to thermal effects, however, cause gradual loss in resonance.

Active mode locked lasers are very sensitive to detuning between the applied modulating frequency and the round trip frequency $c/2d$ of the laser cavity. Commercial systems use complicated temperature compensation techniques for stabilizing both the laser cavity and the mode locker separately. Temperature stabilization of the frequency selecting component and mode locker is required within a small fraction of a degree. Alternatively the mode locked frequency can be actively controlled to follow the laser $c/2d$ axial mode spacing frequency. It requires the system to sense the difference between mode locked frequency and axial mode spacing $c/2d$. Once this difference is detected, the modulator drive frequency is controlled to track the $c/2d$ frequency and highly stable pulses are obtained.

The operation of an active mode locker can be described as follows. Assume that Fabry Perot mode ν_o nearest to the peak of the laser gain profile begins to oscillate first. If an amplitude or phase modulation operating at a frequency $\delta\nu$ is inserted into the laser's feedback, the carrier frequency ν_o will develop sidebands at $\pm\delta\nu$. If the modulating frequency $\delta\nu$ is chosen to be commensurate with axial mode frequency separation $\delta\nu = c/2d$, the coincidence of the upper $(\nu_o+\delta\nu)$ and the lower $(\nu_o-\delta\nu)$ sidebands with the adjacent axial mode resonances will couple the $(\nu_o-\delta\nu)$, ν_o and $(\nu_o+\delta\nu)$, modes with a well defined amplitude and phase. As the $(\nu_o+\delta\nu)$ and $(\nu_o-\delta\nu)$ oscillations pass through the modulator, they will also become modulated and their sidebands will couple the $(\nu_o\pm2\delta\nu)$ modes to the previous three modes. This process continues until all axial modes falling within the oscillating line width are coupled. The constructive and destructive interference of these simultaneous phase locked modes can be described by the interference of Fourier series components in the construction of a repetitive pulse train.

Mode locking with active modulator requires critical adjustment of mirror spacing and modulating frequency as well as compensation for any perturbations in optical length of the feedback. Such compensation is of particular importance in the mode locking of solid state lasers because of variation in optical path length of laser rods during optical pumping. The use of saturable absorbers as passive modulators eliminates the need for such critical adjustments.

An active mode locker has the general arrangement shown in Fig. 2.5. The essential element of the arrangement is an internal time varying perturbation producing device which establishes coherence between the laser modes. This also enables stabilization of the modes. Internal time varying perturbation is caused by a variation of the optical path length, or the loss, by an externally applied modulating signal to the perturbation device. The perturbation is thus either of phase (variation of optical path length) or of loss. Both types of perturbations may be achieved by means of either the linear electro-optic effect in crystals (Pockels cell) or in certain cases by acoustic means.

If we consider an internal loss perturbation weakly driven at frequency approximating the axial mode interval c/2d, its effect is to produce sidebands at all frequencies which are spaced by the modulation frequency from each of the original free running axial modes. The perturbation makes the modes of the free running laser slightly pulled towards the generated sidebands. If the drive strength of the perturbation is increased, a point is reached when one of the free running modes is pulled into lock with the side band from an adjacent mode. With further increase of drive strength, the last free running mode of the laser is pulled into synchronism and a complete phase locking of the laser occurs. The time domain output corresponding to such a phase locked spectrum is a repetitive series of narrow pulses having a repetition frequency equal to that of the perturbation.

When a fairly strong internal phase perturbation is considered, again at frequency near the axial mode interval c/2d, the effect of perturbation is again to associate a set of sidebands with each of the previously free running laser modes. Each of the free running laser modes then becomes the

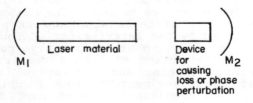

Fig.2.5. Schematic of an active mode locker

centre frequency, or carrier, of an FM signal. For large phase perturbation strength and a certain range of the frequency difference between driving frequency of phase perturbation and the axial mode interval, FM laser oscillator may be obtained. Such FM laser oscillation has been demonstrated [2.5b]. Alternatively, very small difference between driving frequency, the axial mode interval and lower perturbation strengths yield a time domain pulsing behaviors which is very similar to that obtained by internal loss perturbation.

A certain minimum perturbation strength is required to achieve mode locking. If the cavity modes are equally spaced, then in the limit of perfect tuning of the modulation frequency, a very small perturbation would be required. In the presence of mode pulling however, the free running modes are unequally spaced and the perturbation must be sufficiently strong to pull these modes until their frequency spacing is equal to that of the modulator drive frequency.

2.3.3. Acousto-optic and Electro-optic Modulators as Mode Locking Devices

Acousto-optic mode lockers are used to loss modulate the longitudinal modes in a laser cavity. Strong mode coupling results and short pulses of high peak power are emitted. The mode locker is formed with a high quality quartz as interaction medium and consists of an acoustic standing wave cavity with piezoelectric transducer bonded to one of the cavity surfaces (Fig. 2.6). Optimum loss modulation is provided when the optical polarization of the laser is perpendicular to the direction of acoustic propagation. Longitudinal acoustic waves generated by the transducer traverse the quartz cavity and are reflected back and forth between the cavity surfaces. Acoustic waves travelling in opposite direction, after many reflection, combine and form standing waves between the faces of the cavity. The acoustic standing waves produce a periodic variation in the refractive index of quartz and a stationery phase grating is formed. Since the intensity of the acoustic standing wave is amplitude modulated at twice the acoustic resonant frequency, the optical modulation frequency f_o will be twice the acoustic resonant frequency f_m so that

$$f_o = c/2d = 2f_m. \qquad (2.12)$$

The acoustic resonant frequency f_m is given by

$$f_m = N\,v/2l, \qquad (2.13)$$

where N is number of half acoustic wavelengths between

106   LASERS AND HOLOGRAPHY

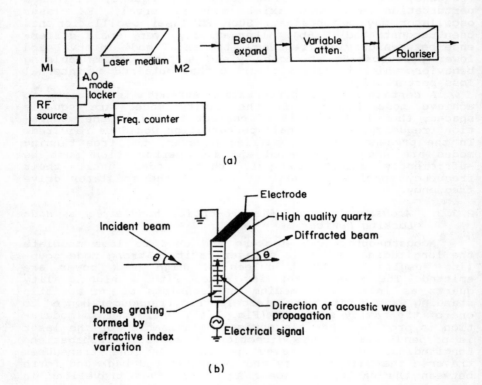

Fig. 2.6. (a) Experimental setup for active mode locking using acousto-optic modulator. (b) Acousto-optic modulator in Bragg mode.

parallel faces of the resonator of length l. v is the acoustic velocity in the interaction medium (in this case quartz). The fundamental mode spacing frequency is $v/2l$ (N=1). The tuning of the resonant frequency is possible by varying the temperature of the mode locker. Laser beam passing through the acoustic standing wave phase grating will be diffracted into higher order beams obtained by Bragg diffraction. For maximum diffraction efficiency in loss modulation, the light beam must enter the acoustic grating at the Bragg angle

$$\theta = \lambda f_m/2v, \qquad (2.14)$$

where λ is the wavelength of light beam.

Loss modulation could also be obtained by passing a beam through an electro-optic device like $LiNbO_3$ in which the beam suffers a polarization rotation through the crystal given by $\pi r_{22} n_o^3 l E_x/\lambda_o$ where n_o is the refractive index of ordinary ray, r_{22} is the electro-optic coefficient of $LiNbO_3$, l is the length of crystal and λ_o is the free space wavelength of light [2.6]. The light is assumed to propagate along z direction while modulating field E is applied in x direction. The light is also polarized in x direction. The peak single pass power loss to the incident polarization for $V_x = V_o \cos\omega_m t$ is

$$\delta = \sin^2[\pi r_{22} n_o^3 V_o l/\lambda_o l_z], \qquad (2.15)$$

where l_z is the dimension of crystal in z direction.

2.3.4. Self Mode Locking

Under certain conditions, the nonlinear effects of the laser medium itself may cause a fixed phase relationship to be maintained between the oscillating modes. We may therefore obtain a train of narrow pulses in the absence of any driving perturbation. The laser is then said to be self locked or self pulsing [2.6]. Attempts to explain self pulsing have been made based on combination tone theory of Lamb [2.7] but this analysis becomes prohibitively complex if a large number of longitudinal modes are lasing above threshold as is often the case. A simple approach is to consider the problem in time domain [2.8]. In general it is not possible to predict the exact condition for self pulsing, but the following conditions are usually necessary:

(a) The oscillation must be confined to a single transverse mode, TEM_{oo}.

(b) The round trip time for radiation in the resonator must be of the order of, or greater than, the atomic decay time.

(c) The laser must not be operated too much above threshold.

In view of somewhat uncertain nature of self locking, it is better to use a driven cavity perturbation for mode locking the laser.

2.3.5. Experimental arrangement of an Active Mode Locked Laser

Loss modulation mode locking can be accomplished by inserting a modulator [2.9] (e.g. acousto-optic Bragg modulator) inside the laser cavity adjacent to one of the laser mirrors. An experimental set up used for mode locking is shown in Fig. 2.6 [2.10]. The arrangement is placed on an isolation table to avoid disturbance due to vibration. The operating wavelength can be changed by adjusting the frequency selective element, such as prism, in the cavity of the laser. In the acousto-optic modulator used in the cavity, the resonant mode spacing frequency is determined by the thickness of the interaction medium. It produces a modulation at twice the applied acoustic frequency. The mode locker modulation frequency is adjusted equal to the laser cavity resonant frequency $c/2d$. The output of RF signal generator is amplified by a tunable power amplifier and is fed to the mode locker. The expanded laser beam is passed through a variable alternator for varying the intensity of the laser beam. Polarizer is used for selecting the horizontal or vertical component of the polarized light.

2.3.6. Stabilization of Mode Locked Laser

Active mode locking of a laser can be achieved by modulating the laser loss, with a tunable RF source and adjusting the modulating frequency to the $c/2d$ mode spacing frequency or by selecting a fixed frequency and adjusting the cavity length (mirror spacing) accordingly. Slow shifts in the modulating frequency, or the changes in cavity length due to thermal effects, will cause gradual loss of resonance. In case of a loss modulated laser the pulses pass through the modulator at a time when the polarization rotation (in an electro-optic $LiNbO_3$ modulator for example) is nonzero and the average loss introduced by the modulator is increased. In the case of a phase modulated laser the pulses pass through the modulator at a time when the instantaneous phase retardation is changing with time. Actively mode locked lasers are very sensitive to even very small changes in cavity length or the modulating frequency. Very small detuning between the applied modulating frequency and exact round trip repetition frequency $c/2d$ of the laser cavity can cause loss of mode locking. These two frequencies must therefore be very accurately synchronized. To maintain stable mode locking operation some form of electronic feedback is required to maintain the driving frequency in exact synchronism with the cavity resonant frequency $c/2d$.

Phase lock loops (PLL) have been used for mode locking the laser for long term operation. Block diagram for one such arrangement [2.6] is shown in Fig. 2.7. It requires a high speed photodetector to sense the first order beat

LASER CHARACTERISTICS

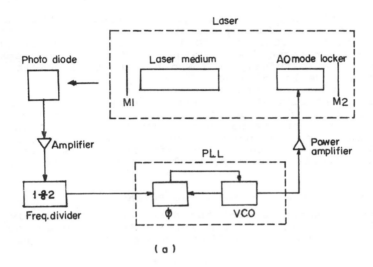

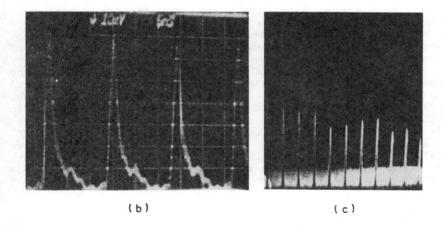

Fig. 2.7. (a) Block diagram of the stabilized mode locked laser. Mode locked output in time domain (b) and in frequency domain (c). Frequency spacing is 70 MHz.

frequency c/2d. This beat frequency signal is amplified and applied to a 'divide by two' frequency divider network. The optical modulation frequency is twice the acoustic resonant frequency.

The difference of the frequency divider output and the VCO is detected by a phase comparator and is fed back to the control point of VCO. The VCO is locked onto the frequency of the divider network and changes in the c/2d mode spacing frequency are faithfully followed by VCO. This results in a stabilized mode locked laser.

This type of stabilization obviates the need of temperature stabilization of laser cavity, mode locker and the mode locker drive source individually or collectively. Though the arrangement has been shown to work with acousto-optic modulator, the procedure would work equally well for the electro-optic phase or loss modulators also. The frequency divider, however, would not be required in that case.

2.4. Measurements on Mode Locked Pulses

Direct measurement of the duration of relatively long optical pulses is most often made by displaying on oscilloscope, the output of a suitable photo detector illuminated by the optical radiation.

The measurement of time duration of light pulses of less than several hundred picoseconds, however, requires special techniques. Since the first successful generation of picosecond optical pulses a lot of progress has been made in the measurement techniques for such short duration pulses. Several techniques have been used for measuring the width of picosecond pulses.

Nonlinear measurements based on autocorrelation of the pulse intensity can provide information about the actual time duration of the pulse. Linear measurement, on the other hand, measures the autocorrelation function of the pulse amplitude. Since the power density spectrum and the amplitude autocorrelation function are a Fourier transform pair, knowledge of one uniquely specifies the other. The power density spectrum is usually measured with a spectrometer. The linear system can also be used to measure the coherence length l_c and the coherence time t_c related to the spectral bandwidth $\Delta \nu$ by

$$l_c \simeq c/2\Delta \nu \simeq ct_c/2. \qquad (2.16)$$

Since the relation $t_c \Delta \nu \gtrsim 1$ provides only a lower limit to a pulse duration for given $\Delta \nu$, the measurement with a linear system can provide only a lower limit to the time duration of a pulse. In other words, coherence time t_c equals the duration of the pulse only in the special case where entire spectral content of the pulse is due to the short duration of its envelope.

In the linear measurement device like the Michelson interferometer (Fig. 2.8) an incident pulse having amplitude $E(t)$ is split into two pulses each with an amplitude $E(t)/2^{1/2}$. Each of these pulses is made to traverse a separate orthogonal arm of the interferometer. After traversing their respective paths $2l_1$ and $2l_2$ the pulses are recombined on a square law detector. The intensity on the detector is given by

$$I(t,\tau) = (1/2)|E_0(t) + E_0(t-\tau)|^2, \qquad (2.17)$$

where $\tau = 2(l_2-l_1)/c$ and $E_0(t)$ is the slowly varying envelope of the pulse with respect to ν. The response of the detector being slow compared to pulse duration or the delay τ, the detector output signal is given by

$$S(\tau) = \int_{-\infty}^{\infty} I(t,\tau)dt = W[1+A(\tau)], \qquad (2.18)$$

where $I(t,\tau)$ is defined by Eq. 2.17, W is the pulse energy

$$W = \int_{-\infty}^{\infty} E_0^2(t)dt$$

and $A(\tau)$ is the autocorrelation function of the pulse amplitude given by

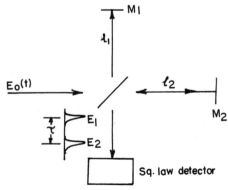

Fig. 2.8. Michelson interferometer as a linear optical measurement device.

$$A(\tau) = \frac{\int_{-\infty}^{\infty} E_0(t) E_0(t-\tau) dt}{\int_{-\infty}^{\infty} E_0^2(t) dt} \qquad (2.19)$$

when $\tau=0$, $S(\tau)/W = 2$. When τ is large enough so that no overlap between E_1 and E_2 exists, $S(\tau)/W=1$. From Fourier analysis,

$$A(\tau) = (2\pi)^{-1/2} \int_{-\infty}^{\infty} \exp(j\omega t) |E(\omega)|^2 d\omega,$$

where $|E(\omega)|^2$ is the power density spectrum of the original laser pulse. [$E(\omega)$ is the spectral amplitude and forms the Fourier transform pair with $E(t)$]. Since power density spectrum and amplitude autocorrelation functions are a Fourier transform pair, knowledge of one uniquely specifies the other. Power density spectrum is measured with a spectrometer and $A(\tau)$ with an interferometer. The two are equivalent since both are linear optical systems.

In the case of nonlinear optical system suppose one passes the two output pulses $E_1(t)$ and $E_2(t)$ of the linear interferometer of Fig. 2.8 through a nonlinear optical crystal. The second harmonic output from nonlinear crystal will be $E_{2\nu} \propto |E_1(t)+E_2(t-\tau)|^2$. The output signal $S(\tau)$ from a detector, having slow response with respect to ν and τ, is

$$S(\tau) = \int_{-\infty}^{\infty} |E_{2\nu}(t)|^2 dt$$

$$= W_{2\nu}[1+2G(\tau)], \qquad (2.20)$$

where $W_{2\nu}$ is the second harmonic pulse energy and $G(\tau)$ is the autocorrelation function of the pulse intensity. Here we have neglected the constant of proportionality.

$$W_{2\nu} = \int_{-\infty}^{\infty} E^4(t) dt,$$

$$G(\tau) = \frac{\int_{-\infty}^{\infty} E^2(t) E^2(t-\tau) \, dt}{\int_{-\infty}^{\infty} E^4(t) \, dt} \qquad (2.21)$$

when $\tau=0$, $S(\tau)/W_{2\nu}=3$ and when τ is large enough so that no overlap between $E_1(t)$ and $E_2(t)$ exists, $S(\tau)/W_{2\nu}=1$. Measurement of the variation of $S(\tau)/W_{2\nu}$ as a function of τ gives the time duration over which the energy of the pulse is distributed.

In the arrangement of Fig. 2.9 for the nonlinear system, an optical pulse is passed through a birefringent crystal of length l and the crystal resolves the single pulse into two pulses of equal amplitude and orthogonal polarization (o and e components). The delay so introduced is $\tau = l(n_e - n_o)/c$, where n_e and n_o are the extraordinary and ordinary indices of refraction. The two orthogonally polarized pulses are then allowed to interact in a nonlinear crystal to produce second harmonic. The crystal is adjusted so that no second harmonic is produced by either component acting alone but only when both are present. The method thus measures overlap of the pulse with a delayed replica. When a delay is introduced such that no overlap of the two pulses occurs, the second harmonic output drops to zero. If we expressed the two orthogonally polarized components by E_H and E_V and the input signal as E_i, then

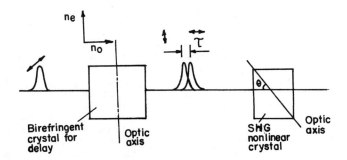

Fig. 2.9. Schematic arrangement for measuring picosecond laser pulses by second harmonic generation.

$$E_i = E_H(t)\exp(2\pi j\nu t) + E_v(t-\tau)\exp 2\pi j\nu(t-\tau). \quad (2.22)$$

The second harmonic signal is

$$E_{2\nu}(t) = E_H(t)E_v(t-\tau)\exp 2\pi j\nu(2t-\tau), \quad (2.23)$$

where $E_H(t)$ and $E_v(t-\tau)$ are the slow varying envelope of the horizontal and vertical polarized electric fields of frequency ν. The constant representing SHG efficiency is again neglected. If now $\tau \gg t_c$, then $E_{2\nu}=0$ and if $\tau=0$, $E_{2\nu}=$max. The output signal from a detector having a slow response time with respect to 2ν is given by

$$S(\tau) = \int_{-\infty}^{\infty} E_H^2(t)E_v^2(t-\tau)\,dt. \quad (2.24)$$

The experimentally measured quantity is $S(\tau)/W_{2\nu}=I(\tau)$ as a function of τ by selecting different lengths of delay crystals. When $\tau=0$, $S(\tau)/W_{2\nu}=1$ and when τ is large enough so that no overlap between E_H and $E_v(t-\tau)$ exists, $S(\tau)/W_{2\nu}=0$.

The efficiency of second harmonic generation (SHG) in nonlinear crystal is enhanced if the fundamental is in the form of a train of narrow pulses [2.11]. This enhancement is used to estimate the width of the mode locked pulses. Alternatively, intensity correlation of SHG is used in which comparison with the free running laser is required [2.12]. Another technique using a dye, exhibiting two photon absorption fluorescence (TPF) as a nonlinear process, was proposed by Giordmaine et al. [2.13]. This technique, which is also an intensity correlation system, has been widely used due to its simplicity (Fig. 2.10). The correlation of one pulse with succeeding pulse (cross correlation) or with itself (autocorrelation) can be used. However, there is a problem in the interpretation of fluorescence pattern if the contrast ratio is not the theoretical maximum (ratio of the intensity of fluorescence in the region where pulses overlap to the fluorescence intensity in between the maxima). A contrast ratio of 3:1 indicates complete mode locking with zero phases i.e. narrowest laser pulses, while a contrast ratio of 1.5:1 represents a completely random phase relationship. It has, however, been shown that much more information on pulse structure can be obtained with the method of third harmonic generation than by the TPF technique [2.14].

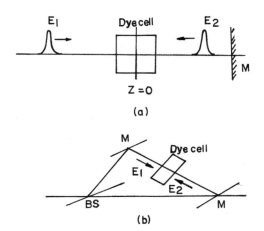

Fig.2.10. Arrangement for TPF method for measurement of picosecond pulses by (a) intensity cross-correlation and (b) intensity autocorrelation.

Multiphoton conductivity effect in a semiconductor is also used for measuring subnanosecond optical pulses [2.15]. In the TPF method the fluorescent signal from a dye is measured while in the two photon (or three photon) conductivity methods the photoconductive signal from a biased semiconductor is measured.

Autocorrelation by SHG technique is not applicable in uv laser pulses since there is no nonlinear crystal capable of generating SHG of uv radiation. Streak cameras have been used to evaluate short duration pulses up to 1 ps (section 2.5.4). UV short pulse width measurement by multiphoton auto correlation has also been demonstrated [2.16].

A lot of work has been done in the development of high speed photodetectors with response time in the picosecond range. The limiting time resolution in such cases is determined by the photodetector [e.g. 200-400 ps from photomultiplier tubes (PMT) and 100-200 ps from PMTs with micro channel plate multiplier (MCP)]. The application of MCP-PMT is limited by the degradation of photocathode quantum efficiency resulting in lower working life of the device. A metal oxide-silicon oxide-metal junction photodetector having 50 ps rise and fall time has been reported [2.17].

2.5. Generation and Measurement of Ultrashort Pulses

In the two decades following the discovery of mode

locked glass lasers in 1966, techniques for generation of shorter pulses advanced dramatically. An important example of ultrafast laser technology is the discovery in 1970 of self phase modulation which has turned out to be a key process for femtosecond pulse generation. The development of the colliding pulse mode locking technique in 1981 pushed the limit of attainable pulse widths to less than 100 fs. Novel shaping and pulse compression techniques have led to the shortest reported optical pulse width of 8 fs in 1985. The chronological development of ultrafast laser technology has been described at length [2.18], however, the milestones in the generation of ultrashort pulses could be listed as mode locking of Nd glass laser in mid sixties, the flash lamp and CW dye laser in seventies and the colliding pulse technique followed by pulse compression in eighties. Pulse width shortening has varied from subnanosecond to subpicosecond, the shortest pulse width obtained is just under 2 μm in length or about 3 wavelengths of visible light.

2.5.1. Synchronous Mode Locking

In this technique, the gain medium (usually a dye) is pumped with a continuous train of ultra short pulses. This modulates the gain of the medium. If the round trip time of the dye laser cavity is equal to, or a submultiple of the cavity length of the pumping laser, the gain of the dye laser is impulsively driven in synchronism with the round trip repetition rate, and a train of ultrashort light pulses is generated from the dye laser. It has been shown that the temporal duration of the output laser pulses is proportional to the square root of the ratio of the pump pulse duration to the intra-cavity bandwidth. Thus to produce short pulses the pulse duration of the pumping source is the most crucial parameter of the system. Matching the cavity length with submicron accuracy is also required.

2.5.2. Colliding Pulse Mode Locking

This technique has produced pulses as short as 27 fs without employing any external pulse shaping or pulse compression scheme. This is the most successful method for generating the shortest pulses.

In this method, a series of mirrors form a ring cavity whose central elements are a saturable gain dye and a saturable absorber dye. The choice of gain medium depends on the wavelength of interest. A particular design uses a rhodamine 6G as the gain medium and diethyloxacarbocyanin iodide as the saturable absorber. The two dyes are placed at two focal points in the cavity (Fig. 2.11). A CW laser is used to pump the gain medium. As the lasing process builds up, two counter propagating beams are formed in the cavity.

The pulse shortening mechanism is similar to that in

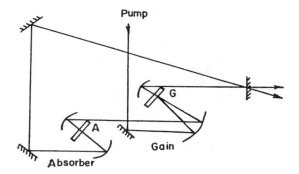

Fig. 2.11. Schematic diagram of a colliding pulse mode locked cavity. A and G are dye cells for saturable absorption and gain respectively.

passively mode locked dye laser. A slowly relaxing saturable absorber acts to steepen only the leading edge of the pulse. Saturation of the gain of the laser medium in combination with linear loss discriminates against the trailing edge. The absorber should saturate more easily than the gain. This provides a net gain at the peak of the pulse and a loss on either side. Shortening continues until dispersion effects prevent the pulse from becoming narrower.

The separation between the saturable absorber dye and the saturable gain dye and their proper positioning in the cavity are crucial to the stable operation of a colliding pulse ring dye laser. A separation of d/4, where d is the optical length of the cavity, is a good choice. In this arrangement, the two counter propagating pulses see the same gain and collide in the absorber with equivalent power. This d/4 spacing also reduces the formation of extra pulses in the cavity.

For achieving the best results, a careful balancing of the four basic pulse shaping mechanisms, namely, self phase modulation, group velocity dispersion, saturable absorption and saturable gain, within a single resonator has to be obtained. Further shortening may be done by compressing the output from the colliding pulse arrangement with various pulse compression techniques. This dual arrangement of combining colliding pulse technique with pulse compression has enabled the ultra fast pulses to be as short as 6-8 fs.

The peak power of output pulses from a colliding pulse laser oscillator is typically in the kilowatt range. However, this can be amplified to generate femtosecond light pulses at gigawatt levels. In multiple amplifying stages the amplified pulse from each stage is passed through a saturable absorber dye before it enters the next stage. This effectively clips the low energy wings of the pulse and any accompanying spontaneously emitted light from amplifier dye cell. Considerable temporal broadening of the pulse occurs during amplification process due to group velocity dispersion in the solvent of the amplifying dye and other optical components. This may be compensated by employing a pulse compressor after the final amplifier stage.

2.5.3. Self Phase Modulation and Pulse Compression

Self phase modulation is an important mechanism that enables optical pulse compression. It provides the necessary bandwidth and frequency sweep needed for pulse compression. The phase modulation is due to intensity dependent index of refraction. An intense optical pulse travelling through the medium distorts atomic configuration of medium resulting in change of refractive index via nonlinear coefficients.

$$n = n_o + n_2 E^2$$

$$E(t) = (1/2) E_o(t) \exp[-2\pi j \{\nu t - n(t) z \nu/c\}]. \qquad (2.25)$$

The intensity dependent term in the index of refraction modulates the spectral intensity by modulating the instantaneous phase. The broadening $\Delta\nu$ at the laser frequency ν is given by

$$\Delta\nu(t) = -(\nu z/c) \; \partial(n_2 E^2)/\partial t. \qquad (2.26)$$

The pulse modifies its own spectra through a change in phase and envelope through process of self phase modulation and self steepening respectively. The maximum frequency spread is given by

$$|\Delta\nu_{max}| \simeq \nu n_2 E_0^2 z/c\tau, \qquad (2.27)$$

where τ is the input pulse duration (FWHM) and z the interaction length in the medium.

Pulse compression is a technique to reduce the pulse width of ultra short lasers down to bandwidth limited region. It also increases the attainable pulse peak power simultaneously as it narrows down the pulse. The ultimate duration of a laser pulse is dictated by the uncertainty relation

$$\Delta t \; \Delta\nu \geq K,$$

where K is a constant of the order of unity and Δt and $\Delta\nu$ are the FWHM pulse widths of the profiles $I(t)$ and $I(\nu)$ in the time and frequency domains respectively which form a transform pair. Generation of increasingly shorter pulses, therefore, requires corresponding spectral broadening.

The idea behind pulse compression is to send a chirped pulse (a pulse in which the frequency of the carrier varies continuously through out the pulse) through a linearly dispersive delay line. The group velocity being dependent on instantaneous frequency, different spectral portions of the pulse travel at different speeds through the delay line. In principle the length of the delay line may be adjusted such that the leading edge is delayed by just the right amount to coincide with the trailing edge at the output of delay line. The width of the resulting output pulse may be as short as the inverse of the bandwidth of the frequency sweep. The process of pulse compression thus involves two steps; first impressing a chirp on the pulse and then propagating the chirped pulse through a dispersive delay line. Optical pulses can be chirped by self phase modulation in a non-linear optical material. To first order, the magnitude of frequency sweep induced by self phase modulation is proportional to the first time derivative of the optical pulse shape. As a result only the central portion of the pulse has the approximate linear chirp suitable for compression. The wings of the pulse have opposite frequency sweep which leads to temporal broadening. The leading and trailing edges of the pulse thus give rise to wings on a compressed pulse (Fig. 2.12).

A single mode fibre can be used as a nonlinear medium for chirping and pulse compression. However, a more successful method of generating high quality compressed pulses is by chirping in a fibre with the combined action of self phase modulation and positive group velocity dispersion followed by compression in an external compressor. However, due to group velocity dispersion (GVD) the pulse is not as narrow as it would be without GVD. A matched pair of gratings has been considered a versatile compressor and used in a variety of pulse compression applications (Fig. 2.13). The grating compressor consists of a pair of plane ruled diffraction gratings arranged with their faces and rulings parallel. Each frequency propagating through the grating pair is diffracted in a different direction and follows a different route giving rise to a time delay that is a decreasing function of frequency. The grating spacing can be adjusted properly to provide the right amount of group delay, generating the compressed pulse. The variation of the group delay with wavelength for a grating pair compressor is given by

$$\frac{\Delta\tau}{\Delta\lambda} = \frac{1(\lambda/a)}{ca\,[1-\{(\lambda/a)-\sin r\}^2]},\qquad(2.28)$$

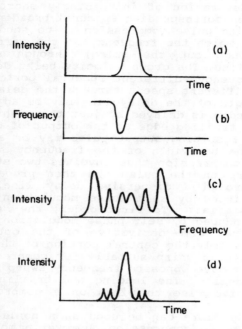

Fig. 2.12. Pulse compression by self phase modulation (SPM). (a) input pulse, (b) frequency modulation by (SPM), (c) Fourier transform of pulse and (d) compressed pulse.

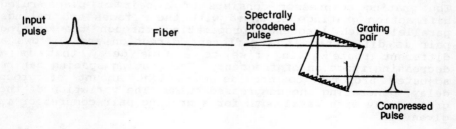

Fig. 2.13. Experimental arrangement for compressing a positively chirped laser pulse.

LASER CHARACTERISTICS   121

where l is the slant distance, a the grating constant, c the velocity of light and r the angle of incidence. In terms of phase function, $\phi(\nu)$, the effect of the grating compressor is to generate a phase function

$$\phi(\nu) = \phi_0 - K\nu^2, \qquad (2.29)$$

where K is an experimentally adjustable compressor constant. The output pulse from a grating compressor can be used as the input pulse for another fibre and compressor. In the arrangement for pulse compression shown in Fig. 2.13 input pulses 40 fs in duration were compressed down to 8 fs with a fibre of 3 to 8 mm in length.

2.5.4 Measurement of Ultrafast Pulses

The technique most widely used for picosecond measurements uses a streak camera. In this technique, light is focussed onto a photocathode and photoelectrons are emitted. The flux of emitted electrons is proportional to the light intensity hitting the photocathode. These electrons are accelerated and then deflected by an applied voltage that sweeps the electrons across the phosphor screen. The electrons released at different times from the photocathode strike the phosphor screen at different positions. This causes a track, or streak, which has a spatial intensity profile directly proportional to the incident temporal profile of the light. This phosphorescent streak is then analyzed electronically by a video system. In one shot, a complete fluorescence profile in time can be measured. The temporal resolution of streak cameras commercially available approaches 1 ps and with UV or IR spectral sensitivity.

2.6. Mode Locking of Transverse Modes

When we consider a set of transverse modes with the same longitudinal mode number but with transverse mode numbers differing by one in the y coordinate, these modes are approximately equally spaced in frequency. If we consider a set of modes locked with equal frequency spacing and zero phase difference with field amplitudes A_n where

$$|A_n|^2 = [(\bar{n})^n/n!] \exp(-\bar{n}), \qquad (2.30)$$

in which $\bar{n}$ is a parameter that determines the number of oscillating transverse modes, the intensity distribution of the optical field is given by

$$I(\xi,t) = \pi^{-1/2}\exp[-(\xi-\xi_0\cos\Omega t)^2], \qquad (2.31)$$

where ξ (= y/w) is the transverse coordinate in the y direction normalized with respect to the fundamental spot size, $\xi_0 = (2\bar{n})^{1/2}$ and Ω is the transverse mode frequency spacing. Eq. 2.31 represents a scanning beam of width equal to the fundamental spot size that moves back and forth in the transverse plane with maximum excursion equal to ξ_0. Fig. 2.14 illustrates this motion over a complete period of $2\pi/\Omega$. The maximum number of resolvable spots is roughly equal to the number of oscillating modes.

Simultaneous locking of the modes of a laser oscillating in several longitudinal and transverse modes has also been studied. If a set of transverse modes corresponding to each longitudinal mode number is locked together to form a scanning beam and each of the sets of longitudinal modes is also locked with zero phase difference to form a pulse in the resonator, the light will be confined to a small region of space both in axial and transverse directions. This blob of light bounces back and forth in the laser resonator following the zig zag path to be expected from geometrical

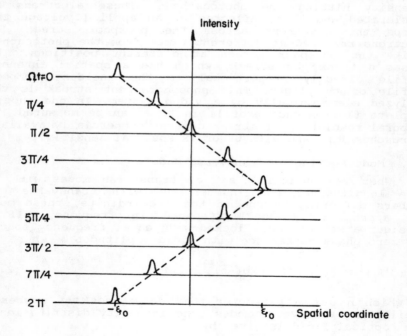

Fig. 2.14. Plot of spatial intensity distribution as a function of time in units of $1/\Omega$ for a transverse mode locked laser.

optics. In most situations existing in practice, the laser is operated in the lowest order transverse mode TEM_{ooq} and therefore multitransverse mode operation is rare. Transverse mode locking therefore is only of academic interest [2.19].

2.7. Q-Switching and Cavity Dumping

Although mode locking can be extended to low repetition rates by using long optical cavities (repetition frequency proportional to c/2d) it is somewhat impractical at repetition rates below 30 MHz. Up to 100 kHz, Q-switching is effective with a laser such as Nd:YAG; for 100 kHz to 30 MHz cavity dumping is preferred. The latter two techniques can be described with the aid of Fig. 2.15 where the transmission of the output mirror of the oscillator is assumed to be time dependent in response to a drive signal. In practice this can be effectively accomplished using an intra-cavity modulator when the mirror transmission is time independent.

2.7.1. Cavity Dumping

In cavity dumped operation (Fig 2.15b) the oscillator stores energy in the optical cavity during the interpulse period when the output mirror transmission is zero. To achieve maximum power output the transmission is switched to 100 percent (for a time interval>2d/c) at a time when the optical cavity has maximum stored energy. As a result, all the stored energy of the cavity is dumped and the output is a narrow pulse whose peak power is equal to the maximum internal circulating power. Restoration of the mirror transmission to zero (full reflection) permits a repetition of the cycle. In general, to maximize the average output power for a given repetition rate the mirror transmission in the high transmission state is less then 100 percent and is held for times longer than 2d/c. The technique of cavity dumping was first proposed for pulse pumped lasers [2.20] and then applied to CW gas lasers [2.21].

There are many applications where only a single ultra short pulse is required rather than a train. A schematic diagram of an arrangement for selecting a single pulse based on cavity dumping technique is shown in Fig. 2.16. The reflectivity of both mirrors is high. In addition to dye cell, a Glan polarizer and a polarizing switch such as a Pockels or Kerr cell are inserted into the feedback path of the laser. The polarizing switch is initially unenergised and the polarizer is adjusted for maximum transmission. A pulse of radiation then bounces back and forth between the two reflectors. The leakage radiation from one of the mirrors (M_1 in Fig. 2.16) is used to trigger a high voltage pulse at a predetermined optical pulse amplitude. The high voltage pulser energizes the polarization switch to its

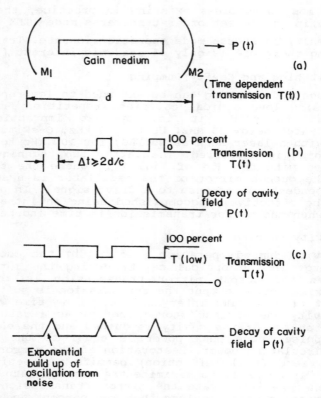

Fig. 2.15. Repetitive cavity dumping and Q switching of laser. (a) Laser oscillator which has time dependent mirror transmission. In practice this is effectively accomplished using an intra cavity modulator and a mirror where transmission is time dependent, (b) cavity dumping cycle and (c) Q switched cycle.

quarter wave voltage. After the pulse has made one traversal through the energised polarizing switch, the polarization of the pulse is rotated through 45°. For a two trip traversal of the polarizing switch, the polarization of the pulse is rotated through 90°. The Glan prism prevents the propagation of the pulse with this polarization in the system and the propagation direction of the pulse is redirected (as shown by the arrow) to output. In effect, the high reflectivity of the cavity is suddenly changed to a low reflectivity and the pulse stored within the cavity is suddenly dumped.

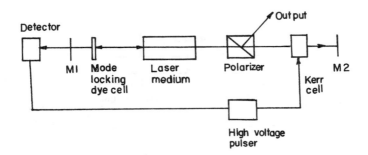

Fig. 2.16. Schematic arrangement for selecting a single pulse from a train of ultra short pulses.

The main difficulty in selecting a single ultra short pulse lies in the high voltage pulser. The rise time of the high voltage supplied to the polarizing switch must be faster than the time required for the light pulse to make a round trip traversal between the two mirrors.

2.7.2. Q-Switching

Referring back to Fig. 2.15 we consider the operation cycle for Q-switch as shown in Fig. 2.15c. Here the mirror transmission is 100% during most of the cycle so as to prevent oscillation and to allow the energy to be stored in the laser medium (in contrast to cavity dumping in which the flux is stored in the cavity). During this period, the single pass laser gain builds up to a high value. The mirror is then switched to its low transmitting (high reflection) state for a period sufficient to allow the oscillator to build up from noise. This period depends on the round trip transit time of light in the cavity and the maximum single pass unsaturated gain achieved. Q-switching was first proposed in 1961 [2.22], applied to pulsed solid state lasers [2.23] and then investigated experimentally and theoretically for CW pumped lasers [2.24, 2.25].

Let us assume a simple laser geometry with some sort of a closed shutter to spoil the cavity Q (Q-switching is also termed Q-spoiling) and thereby prevent oscillation (Fig. 2.17). Under this situation one can continuously pump energy to invert population until some sort of equilibrium is reached between the pump and the spontaneous decay processes of the system. This initial population inversion may be many times that required for CW oscillation in the absence of the shutter. Let us consider t=0 as the point of time when the shutter is opened (i.e. Q is switched to high value). At this moment the system is far above threshold,

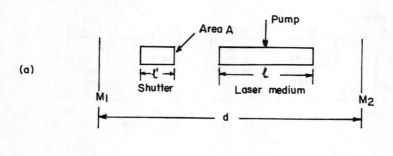

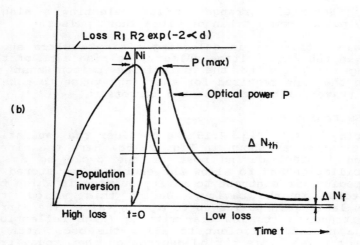

Fig. 2.17. Schematic of a Q-switched laser. (a) Q-switched cavity, (b) sequence of events during Q-switch. Q is switched at t=0.

the spontaneous emission along the axis of the cavity is greatly amplified and with the feedback provided by the cavity, the radiation builds up high enough to start depleting the population inversion (Fig. 2.17b). Let N_p be the number of photons inside the cavity. In one round trip, this number will increase to

$$[R_1 R_2 T^2 \exp(2\gamma 1 - 2\alpha 1')]N_p,$$

so that

$$\frac{dN_p}{dt} = \frac{[R_1 R_2 T^2 \exp(-2\alpha l') \exp(2\gamma l) - 1] N_p}{2(l_a + n'l' + nl)/c},$$

$$= [\exp\{2(\bar{\gamma} - \bar{\alpha}_t)d\} - 1] N_p / \tau. \qquad (2.32)$$

In Eq. 2.32

$\bar{\gamma} = \gamma l/d$
  = gain coefficient per unit length of the cavity,

$\bar{\alpha}_t = \alpha l'/d + (1/2d)\ln(1/R_1 T^2) + (1/2d)\ln(1/R_2)$
  = loss coefficient per unit length of the cavity,

$\tau = 2(l_a + n'l' + nl)/c$ = round trip time,
$d = l + l' + l_a$ = cavity length,
l_a = length of air path,
l = length of gain medium,
l' = length of shutter path,
R_1, R_2 = the reflectivity of mirrors M_1, M_2 respectively,
n, n' = refractive index of gain medium and shutter medium respectively, and
$T = T_1 T_2,$

where T_1 is the product of Fresnel transmission of two faces of the gain medium and T_2 that of the shutter medium. When the effective gain is small,

$$\exp 2(\bar{\gamma} - \bar{\alpha})d \approx [1 + 2(\bar{\gamma} - \bar{\alpha})d]$$

so that

$$dN_p/dt = 2(\bar{\gamma} - \bar{\alpha}_t) dN_p/\tau. \qquad (2.33)$$

The photon life time of the passive cavity, with $\bar{\gamma} = 0$, can be expressed as

$$1/\tau_p = [1 - \exp(-2\bar{\alpha}_t)d]/\tau = 2\bar{\alpha}_t d/\tau \qquad (2.34)$$

so that

$$dN_p/dt = [(\bar{\gamma}/\bar{\alpha}_t) - 1] N_p/\tau_p. \qquad (2.35a)$$

Further, we can write

$$\bar{\gamma} = \gamma_1/d = \sigma(N_2 - N_1)_1/d$$

$$\Delta N = (N_2 - N_1) A_1,$$

where σ is stimulated emission cross-section and ΔN is the total inverted population in the cavity of volume A_1. Also $\bar{\gamma} = \bar{\alpha}_t$ at the threshold, so that

$$\bar{\alpha}_t d = \sigma(N_2 - N_1)_{th} 1$$

$$\Delta N_{th} = (N_2 - N_1)_{th} A_1$$

which lead to

$$\bar{\gamma} = \sigma \Delta N/Ad,$$

and $\bar{\alpha}_t = \sigma \Delta N_{th}/Ad$.

Thus, $\bar{\gamma}/\bar{\alpha}_t = \Delta N/\Delta N_{th}$ (2.35b)

Using Eq. 2.35b, Eq. 2.35a becomes

$$dN_p/dt = [(\Delta N/\Delta N_{th}) - 1] N_p/\tau_p \qquad (2.36)$$

We now consider a generalized model of the two atomic states involved in a laser (Fig. 2.18) and write the differential equations describing the dynamics of the population (Rate equations) in terms of the life time for the various processes.

$$dN_2/dt = W_2 - (N_2/\tau_2) - (\sigma I/h\nu)(N_2 - N_1)$$

$$dN_1/dt = W_1 - (N_2/\tau_{21}) - (N_1/\tau_1) + (\sigma I/h\nu)(N_2 - N_1). \qquad (2.37)$$

Here the various emission and decay rates and other parameters are defined as follows:

(a) The sum of the spontaneous emission rate and any other collision process which decreases the population in 2

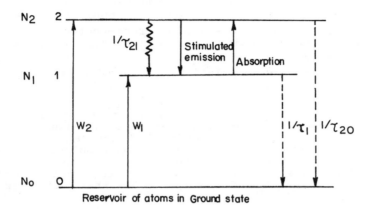

Fig. 2.18. Generalized pumping scheme of a laser.

and simultaneously increases that in 1 is denoted by $1/\tau_{12}$.

(b) The rate of loss of state 2 that does not result in an atom in state 1 is denoted by $1/\tau_{20}$.

(c) The total rate of decay of state 2 is the sum of the above rates and defines the life time of state 2 by

$$1/\tau = 1/\tau_{21} + 1/\tau_{20}.$$

(d) The decay rate of state 1 due to any and all causes is denoted by $1/\tau_1$.

(e) W_1 includes direct excitation from ground state to 1 state and also any indirect routes such as excitation to a higher state followed by spontaneous emission from the higher state back to state 1. It does not include the spontaneous decay of state 2 into 1 nor the stimulated processes. State 2 is pumped directly from ground at a rate W_2 that includes indirect paths to higher levels followed by a decay to 2.

(f) State 2 can radiate a photon of energy $h\nu_{12}$ spontaneously. The rate of decay of state 2 is proportional to the density in 2 times a rate constant. Atoms in state 1 behave similarly; they can radiate spontaneously to another level, be deactivated by a collision, or be swept out by mass motion from the volume of interest. The stimulated emission rates are in addition to the above natural decay processes.

(g) σ is the stimulated emission cross-section and I is the radiation intensity (i.e. power per unit area) at frequency ν.

(h) N_1, N_2 are the populations in level 1 and 2 respectively.

The total evolution of the giant laser pulse by Q-switching process is typically completed in about 20 ns. We can therefore neglect the effect of population relaxation and pumping that take place during the pulse. If we also assume that the switching of the Q from low to high state is accomplished instantaneously, we need to consider only the stimulated emission and absorption so that rate Eqs. 2.37 reduce to

$$dN_2/dt = (-\sigma I/h\nu)(N_2-N_1)$$

$$dN_1/dt = (\sigma I/h\nu)(N_2-N_1)$$

or

$$\frac{d(N_2-N_1)Al}{dt} = \frac{-2\sigma(N_2-N_1)l \; IAd(d/\tau)}{dh\nu(d/\tau)}. \qquad (2.38)$$

Observing that

$$\Delta N = (N_2-N_1)Al,$$

$$\bar{\gamma} = \gamma l/d = \sigma(N_2-N_1)l/d$$

and $N_p = IAd/(h\nu d/\tau)$,

we can write

$$d(\Delta N)/dt = -2\bar{\gamma}(2d/\tau)N_p. \qquad (2.39)$$

Using Eq. 2.34 and Eq. 2.35b, Eq. 2.39 becomes

$$d(\Delta N)/dt = (-2\bar{\gamma}/\bar{\alpha}_t)(\bar{\alpha}_t 2d/\tau)N_p$$

$$= -2(\Delta N/\Delta N_{th})N_p/\tau_p. \qquad (2.40)$$

If now we define a time scale normalized to τ_p i.e. $T=t/\tau_p$ [2.26], then Eqs. 2.36 and 2.40 become

$$dN_p/dT = [(\Delta N/\Delta N_{th})-1]N_p. \qquad (2.41)$$

$$d\Delta N/dT = -2(\Delta N/\Delta N_{th})N_p. \qquad (2.42)$$

The Eqs. 2.41 and 2.42 are nonlinear and coupled and cannot be solved for N_p and ΔN in terms of elementary functions of time. However, we can find an elementary solution from the photon number N_p in terms of the inversion ΔN. Eliminating time from the Eqs. 2.41 and 2.42, we get

$$dN_p/d\Delta N = [(\Delta N/\Delta N_{th})-1]/2. \qquad (2.43)$$

Further, we note that photon number N_p reaches a peak when the inversion crosses the threshold value. Therefore we integrate from ΔN_i to ΔN_{th} to get

$$\int_{N_p \approx 0}^{N_p(max)} dN_p = (1/2)\int_{\Delta N_i}^{\Delta N_{th}} [(\Delta N_{th}/\Delta N)-1]d(\Delta N),$$

$$N_p(max) = [(\Delta N_i - \Delta N_{th})/2] - (\Delta N_{th}/2)\ln(\Delta N_i/\Delta N_{th}). \qquad (2.44)$$

$N_p h\nu$ represents the optical energy stored in the cavity. The fractional loss of this energy per round trip, as a result of coupling, divided by the time for a round trip, equals The power emerging from the output mirror M_2 of reflectivity R_2. Therefore,

$P = h\nu\, N_p$(fractional loss per round trip/round trip time)

$$= h\nu\, N_p\, \bar{\alpha}_{ext} 2d/\tau, \qquad (2.45)$$

where

$$\bar{\alpha}_{ext} = (1/2d)\ln(1/R_2) \approx (1/2d)(1-R_2), \text{ when } R_2 \text{ is high.}$$

Again substituting for $2d/\tau$ in terms of τ_p from Eq. 2.34

$$P(\max) = (\bar{\alpha}_{ext}/\bar{\alpha}_t) h\nu N_p(\max)/\tau_p. \qquad (2.46)$$

If we integrate Eq. 2.43 over the complete time interval of the pulse in order to estimate the fraction of the inversion which is converted to photons,

$$\int_0^0 dN_p = (1/2) \int_{\Delta N_i}^{\Delta N_f} [(\Delta N_{th}/\Delta N) - 1] d\Delta N$$

giving

$$\Delta N_f/\Delta N_i = \exp[-(\Delta N_i - \Delta N_f)/\Delta N_{th}]. \qquad (2.47)$$

Eq. 2.47 determines ΔN_f implicitly in terms of initial value ΔN_i and the threshold value ΔN_{th}. This can be solved numerically (or graphically). The fraction of initial inversion converted to photons is

$$\eta = \frac{(\Delta N_i - \Delta N_f)}{\Delta N_i} \qquad (2.48)$$

and tends to unity as $\Delta N_i/\Delta N_{th}$ increases (Fig. 2.19).

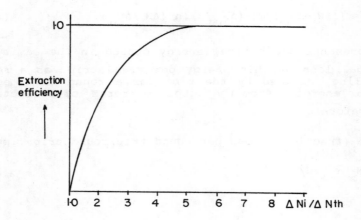

Fig. 2.19. Variation of extraction efficiency η as a function of n_i/n_{th} for a Q-switched pulse.

The total energy converted to photons is

$$E_p = h\nu(\Delta N_i - \Delta N_f)/2$$

$$= \eta h\nu \Delta N_i/2, \qquad (2.49)$$

and the output energy is

$$E_{out} = (\bar{\alpha}_{ext}/\bar{\alpha}_t) E_p. \qquad (2.50)$$

The pulse width can be estimated from the relation

$$P(max) \Delta t \approx E_{out} \qquad (2.51)$$

The basic equations of Q-switching, Eqs. 2.41 and 2.42, can be solved by numerical techniques, given that a few photons are present within the cavity to start the stimulated emission. The evolution of Q-switch pulse with time is shown in Fig.2.20. It is seen that as inversion ratio $\Delta N_i/\Delta N_{th}$ is increased, the time required to develop a more

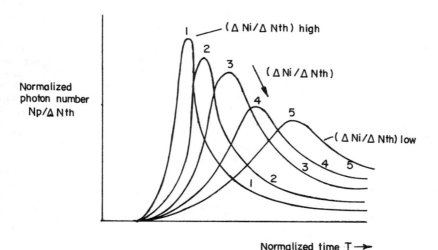

Fig. 2.20. Time evolution of Q-switched pulse for different inversions. The nature of curves 1 to 5 correspond to decrease of n_i/n_{th}.

intense pulse decreases. This is because firstly, higher inversion leads to higher gain and hence faster growth rate and secondly, higher inversion ratio implies more atoms in the upper state which contributes spontaneous power to start the oscillation.

One cannot however increase the initial inversion too high and obtain arbitrary high peak powers. This is because, as the initial inversion is increased the spontaneous emission rate increases proportionately which is then amplified by the remaining section of the gain medium. Since the single pass gain is very high, the amplifier spontaneous emission may be sufficient to saturate the gain, thereby limiting ΔN_i. This limits the amount of energy that can be stored in the population inversion in any laser. The switch must have enough loss, or hold off capability, so that the round trip gain is <1. For a given switch, this limits the initial inversion ΔN_i.

A variety of methods have been used for Q-switching. They range from spinning the mirrors or retroprisms and using saturable absorbers or electro-optic switches in the cavity resonator.

Giant laser pulses obtained by Q-switching technique are used extensively in applications that depend on high peak powers and short duration. These applications include nonlinear optics, ranging, material processing, initiation of chemical reaction and plasma diagnostics.

A typical arrangement for obtaining a Q-switched pulse for ranging experiment is shown in Fig.2.21, which was used with Nd glass rod developed by CGCRI, India. Q-switching was obtained by rotating one end reflector, i.e. a 90° total reflecting prism rotated at 20,000 rpm [2.27]. The laser rod was kept at one focus of an elliptical reflector while the pump flash tube was kept at the other. A glass tube was used to shield the laser rod from the harmful effects of the uv radiation from the flash lamp. The flash tube was operated by shunt triggering. A preset delay was adjusted by a synchronizing lamp and photo diode arrangement placed near the rotating prism and a delay generator. The flash tube discharge circuit used an aircore inductance to shape the flash tube current pulse. The time delay provided by the electronic delay generator is necessary in order to fire the flash tube in such a way that optimum population inversion is built up at the instant when resonant cavity reflectors are aligned for high Q condition. The required time delay can be estimated experimentally by first noting the onset of normal mode oscillation with respect to the initiation of the flash pulse. For this, the delay is set to zero and the prism is kept stationary and made to align with the output mirror for obtaining laser action.

When Q-switched output is measured for various values

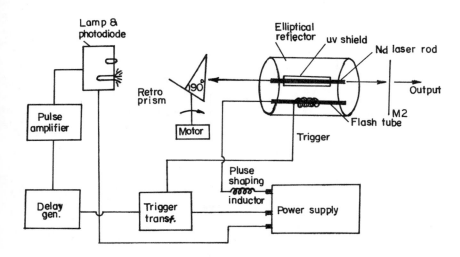

Fig. 2.21. Experimental arrangement of the rotating prism Q-switched Nd glass pulsed laser [2.27].

of delay, an optimum value for delay is found for which output is maximum. An important aspect of the delay vs output curve is the appearance of double pulse for certain values of delay for given value of end reflectivity and input energy. This multiple pulsing is an expected feature of the slow type of Q-switching in the rotating prism arrangement (Fig. 2.22). Fast switching is done by electro-optic means like the Kerr cell or the Pockels cell. To store maximum energy, delay should be kept to optimum value. This provides peak emission. Keeping delay below this optimum value gives less output because population is not built sufficiently. Increasing delay more than the optimum value will again decrease the output because then spontaneous emission losses increase while population inversion can not increase beyond the saturation value. The optimum value of delay is approximately equal to the life time of the lasing level.

If Q-switching is done by a passive Q-switch, e.g. the Eastman 9740 saturable absorption dye for Nd:YAG laser, it is observed that both mode locking and Q-switching is achieved simultaneously. Within the envelope of Q-switched pulse there are mode locked pulses. The time scales of the mode locked and Q-switched pulses are orders of magnitude

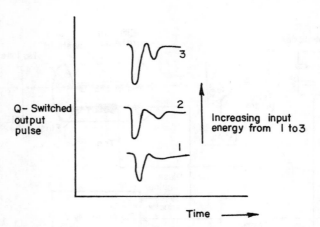

Fig. 2.22. Appearance of double pulsing with increasing energy in slow Q-switching, other parameters remaining fixed.

different (ps and ns respectively).

Another case of time varying output is of relaxation oscillation which is observed in many types of lasers. This oscillation takes place characteristically with a period that is considerably longer than the cavity decay time or the resonator round trip time. Typical values range from 0.1 to 10 μs.

2.8. Relaxation Oscillation

The basic physical mechanism of relaxation oscillation is an inter play between the oscillation field in the resonator and the atomic inversion. An increase in the field intensity causes a reduction in the inversion due to the increased rate of stimulated transition. This causes a reduction in the gain that tends to decrease the field intensity.

Relaxation oscillation occurs when the population inversion from pumping builds up faster than the photon density in the cavity. This happens in a laser system with a very low probability for spontaneous emission and non availability of sufficient photons to start the oscillation process. As a consequence of the low spontaneous rate, it takes many round trips before the photon density can build up to a level sufficient to affect the population inversion in a significant way.

Relaxation oscillator involves gain switching as against Q-switching and mathematical investigation [2.28] requires time dependence of the pumping process which we neglected while analyzing Q-switching. We consider the laser cavity configuration of Fig. 2.23 and assume that

(a) pump is suddenly applied at t=0 and held constant there after,

$u(t) = 0$, for $t<0$

$= 1$, for $t>0$

(b) both states 2 and 1 decay back to ground state with the same time constant τ.
(c) Only state 2 is pumped.

The population densities then obey the rate equations

$$dN_2/dt = W_2 u(t) - (N_2/\tau) - (N_2-N_1)\sigma I/h\nu$$

$$dN_1/dt = -(N_1/\tau) + (N_2-N_1)\sigma I/h\nu. \qquad (2.52)$$

We now follow the same steps that we did for the case of Q-switching in section 2.7.2; namely, subtract dN_1/dt from dN_2/dt, multiply by the volume Al of the active medium, equate $IAd/h\nu c$ to N_p, the number of photons in the cavity,

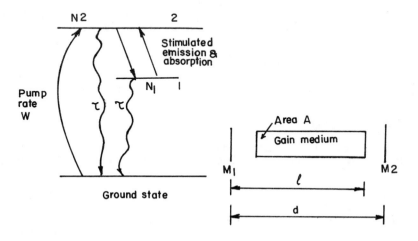

Fig. 2.23. Configuration of a gain switched laser for analysis of relaxation oscillation.

and $(N_2-N_1)Al$ to the inverted population ΔN. We then get

$$d\Delta N/dt = W_2 Al u(t) - (\Delta N/\tau) - 2c'\gamma_1 N_p/d, \qquad (2.53)$$

where

$$\gamma = (N_2-N_1)\sigma.$$

The number of photons within the cavity is treated again in the same way as for Q-switching so that

$$dN_p/dt = [(\Delta N/\Delta N_{th})-1] N_p. \qquad (2.54)$$

Again, time is normalized to that of the photon life time of the passive cavity ($T=t/\tau_p$), the inversion density ΔN by the threshold value ΔN_{th} and the photon number N_p by the CW or steady state value N_{po}, so that

$$(d/dt)(N_p/N_{po}) = [(\Delta N/\Delta N_{th})-1](N_p/N_{po}) \qquad (2.55)$$

$$(d/dt)(\Delta N/\Delta N_{th}) = (W_2 Al \tau_p/\Delta N_{th}) u(t) - (\Delta N/\Delta N_{th})(\tau_p/\tau)$$

$$-2(\Delta N/\Delta N_{th})(N_{po}/\Delta N_{th}) N_p/N_{po}. \qquad (2.56)$$

Eqs. 2.55 and 2.56 can only be solved numerically. However, we can, under some limitations, obtain a few simple forms. If we prevent stimulated emission by destroying the Q of the cavity,

$$\Delta N = W_2 \tau [1-\exp(-\tau_p/\tau)T] \qquad (2.57)$$

and by putting $\Delta N = \Delta N_{th}$, $N_p = 0$

$$\Delta N_{th} = W_{20} Al \tau. \qquad (2.58)$$

When all transients have died down in the steady state, $\Delta N/\Delta N_{th}=1$, $N_p=N_{po}$ and all time derivatives are set to zero,

$$\frac{N_{po}}{\Delta N_{th}} = \frac{\tau_p(K-1)}{2\tau}, \qquad (2.59)$$

where $K=W_2/W_{20}$ is the ratio of the actual pumping rate to threshold value.

The solution of the coupled differential Eqs. 2.55 and 2.56 would give a large initial pulse (similar to Q-switching situation) followed by a damped oscillation approaching the steady state value (Fig. 2.24). This damped oscillatory behaviour is characteristic of the response of many types of lasers and is a result of the interchange of energy between the photons and the inversion driven by the pumping. In order to understand the oscillatory nature of response, it is desirable to find the change in photon number in response to a small change in pumping rate.

We learn from Eq. 2.56 that as the photons increase, the inversion decreases, which then causes a decrease in the photons by Eq. 2.55 and in turn increases the inversion and so on.
Let

$$W_2/W_{20} = 1+p(t), \tag{2.60}$$

$$N_p/N_{po} = 1+q(t), \tag{2.61}$$

$$\Delta N/\Delta N_{th} = 1+r(t), \tag{2.62}$$

where $p(t)$ is the slight change in pumping and q, r are the corresponding small changes in photons and inversion respectively. Inserting Eqs. 2.60-2.62 in Eqs. 2.55 and 2.56,

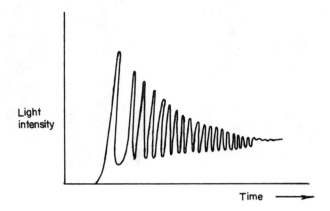

Fig. 2.24. Relaxation oscillation in a laser. Damped sinusoidal perturbation of intensity is shown as function of time.

neglecting the products of q and r because of smallness, and removing the normalization on time and substituting for ΔN_{th} from Eq. 2.58 and also noting that $K=W_2/W_{20}$, we get

$$dq/dt = r/\tau_p \tag{2.63}$$

$$dr/dt = (Kp/\tau)-(Kr/\tau)-(K-1)q/\tau. \tag{2.64}$$

One can combine Eqs. 2.63 and 2.64 to obtain a single equation for the change in photon number in terms of the change in pumping rate,

$$d^2q/dt^2 +(K/\tau)dq/dt +[(K-1)/\tau\tau_p]q = Kp/\tau\tau_p. \tag{2.65}$$

When the change in pumping rate p is put to zero, the right hand side of Eq. 2.65 becomes zero, so that

$$d^2q/dt^2+(K/\tau)dq/dt+[(K-1)/\tau\tau_p]q = 0. \tag{2.66}$$

Eq. 2.66 is the differential equation describing a damped harmonic oscillator. However when p, the excess pumping, is positive and finite, such that

$$p(t) = \Delta p\ u(t),$$

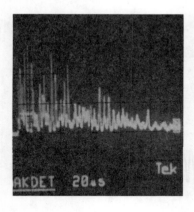 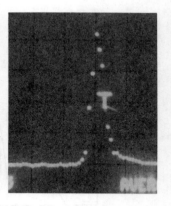

(a)   (b)

Fig. 2.25. Output from a Nd:YAG laser. (a) Intensity fluctuation of undamped relaxation oscillation, (b) Q-switched pulse.

the solution of Eq. 2.66 leads to the time evolution of photon change given by

$$q \approx (K/K-1)\Delta p\,[1-\exp(-t/\tau_d)\cos\omega_o t], \qquad (2.67)$$

where $1/\tau_d = K/2\tau$ = damping constant

$$W_o = (K/2\tau)[\{4(K-1)\tau/K^2\tau_p\}-1]^{1/2}.$$

It is to be noted that the oscillatory behaviour shows that the population inversion on a transient basis can exceed the threshold value. In Q-switching, one adds a controllable loss, whereas in gain switching leading to relaxation oscillation, the inversion builds up faster than the photons.

Although some lasers display the damped sinusoidal perturbation of intensity [solution of Eq. 2.65 for p=0 represented typically in Fig. 2.24], in many other lasers the perturbation is undamped [solution of Eq. 2.65 with p≠0 exemplified in Fig. 2.25]. The undamped relaxation in frequency domain shows spectral broadening and a shift to higher frequencies. This spectral broadening results from increase in pumping strength. For spectral purity, it is desirable to keep the pumping level only very slightly above threshold.

References

2.1 P.W. Smith, *Proc. IEEE* **60** (1972) 422.

2.2 P.W. Smith, *Proc IEEE* **58** (1970) 1342.

2.3 C.C. Cutler, *Proc. IRE* **43** (1955) 140.

2.4 A.J. Demaria, W.H. Glenn Jr, M.J. Brienza and M.E. Mack, *Proc. IEEE* **57** (1969) 2.

2.5 (a) V.J. Fowler and J. Schlafer, *Proc. IEEE* **54** (1966) 1437.

  (b) S.E. Harris, *Proc. IEEE* **54** (1966) 1401.

2.6 P.W. Smith, *IEEE J. Quant. Electron.* **QE-3** (1967) 627.

2.7 W.E. Lamb Jr, *Phys. Rev.* **134** (1964) A 1420.

2.8 T. Uchida and A. Veki, *IEEE J. Quant. Electron.* QE-3 (1967) 17.

2.9 W. Koechner, *Solid State Engineering* (Springer Verlag, 1976).

2.10 S.C. Jain, V.V. Rampal and U.C. Joshi, *Opt. Eng.* **31** (1992).

2.11 M. DiDomenico, H.M. Marcos, J.E. Geusic and R.E.

Smith, *Appl. Phys. Lett.* **8** (1966) 180.

2.12 J.A. Armstrong, *Appl. Phys. Lett.* **10** (1967) 16.

2.13 J.A. Giordmaine, P.M. Rentzepis, S.L. Shapiro and K.W. Wecht, *Appl. Phys. Lett.* **11** (1967) 216.

2.14 A.A. Malyutin, M.Y. Schelev, *JETP Lett.* **9** (1969) 266.

2.15 C.K.N. Patel, *Appl. Phys. Lett.* **18** (1971) 25.

2.16 D.M. Rayner, P.A. Hachett and C. Willis, *Rev. Sci. Instru.* **53** (1982) 537.

2.17 S. Thaniyavarn and T.K. Gustafson, *Appl. Phys. Lett.* **40** (1982) 2551.

2.18 *Encyclopedia of Lasers and Optical Technology,* ed. R.A. Meyers (Academic Press, 1991).

2.19 D.H. Auston, *IEEE J. Quant. Electron.* **QE-4** (1968) 420, 471.

2.20 A.A. Vuylsteke, *J. Appl. Phys.* **34** (1963) 1615.

2.21 W.H. Steier, *Proc. IEEE* **54** (1966) 1604.

2.22 R.W. Hellworth in *Advances in Quantum Electronics,* ed. J.R. Singer (Columbia Univ Press, 1961) p.334.

2.23 F.J. Mcclug and R.W. Hellworth, *J. Appl. Phys.* **33** (1962) 828.

2.24 V. Evtuhov and J.K. Neeland, *IEEE J. Quant. Electron.* **QE-5** (1969) 207.

2.25 R.B. Chesler, M.A. Karr and J.E. Geusic, *IEEE J. Quant. Electron.* **QE-5** (1969) 345.

2.26 W.G. Wagner and B.A. Lengyl, *J. Appl. Phys.* **34** (1963) 2042.

2.27 N. Mansharamani, V.V. Rampal and K.P. Srivastava, *J. Instrum. Soc. India* **6** (1976) 39.
M.K. Basu, V.S. Bhatnagar, N. Mansharamani and V.V. Rampal, *J. Instrum. Soc. India* **7** (1977) 23.

2.28 J.T. Verdeyen, *Laser Electronics* (Prentice Hall, 1989).

CHAPTER 3

SPECIFIC LASER SYSTEMS

The laser is an extension of maser action at optical frequencies. The words maser and laser are acronyms signifying amplification by stimulated emission of radiation at microwave (maser) and light (laser) frequencies. The conditions for the amplification and oscillation at light frequencies were first brought out by Schawlow and Townes in 1958 [3.1]. Two years later, the first experimental observation of laser action was achieved by Maiman in a crystalline solid ruby, $Al_2O_3:Cr^{3+}$ [3.2]. Since then, much effort has been spent in finding new laser transitions in various media; solid, liquid, gas, and semiconductors. Today laser frequencies run into thousands covering the spectrum from UV to far infrared. Some of the milestones in the search for new laser systems after Maiman's success are the following:

1. Gas laser action in a mixture of helium and neon by A. Javan, W. Bennett and D. Herriot of Bell Laboratories in 1961.
2. Demonstration of neodymium laser by L.F. Johnson and K. Nassau in 1961.
3. Semiconductor laser demonstrated in 1962 by R. Hall of General Electric Research Laboratories.
4. Carbon dioxide laser discovered by C.K.N. Patel of Bell Laboratory in 1963.
5. Argon ion laser developed by W. Bridges of Hughes Research Labs in 1964 and later in 1966, the He-Cd, metal vapour and ion laser by W.T. Silfvast, G.R. Fowles and B.D. Hopkins at University of Utah.
6. Liquid laser in the form of a fluorescent dye discovered in 1966 by P.P. Sorokin and J.R. Lankard of the IBM Research Laboratory. This provided a broadly tunable laser.
7. Rare gas halide excimer laser in Xenon fluoride discovered in 1975 by J.J. Ewing and C. Brau of the Avco Everett Research Laboratory.
8. Free electron laser amplifier in 1976 by J.M.J. Madey and coworkers at Stanford University.

9. Soft X ray laser demonstration in 1985 in highly ionized plasma by D. Mathews and coworkers at Lawrence Livermore Laboratory.

In addition there have been important developments in the areas of gas dynamic laser, chemical laser, plasma recombination laser, quantum well and hetrostructure semiconductor lasers, guided wave and distributed feedback lasers. A number of pumping schemes including optical, electric discharge and current injection have been adopted. We shall now discuss some representative systems.

3.1. Solid State Lasers

Solid state lasers, as a class, concern those laser systems which are based on crystalline or glassy insulating media in which stoichiometric component, or extrinsic dopants incorporated into the material, serve as the laser species. The number of potential impurity host systems is virtually limitless. Some of the reported laser ions and host materials are:

Laser ions

Transitional metal ions	Ti^{3+}, V^{2+}, Cr^{4+}, Co^{2+}, Ni^{2+}, Cr^{3+}
Trivalent Rare Earths	Ce^{3+}, Pr^{3+}, Nd^{3+}, Pm^{3+}, Sm^{3+}, Eu^{3+}, Gd^{3+}, Tb^{3+}, Dy^{2+}, Ho^{3+}, Er^{3+}, Tm^{3+}, Yb^{3+}
Divalent rare earths	Sm^{2+}, Dy^{2+}, Tm^{2+}
Actinide	U^{3+}

Laser hosts

Fluoride crystals	MgF_2, CaF_2, LaF_3, $LiYF$, $LiCaAlF_6$
Oxide crystals	MgO, Al_2O_3, Y_2O_3, $BeAl_2O_4$, $YAlO_3$, $CaWO_4$, YVO_4, $Y_3Al_5O_{12}$, $Gd_3Sc_2Ga_3O_{12}$ $LiNdP_4O_{12}$
Glasses	ZrF_4-BaF_2-LaF_3-AlF_3, SiO_2-Li_2O-CaO-Al_2O_3, P_2O_5-K_2O-BaO-Al_2O_3

The listed ions lase with varying degrees of proficiency and may have both advantages and disadvantages in their use e.g. Ti^{3+} has wide tuning range but lacks ability to store energy, Sm^{2+} lacks good chemical stability etc. Each of the laser host also has plus and minus points e.g. ease of crystal growth, ability to form a melt in glassy materials and compatibility of impurity ion with particular host. Generally, the size and charge of the substitutional host metal ion must be similar to that of the impurity ion.

3.1.1. The Ruby Laser

This is a representative of the crystalline solid state lasers [3.3] and was the first to be demonstrated. Ruby is still one of the most useful laser materials and gives laser output in the red region of the visible spectrum at $\lambda = 0.6943$ μm. The chemical composition of ruby consists of a crystal of sapphire Al_2O_3 in which a small amount of aluminium is replaced by chromium by adding Cr_2O_3 (~5% by wt.) to the melt in the growth process (giving Cr^{3+} concentration of about 10^{19} cm^{-3}). The host crystal has a rhombohedral unit cell where the axis of symmetry is the C-axis. The crystal is uniaxial with refractive indices $n_o = 1.763$ and $n_e = 1.755$. In the ordinary wave the electric field E is perpendicular to C-axis and in the extraordinary wave it is parallel to C-axis. The chromium atoms are active in the lattice as triply ionized Cr^{3+}, and give rise to the energy level diagram of Fig. 3.1a. Because of the energy level scheme relevant to this laser, it is classified as a three level laser (as against 4 level schemes of Nd laser to be described later). The three level scheme is shown in Fig. 3.1b for comparative understanding.

The pumping of ruby is performed usually by subjecting it to the light of intense flash lamps (similar to flash photography). A portion of this light, corresponding in frequency to the two absorption bands $^4F_2$ and $^4F_1$, is absorbed; thereby causing Cr^{3+} ions to be transferred into these levels. The ions proceed to decay fast into the upper laser level ^2E. The level ^2E is composed of two levels $2\bar{A}$ and $\bar{E}$ separated by 29 cm^{-1}. Transitions originating from the $2\bar{A}$ are denoted by R (0.6929 μm) those from $\bar{E}$ are called R_1 (0.6943 μm). The lower laser level is the ground state. The life time of the E states, because of spontaneous emission, is 3 ms which is long by atomic standards. The E levels are therefore sometimes classified as metastable. Because of the anisotropy, the spontaneous emission (and thus stimulated emission) depends on the orientation of the optical electric field with respect to C-axis. The degeneracy of each of the upper laser levels is 2, whereas that of the ground state is 4; hence the population of $\bar{E}$ level is required to be half of that in $^4A_2$ in order to obtain optical transparency. This is because absorption is proportional to $[(N_1 g_2/g_1) - N_2]$ and for

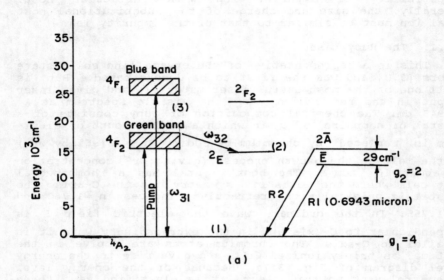

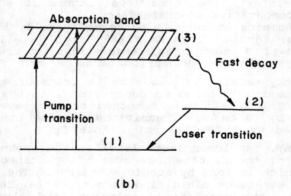

Fig. 3.1. Energy level diagram relevant to (a) ruby laser (b) three level scheme.

this to be zero, $N_1/2 = N_2$.

This means 2A level will have an almost equal population to that of the $\bar{E}$ state and thus roughly half of the doped chromium atom must be promoted to the ^2E manifold to obtain transparency on the R_1 line. Further pumping yields gain and oscillation if proper feedback is supplied. Fortunately ruby has very strong absorption bands corresponding to the $^4A \rightarrow {}^4F$, ^4F transitions. These bands nearly 1000 Å wide, are located in the green (18000 cm^{-1}) and violet (25000 cm^{-1}) portion of the spectrum and provide ruby its characteristic colour. The absorption bands absorb a significant fraction of the light emitted by an incoherent flash lamp that radiates more or less as a black body source at an elevated temperature of about 7000 to 9000°K. The life time of the pumped states $^4F_2$ and $^4F_1$ are very short (~50 ns), with most of the atoms returning to the ^2E levels rather than back to the ground state. About 70% of the atoms elevated to ^4F levels appear in the ^2E levels. The stimulated emission cross-section is obtained from the absorption cross-section from the relation

$$\sigma_s = \sigma_{abs}(g_1/g_2) \tag{3.1}$$

with $g_1=4$ and $g_2=2$, σ_{abs} for R_1 line=1.22x10^{-20}cm^2, σ_s is obtained as 2.5x10^{-20}cm^2.

The width $\Delta\nu$ of the laser transition is a function of temperature and at 300°K it is 11cm^{-1} (from 250°K to 350°K it increases linearly with a slope of about 0.1cm^{-1}/°K).

Fig. 3.2 shows the pumping arrangement of ruby laser which is typical of optically pumped solid state lasers. The flash lamp (helical) may surround the ruby rod (Fig. 3.2a) or the flash radiation may be focussed onto the ruby rod in a single elliptical reflector (Fig. 3.2b) or a double elliptical system (Fig. 3.2c). The flash excitation is provided by the discharge of a capacitor bank. The resonant cavity is formed by the mirrors M_1 and M_2 and output is obtained from the partially reflecting mirror (say M_2).

In the three level scheme of Fig. 3.1b, certain clarifications are in order:

(a) Fast decay from level 3 to 2 makes the emission from 3 to 1 insignificant in comparison.

(b) Fast relaxation from 3 to 2 and slow emission from 2 to 1, enables accumulation of atoms in ^2E state. With fast enough pumping, this provides population inversion on the 2→1 transition.

(c) The decay from 3 to 2 is nonradiative and the energy goes in heating the crystal lattice.

(d) Pump energy absorption at R_1, R_2 transition is insignificant due to large difference in absorption cross-section at the laser wavelength (0.694 μm) and the peak emission from flash lamp (0.55 μm).

(e) Amplification of pump radiation at R_1, R_2 lines is small since path of pump amplification is the diameter of rod (in transverse pumping). This is unlike the amplification in laser action where length of the laser rod forms the amplifying medium.

(f) The spectral emission characteristics of a flash lamp depend on the current density. It is thus desirable to operate it such that its peak emission band spectrally matches with the absorption band of the laser medium (in the present case $^4F_1$, $^4F_2$ bands of ruby). The type of gas filled in the flash lamp also determines the spectral output. Normally Xenon or Krypton filled lamps are used. Typically, the flash pulse is about 500 μs long.

To estimate the population N_2 in level 2 due to the flash pulse of duration T, let us assume that the flash pulse is rectangular in time and gives an optical flux at the crystal surface $W(\nu)$ watts per unit area per unit frequency at the frequency ν. If the absorption coefficient of the crystal is $a(\nu)$ then the amount of energy absorbed by the crystal per unit volume is

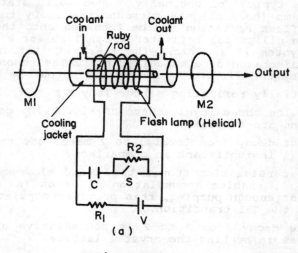

Fig. 3.2a

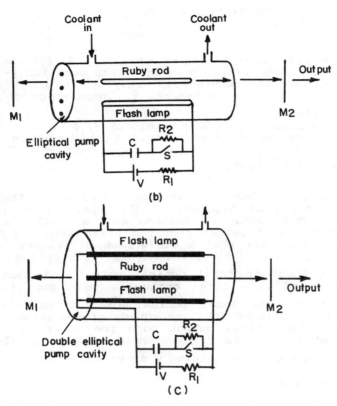

Fig.3.2. Schematic of an optically pumped solid state laser with ruby as gain medium and pump cavity with (a) helical flash lamp (b) single elliptical reflector (c) double elliptical system.

$$T \int_0^\infty W(\nu) a(\nu) d\nu$$

provided of course $a(\nu)$ is taken to be independent of the distance travelled by pumping radiation through the crystal. If the absorption quantum efficiency (the probability that absorption of a pump photon at ν results in transferring one atom into the upper laser level) is $\eta(\nu)$, the number of atoms pumped into level 2 per unit volume is

$$N_2 = T \int_0^\infty (1/h\nu) W(\nu) a(\nu) \eta(\nu) d\nu. \qquad (3.2)$$

Since the life time of atoms in level 2 (~3 ms) is an order of magnitude larger than the flash duration (~ 0.5 ms) we may neglect the spontaneous decay out of level 2 during the time T so that N_2 represents the population of level 2 after the flash.

Considering the average values of W, a, η over the useful absorption region of width $\Delta \nu$, we could express

$$N_2 = T \bar{W} \bar{a} \bar{\eta} \Delta\nu/h\bar{\nu}, \qquad (3.3a)$$

where bar represents the average value over the narrow spectral region of absorption. The spectral region over which absorption occurs is however only a small fraction of the total spectral output from the flash lamp and thus the electrical to laser photon conversion efficiency is quite low. First, there is efficiency of conversion ε_1 from electrical input to total optical output and then only a fraction ε_2 of this total optical output goes as useful absorption for the laser medium. In this calculation of conversion efficiency, the inadequacy of the pump cavity to focus all the optical output onto the laser medium by a factor say ε_3 is also to be considered. Thus, electrical input E required for obtaining $\bar{W}$ is given by,

$$E = W/\varepsilon_1 \varepsilon_2 \varepsilon_3. \qquad (3.3b)$$

To estimate the threshold energy for laser action, one can first calculate the population N_2 required at threshold from applying condition of oscillation for given losses in the cavity (Eqs. 1.145, 1.146) and then estimate electrical input required to achieve this value of N_2 from Eqs. 3.3a, 3.3b. For this of course one must remember that

$$N(2\bar{A})/N(\bar{E}) = \exp(-\Delta E/kT) = 0.87 \text{ for } \Delta E = 29 \text{ cm}^{-1}$$

$N(^4A_2) + N(\bar{E}) + N(2A) =$ chromium ion density (dependent on doping level).

3.1.2. Neodymium Lasers

The doping of a rare earth nodymium as trivalent ions in a variety of hosts such as ytrium aluminium garnet

$Y_3Al_5O_{12}$(YAG) [3.3, 3.4] and glass [5.5] provides a class of 4 level laser system. In each case, the active atoms behave as triply ionized Nd^{3+}, with energy levels and broadening of the states dependent on the host lattice. (The crystal is of course charge neutral, the three electrons being bonded to the neighbouring atoms of the host). The dopant (1 to 2%) goes into the amorphous glass at random sites and each Nd^{3+} ion sees a slightly different environment. In the YAG (a cubic crystal) the Nd^{3+} substitutes for Y^{3+} and each of these dopant atoms sees more or less identical environment. The glass laser transition is inhomogeneously broadened (though not Doppler broadened) with a comparatively wide line. The YAG transition line widths are much smaller, though still inhomogeneously broadened by lattice vibrations.

Nd:YAG Laser

YAG ($Y_3Al_5O_{12}$) is an optically isotropic crystal with cubic structure. The active atoms are in a well-defined environment and the energy levels are well defined and narrow. The dominant transition is 1.0641 μm between the upper state $^4F_{3/2}$ (11507 cm^{-1}) and the lower state $^4I_{11/2}$ at 2110 cm^{-1}. The energy levels are shown in Fig. 3.3. The

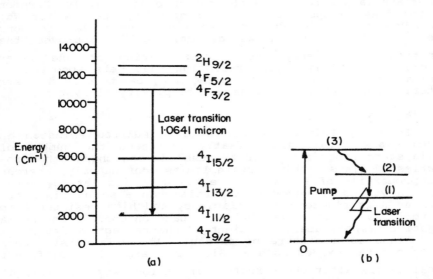

Fig. 3.3. Energy level scheme for (a) Nd:YAG (b) 4 level laser.

lower laser level $^4I_{11/2}$ decays with a life time of ~30 ns but that radiation is strongly absorbed by the host lattice and the energy shows up as heat. YAG has a high thermal conductivity and the heat can be removed by conduction. The system can be operated either CW or pulsed, mode locked or Q-switched. Because of the narrow line widths the stimulated emission cross-section is large and the pumping threshold is low. The absorption bands are also narrow and lie between 1300 and 2500 cm^{-1}. The lower laser level is at $E_1 \sim$ 2111 cm^{-1} from the ground state so that at room temperature its population is down by a factor of $\exp(-E_2/kT) \approx e^{-10}$ from that of the ground state and can be neglected. We thus have $(N_2)_{th} \sim (\Delta N)_{th}$, a property of four level lasers.

The gain line width at room temperature is ≈ 6 cm^{-1} and the spontaneous lifetime from the laser transition is 0.55 ms. The room temperature emission cross-section at centre of laser transition is $\sigma = 9 \times 10^{-19}$ cm^2. Compared to ruby ($\sigma = 1.22 \times 10^{-20}$ cm^2) the gain constant in Nd^{3+} YAG at given inversion is nearly 75 times that for ruby. This makes the oscillation threshold to be very low for Nd:YAG and explains the easy CW operation of Nd:YAG compared to ruby. For the narrow absorption, it is necessary that the emission from the flash lamp matches the absorption bands exactly. Krypton filled flash lamps are therefore more appropriate for Nd:YAG. Recently semiconductor lasers have been used to pump the $^4F_{3/2}$ level of Nd:YAG directly. This improves the conversion efficiency and avoids heating of the laser material by unwanted pump radiation that gets absorbed in the material. The use of Nd:YAG as low threshold laser material has reduced the size and complexity of system.

Nd Glass Laser

In 1961, Snitzer demonstrated stimulated emission in doped glass [3.6]. This greatly increased the number of solid state laser materials because one can change the properties of glass within certain limits continuously, whereas the properties of a crystal are fixed by its lattice type. Certain values of several properties can also be aimed at, such as high fluorescence efficiency and high resistance to thermal shocks, or a low thermal expansion coefficient and a negative temperature coefficient of the refractive index.

Glass is a mixture of oxides. Its main constituents are non metal oxides such as SiO_2, B_2O_3 and P_2O_5. Different metal oxides alter the structure in various ways and make it possible to obtain a large variety of properties. The dopants, such as Nd, are also added to the batch as oxides (e.g. Nd_2O_3). The components are mixed before melting. An advantage of doped glass as laser material is that glass of extremely good optical homogeneity can be produced in large volumes and in different shapes. It is even possible to produce fibres of laser glass because of its very good fusibility.

Some of the types of doped glasses which have been tried for laser action are K-Ba-Si [3.6], Na-Ca-Al-Si, La-Ba-Th-B [3.7], SiBaRb, $Ba(PO_3)_2$, LaBBa, SiPbK and LaAlSi [3.8]. The most common dopant is neodymium though other dopants like Tb^{3+}, Ho^{3+}, Er^{3+}, Tm^{3+}, Yb^{3+} have also been tried. The wavelengths of the induced emission are practically all in the near infrared (0.9 to 2 μm). Glass doped with neodymium shows laser action at room temperature. The energy levels involved in the laser transition in a typical Nd doped glass are shown in Fig. 3.4. The laser emission wavelength is at 1.059 μm and the lower level is approximately 1950 cm^{-1} above ground state. This is again a 4 level system with negligible thermal population of lower laser level. The fluorescent line width is around 300 cm^{-1} which is more than order of magnitude larger than that of Nd^{3+} in YAG. This is due to amorphous structure of glass which causes different Nd^{3+} ions to see slightly different surroundings. This makes different ions radiate at slightly

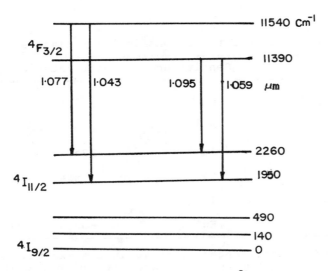

Fig. 3.4. Energy level diagram for Nd^{3+} in RbKBa silicate glass [3.8].

different frequencies, causing a broadening of the spontaneous emission spectrum.

When glass is doped with Nd as lasing ions alongwith another dopant ion called the sensitizer, the sensitizer absorbs radiation at wavelength other than that of the laser ion and then radiates within the pump band of the laser ion. Alternatively, the sensitizer ion transfers its excitation energy directly to the laser ion. The ratio of the concentrations of laser and sensitizer ions is very critical. Quenching of the fluorescence may occur at higher concentrations of sensitizer. Some of the sensitizers for Nd^{3+} laser ion are given in Table 3.1.

Glass doped with erbium as laser ion is of special importance. Its radiation at 1.55 μm does not penetrate the lens of the human eye and, therefore, cannot destroy the retina.

The lifetime of the upper laser level depends on the host glass and on the Nd^{3+} concentration. The highest life times (upto 1 ms) have been observed in silicate glass; smaller ones in fluoride glass and phosphate glass. Borate glass doped with neodymium exhibits the shortest life time (50 μs). The spectral width of the fluorescence band is small in phosphate glass and in fluoride glass. Therefore, these types of glass show the lowest threshold for laser action [3.9]. The silicate glass and even more the borate glass, emit radiation within relatively wide fluorescence bands. Though phosphate glass has advantages of narrower

Table 3.1 Sensitizer ions in laser glasses

Laser ion	Sensitizer	Host material
Nd^{3+}	UO^{2+}	KBaSi
	Mn^{2+}	Phosphate
	Ag	KBaSi
	Ce^{3+}	LiMgAlSi
	Tb^{3+}	Borosilicate
	Eu^{3+}	Borosilicate

fluorescence band, longer lifetime in the excited state and better optical pumping efficiency with higher gain coefficient, it has some problems in chemical durability (some phosphate glasses are hygroscopic) and in wavelength mismatch with Nd:YAG oscillators. Another difficulty is thermal expansion, almost double that of silicate glass.

Besides a high fluorescence efficiency, other properties of the glass are also desirable e.g. high resistance to intense giant pulses, very good optical homogeneity, chemical stability, a high coefficient of thermal conductivity and a temperature independent optical path length. These properties cannot be optimized at the same time.

Unlike many crystals, the concentration of the dopant ions can be very high in glass. But one has to take into account that the decay time of fluorescence and, therefore, the efficiency of stimulated emission, decreases with higher concentrations. In silica glass, this decrease becomes noticeable at a concentration of 5% Nd_2O_3, in borate glass even at 1%.

The threshold energy of stimulated emission decreases and the emitted energy increases with growing fluorescence efficiency and also with decreasing losses in the glass. These losses are caused by absorption and dissipation. The most important impurities responsible for absorption losses are iron, cobalt and nickel. Iron in the bivalent state is mostly undesirable for neodymium glasses, since it has a wide absorption band at approximately 1.1 µm which causes losses at the main emission wavelength of neodymium at 1.06 µm. Optical losses are reduced to < 0.001 cm^{-1} and refractive index variation below 2.10^{-6} over several centimeters.

The ultra violet portion of the pumping light can generate in the glass centers which lower the fluorescence efficiency. This undesired effect, known as solarisation, is eliminated by incorporating UV absorbing substances. Typical absorption spectra is shown in Fig. 3.5. An alternative scheme is to use a protective covering of a fluorescent dye over the glass in the pump cavity which absorbs the UV and fluoresces in the desired pump band of the active ion. An example of such a dye is Rhodamine 6G. This not only protects from solarisation but also improves the pump efficiency.

An essential characteristics of the glass for giant pulse laser is its resistance to intense laser radiation. There are several reasons for destruction occurring within the glass. One possible reason is the heating of absorbing inclusions to high temperature, which causes cracks in the glass. Another reason is self focussing of the laser beam which produces electric field strengths at which an electric discharge occurs. This causes thin filaments of small bubbles, an effect which can also occur in crystals such as ruby. An analysis of light emitted during glass damage indicates the occurrence of temperatures of about $1000°C$ [3.10]. Several explanations are given for self focussing of light e.g., electrostriction, thermal effect and nonlinearity of the dielectric constant. The destruction by self

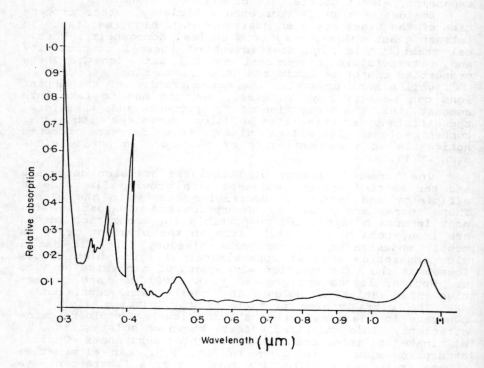

Fig. 3.5. Absorption spectra of a UV absorbing Sm doped filter glass.

focussing in good laser glass occurs at 20 to 50 J/cm^2 for giant pulse of 30 ns duration at power density of several gigawatts [3.11].

The pumping light causes nonuniform heating of the laser rod, resulting in a distortion of the wavefront of the passing laser radiation. This effect can be reduced by using glass with a negative temperature coefficient of the refractive index. Again, it is also advantageous to keep the infrared portion of the pumping light away from the laser rod by means of filters. Spectroscopic properties and gain coefficients of some commercial Nd laser glass at room temperature have been studied [3.12] and are listed in Table 3.2.

Data of laser glass developed at CGCRI, India in the early seventies [2.27] is briefly summarized as follows.

Table 3.2 Properties of Nd laser glasses at ambient temperature.

Glass types	Wavelength (nm)	Emission Cross section (10^{-20} cm^2)	Bandwidth of fluorescence (nm)	Fluorescence decay (μs)	Nd wt. (%)	Gain coeff. (cm^{-1})
1	2	3	4	5	6	7
Silicates						
ED	1061	2.7	–	490	3.6	0.115
LG650	1057	1.0	22	650	5.0	0.079
LG630	1060	–	22	640	3.0	–
LGN55	1060	–	22	600	5.0	–
LG57	1060	–	21	600	0.8	–
LG56	1060	–	22	650	3.0	–
Phosphates						
EV-2	1053.5	4.7	–	370	2.5	0.185
LHG5	1053	4.4	–	350	3.3	0.168
LHG7	1053	3.7	–	370	3.4	0.140
Q-88	1054	4.2	–	360	3.0	0.193
LG718	1053.5	4.5	–	296	8.0	0.19
LG750	1054	–	–	350	–	–
Flouro-phosphate						
FK50	1054.5	2.7	–	444	3.0	0.141

Glass types	Coefficient of thermal expansion ($\alpha \cdot 10^7 /°C$)	Hardness (kp/mm^2)	Temperature coefficient of refractive index ($\Delta n \cdot 10^6 /°C$)	Thermal conductivity (kcal/mh°C)
	8	9	10	11
LG650	95.1	418	-1.9	0.83
LG630	97	365	-2.2	0.85
LGN55	91	372	-2.1	0.80
LG57	95.8	391	-2.1	1.0
LG56	93	360	-2.4	0.8

The glass sample had transmission spectra as shown in Fig. 3.6, a fluorescence decay of about 250 μs at 1/e point

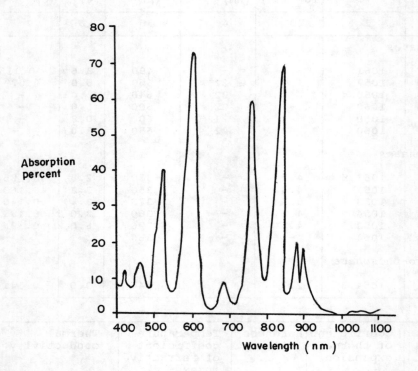

Fig. 3.6. Absorption spectra of a 40 mm dia.x 3.5 mm thick Nd glass sample from CGCRI.

and stimulated emission cross-section 3.2×10^{-20} cm^2. The material when formed into a rod of 6.35 mm dia.x76 mm long and pumped in a single ellipse cavity lased in normal mode as well as in Q-switched mode. In the arrangement of Fig. 3.7 it gave a Q-switched threshold of 90 J. The fluorescence emission bandwidth was 22.5 nm. The laser output versus the electrical input to flash lamp is shown in Fig. 3.8. The measurements on this silicate glass showed it to be somewhat similar in performance to ED2.

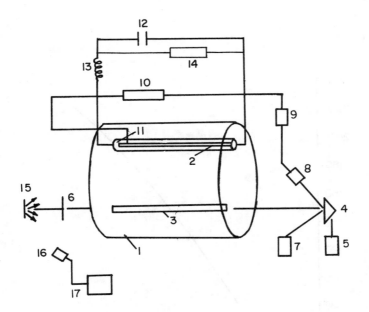

Fig. 3.7. Experimental arrangement for measurement of Q-switch threshold. 1 pump cavity, 2 flash lamp, 3 laser rod, 4 90° prism, 5 high speed motor, 6 output mirror, 7 lamp, 8 photodetector, 9 delay network, 10 trigger circuit, 11 UV shield, 12 energy storage capacitors, 13 pulse shaping inductance and 14 HV power supply.

3.1.3. Tunable Solid State Lasers

Recently a new class of tunable solid state lasers has appeared on the commercial scene which can be characterized by a vibronic laser medium in which the lasing transitions occur between vibrational levels of different electronic states. This class includes the Alexandrite ($Cr^{3+}:BeAl_2O_4$), titanium doped sapphire ($Ti:Al_2O_3$) and the cobalt doped magnesium fluoride ($Co:MgF_2$). The colour center lasers, though belonging to the tunable solid state category, do not belong to the vibronic class of lasers. The energy level diagram for the vibronic laser material shows a band of possible upper state levels rather than the discrete levels of say, an argon laser. Thus, instead of having a discrete energy difference between the upper and lower levels,

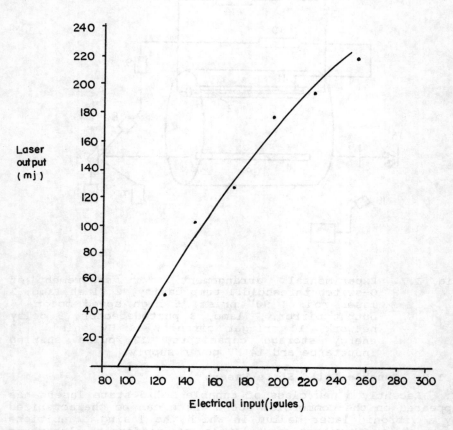

Fig. 3.8. Laser output vs. electrical input for CGCRI Nd glass.

corresponding to a fixed wavelength, a range of energy differences is possible with vibronic lasers. This corresponds to a multiplicity of available wavelengths.

Alexandrite Laser

Alexandrite *laser* was first developed as a commercial laser in the late seventies. The material lases at wavelengths ranging from approximately 720 nm to approximately

800 nm, while the peak wavelength falls near 750 nm. Alexandrite is capable of CW operation from approximately 700-800 nm but this region is now well covered by Ti sapphire. The upper laser level lifetime is around 250 µs and flash lamp pumping enables pulsed operation.

Titanium Sapphire Laser

Ti:sapphire lasers were first developed in the mid 1980s and subsequently commercialized. Introduction of CW Ti:sapphire has been one of the most significant developments in tunable lasers in respect of capability and ease of operation. Tunable from 660 to 1100 nm, with over 5 W of output power around 800 nm, Ti:sapphire has been established as the gain medium of choice in the near IR region of spectrum. CW Ti sapphire lasers are now available in a range of configurations and sophistication. Even in single frequency configuration, Ti: sapphire laser can be efficiently pumped by multiline output of cw argon ion laser. To cover the full tuning range, more than one set of cavity mirrors are required.

In the energy level scheme for Ti:sapphire, the 3d electron is split into two states, the ^2E and $^2T_2$, by the six nearest neighbour oxygen anions surrounding the Ti^{3+} ion. The displacement between the $^2T_2$ and ^2E states results in broad absorption and emission features. The absorption and emission spectra of Ti^{3+} doped Al_2O_3 at $300^\circ K$ is shown in Fig. 3.9. The material Ti:sapphire may be optically pumped with second harmonic of Nd:YAG (0.53 µm) or an Ar^+ laser (to be described later).

Flash lamps, frequency doubled Nd:YAG, or Nd:YLF lasers serve as pump sources for a variety of pulsed Ti:sapphire lasers. Such lasers are capable of tens of millijoules upto about 120 mJ of pulse energy at repetition rates that depend on the pump lasers repetition rate. As in CW case, pulsed Ti:sapphire lasers can be tuned from roughly 700 to 1080 nm by interchange of several sets of cavity optics. A few millijoules per pulse at repetition rates upto 1 kHz and 100 mJ per pulse at repetition rates of 10 to 20 pulses per second are commercially available.

In general, high pulse energies are obtainable from Alexandrite and substantial pulse energies and good beam quality from Ti:sapphire, provided spectral constraint is not present. Mode locked Ti:sapphire options are also available.

$MgF_2:Co^{2+}$ Laser

Co:MgF_2 lasers were commercialized in 1989. At room temperature, they are tunable in pulsed operation from 1750

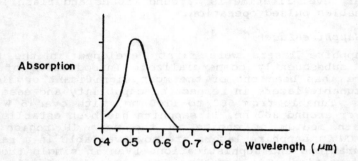

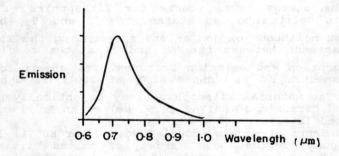

Fig. 3.9. Absorption and emission spectra of Yi^{3+} doped Al_2O_3 at 300°K.

to 2500 nm, with peak wavelength falling near 2050 nm [3.13].

Other Developments

Considerable efforts have been made to develop new tunable solid state materials for CW operation on the blue and red end of Ti:sapphire. Some of these are

(a) Tm:YAG, 1.85-2.16 μm, CW room temperature operation,
(b) $Cr:Mg_2SiO_4$ in the 1170-1340 nm region,
(c) $Cr^{3+}:LiCaAlF_6$, $Cr^{3+}: LiSrAlF_6$ at 700-900 nm and

(d) Colour centre lasers in the 1400-1700 nm and 2300-3500 nm range.

3.1.4. Sensitized Solid State Laser Materials

The performance of a laser material may be enhanced by the presence of additional impurity ions. Fig. 3.10 shows the two ways in which these extra ions can work. A sensitizer increases the level of pump light absorption by absorbing in spectral regions where the laser ion normally does not absorb. The sensitizer then efficiently transfers energy to the upper level of the laser ion.

Another likely role of the additional ion may be to deactivate the lower laser level. The lower laser level remaining unpopulated provides an advantage in a four level laser. If the lower laser level does not rapidly drain following the laser action it becomes populated. The deactivator avoids this situation by funnelling the energy away from the laser ion and depopulating the lower laser level. There are several sensitized laser materials that have proved to be useful. Two of these are CrNd:GSGG ($Gd_3Sc_2Ga_3O_{12}:Cr^{3+},Nd^{3+}$) lasing at 1.061 μm and Cr^{3+}, Tm^{3+}, Ho^{3+}:YAG operating at 2.1 μm.

GSGG

In doubly doped GSGG, Cr^{3+} acts as the sensitizer and Nd^{3+} as the laser ion. Cr^{3+} ion provides greatly enhanced absorption for the flash lamp pumped system. The Cr-Nd

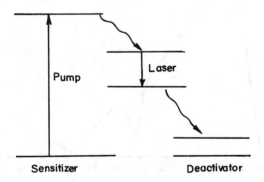

Fig. 3.10. Role of sensitizer to excite the upper laser level and that of deactivator to depopulate the lower laser level.

energy transfer is extremely efficient (>90%). GSGG:Cr^{3+}, Nd^{3+} and related systems have provided the highest flash lamp pumped efficiencies to date.

3.1.5. Eye Safe Solid State Lasers

Fig. 3.11 shows transmission of human eye as a function of wavelength. It is seen that both ruby and Nd glasses have some transmission which for high power lasers can be a source of damage to the eye. This is particularly significant since the damage threshold of eye retina is low (of the order of a few $\mu J/cm^2$) and the eye lens focusses the radiation on the retina. Further, any spectacles, or binoculars, increase the collecting aperture of the eye and enhance the risk of damage. Where high power lasers are used in training devices, particularly in military establishments, the risk of eye damage is a serious drawback. Efforts have therefore been made to use solid state laser materials which have emission peak lying beyond 1.4 μm. Several candidate materials, similar to Nd:YAG, have been tried. One such material is holmium in yttrium lanthanum fluoride (Ho:

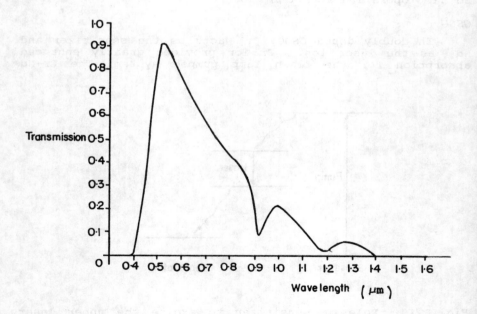

Fig. 3.11. Transmission of human eye as a function of wavelength.

YLF) which lases at 2.06 μm. Another is erbium in YLF which lases at 1.73 μm. Both of these materials however suffer from lack of sensitive detectors at these wavelengths. The two systems that have found favour as eye safe lasers are the following:

i) An eye safe laser using a Nd:YAG as laser source and Raman shifting its wavelength to 1.54 μm. The Raman shifter is a cell filled with methane at high pressure [3.14].

ii) Use of erbium in glass host to emit at 1.54 μm. The main advantage of the Er:glass laser is that it uses direct flash lamp pumping. The comparative damage threshold levels for Nd:YAG and Er:glass are shown in Table 3.3.

Table 3.3 Permissible exposure levels per pulse (J/cm^2)

Pulse rate	Nd:YAG	Er:glass
1 Hz	5.10^{-6}	1
10 Hz	$1.6.10^{-6}$	1.10^{-2}

3.1.6. Diode Laser Pumping of Solid State Lasers

The basic elements of a diode-pumped solid state laser are quite similar to the corresponding elements of the lamp pumped type. (The diode laser is described in the section on semiconductor laser). Fig. 3.12 shows a schematic arrangement of a diode pumped solid state laser in the end pumped configuration. The main difference between the diode pumped and lamp pumped arrangements is the source of optical excitation, the diode laser array. The lamp pumping involves a

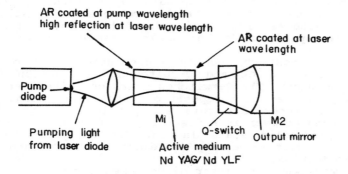

Fig. 3.12. Laser pumped solid state laser with end pumping.

broad band emission and consequently a lot of energy is wasted in the form of heat which makes cooling of laser medium essential. The laser diode on the other hand emits a narrow band spectrum which can be temperature tuned to overlap the strong absorption band of the lasing crystal. This makes the pumping process more efficient and greatly reduces cooling requirements [3.15].

The main advantage of diode pumping lies in providing higher efficiency, longer life time and higher stability. Higher efficiency, arising from matching of emission and absorption bands of laser diode and laser material respectively, obviates the need for water cooling and makes the system compact and light weight. The high efficiency of semiconductor laser also contributes to higher efficiency of diode pumped solid state laser. Higher efficiency also leads to reduced power supply requirements thereby resulting in smaller power supplies and overall reduction of size and weight of the system.

Greater life of diode pumped laser arises from long operating life (>30000 hrs) of semiconductor diodes as against limited life (<1000 hrs) of an arc lamp or a flash lamp). Greater reliability implies a large value of MTBF (mean time between failure). In the diode laser pumped solid state lasers, MTBF of over 20,000 hrs is possible. Reduction in heat automatically implies longer life and higher reliability. Good heat sinking is still required in pump diodes but on a smaller scale. Even though the laser diodes are much more efficient than lamps as pumping source, they can only generate a small fraction of the power that a lamp can. Whereas lamp based systems in CW mode reach over 30 W in TEM_{oo} mode, most single diode pumped systems generate less than 1 W at 1.06 μm (typically 500 mW by end 1990). Use of diode arrays however enables higher powers which is likely to improve in the years to come. The technology of diode pumped lasers has yet to catch up in maturity compared to lamp pumped systems though progress is fast. Newer approaches such as side pumping in tightly folded resonator scheme are capable of higher powers than end pumping (Fig. 3.13).

The two most frequently implemented types of diode based systems are CW and Q-switched. CW version is for narrow line width. The Q-switched version gives very short duration pulses at high repetition rates (nanosecond pulses at kilohertz repetition rate are possible). High cost of diode systems however is a discouraging factor in the use of diode pumped solid state lasers. But as with other new devices, the cost generally decreases with time and higher volume production.

The wavelength of laser diodes used for pumping Nd:YAG is arranged to coincide with the strong absorption band of Nd:YAG at 808 nm.

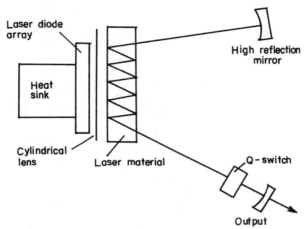

Fig. 3.13. Side pumping of a solid state laser with laser diodes in tightly folded resonator scheme.

3.1.7. The Slab Laser

A recent development in the area of solid state lasers is the slab laser (Fig. 3.14) in which the active medium used is in the form of a slab as against the more conventional cylinderical rod. The slab laser is particularly relevant to the industrial applications such as material processing, welding, cutting, drilling etc.

Thermal lensing is a serious problem in Nd lasers. This is due to the absorption of radiation in the material which gets converted to heat. This heat in the laser rod causes refractive index variation resulting in a thermal lens, the focal length of which depends on the pump power. The slab laser avoids this thermal lensing problem by zig zag propagation of beam through the slab as shown in Fig. 3.14. Since the laser material forms part of the resonant cavity, a pump power dependent thermal lens in the cavity will change the beam profile. The slab laser provides an alternate means of providing high power without changes in the beam profile. Also, slab laser can generate increased output by employing large volume of the active medium. Reduction in thermal distortion of the beam enables tight focussing.

The beam in slab laser enters and exits through tilted surfaces (Brewster ends) and zig zags through the active medium. The crystal is cooled along the pump faces. It is called a total internal reflection (TIR) face pumped laser. The heat removal from the slab is significantly more efficient than from the rod. The laser medium can thus be pumped harder yielding more laser power without sacrificing beam quality.

168 LASERS AND HOLOGRAPHY

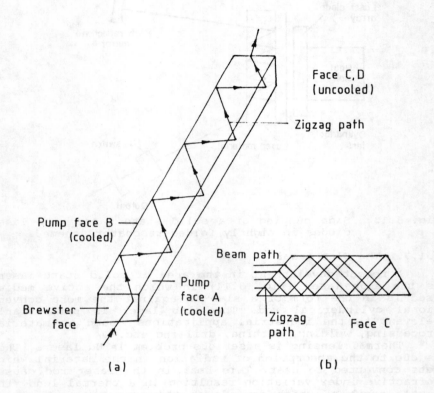

Fig. 3.14. The active medium in slab laser. (a) Pumping is from faces A and B which are cooled. The ray passes in a zig zag manner by total internal reflection from faces A and B, (b) the view from top of Fig. 3.14a showing beam path.

3.2. Colour Centre Lasers

Certain colour centres in the alkali halides can be used to create broadly tunable, optically pumped lasers in the near infrared [3.16-3.18]. Colour centres are simple point defects in crystal lattices, consisting of one or more electrons trapped at an ionic vacancy in the lattice. These point defects are common in many crystalline solids and have been studied rather extensively in alkali halide crystals such as NaCl and KCl. Certain colour centers have optical absorption and emission bands that make them suitable as laser gain media. Lasers based on these colour centres are closely analogous to organic dye lasers (to be described later); they are optically pumped can be broadly tuned, share a similar cavity design, and can generate ultra short pulses. The significant difference between colour centre lasers and dye lasers is the tuning range. Dye lasers operate in the visible and near IR (0.4 to 0.9 µm) while through the use of several different types of centres and host lattices (like different dyes and solvents in dye lasers) the colour centre lasers cover the entire region between 0.8 and 4.0 µm. These lasers may be operated in CW mode with output powers of the order of 1 W, or mode locked to give pulses of less than 100 fs duration. In pulsed mode, output powers in 1 MW range have been achieved. It is the unique combination of broad wavelength tuning range and continous wave power that makes the colour centre laser useful. The spectral purity of single frequency colour centre laser is noteworthy (<4 kHz for single mode CW operation). Laser active colour centres generally provide long lived operation only when operated at cryogenic temperature, although many of the centres can be stored at room temperature with no degradation.

Due to the relatively short radiative life time of the colour centre, intense optical pumping from a laser source is necessary to achieve efficient laser operation. The pump laser depends on the crystal, but usually a Nd:YAG laser operating at 1.06 or 1.32 µm or an Ar or Kr ion laser operating in the visible is used. The cavity of colour centre laser must contain dispersive elements, such as a prism, in order to facilitate tuning and line narrowing. When properly designed, a colour centre laser is usually capable of tuning over a range exceeding 25% of its central wavelength.

While colour centres exist in many different crystal lattices, most work has been done on point defects in alkali halide crystals such as KCl. A representative sample of colour centres in alkali halide crystals is shown in Fig. 3.15. All laser active colour centres involve an ion (halide ion) vacancies. The F centre has the simplest structure, consisting of a single electron trapped at a vacancy surrounded by an essentially undisturbed lattice. If one of the neighbouring alkali ions is a substitutional alkali impurity, say Li^+ in a KCl crystal, the centre is called an F_A centre. Similarly the F_B centre consists of an F centre beside two substitutional impurities. Two adjacent F centres along a (110) axis of the crystal form the F_2 centre. Its ionised counterpart is called the F_2^+ centre which is an F_2 centre with only one electron. The $(F_2^+)_A$ centre consists of an F_2^+ centre adjacent to a substitutional alkali ion. F_A,

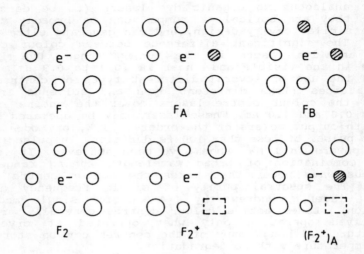

Fig. 3.15. A representative sample of colour centres in alkali halide crystals. The large and small circles represent negative and positive ions respectively. Filled circles represent alkali impurity. Dotted rectangle represents vacancy.

F_B, F_2^+ and $(F_2^+)_A$ defects form colour centre lasers. There are some other laser active centres like $F_2^+:O_2^-$ which consists of F_2^+ centre associated with an oxygen ion and forms one of the most stable and powerful colour centre lasers. N_2 centre consists of three F-centres in a trigonal arrangement. If one considers F centre as essentially electrons trapped in a three dimensional square well of dimension **a** formed by the electrostatic potential of the surrounding positive ions, the F band energy can be related to **a** for most alkali halides by

$$E_F = 17.7 \, a^{-1.84} \tag{3.4}$$

where **a** is in angstroms and E_F is in electron volts. Many colour centres follow a similar relation between the lattice constant and the energies of their absorption and emission bands. A strong coupling between the lattice dimension and the transition energy is also predicted by this model. Lattice vibrations harmonically vary the actual dimensions of the square well in a period $<10^{-13}$ s causing the energy levels to vary on this time scale. Since this perturbation is random and occurs at all F centres at a rate faster than the excited state life time of the centre, the absorption and emission bands of the centre are homogeneously broadened.

To generate laser active centres it is required to create a population of ordinary F centres, either through additive colouration or radiation damage. Additive colouration is the preferred technique because F centres produced are very stable, whereas radiation damaged crystals usually require cryogenic storage at all times. Some centres like the F_2^+ however can be coloured only through radiation damage.

Colour centres form an ideal 4 level system for laser operation. The energy level diagram of F_2^+ centre is shown in Fig. 3.16. 0 denotes the ground state, 1 represents the first excited state and 2 the relaxed excited state. 3 denotes the lower laser level. Laser emission is between 2 and 3. Performance data of some colour centre lasers is shown in Table 3.4.

Table 3.4 Some colour centre lasers

Host lattice	Centre	Pump wavelength (μm)	Tuning range (μm)	Maximum power (W)	Remarks
LiF	F^{2+}	0.647	0.82-1.05	1.8	operational life short
NaCl	F_2^+	1.06	1.4-1.75	1	
NaCl:OH	$F_2^+O^{2-}$	1.06	1.42-1.85	3	Type I and II of F_A and F_B differ in their relaxation behaviour.
KCl:Li	$(F_2^+)_A$	1.32	2.0-2.5	0.4	
KCl:Na	$F_B(II)$	0.514	2.25-2.65	0.05	
RbCl:Li	$F_A(II)$	0.647	2.6-3.6	0.1	
KCl	N_2	1.064	1.27-1.35	0.04	Pulsed

Due to the relatively short radiative life time of the colour centre, the pump intensity required to achieve a useful population in level 2 must be of the order of 100

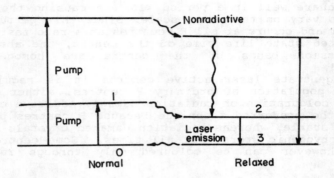

Fig. 3.16. Energy level diagram of F_2^+ centre.

kW/cm^2. Such large intensities can be obtained by focussing the pump laser beam. Most colour centre lasers are pumped by the Ar ion or Nd:YAG laser. Like the 1→2 transition the 3→0 transition is nonradiative and occurs within 10^{-12} s, the residual population in 3 is thus essentially zero.

Typical values of radiative life time τ, the emission bandwidth $\Delta \nu$, wavelength of emission λ and the stimulated emission section σ for some colour centres are shown in Table 3.5.

Table 3.5 Typical values of λ, τ, $\Delta \nu$ and σ for some colour centres

Centre	λ (μm)	τ (ns)	$\Delta \nu$ (THz)	σ (10^{-16}cm^2)
FA(II)	2.7	200	15	1.7
F_2^+	1.5	80	30	1.6
$(F_2^+)_A$	2.3	170	20	2.7
$F_2^+:O^{2-}$	1.6	160	45	0.9
N_2	1.3	210	20	0.4

A colour centre laser cavity is schematically shown in Fig. 3.17. The crystal is oriented at Brewster's angle to minimize reflective losses. The tight focus at the crystal ensures high intensity needed for population inversion.

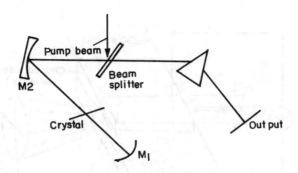

Fig. 3.17. Schematic of a colour centre laser. Beam splitter reflects pump wavelength but transmits colour centre wavelength. Tuning is by rotating the output coupler.

Typical spot size at beam focus is in the range of 20 to 30 μm. Pump absorption is given by

$$P = P_o[1-\exp(-\alpha N_o l)], \qquad (3.5)$$

where α ($\approx \sigma$) is the ground state absorption cross-section, N_o the ground state population, l the crystal thickness, P_o the input power and P is the absorbed power. Any local defects or scratches on the crystal are to be avoided in the path of the beam.

3.3. Semiconductor Lasers

So far we have considered the solid state laser materials, crystalline and glassy, which are characterized as insulators. We now discuss the solid state laser materials which are semiconductors. These semiconductor lasers have many properties that are quite similar to conventional lasers but they also differ in many respects. The most obvious differences are that they are quite small (less than a mm in dimension) and laser action is produced simply by passing a current through the semiconductor diode itself. The pumping energy is directly converted to populating the upper levels and the system is highly efficient. Direct excitation by a current makes it easy to modulate the output beam and the short life time of levels permits high frequency modulation. The divergence of the laser beam is however relatively large. This apparent disadvantage is due to the smallness of the active region. The properties of p-n junc-

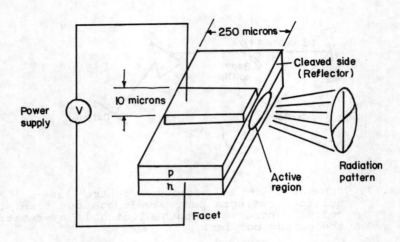

Fig. 3.18. A p-n junction laser

tion lasers are the result not of transitions between discrete energy levels but of transitions associated with the band properties of semiconductors. The pumping properties are determined by the characteristics of the junction.

Semiconductor lasers have a very simple structure; two leads connected to a power supply, a p-n junction and cavity mirrors formed by cleaving the crystal (Fig. 3.18). The output radiation is not a simple TEM_{oo} mode. Because of the small dimension (μm size) of the emitter the beam diverges with angles on the order of a few degrees; and because of unequal dimensions of emitter in orthogonal planes the beam spreads unequally in the two directions.

Semiconductor lasers can be mass produced by the same photolithographic techniques as electronic circuits and can be integrated monolithically with the latter. This combination has proved helpful to the field of integrated optoelectronic circuits.

The energy level scheme for semiconductors differs from the discrete level transitions discussed earlier. There is a continuous spectrum of electronic states and the band structure involves valence band and conduction band separated by forbidden gap energy E_g. The highest band filled with electrons is the valence band. The next higher band, the conduction band, is separated from the valence band by a region of energy, the energy gap E_g, where there

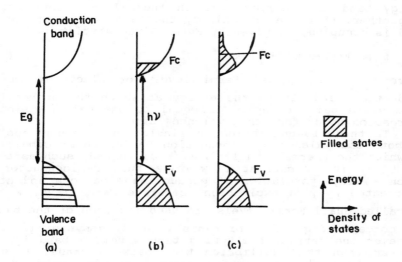

Fig. 3.19. Energy vs. density of states for an intrinsic semiconductor. (a) equilibrium state (b) inverted at $T=0°K$ and (c) inverted at $T>0°K$.

are no allowed states. Fig. 3.19a shows the situation in equilibrium for an intrinsic semiconductor (with no impurity doping). If a light with photon energy $h\nu > E_g$ is incident on the crystal, photons will be absorbed by electrons making tansitions from the valence band to the conduction band. Fig. 3.19b shows the situation for an inverted population at $0°K$. The valence band is empty of electrons (or filled with holes) down to an energy F_v and the conduction band filled upto F_c. Photons with energy $h\nu$ lying in the range $E_g < h\nu < (F_c-F_v)$ will cause downward transitions and hence stimulated emission. Thus,

$E_g < h\nu < (F_c-F_v)$ amplification

$(F_c-F_v) < h\nu$ absorption

At finite temperature, there will not be a sharp energy distinction between filled and unfilled states. Rather, the distribution of carriers will be smeared out in energy as shown in Fig. 3.19c. In this case overall thermal equilibrium does not exist, but if we assume that in a given energy band the carriers are in thermal equilibrium with each other, then the probability that a state in conduction band is occupied, is given by Fermi Dirac statistics

$$f_c = 1/(1+\exp(E-F_c)/kT), \qquad (3.6)$$

where F_c is the quasi Fermi level for electrons in the conduction band. It is the energy at which the probability of a state being occupied is equal to one half. A similar expression holds for the valence band.

In thermal equilibrium, a single Fermi energy applies to both the valence and conduction bands. Under conditions in which the thermal equilibirum is disturbed, such as a p-n junction with a current flow or a bulk semiconductor in which a large population of conduction electrons and holes is created by photoexcitation, separate Fermi level F_c and F_v called quasi Fermi levels are used for each of the bands. The concept of quasi fermi bands in excited system is valid whenever the carrier scattering time within a band is much shorter than the equilibrium time between bands. This is usually true at the large carrier densities used in p-n junction lasers.

The number of quanta of radiation absorbed depends on the electron in the valence band that can absorb the quantum and on the presence of an empty state in the conduction band, i.e. the number is proportional to $f_v(1-f_c)$. Similarly for quanta to be emitted, an electron must be in the upper state and a hole in the lower state i.e. the number is proportional to $f_c(1-f_v)$.

For emission to be higher than absorption, therefore,

$$f_c(1-f_v) > f_v(1-f_c) \qquad (3.7)$$

which leads to

$$(F_c-F_v) > h\nu. \qquad (3.8)$$

We have considered above the intrinsic semiconductor. For a doped semiconductor however it is only necessary to use the quasi Fermi levels for the impurity energy states. A highly doped (degenerate) semiconductor behaves like a metal where the conductivity does not disappear at low temperature. The unoccupied states in the valence band are referred to as holes and they are treated like the electrons except that their charge, corresponding to an electron deficiency, is positive and their energy is measured downward.

When an electron is excited from valence band into the conduction band, it leaves a vacancy (hole) in the valence band. The process of emission of radiation is the recombination radiation emitted on recombination of an electron with hole in downward transition. The thermalisation time, i.e. the time required for the electrons (holes) to come to equilibrium in the conduction (valence) band, is short compared to the recombination time. Thus carriers go to states close to their band edges before recombining. In analogy to the 4 level scheme of laser action (Fig. 3.3b) level 1 corresponds to states close to the top of the valence band. Level 2 corresponds to states close to the bottom of the conduction band. Level 3 is less precisely defined but can be thought of as higher states in the conduction band. When an electron combines with a hole, intrinsic recombination radiation, corresponding approximately to the band gap energy, is given off.

The energy band structures of semiconductors are of two types, direct and indirect. For the direct band semiconductors, the lowest conduction band minimum and the highest band maximum are at the same wave vector in the Brillouin zone (Fig. 3.20). In the indirect semiconductors the extrema are at different wave vectors. Since the wave vector of the light is much smaller than the electronic wave vector, first

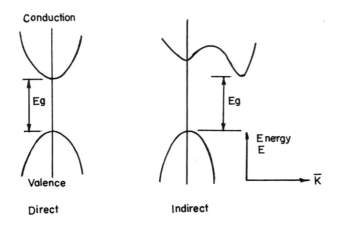

Fig. 3.20. Energy vs. wave vector for direct and indirect semiconductor.

order radiative transitions occur between states with the same electronic wave vector. Optical gain in direct semiconductors is thus quite high and laser action can occur easily. On the other hand, in the indirect materials the change in wave vector must be taken up by some other agent such as lattice vibration or impurity. The radiative transitions in indirect material are weaker and available gain is usually small, so that losses cannot be easily overcome. For this reason most of the semiconductor lasers have been developed with direct band materials.

3.3.1. The p-n Junction Laser Diode

To achieve laser action it is necessary to have a method for inverting the population and a resonant structure. A commonly used method for semiconductor laser is the current injection in a p-n junction diode (Fig. 3.18). To understand the mechanism of a p-n junction for providing stimulated emission, consider the energy diagram of p-n junction (Fig. 3.21). In the energy vs. distance plot, E_c is the conduction band edge, E_v the valence band edge. The left and right parts correspond to the n and p doped degenerate

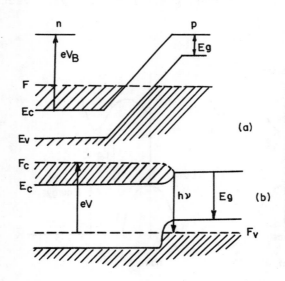

Fig. 3.21. Energy vs. distance for a p-n junction. (a) V=0 (b) V >0.

semiconductor in which there are enough electrons added by impurity doping to fill the conduction band upto the Fermi level F (n type) and enough holes added by acceptor impurity which deplete electrons from the valence band (p type).' At the junction, at zero applied voltage, (Fig. 3.21a) electrons flow from the n side to the p side until an electrical potential barrier V_B is built up which prevents further current flow. Now, if a voltage is applied to reduce the barrier or to raise the n side relative to the p side (Fig. 3.21b), electrons can flow over the top of the barrier to the p side where they make a transition to an empty state in the valence band and emit photons with energy approximately equal to E_g. It is also possible to have holes flow to the n side where they recombine with electrons. The predominant process is determined by the relative impurity densities, the carrier life times and the carrier mobilities. In either case, for high enough applied voltage, there can be a region in the vicinity of the junction where the population is inverted. At low currents there is spontaneous emission in all directions. As the current is increased, the gain increases until the threshold for laser action is reached.

To estimate the threshold current density, we proceed as follows:

From continuity equation, it is noted that if I is the current density (charge flow per second per unit area), and N, the electron density (i.e. number per unit volume),

$$I/b = eN/\tau, \qquad (3.9)$$

where b is the thickness of active region, e the electronic charge and τ is the life time due to both radiative and non radiative processes. Eq. 3.9 follows by noting that

$$I = Ne(VA)/A$$

and $V = b/\tau$.

V is velocity with which the charges diffuse in time τ through the active region of thickness b and area A. Recalling Eq. 1.144 and writing the condition of oscillation at threshold as

$$R \exp(\alpha_t - \beta)d = 1, \qquad (3.10)$$

where α_t is the gain coefficient at threshold and β represents loss coefficient.
Further, expressing Eq. 1.89 for the gain coefficient as

$$\alpha = c^2(N_2 - N_1 g_2/g_1)/8\pi\nu^2 n_0^2 \tau_{sp} \Delta\nu \tag{3.11}$$

where $n_0 = (\varepsilon/\varepsilon_0)^{1/2}$ is the refractive index, $g(\nu) = 1/\Delta\nu$, $c/\nu = \lambda$.

Combining Eq. 3.9- 3.11, assuming $N_2 \to N$ and $N_1 = 0$, we can show that threshold current density, I_t is given by

$$I_t = [8\pi e n_0^2 \nu^2 b \Delta\nu/\eta c^2][(1/d)\ln(1/R) + \beta], \tag{3.12}$$

where η is the quantum efficiency and is equal to τ_{sp}/τ. Eq. 3.12 is valid only at T= 0°K, since lower band has been assumed to be unoccupied ($N_1 = 0$). The functional form of Eq. 3.12 has been found experimentally valid in GaAs upto 300°K. In a typical p-n junctin laser both the recombination process and the light are confined to a narrow region near the junction. Also, in most lasers the stimulated emission occurs at photon energies close to the energy gap. In order to satisfy Eq. 3.7, the built in voltage V_B must be greater than $h\nu$. This requires that the semiconductor be heavily doped with impurities on both sides of the junction so that F will be close to E_c on the n side and to E_v on the p side.

GaAs was the first material to lase [3.19] and since then the laser action has been observed in alloys of several compounds, especially the III-V compounds, with different pumping mechanisms such as optical, electron beam, avalanche breakdown and of course the more widely used injection current [3.20]. One of the primary advantages of the junction laser is the ease with which it can be amplitude modulated.

The nature of the temperature dependence of I_t is shown in Fig. 3.22. At high temperatures a dependence close to T^3 is observed and at low temperatures I_t is fairly constant. The temperature T_c at which the T^3 dependence begins, increases with increasing substrate carrier concentration N.

3.3.2. Heterojunction Lasers

There are many problems associated with the simple p-n junction lasers which are attributed to using the same material GaAs for both the p and n regions. The two main drawbacks are

(a) The injected minority carriers are free to diffuse and thereby dilute the spatial distribution of recombi-

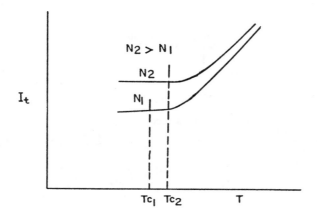

Fig. 3.22. Threshold current density vs. temperature.

nation and in turn the gain.

(b) There is very little guiding and confinement of the electromagnetic wave being amplified.

Improvement in both these problem areas can be effected by the use of heterostructures to form the active region of the laser. These are junctions between two dissimilar materials such as GaAs with $Al_xGa_{1-x}As$ where x is the fraction of gallium being replaced by aluminium. GaAs and AlAs semiconductors have almost identical lattice constants (5.65-5.66Å) and can be mixed and grown on top of each other with little strain involved and very small density of traps at interface. As x increases the band gap increases and index of refraction goes down. The heterostructure configuration of junction lasers results in reduction of threshold current for the semiconductor laser.

The two common constructions of heterostructure lasers are GaAs with $Ga_{1-x}Al_xAs$ and $Ga_{1-x}In_xAs_{1-y}P_y$ as active regions. The former emits in the range 0.75 to 0.88 μm and the latter in the range 1.1 to 1.6 μm. Both types are useful for communication in optical fibres. Fig. 3.23 shows a generic form of $GaAs/Ga_{1-x}Al_xAs$ laser [3.21]. The essential features of this construction are

(a) Thin region of GaAs sandwitched between two regions of $Ga_{1-x}Al_xAs$ of opposite doping forming a double heterojunction (DH) laser.

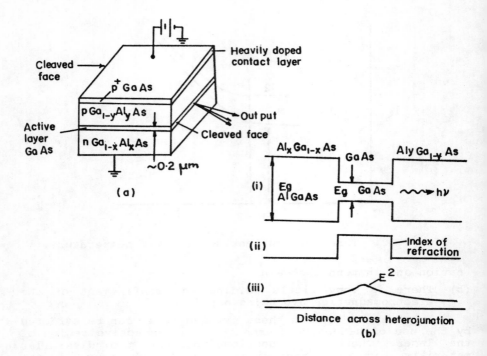

Fig. 3.23. (a) Hetrostructure construction of GaAs/$Ga_{1-x}Al_xAs$ laser. (b) (i) band diagram for forward bias, (ii) refractive index and (iii) light intensity.

(b) A dielectric wave guiding effect by the sandwitch brought about by lowering of the energy gap causing an increase in the index of refraction. (E_g and Δn are proportional to the molar fraction x)

The basic layered structure of a quarternary GaInAsP/InP laser used for optical communication in 1.0-1.6 μm range is shown in Fig. 3.24 [3.22].

Another construction called the buried heterostructure (BH) laser utilises transverse confinement for lowering the threshold current (Fig. 3.25). In this configuration a rectangular GaAs active region prism (~2 μm x 0.1 μm) is embedded in a lower index material $Ga_{1-y}Al_yAs$. The active

SPECIFIC LASER SYSTEMS 183

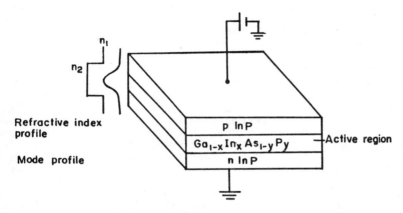

Fig.3.24. The basic layer structure of GaInAsP/InP DH laser.

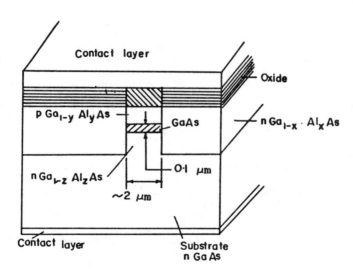

Fig. 3.25. Buried hetrostructure laser : construction features.

region is surrounded on all sides by the lower index GaAlAs so that electromagnetically the structure is that of a rectangular dielectric wave guide. The transverse dimension of the active region and the index discontinuities (or the molar fraction x,y) are chosen so that only the lowest order transverse mode can propagate in the wave guide. BH lasers have been fabricated with threshold currents of less than 1 mA though typical thresholds are a few milliamperes.

3.3.3. Recent Advances

Because of the relatively large diffusion length of the early homojunction devices (forward biased p-n junctions), the diode lasers needed large current for laser action. The development of double heterostructure (DH) devices provided low current thresholds. In the DH structure, injected electrons are trapped in a material thinner than the diffusion length (1000-3000 Å) with a lower band gap energy. The gain is higher and so is the index of refraction in the active layer. The higher refractive index creates a planar waveguide. The guide is used on the device to produce a single transverse mode. Spatial variation of the gain (or index of refraction) provides guidance of propagating mode. The cavities used for the semiconductor lasers are characterised by transverse dimensions comparable to the operating wavelength. Hence a waveguide approach is most often used in analysis.

The buried heterostructure (BH) devices provide better transverse mode operation than stripe geometry devices. In the geometry of a typical stripe laser, the injected current is usually confined to a stripe where width is much larger than the depth over which the recombination of the carriers takes place. More recent manufactoring techniques such as organometallic vapour phase epitaxy and multiple quantum well structuring have enabled active layers thinner than 100 Å.

Current areas of exploration are in single frequency lasers with very small frequency chirping under fast modulation and wavelength tunable lasers with narrow line-width [3.23]. The distributed feedback laser is the most favoured structure for single frequency operation. Tuning the wavelength can be done either by using (i) external cavity with its own filter or (ii) with monolithically integrated multi-electrode distributed feedback (DFB) or distributed Bragg reflection (DBR). DFB and DBR differ structurally in the placement of the grating structure. Holographic gratings can be used for the tuning of a DFB laser.

3.3.4. Quantum Well Lasers

In the conventional DH laser, though the active layer is thin enough (1000-3000 Å) to confine electrons and the optical field, the electronic and the optical properties remain the same as in the bulk material. In a quantum well laser [3.24] on the other hand active layer is made thinner than 100 Å so that the electronic and the optical properties change drastically due to the reduced dimensionality of free electron motion from three to two dimensional. The confinement of electrons causes a quantization in the allowed energy level (Fig. 3.26) and the formation of sub bands of energy.

$$E_n = (\hbar^2/2m)(n\pi/L_z)^2, \qquad (3.13)$$

where h is Planck's constant, m is electron effective mass, L_z is the thickness of the quantum well and n is an integer. Expressions similar to Eq. 3.13 hold for the coduction and valence bands. The density of states changes from the parabolic dependence of energy to a step like structure (Fig.3.27).

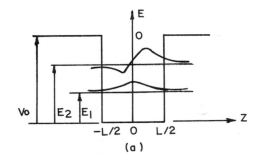

(a)

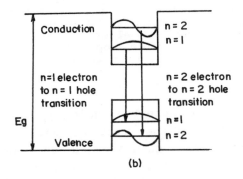

(b)

Fig. 3.26. (a) Wave function and energy sub bands of an infinitely deep quantum well, (b) first two, n=1 n=2, quantized electron and hole states and their eigen function in an infinite potential well.

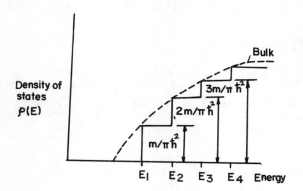

Fig. 3.27. Density of states in a two dimensionally confined quantum well compared to that of a bulk semiconductor.

$$\rho(E) = \sum_{n=1}^{\infty} (m/\pi\hbar^2) H(E-E_n),$$

where $H(x)$ is the Heaviside function.

Since the density of states is constant, rather than gradually increasing from zero, there is a group of electrons of nearly the same energy available to recombine with a group of holes of the same energy. Maximum gain can thus be larger than in conventional DH lasers.

When sufficient number of electrons are injected into the conduction band to provide population inversion, the active region exhibits optical gain given by

$$\alpha = A(N_2-N_1) \tag{3.14}$$

where $A=d\alpha/dn$ is the differential gain, N_2 is the injected electron density and N_1 is the electron density required to reach population inversion. Both A and N_1 are material constants depending on the band structure (Typically, for InGaAsP DH laser $A \sim 1-2.10^{-16}$ cm^2 and $N_1 \sim 1-1.5.10^{18}$ cm^{-3}, depending on laser wavelength and doping levels). Using quantum confinement of very thin active layers, it is possible to alter the band structure so that A becomes optical gain.

The reduced density of states in the step like structure require fewer electrons to reach optical transparency leading to a low threshold current density. As stated above, the steplike structure provides higher gain. For given values of A and N_2, this leads to smaller values of N_1 from Eq. 3.14. The semiconductor is said to be transparent when enough electrons are injected into the conduction band to provide significant population inversion (this implies high gain by electron hole recombination at the p-n junction). The differential gain A is much larger than that of bulk material [3.25] and gives rise to high speed and narrow line width. An increased value of gain bandwidth can be achieved by optimizing the laser cavity design. This will extend the lasing wavelength range from the lowest energy state to the higher energy states.

Quantum well lasers have been successfully made in GaAs/AlGaAs materials. Threshold current densities lower than 100 A/cm^2 and submilliampere threshold currents, have been reported. Compare this with threshold current density of several kA/cm^2 for DH laser. More recently, InGaAs/ InGaAsP multiple quantum well (MQW) lasers for the 1.3 and 1.5 μm fibre communication systems have been made. Typical MQW structures are shown in Fig. 3.28.

To summarise a thin region of GaAs in a DH laser acts as a trap for electrons and holes. If the thickness of active GaAs is reduced to say <200 Å, the confined electrons and holes display quantum effects. Their energies and wave functions are determined mostly by the confinement distance and the resulting lasers are called quantum well (QW) lasers. The major advantage of thining of the active

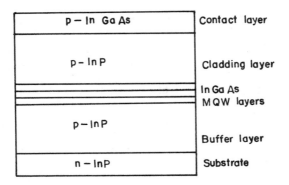

Fig. 3.28. Multiple quantum well structure.

region is the reduction of the total number of carriers needed to achieve transparency in the active region relative to conventional laser. This reduction is in the ratio of the active region thickness. The free carrier absorption is also reduced by the same factor. The result is a greatly reduced threshold current.

3.3.5. Distributed Feedback lasers

A significant achievement of the 1980's has been the realisation of high performance single frequency lasers. The distributed feedback (DFB) concept, first applied to dye lasers [3.26], was extended to semiconductor lasers. Distributed feedback lasers [3.27] and distributed Bragg reflector lasers [3.28] were demonstrated in AlGaAs/GaAs lasers. The same concept has also been applied to InGaAsP/InP lasers [3.29, 3.30]. Work on single frequency lasers with very small frequency chirping under high speed modulation and wavelength tunable lasers with narrow line width, has recently received attention due to the needs of high speed and coherent light wave systems. Single frequency oscillation can be obtained by integrating a frequency selective grating element inside the laser cavity. In the distributed feedback (DFB) laser, the grating region is built-in a waveguide layer adjacent to the active layer. In a distributed Bragg reflector (DBR), on the other hand, the grating region is outside the active region along the length of the cavity. In the DBR laser, the reflection is enhanced at the Bragg wavelength λ_B, related to the period of the grating, Λ, by

$$\lambda_B = 2n\Lambda/\text{l},$$

where n is the effective index of the mode and l is the integer order of the grating. In the DFB laser, the grating region has a periodically varying index of refraction, which couples two counter propagating travelling waves. The coupling is maximum for wavelength close to the Bragg wavelength. Because of the periodic index variation due to grating structure, there is a wavelength band known as the stop band within which the transmission through this periodic structure is zero. Thus within the stop band the reflection is the largest.

In a normal laser, the feedback is provided by the cavity reflectors at the two ends of the cavity. However it is possible to distribute this feedback along the length of the cavity in order to obtain the advantages of high frequency selectivity. Consider the cavity structure, shown in Fig. 3.29 which can be realized in a semiconductor by etching a periodic structure along the axis of the diode and then regrowing a different material on top of the array. The index of refraction goes down as aluminium fraction goes

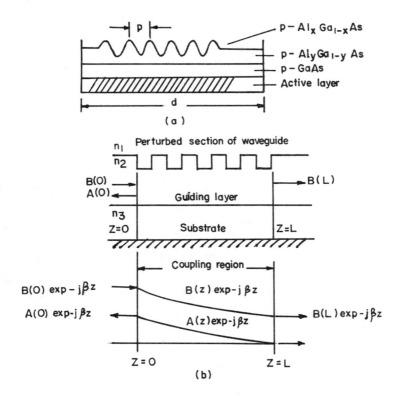

Fig. 3.29. (a) Idealised version of corrugated guide for analysis in which only passive cavity without gain is considered and (b) the incident and reflected fields.

up and thus the guiding properties and the effective width of the guide are both periodic functions of z.

To understand the effect of the corrugated guide on the incident electromagnetic field, consider a corrugated section of length L and a wave with an amplitude $B(0)$ incident on the corrugated section from the left. The coupling of the forward and backward modes of the electromagnetic field are given by the coupled mode equations

$$dA/dz = KB \exp(-2j\Delta\beta)z$$

$$dB/dz = K^*A \exp(2j\Delta\beta)z, \qquad (3.15)$$

where $\Delta\beta = \beta - \beta_0$, $\beta_0 = 2\pi/\lambda_0 = \pi l/p$, $l = 1, 2, 3, ..$

In Eq. 3.15 K^* is complex conjugate of K and K is the coupling constant. A and B are the amplitudes of the forward and backward travelling modes. β is the propagation constant ($=2\pi n/\lambda$) in the medium of refractive index n. β_0 is the value of β in free space.
The solution of Eq. 3.15 gives

$$A(z) = B(0)K \sinh|K|(z-L)/|K|\cos|K|L$$

$$B(z) = B(0) \cosh|K|(z-L)/\cosh|K|L, \qquad (3.16)$$

where $A(L) = 0$ and the fields for $z > L$ and $z < 0$ are as shown in Fig. 3.29b. The plot of mode powers $|B(z)|^2$ and $|A(z)|^2$ (Fig. 3.29b) shows that incident mode power drops off exponentially along the perturbation region. This is however not due to absorption but to the reflection of power into the backward mode A.

If now the periodic medium is provided with sufficient gain at frequencies near the Bragg frequency, oscilation can result without the benefit of end reflectors. The feedback is provided by the continuous coherent back scattering from the periodic perturbation.

In the case of the guiding medium possessing gain the coupled Eq. 3.15 are modified so that when $|K|=0$ the two independent solutions $A(z)$ and $B(z)$ correspond to exponentially growing waves along the -z and +z direction, respectively. Eq. 3.15 are thus replaced by

$$dA/dz = KB \exp(-2j\Delta\beta)z - \alpha A$$

$$dB/dz = K^*A \exp(2j\Delta\beta)z + \alpha B, \qquad (3.17)$$

where α is the exponential gain constant of the medium.

For frequencies very near the Bragg frequency and for sufficiently high gain, the guide acts as a high gain amplifier. The amplified output is given in reflection with a gain

$$\frac{E_r(0)}{E_i(0)} = \frac{K\sinh(\delta L)}{(\alpha - j\Delta\beta)\sinh(\delta L) - \delta\cosh(\delta L)} \qquad (3.18)$$

and in transmission with a gain

$$\frac{E_i(L)}{E_i(O)} = \frac{-\delta\exp(-j\beta_0 L)}{(\alpha-j\Delta\beta)\sinh(\delta L) - \delta\cosh(\delta L)} \qquad (3.19)$$

where $\delta^2 = |k|^2 + (\alpha-j\Delta\beta)^2$

$\Delta\beta = \beta - \beta_0 + j\alpha$.

The behaviour of the incident and reflected field for a high gain case is shown in Fig. 3.30. The difference between this case, and the passive one shown in Fig. 3.29c, is apparent. The oscillation condition can be written as

$$\frac{\delta - (\alpha - j\Delta\beta)}{\delta + (\alpha - j\Delta\beta)} \exp(2\delta L) = -1. \qquad (3.20)$$

In general, one has to find a numerical solution to

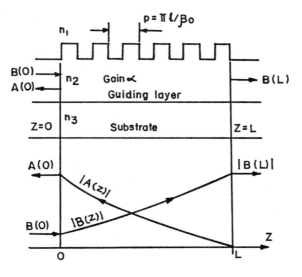

Fig. 3.30. The incident and reflected fields inside an amplifying periodic waveguide near Bragg condition $\beta = \pi_1/p$.

obtain the threshold values of $\Delta\beta$ and α for oscillation [3.31]. In some limited cases however, one can find approximate solutions. In the high gain situation, the oscillation condition becomes

$$\frac{4(\alpha-j\Delta\beta)^2}{|K|^2} \exp(2\delta L) = -1, \qquad (3.21)$$

which on equating amplitudes gives the threshold gain value α_m through

$$\frac{\exp 2\alpha_m L}{\alpha_m^2 + (\Delta\beta)_m^2} = \frac{4}{|K|^2} \qquad (3.22)$$

indicating an increase in threshold with increasing mode number m. Equating phases in Eq. 3.21 gives the oscillating mode frequencies for the limit $\alpha_m >> (\Delta\beta)_m, K$

$$(\Delta\beta)_m L \approx -(m+1/2)\pi. \qquad (3.23)$$

The two lowest order modes correspond to m=0 ($\Delta\beta = -\pi/2L$) and m=-1 ($\Delta\beta = +\pi/2L$). From Eq. 3.20 it follows that the threshold gain depends only on the magnitude of $\Delta\beta$ so that the lowest order modes m=0 and m=-1 posses identical thresholds. The mode frequency spacing from Eq. 3.23 is

$$\nu_m - \nu_{m-1} \approx c/2nL \qquad (3.24)$$

and is approximately the same as in a two reflector resonator of length L where n is the effective refractive index of the medium.

The distributed feedback laser thus has a strong built-in discrimination against all modes except those with m=±1 that are equispaced about the Bragg frequency.
Distributed feedback semiconductor lasers that incorporate corrugated interface in their epitaxial layered structure, have assumed great importance in the optical fibre communication for high data rate systems. This is because they can be made to oscillate in a single mode. Single mode laser is required to reduce dispersion with fibre, and hence improve the data rate. A slight asymmetry in the structure can favour one of the two equal threshold modes and thereby make the laser operate in a single mode. Various high speed DFB laser structures have been developed [3.23]. Line widths of 3 MHz for 300 μm cavity length and 500 kHz for 1.2 mm cavity length have been achieved. Compared to this a line width of 250 kHz has been obtained for a

1500 μm long MQW laser with an optimized 3 well structure.

3.4. Dye Lasers

Dye laser is not just another laser. It is important because it is tunable over a wide range of wavelengths. The active medium is a dye which can be used in solid, liquid, or gas phases (more commonly in liquid) and its concentration, and hence absorption and gain, is readily controlled. The active medium can have high optical quality and cooling is achieved by a flowing cycle through a reservoir.

Laser action in dye was first obtained in chloroaluminium pthalocyanine [3.32] and then extended to many dyes with different solvents [3.33]. Dye laser wavelengths now cover the whole visible spectrum with extension into the near ultra violet and infrared.

An important step in the development of dye lasers was the use of diffraction grating as one of the resonator mirrors to introduce wavelength dependent feedback [3.34]. A continuous tuning range with spectral narrowing to 0.06 nm was obtained. Since then many different schemes have been developed for tuning the dye laser wavelength.

Flash lamp pumping with a fast rising pulse provided a convenient incoherent pumping source for pulsed dye laser-single shot as well as repetitively pulsed. Use of triplet quenchers enabled long pulse operation leading to the possibility of CW operation [3.35].

CW operation was an important step towards the full utilisation of the dye lasers potential, first achieved in 1970 [3.36]. An argon ion laser was used to pump solution of rhodamine 6G in water with some detergent added to it. Water, because of its high heat capacity, reduced temperature gradients. The detergent acted as triplet quencher and prevented formation of non fluorescing dimers of dye molecules. Dye lasers, because of their broad spectral bandwidth provide ultrashort pulses by mode locking.

A large number of organic dyes exist (more than 200 actively used) each fluorescing in a specific wavelength region. One can find a formal listing of such dyes [3.37]. However, as examples, four dyes are shown in Table 3.6 which cover major part of visible range.

3.4.1. The Dye as Laser Medium

Laser action in dyes takes place between the broad, diffuse energy bands characteristic of complex molecules in condensed liquid solvents. The diffuseness of the energy bands is evidence of the short life times of the laser states, due both to the strongly allowed nature of the laser electronic transition and the frequent collision of each dye molecule with its neighbours. To invert such states, large pumping rates obtained by high intensity flash lamp or a

Table 3.6 Molecular structure, laser wavelength and solvents for some laser dyes

Dyes	Solvent	Wavelength (nm)
7 Hydroxycoumarine	Water (pH~ 9)	Blue 450-470
Na Fluorescein	Ethyl alcohol Water	Green 530-560
Rhodamine 6G	Ethyl alcohol Methyl alcohol Water, DMSO Polymethyl methacrylate	Yellow 570-610
Rhodamine B	Ethyl alcohol Methyl alcohol Polymethyl methacrylate	Red 605-635

pumping laser are required.

The lasers that have been used as pump for the dye lasers are

(a) argon ion laser (0.514 μm),
(b) second harmonic of the Nd:YAG laser (0.53 μm),
(c) the nitrogen (N_2) laser (0.337 μm),
(d) the near UV excimer laser and
(e) copper vapour laser with emission in green.

The flash lamp pump pulse, for reasons of triplet state population (described later), needs to be a fast rising one. This is generally obtained from the discharge of a capacitor at high voltage and negligible inductance in the circuit. Pumping pulses of a rise time less than 100 ns are desirable.

Organic dyes are those substances which contain conjugated double bonds in their molecule. Their thermal and photo chemical stability is important and determines the long wavelength limit of dye lasers. It is difficult to find stable dyes beyond one micron. Many dyes like fluorescein, can exist as cationic, neutral and anionic molecules depending on the pH value of the solution. Maximum concentration of the dye in the solution is governed by concentration quenching of fluorescence which usually sets in whenever the dye molecules approach each other closer than about 10 nm. Dyes can also be doped in host crystals and solid solutions of dyes can be used. A dye can be dissolved in the liquid monomer of a plastic material (such as polymethyl methacrylate) and then polymerized. One can also dissolve it in glass or gelatin.

3.4.2. Spectra of Organic Dyes

A peculiarity of the spectra of organic dyes, as opposed to the atomic and ionic spectra, is the width of the absorption bands which usually covers several tens of nanometers. This is because a typical dye molecule may posses fifty or more atoms, giving rise to about 150 normal vibrations of the molecule. These vibrations together with their overtones, densely cover the spectrum between a few wave numbers and 3000 cm^{-1}. Many of these vibrations are closely coupled to the electronic transitions by the change in electron densities over the bonds constituting the conjugated chain. After the electronic excitation has occurred, there is a change in bond length due to the change in electron density. Quantum mechanically this means that transitions have occurred from the electronic and vibrational ground state S_0 of the molecule to an electronically and vibrationally excited state S_1.

In the general case of a large dye molecule, many normal vibrations of differing frequencies are coupled to the electronic transition. Also collision and electrostatic perturbation, caused by surrounding solvent molecules broaden the individual vibrational lines. Further, every vibronic sub level of every electronic state, including the ground state, has superimposed on it a ladder of rotationally excited sublevels. These are extremely broadened because of the frequent collision with solvent molecules which hinder the rotational movement, so that there is a quasi continuum of states superimposed on every electronic level. The population of these levels is determined by a Boltzmann distribution. After an electronic transition, the approach to thermal equilibrium is very fast in liquid solutions at room temperature. This is because a large molecule experiences at least 10^{12} collisions per second with solvent molecules, so that equilibrium is reached in ~1 ps. Thus, the absorption is practically continuous all over the band. The same is true for the fluorescence emission corresponding to the transition from the electronically excited state of the molecule to the ground state. This results in a mirror image of the absorption band displaced towards lower wave numbers by reflection at the wave number of the purely electronic transition. This happens because the emissive transitions start from the vibrational ground state of the first excited electronic state S_1 and end in vibrationally excited sublevels of the electronic ground state. The resulting typical form of the absorption and fluorescence spectrum of an organic dye is shown in Fig. 3.31. The

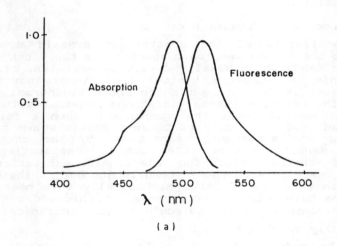

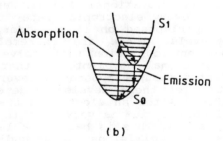

Fig. 3.31. (a) Absorption and fluorescence spectrum of a typical dye molecule. (b) Energy levels responsible for absorption and fluorescence in a dye molecule. The difference in absorption and fluorescence band arises due to Frank Condon shift.

difference in the wavelength region between absorption and fluorescence arises due to Frank Condon shift (Fig 3.31b).

Temperature, concentration and pH value dependence further modify the dye spectra. If the temperature of dye solution is increased, higher vibrational levels of the ground state are populated according to a Boltzmann distribution and more and more transitions occur from there to higher sublevels of the first excited singlet state. Consequently, the absorption spectrum becomes broader and the superposition of so many levels blurs most of the vibrational fine structure of the band while cooling of the solution usually reduces the spectral width and enhances any vibrational features that may be present.

The concentration dependence of dye spectra is most pronounced in solutions when the solvent consists of small, highly polar molecules, notably water. The dimer formation affects the fluorescence yield. The dimers show a very slow decay of fluorescence. This makes them susceptible to competing quenching processes, which in liquid solution are generally diffusion controlled and hence very fast. Consequently, in most of these cases the fluorescence of the dimers is completely quenched and cannot be observed. This is why dimers constitute an absorptive loss of pump power in dye lasers and must be avoided by all means. This can be done by using less polar solvent like alcohol or chloroform. Another possibility is to add a detergent to the aqueous dye solution which then forms miscelles that contain one dye molecule each.

For every excited singlet state there exists a triplet state of somewhat lower energy and the transitions between singlet ground state S_o into triplet states are spin forbidden. The eigen states of a typical dye molecule are shown in Fig. 3.32. There is a ladder of singlet states S_i (i=0,1,2,3,..) containing the ground state S_o. Somewhat displaced towards lower energies there is the ladder of triplet states T_i (i=1,2,3..). The longest wavelength absorption is from S_o to S_1, the next absorption band from S_o to S_2 etc. The absorption from S_o to T_i is spin forbidden. The absorption $S_o \rightarrow S_1$, $S_o \rightarrow S_2$, etc., usually have different transition moments.

There are many processes by which an excited molecule can return directly or indirectly to the ground state. Some of these are schematically shown in Fig. 3.32. It is the relative importance of these which mainly determines how useful a dye will prove in dye lasers. The process directly used in dye lasers is the radiative transition from the first excited singlet state S_1 to the ground state S_o. If this emission, termed fluorescence, occurs spontaneously,

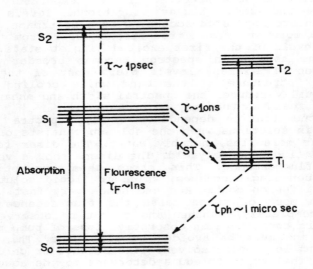

Fig. 3.32. Energy levels of a typical dye molecule with radiative (solid line) and nonradiative (broken lines) transitions.

its radiative lifetime τ_F is connected with Einstein coefficient for spontaneous emission, $A=1/\tau_F$. Generally the fluorescence spectrum is broad and shows considerable Stokes shift.

The internal conversion between S_2 (and higher excited states) and S_1 is usually very fast ($<10^{-11}$ s). The radiationless transitions from an excited singlet state to a triplet state can be induced by internal perturbation as well as by external perturbations. These radiationless transitions are usually termed as inter-system crossing with the crossing rate say K_{sT}. The radiating transition from $T_1 \rightarrow S_0$ is termed phosphorescence. Some of the paths of deactivation of excited molecule include reversible and irreversible photochemical reactions. The absorptive transitions $S_1 \rightarrow S_n$

and $T_i \to T_n$ can constitute a loss mechanism for the pump as well as for the dye laser radiation. It is, therefore, very important for dye laser operation.

3.4.3. Requirements for Starting Oscillation

In principle, one could use either the fluorescence or phosphorescence emission for laser action. However, in phosphorescence, due to strongly forbidden transition, a very high concentration of active species is required to obtain amplification large enough to overcome the cavity losses. Very high concentration, however, is not practicable since, for many dyes, this may be higher than the solubility in any solvent. Also, triplet-triplet (T-T) absorption would present high losses, because T-T absorption bands are broad and diffuse and overlap with phosphorescence band. If fluorescence emission is used for laser action, the allowed transition from the lowest vibronic level of S_1 to some higher vibronic level of S_0 will give a high amplification factor even at low concentration. However, the inter-system crossing to T_1 is high enough in most molecules to reduce the quantum yield of fluorescence to a value less than unity. This inter-system crossing reduces the population of S_1 and hence the amplification factor, secondly it enhances the T-T absorption losses by increasing the population of T_1.

In pulsed pumping, a slowly rising pump pulse would transfer most of the molecules to the triplet state and deplete the ground state correspondingly. The population of state can therefore be held small if pumping light flux density rises fast enough i.e. threshold is reached in a time t_r small enough compared to inter system crossing rate K_{sT} so that

$$t_r \ll 1/K_{sT}, \qquad (3.25)$$

where t_r is the time of pump light power during which it rises from zero to threshold level. For $K_{sT} \approx 10^7/s$, $t_r < 100$ ns. This requires a giant (Q switched) laser pulse. In laser pumped dye lasers therefore, one may neglect all triplet effects in a first approximation.

In its simplest form, a dye laser consists of a cuvette of length l cm with dye solution of concentration N (cm^{-3}) and of two parallel end windows carrying a reflective layer each of reflectivity R for the laser resonator. With N_1

molecules per cm^3 excited to the singlet state S_1, the dye laser will start oscillating at frequency ν if the overall gain is equal to, or greater than, unity. Thus

$$\exp[-\sigma_a(\nu)N_0 l] R \exp[\sigma_f(\nu)N_1 l] \geq 1 \qquad (3.26)$$

is the condition of oscillation where $\sigma_a(\nu)$ and $\sigma_f(\nu)$ are the cross-sections for absorption and stimulated fluorescence at frequency ν respectively and N_0 is the population of the ground state. The first exponential term gives the attenuation due to reabsorption of fluorescence by the long wavelength tail of the absorption band. This attenuation becomes more important the greater the overlap between the absorption and fluorescence bands. Since the fluorescence band usually is a mirror image of the absorption band, the maximum value of the cross-sections in absorption and emission are found to be equal, i.e.

$$(\sigma_a)\max = (\sigma_f)\max$$

3.4.4. Cavity Arrangements

Some of the practical arrangements for pumping and forming a dye laser resonator are shown in Fig. 3.33. A simple structure (Fig. 3.33a) consists of a cuvette filled with dye solution excited by a beam from a suitable laser. The resonator is formed by the two glass air interfaces of the polished sides of the cuvette. The pump laser and the dye laser beams are at right angles. Reflective coatings are used on the windows. It is also possible to use AR coatings, or Brewster windows, at the cuvette and separate resonator mirrors. The arrangements of Figs. 3.33b,c are self explanatory. In the arrangement of Fig. 3.33d, the coating of mirrors (M_1, M_2) with required reflection and transmission values poses problems, particularly for high power lasers. However, in the longitudinal pumping arrangement shown, there is a possibility of using an extremely small depth of dye solution.

The arrangement of Fig. 3.33e, due to Hänsch [3.38], is particularly attractive since it provides spectrally narrow tunable output for spectroscopic applications. The dye solution is transversely circulated by a centrifugal pump. The nitrogen laser (pump) radiation is focussed by a lens into a line at the inner cell wall. To provide a near circular active cross-section, the dye concentration is adjusted so that the penetration of the pump light is equal to the width of active medium. A plane output mirror is mounted at one end of the optical cavity. The other end of

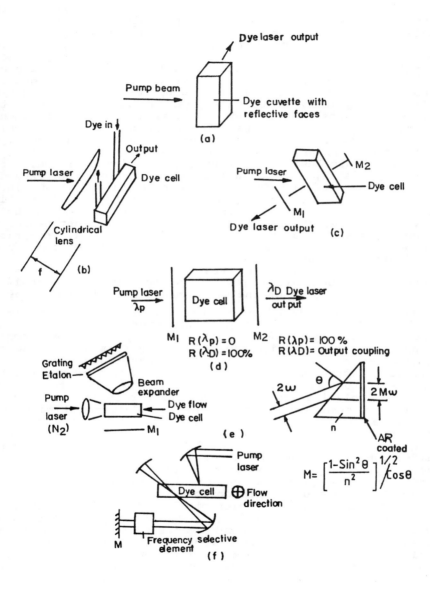

Fig. 3.33. Cavity arrangements for a dye laser.

the cavity contains an inverted telescope corrected for spherical aberration and coma. The initial waist size is enlarged and the beam divergence is reduced. This is especially important if a diffraction grating in Littrow mounting is used instead of a second mirror. If the beam remains unexpanded, it would eventually burn the grating and the spectral resolution would be very low. Also, the grating resolution is determined by the number of lines illuminated, times the grating order. The beam expander increases the number of lines illuminated. The etalon acts as a frequency filter. It is a pair of parallel reflecting surfaces, which makes a wavelength filter by interference of the multiply reflected beams as in a Fabry Perot resonator.

The basic design of Hänsch has a variation in which the telescope is replaced by a set of half prism beam expander (Fig. 3.33e inset). The magnification given by a single prism is large only at very high incidence angles, creating a large reflection loss at the entrance face. By using several prisms, the incidence angle can be brought nearer to Brewster's angle to drastically reduce this loss. The prisms automatically polarize the dye laser beam and are more economical than an achromatic telescope.

Fig. 3.33f shows an arrangement for a CW dye laser. Here the dye solution flows across the cavity at a position where the spot size of the TEM_{oo} mode is a minimum so as to enhance the stimulated emission rate. The continuous stream of dye solution is oriented at Brewster's angle to enhance the Q of the dye laser cavity. The angle of the incoming pump laser beam is adjusted to give maximum gain along the region where the pump and the dye laser beams overlap. The dye is collected and recirculated. The dye flow replenishes the dye before the triplets can quench the laser action.

The polarization of the dye laser beam in the longitudinal and transverse arrangements is determined mainly by the polarization of the exciting laser beam, the relative orientation of the transition moments in the dye molecule for the pumping and laser transitions and the rotational diffusion-relaxation time. The latter is determined by solvent, viscosity, temperature and molecular size. In addition, the polarization of dye laser beams can, of course, be manipulated in obvious ways by introducing into the resonator polarizing elements like Brewster windows.

The above arrangements (Figs. 3.33 a-e) have been shown with liquid dye solutions but even simpler arrangements are possible with solid solutions of dyes or dye crystals. For pulsed dye lasers, if the repetition rate of the pump pulses is increased above about one shot per minute, the heat generated via radiationless transitions in the dye molecules and energy transfer to the solvent (for liquid solutions) can cause Schlieren in the cuvette which reduce the dye laser output. It is thus advisable to use a flow system so that the cuvette contains fresh solution for every shot. Obviously this is also the way to deal with problems of photo chemical instability. However, these rarely arise in laser pumped dye lasers, except for some very unstable IR dyes.

3.4.5. Output Characteristics

Most often the pump pulse is of approximately Gaussian shape and its fullwidth at half maximum power is less than the reciprocal of the intersystem-crossing rate-constant K_{sT}. When this is so, then shortly after the pump pulse reaches threshold level, dye laser emission starts. The dye laser output power closely follows the pump power till it drops below threshold, when dye laser emission stops. The dye laser pulse shape thus closely resembles the part of the pump pulse above the threshold level.

The temperature dependence of the dye laser wavelength shows a shift towards shorter wavelengths with decreasing temperature. This is caused by the narrowing of the fluorescence and the absorption band with decreasing temperature which results in higher gain and fewer reabsorption losses near the fluorescence peak.

The solvent has a most important influence on the wavelength and efficiency of the dye laser emission. For a given dye the laser wavelength can shift by as much as 20 nm by the change of solvent [3.33]. The acidity of the solvent relative to that of the dye also has effect on the laser wavelength. As stated earlier, many dyes show fluorescence as cations, neutral molecules and anions. Correspondingly, the dye laser emission of such molecules usually changes with the pH of the solution, since generally the different ionization states of the molecule fluoresce at different wavelengths. In some situations, such as that of the dye rhodanile blue, the dye can exist in two conformations. In this case the absorption spectrum is the same for both conformations but the fluorescence spectrum shows marked difference and hence difference in laser wavelength. The energy conversion efficiency, for same absorption and constant temperature, also strongly depends on the solvent. In an experiment using the dye DTTC [3.39], the efficiency varied from 6.5% for acetone as solvent to 25% for DMSO when concentration of the laser Dye and the pumping conditions were kept the same.

The wavelength coverage of laser pumped dye lasers is determined on the short wavelength side by the shortest available pump laser wavelength of sufficiently high peak power; on the long wavelength side it is given by the stability of the dyes used.

In the case of flash lamp pumped dye lasers, triplet effects become important because of the long rise time, or duration of the pump light pulse. The triplet losses are time dependent and they affect the efficiency as well as the emission wavelength of the dye laser.

The population of triplet state can be quenched by the use of a triplet quencher such as O_2. However, O_2 also increases K_{sT} and thereby also quenches the fluorescence of a dye. The relative importance of triplet and fluorescence quenching then determines whether or not oxygen will improve the performance. The effect of molecular oxygen at different partial pressures on the dye laser emission of a number of dyes has been studied [3.40]. Several dyes were found for which the fluorescence quenching of O_2 was stronger than triplet quenching. It is thus more appropriate not to use a quencher like O_2 which enhances both K_{sT} (fluorescence quenching) and K_{Tso} (triplet quenching), but rather to apply energy transfer from the dye triplet to some additive molecule with a lower lying triplet that can act as accepter molecule. Cyclooctatetrene as accepter molecule has been found effective for the dye rhodamine 6G [3.41].

The dye life time, or stability against photo decomposition, is another important dye property. All dye molecules eventually decompose under repeated pump excitations. In the more stable molecules, this occurs after 10^6-10^7 photon absorption, corresponding to about 10^4 J absorbed energy per cubic centimeter of dye solution. The usable lifetime is about 1% of this value.

In practice, a large volume of dye (about a litre) is circulated through a laser to give several hundred hours of operation between dye changes. Dye concentrations are generally limited to less than 10^{-2} mole/l, where the limit of solubility for some dyes occur and where aggregation (dimer formation) in others alters the dye spectra.

Tunable dye lasers have been operated in pulsed and CW mode and have been mode locked to give a train of ultrashort pulses. The large oscillation bandwidth of dye lasers is very convenient for mode locking. When mode locked with pulse compression, they have provided some of the shortest pulses in the fs regime. To extend the wavelength region of tunable laser radiation they have also been used with nonlinear frequency conversion techniques to extend the range to UV by upconversion, and provide tunable IR and FIR by down conversion [3.42, 3.43]. For nonlinear down conversion, a dual frequency dye laser arrangement shown in Fig. 3.34 seems attractive as it provides orthogonally polarized tunable collinear output for nonlinear mixing [3.43]. Two gratings at grazing incidence with tunable mirrors M are used in the laser cavities of the twin dye laser system. Polarization corresponding to the two wavelengths are made orthogonal with the help of an intra-cavity Glan polarizer.

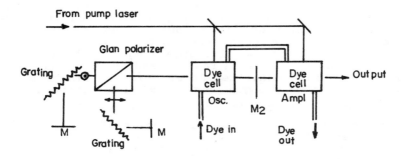

Fig. 3.34. Dual frequency dye laser.

3.4.6. Specific Purpose Developments

We have already discussed distributed feedback semiconductor laser with its special property of high spectral purity. This distributed feedback has also been experimented in dye lasers. The distributed feedback dye laser first demonstrated the tunable and narrow band output from a very compact structure that would be desired in a light source for integrated optical devices. The optical feedback in these mirrorless lasers is provided by Bragg scattering from a periodic spatial variation of the refractive index, or the gain itself, in a thin film gain medium. The first such laser used a holographic phase grating (section 6.6), which was recorded in a dichromated gelatin film on a glass substrate, the film was dyed by soaking in a solution of rhodamine 6G. This was transversely pumped with a nitrogen laser to produce a 630nm beam of 0.05 nm line width from the 0.1x 0.1x10 nm^3 total laser volume.

Tunable dye lasers have also been used for uranium enrichment. Large powers with good spectral purity and ability to tune the wavelength sharply have made the dye lasers a favourite candidate for this application. The raw uranium is vaporised in vacuum with an electron beam and the streaming atomic vapour is exposed to dye laser radiation tuned to selectively photoionize only ^{235}U atoms. These charged ions are deflected out of the stream by electrostatic fields to achieve separation. The technique is applicable to other types of isotope separation as well. CW dye lasers have also been adopted for investigation in the cure of cancer (photodynamic therapy).

3.5. Gas Lasers

Laser emission in gaseous medium has been obtained in the spectrum extending from far infrared (FIR) to the X-ray. Well over 1000 wavelengths have been obtained from gaseous elements in atomic or ionic form. Most elements in the periodic table have been induced to lase when in gaseous form via pulsed or CW discharge pumping. A multitude of molecular gases as well as radical and short lived transient species have also exhibited laser emission. These molecular lasers emit at wavelengths extending from the vacuum UV (H_2 laser, excimer lasers) to the far infrared (HCN laser). CO_2 laser operating at 10.6 μm has proved to be the most useful from commercial point of view because of its wide application in industry. He-Ne laser has occupied the position of a coherent light source in the visible region for many years and still enjoys an important place as an aid to education, instrumentation and alignment needs. Gas lasers are relatively efficient, have homogeneous medium, produce nearly perfect Gaussian beam and can be scaled to large volumes and high powers. However, they are comparatively large in size and require high voltage for discharge. Some of the more well known gas lasers available commercially are (i) He-Ne (1961), (ii) Iodine (1964), (iii) CO_2 (1964), (iv) Nitrogen (1966) (v) Argon ion (1964), (vi) Copper vapour(1966), (vii) HF/DF (1967), (viii) He-Cd (1968) and (ix) Excimer (1975). The year shown in bracket is the year of their discovery. A large number of molecular species have been used for emission in FIR and have been listed [3.44].

Output powers of gas lasers are a function of the mode of excitation, CW or pulsed, as well as of the laser gas. Gas lasers may be pumped electrically, chemically, thermodynamically, or optically by other laser or light sources. Few gas lasers are tunable over an appreciable wavelength range.

3.5.1. He-Ne Laser

Laser oscillation in a mixture of He and Ne provided the first gas laser and also the first CW laser [3.45]. This utilized the neon transition emitting at 1.15 μm. Later, oscillation at two more wavelengths was obtained at 0.6328 μm in the visible and 3.39 μm in the infrared [3.46]. He-Ne laser has become widely available as a compact, reliable commercial product used in the measurement and metrology related applications.

The active medium in the He-Ne laser is a gaseous mixture of He and Ne with proportions of about 10:1 in a sealed pyrex tube at a pressure of several torr. The gas is excited in a positive column discharge. Optimum operating parameters (e.g. discharge current, He-Ne ratio, total gas pressure) are functions of tube diameter and also depend on the output wavelength selected.

Fig. 3.35 shows a simplified energy level diagram of laser transitions in a He-Ne laser. The energetic electrons in the discharge excite helium atoms in a variety of excited states. These excited atoms in coming down to ground state collect in the long lived metastable states 2^3S and 2^1S where life times are 100 and 5 μs respectively. Since these long lived (metastable) levels nearly coincide in energy with the 2S and 3S levels of Ne, they can excite Ne atoms into these two excited states. This excitation takes place when an excited He atom collides with an Ne atom in the ground state and exchanges energy with it. The small difference in energy is taken up by the kinetic energy of the atoms after the collision.

It is seen from Fig. 3.35 that He-Ne laser can be made to oscillate at more than one wavelength. The dominant lines are 632.8 nm in the visible and 3391.3 nm in the infrared. For operation at 632.8 nm (red line of Ne) emission at

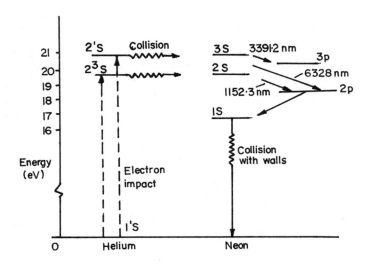

Fig. 3.35. Energy level diagram for He-Ne laser transitions.

3391.3 nm must be suppressed since the two lines share the same upper level. The upper level of 632.8 nm transition is 3S level of Ne whereas the terminal level belongs to the 2p group. The terminal 2p level decays radiatively with a time constant of $\sim 10^{-8}$ s into the long-lived 1S state. This time is much shorter than the lifetime (10^{-7} s) of upper laser level 3S. This makes population inversion in 3S-2p transition possible. The 1S level, because of its long life tends to collect atoms reaching it by radiative decay from 2p level. Atoms in 1S collide with discharge electrons and are excited back into level 2p. This reduces the inversion. Atoms in 1S states relax back to ground state mostly in collision with the wall of the discharge tube. For this reason, the gain in the 0.6328 μm transition is found to increase with decreasing tube diameter.

For the 1.15 μm oscillation, the upper laser level 2S is pumped by resonant collision with metastable 2^3S He level. It uses the same lower level as 0.6328 μm transition and consequently also depends on wall collision to depopulate the 1S Ne level.

The 3.39 μm oscillation involves a 3S-3p transition and uses the same upper level as 0.6328 μm oscillation. Because of the high gain in this transition, oscillation would normally occur at 3.39 μm rather than 0.6328 μm. The 0.6328 μm laser overcomes this problem by introducing losses at 3.39 μm in the optical path through elements such as glass or quartz windows that absorb strongly at 3.39 μm but not at 0.6328 μm.

A typical gas laser setup is shown in Fig. 3.36. The gas envelope windows are normally tilted at Brewster angle to avoid reflection loss at windows. The output is consequently polarized accordingly.

The low gain of 632.8 nm laser line makes the system relatively inefficient with low output power. Commercial

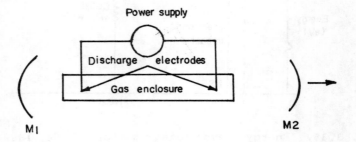

Fig. 3.36. Typical arrangement for a gas laser.

He-Ne lasers provide CW powers between 1 and 50 mW at 0.6328 μm. Low power He-Ne lasers are compact with tube length ~20 cm, beam diameter ~1 mm, beam divergence ~1 mrad, coherence length of about 10-30 cm and emit either linearly polarized or randomly polarized light in a TEM_{oo} mode. High power lasers may be over 1 m long.

The He-Ne laser is a representative of the atomic lasers since the active species is the atomic gas in neutral form. We shall now consider a representative of an ionized species i.e. the rare gas ion lasers in which the active medium corresponds to the ionized states of a rare gas such as Argon or Krypton.

3.5.2. Argon Ion Laser

Argon ion lasers [3.47] belong to the class of rare gas ion lasers. Laser emission in pulsed and CW mode has been observed from Ne, Ar, Kr and Xe in various ionization states. Rare gas ion lasers provide some of the highest CW powers available over the wavelength range 350-700 nm. This is however obtained at low efficiency (<0.1%) thus necessitating large input electrical power and efficient cooling systems.

Energy levels of the Ar^+ laser are shown in Fig. 3.37.

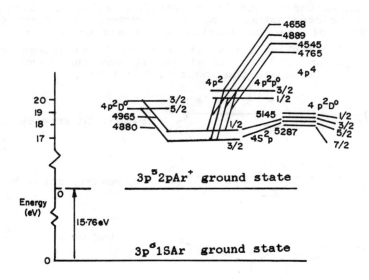

Fig. 3.37. Energy levels of an Ar laser in the blue green region.

Photon emitted by an excited argon ion is the result of the ionization of a neutral atom (15.76 eV above ground state of neutral atom) together with excitation of the resulting ion to states nearly 20 eV above ground state of ion. Electron collision may excite these states directly or indirectly.

The large discharge current required for operation of Ar and Kr ion lasers places strong demands on bore materials. For high power sytems, segmented tungsten disk bores are used. For low power air cooled Ar lasers, however, berylium oxide channels serve the purpose.

Mode locked Ar lasers can provide pulse lengths of 100 ps at a pulse repetition frequency of over 100 MHz.

3.5.3. Carbon dioxide Laser

Laser action from the CO_2 molecule was first observed by Patel in 1964 [3.48]. It is representative of the so called molecular lasers in which the energy levels relate to internal vibration of the molecules i.e. the relative motion of the constituent atoms. The atomic electrons remain in their lowest energetic states and their degree of excitation is not affected.

The CO_2 molecule consists of three atoms and can execute three basic internal vibrations called the symmetric stretch, the bending mode and the unsymmetric stretch (Fig. 3.38). Some of the low vibrational levels of CO_2 are shown in Fig. 3.39. The laser transition at 10.6 µm takes place between the upper laser level (00^01) and the lower laser level (10^00) (shown in the inset of Fig 3.39).

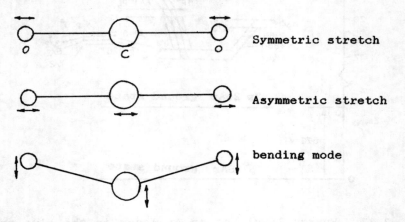

Fig. 3.38. The three normal modes of vibration of CO_2 molecule.

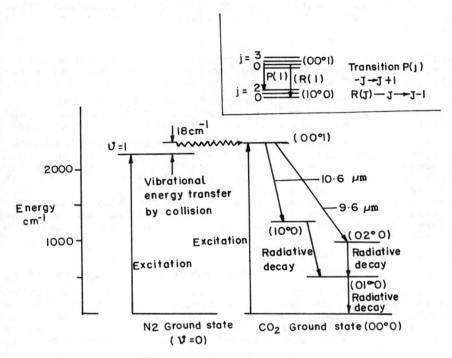

Fig. 3.39. Energy levels pertinent to CO_2 laser.

The excitation is provided to a mixture of N_2, He and CO_2. The overall efficiency is high ~30% which results from the following reasons:

(a) laser levels are all near the ground state and atomic quantum efficiency is ~45%,
(b) large fraction of CO_2 molecules excited by electron impact cascade down to collect in the long lived (00^01) level and
(c) a very large fraction of the N_2 molecules excited by the discharge tend to collect in the v=1 level. Collision with ground state CO_2 molecules result in transferring their excitation to the latter, thereby exciting them to (00^01) state. The slight deficiency in energy of 18 cm^{-1} is made up by a decrease of the total kinetic energy of the molecules following the collision.

Under most excitation conditions, oscillation takes place on P(18) 10.5713 μm, P(20) 10.5912 μm and P(22) 10.6118 μm lines.

The CO_2 laser usually operates on a mixture of CO_2, N_2, and He gases. The optimum proportions of the three components depend on the mode of excitation, but generally for flowing gas CW systems the $CO_2:N_2:$He ratio is in the range 1:1:8 with a total gas pressure of 5-15 torr. The role of He is to quench the population that builds up in $(01^{0}0)$ and other low lying levels of the CO_2 molecules, after laser emission has occurred and to stabilize the glow discharge. Helium atoms are translationally excited during such processes, and heat must be removed from the laser gas for efficient operation. In slow flow devices, the heat is removed by collisions with the walls of the laser tube. Typically power levels are 75 W/m for slow axial flow systems. For fast flow convectively cooled devices, the power increases to 200-600 W/m. CO_2 laser can be operated pulsed, CW or in Q-switched mode.

3.5.4. TEA CO_2 Laser

The development of high pressure CO_2 lasers using transverse, rather than longitudinal, excitation has proved advantageous since the large fields required to achieve breakdown in high pressure gases can be obtained with relatively low applied voltages. The arrangement known as the TEA (transversely excited atmospheric pressure) CO_2 laser (Fig. 3.40) uses a transverse discharge geometry to induce a uniform self sustained avalanche discharge in the laser gas at high pressure [3.49]. Different electrode schemes have been developed to achieve a large area discharge which is more or less uniformly distributed along the entire length of the optical cavity. An important aspect of the transverse excitation is that the discharge impedance is low which allows a very rapid injection of the excitation energy. This enables good efficiencies because the excitation of the laser molecules should take place in a time short compared to the life time of the excited state of the CO_2 molecule. The life time at atmospheric pressure is ~10 μs, hence excitation times of ~ 1 μs are desirable to achieve maximum efficiencies in giant pulses. An added advantage of short excitation time is that no external Q switching accessories such as spinning mirror are needed to obtain giant pulse action in CO_2 lasers. In TEA lasers, giant pulses are automatically obtained due to a gain switching action. In gain switching, the gain reaches a large value before the laser field becomes sufficiently strong to rapidly depopulate the upper laser level (Fig.

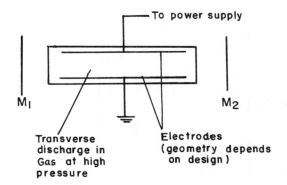

Fig. 3.40. Transverse excitation arrangement for a TEA CO_2 laser.

3.41). One, therefore, has the same condition as in a Q-switching case and a giant pulse results. The theoretical and experimental studies [3.50] indicate that equal or higher peak powers are achieved by this gain switching mechanism than by more conventional Q-switching techniques using rapidly rotating mirrors.

The difference between gain switching and Q-switching has been explained in chapter 2 wherein it is pointed out that in gain switching the Q of the cavity is not changed and remains fixed at high value with time. But in Q-switching it is the Q of the cavity that is changed with time, remaining low initially to prevent oscillation but switched to high value at a time when population inversion is built up to a high value. Both gain switching and Q-switching, however, provide a high peak power pulse.

Figures for the energy per pulse vary from 2.5 to 20 J with only half that energy going in the initial giant pulse which lasts approximately 100 ns. The peak power from TEA CO_2 lasers can be increased by further shortening the laser pulse. The pulsewidth associated with the gain switched or Q-switched giant pulses is determined by the gain per unit length of the laser medium rather than the bandwidth of the CO_2 laser line.(Bandwidth determines the mode locked pulsewidth. At atmospheric pressure this bandwidth is about 3 GHz which implies that subnanosecond pulses could be amplified).

The simplest electrode system for transverse excitation

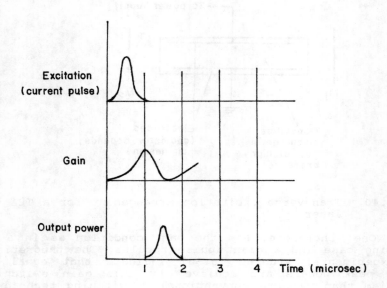

Fig. 3.41. Time sequence of gain switched action in TEA laser.

consists of a long tubular anode and a number of carbon resistors in parallel using their leads as the cathode elements which are distributed linearly along the optical cavity. Lasers using this type of electrode structures have been operated at atmospheric pressure with an efficiency of approximately 5%. A variation of this arrangement is the helixoidal electrode distribution which favours TEM_{oo} mode. As a technique to obtain the required discharge distribution, the resistively loaded electrodes have disadvantage such as reduced efficiency due to the I^2R losses. A number of electrode structures, have been adopted for improving the discharge conditions.

The major factor in the achievement of compact high average power of CO_2 lasers is to achieve a high mass flow through the laser system. Transverse excitation is suited for this purpose since it lends itself readily to transverse gas flows. Also less energy needs to be expanded to move a given gas volume through the large cross-section of transverse flow than the longitudinal one. Operation at atmospheric pressure implies, because of high density, a greater mass flow with subsonic gas velocity than is possible with lower pressure systems.

Large pulse energies can be extracted from electron beam controlled CO_2 lasers. In these devices a high current of ~100 keV electrons is used to preionize the laser gas. When the preionization charge density is sufficiently large, a subsequent main electrical discharge can be operated at a lower field than that required for a self sustained discharge. The discharge field strength may therefore be adjusted to optimize the CO_2 laser excitation and the laser output. In this way CO_2 laser pulse energies of the order of kJ in µs duration can be generated.

In addition to electrically excited gas lasers, there are gas lasers that are either thermally or chemically pumped. But whereas the electrically pumped gas laser benefits from the application of flow only to the extent that waste energy is carried away, the thermal and chemical types depend on the flow for their laser action. We now describe the thermally pumped gas dynamic CO_2 laser.

3.5.5. Gas Dynamic CO_2 Laser

In this type of laser, flow is used to create an inversion from what is initially a completely equilibrated hot gas. A thermally pumped system starts with a hot equilibrium gas mixture in which there is no population inversion. The inversion is produced gas dynamically by rapid expansion through a supersonic nozzle [3.51]. Because a hot gas is the basic energy source, the laser typically operates in the 8-14 µm region. The nitrogen CO_2 gas dynamic laser operates at 10.6 µm in the standard CO_2 laser transition. The CO_2 gas dynamic laser typically involves a gas mixture that is mostly nitrogen and approximately 10% CO_2 and 1% water (mixtures involving helium instead of water are also possible). Nitrogen being a simple diatomic molecule has only one vibrational mode. Energy from this mode can be lost by collisions with nitrogen, CO_2 and H_2O, returning the excited molecule directly to the ground state. CO_2 being a linear triatomic molecule has three basic modes of vibration, asymmetric stretch mode, which forms the upper laser level, symmetric stretch mode which forms the lower laser level and bending mode.

The energy exchange process in the gas dynamic laser includes several transfers i.e. (i) near resonance between nitrogen and the first asymmetric stretch level of CO_2

causes efficient transfer of energy between these modes, (ii) energy can be lost by collisions with N_2, CO_2 and H_2O, most probably transferring into the symmetric stretch and bending modes of CO_2. Lasing however occurs only from first asymmetric stretch mode to the first symmetric stretch, (iii) the Fermi resonance between the bending and symmetric stretch modes of CO_2 tightly couples these modes. Therefore excitation energy is extracted from the lower laser level by deactivation of the bending mode, a process that can occur during collisions with all gas species but principally water.

The basic principle of the gas dynamic laser is to expand the gas rapidly through a supersonic nozzle to a high Mach number. The purpose is to lower the gas mixture temperature and pressure, downstream of the nozzle in a time that is short compared with the vibrational relaxation time of the upper laser level (the asymmetric stretch mode of CO_2 coupled with nitrogen). At the same time, by addition of water (or helium) the lower level relaxes in a time comparable to, or shorter than, the expansion time. Because of the rapid expansion, the upper laser level can not follow the rapid change in temperature and pressure and stays at a population characteristic of the stagnation region. Because of the long relaxation time of the upper level, this population stays for a considerable distance downstream of the nozzle.

Fig. 3.42 shows the basic CO_2 gas dynamic laser. The gas flows from left (stagnation region containing hot gas mixture) to the right. As the gas is expanded through the nozzle (Fig. 3.42a), the random translational and rotational energies are converted into the directed kinetic energy of flow. With expansion through the nozzle, the upper level population drops just a little bit and then remains constant. The lower level population, however, diminishes rapidly within the nozzle, continues to decrease, and virtually disappears a few centimeters downstream. Thus, downstream of the nozzle, the population of the upper level is characterized by a temperature like that of the stagnation region and the population of the lower level is characterized by a temperature like that of the downstream region. Inversion begins approximately one centimeter downstream of the nozzle throat (Fig. 3.42b) and continues for about a meter downstream of the nozzle. Because of the high gas densities involved and the high speed flow down stream of the nozzle, an inversion capable of operation at very high powers is achieved.

Since the optical cavity for the gas dynamic laser is basically short and fat (Fig. 3.42c), geometric angles are large compared to the diffraction angles (geometric angle=

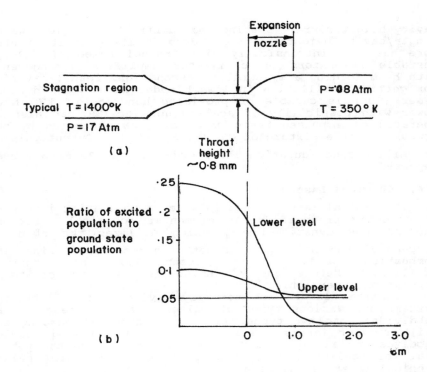

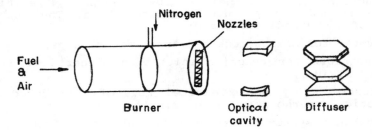

Fig. 3.42. Schematic of gas dynamic laser, (a) nozzle, (b) population inversion and (c) combustion and cavity formation.

cavity height/mirror spacing and diffraction angle = wavelength/cavity height). The gas dynamic laser cavity therefore has an intrinsically high Fresnel number (~ 1000). Unstable resonators or oscillator amplifier arrangements with beam folding are useful for such high Fresnel cavities for getting oscillation in lowest order mode. Gas dynamic laser (GDL) is capable of providing hundreds of kW of CW power with possibility of scaling up. The GDL can also be operated on a pulsed basis with gas flow generated by the combustion of a mixture of CO, O_2, N_2 and H_2. Output pulses of millisecond duration and energy up to 20 J have been reported.

3.6. Chemical Lasers

A chemical laser is one in which a population inversion is directly produced by an elementary chemical reaction. The CO_2 gas dynamic laser, which utilizes the chemical process of burning of gas, is not a chemical laser since its combustion products come essentially into thermal equilibrium before gas expansion is used to create a population inversion.

Chemical lasers are normally divided into CW flowing systems and various types of pulsed devices. These can be subdivided into two types: (i) direct chemical pumping and (ii) transfer lasers, depending on how the lasing species receives its energy. The CW chemical lasers have the potential for scaling to high average powers. The first pulsed chemical laser was discovered in 1965 [3.52].

Most chemical lasers operate on a vibrational rotational transition in a molecule, because many exothermic gas phase reactions liberate their energy through a stretching vibration in the newly formed chemical bond. The common features of a chemical laser are:

(a) a system for mixing together the reagent gases,
(b) a technique for reaction initiation either before or after mixing,
(c) a laser cavity where excited molecules produced by reaction undergo stimulated emission and
(d) an exhaust system to remove expended gases from laser cavity.

A number of mixing gases like H_2F_2, D_2F_2, H_2Cl_2 etc. have been used for laser action by chemical reaction. H_2F_2 is considered an important representative of the class of chemical lasers. Chemical lasers have been operated in a wide range of wavelengths from 2 to over 100 μm.

Each of the chemical reaction involves free atoms,

since the fuels involved do not react rapidly in their molecular forms. The techniques for initiating chemical reaction include UV photolysis ($h\nu + F_2 \rightarrow 2F$), electrical discharge ($e + F_2 \rightarrow e + 2F$), thermal dissociation ($F_2 + heat \rightarrow 2F$) and chemical reaction ($NO + F_2 \rightarrow NOF + F$).

3.6.1. HF/DF Laser

The HF (DF) laser employs one of the two variants for obtaining excitation by chemical reaction; either fuel combustion or electric arc heating.

Combustion or electric discharge produce free atoms of fluorine or hydrogen atoms from the mixture of H_2 and F_2. The following reactions can then take place

$$F + H_2 \rightarrow HF^* + H, \quad \Delta Q = -32 \text{ kcal/mole} \quad (3.27a)$$

$$H + F_2 \rightarrow HF^* + F, \quad \Delta Q = -97.9 \text{ kcal/mole} \quad (3.27b)$$

For DF laser, H is replaced by D in Eq. 3.27. The excited state of HF is denoted by HF^*. In Eq. 3.27a it corresponds to vibrational level denoted by $v \leq 3$ while for Eq. 3.27b the excitation is to vibrational levels denoted by $v \leq 9$. There is a chain reaction with the H atoms from Eq. 3.27a used in the reaction of Eq. 3.27b, its product, in turn, yields the F atom required in Eq. 3.27a. Both reactions are exothermic and the reaction products share the chemical energy released. Some of this excess energy increases the temperature of the gas which, in turn, increases the rate of the reaction. The chain reaction consumes the donor molecules and it will slow down when the densities of H_2 and F_2 are depleted. The important point is that HF molecules are formed in a high vibrational state which leads to selective pumping of the upper laser level.

The lasers based on this chemistry are necessarily low pressure devices due to the relatively short collisional lifetimes for the principal lasing species. Most efficient devices have output energies in the range 100-400 kJ/kg and corresponding chemical efficiencies of 5-20%. The HF laser radiates in the range 2.5-3 μm and the DF laser in the range 3.6-4.2 μm.

The combustion temperatures vary from 1500-2500°K and pressures in the range 30-300 psi. The dissociated fluorine flows through an array of highly cooled small nozzles. The mixing nozzles establish appropriate pressure, temperature and composition for the chemical reactions which proceed rapidly in the laser cavity. The optical cavity is formed by unstable resonator for the reasons described in the discussion for GDL which also has a high Fresnel number and the need for operation in the lowest order mode.

There is another class of chemical lasers, the CW transfer chemical laser. This depends on the transfer of energy from excited reaction products produced in a mixing and reacting flow system to host molecules, or atoms, which can undergo stimulated emission. Most of the systems utilize vibrational rotational transitions in CO_2. $DF-CO_2$ transfer chemical laser is a representative example. This has the reaction

$$F + D_2 \rightarrow DF^* + D, \quad \Delta Q = -32 \text{ kcal/mol}$$

$$D + F_2 \rightarrow DF^* + F, \quad \Delta Q = -100 \text{ kcal/mol}$$

$$DF^* + CO_2 \rightarrow DF(v\sim 1) + \Delta CO_2 (00^01) \quad (3.28)$$

The lasing occurs at 10.6 μm and corresponds to the usual P(20) transition, 00^01-10^00 band of the CO_2 molecule. This CW laser is a versatile chemical laser and has been operated successfully at multi kilowatt powers over a range of cavity pressures of 10 to 250 torr.

Another chemical laser of importance is the oxygen iodine laser (COIL). This is a chemically pumped CW laser utilizing electronic transitions. This has the shortest wavelength (1.3 μm) of any CW chemical laser and has been operated at 40% efficiency.

3.7. Carbon Monoxide (CO) Laser

This is another gas laser which can be operated, like the CO_2 laser, in a number of different configurations such as CW axial flow device, as a gas dynamic laser and under TEA excitation conditions. It can also be run as a chemical laser. Laser emission occurs between high vibrational levels of ground state CO molecules. The output wavelength is line tunable between 5 and 7 μm. Highly excited vibrational levels (up to v=35) are populated in a discharge via collisional process. The laser emission involves the radiative cascade process.

$$CO(v,J) \rightarrow CO(v-1,J+1) + h\nu, \quad (3.29)$$

where J is the rotational quantum number. Output is generally enhanced by cooling the gas. Commercial devices can provide up to 20 W CW. Pulsed output from TEA excitation may yield about 10 mJ energy in about 1 μs pulse at 10 Hz prf. The toxicity of CO gas may present problems in flowing gas system, if proper precautions are not taken.

3.8. Excimer Lasers

Rare gas halide excimer lasers are the most efficient sources of high intensity UV radiation. Strong laser emission is obtained from ArF, KrF, XeF, KrCl, XeCl and XeBr excimers. Output wavelength spans the spectrum from near UV to vacuum UV (Table 3.7). Rare gas halide lasers have been

Table 3.7 Output wavelengths of some excimer lasers.

Excimer	Wavelength (nm)
XeBr	282
XeCl	308
XeF	351
KrCl	222
KrF	249
ArCl	175
ArF	193

operated only in a pulsed mode. Their pulse length extends from picosecond to a microsecond. They may be pumped in pulsed mode by electric discharges, intense electron beams, proton beams, or optical sources. These lasers are wavelength tunable over wavelength ranges of several nanometers. Most commercial table top excimer lasers yield pulsed UV energy ~50 to 100 mJ per pulse in a width of ~10ns with repetition rate of ~ 500 pps at an efficiency of 1-2%. The first lasing of a rare gas halide (RGH) excimer was reported in 1975 [3.53]. In addition to the diatomic RGH lasers, there are triatomic RGH excimers which provide tunable radiation from visible to UV (Table 3.8). In 1986, four atomic RGHs (e.g. Ar_3F) have given broadband emission at 430 nm using electron beam pumped Ar/F_2 and Ar/NF_3 mixtures.

Before the availability of excimer lasers, the role of high intensity UV laser source for pumping dye lasers was performed by the pulsed nitrogen laser operating at 331 nm. But with the appearance of excimer laser in the UV region (XeF 351 nm, XeCl 308 nm, KrF 248 nm) the nitrogen laser is now replaced by excimer laser for pumping dye lasers. The frequency doubled Nd:YAG laser at 0.532 μm and frequency tripled Nd:YAG in the UV region are of course also used extensively for dye pumping.

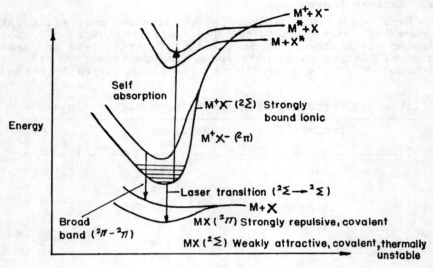

Fig.3.43. Potential energy diagram of rare gas halides [3.53].

Table 3.8 Triatomic RGH excimers

RGH	Wavelength (nm)	RGH	Wavelength (nm)
Ar_2F	285 ± 25	Kr_2Br	318
Ar_2Cl	245 ± 15	Xe_2F	610 ± 65
Kr_2F	420 ± 35	Xe_2Cl	490 ± 40
Kr_2Cl	325 ± 15	Xe_2Br	440 ± 30

excimers. The upper laser level is an ionically bound state while the ground state is covalently bound. The upper laser level is formed through ionic or neutral reactions. At close internuclear separation, the potential energy curve splits into the $^2\Sigma$ and $^2\pi$ states, generally referred to as B and C states respectively. The ground state consists of two states of which the $^2\Sigma$ state has the lowest energy (referred to as X state which is weakly bound). The other is the $^2\pi$ state which is always repulsive and is referred to as A state.

The emission spectrum consists of two bands such as B($^2\Sigma$)→X($^2\Sigma$) and C($^2\pi$)→A($^2\pi$). The B→X transition has a higher stimulated emission cross-section than that of the C→A transition indicating that the B→X transition usually gives intense lasing. The C→A transition consists of relatively broad continuum which is attributed to the repulsive structure of the A state.

Typical gas mixtures of rare gas halide lasers are high pressure (~1-5 atmosphere) rare gas mixtures containing a small amount of halogen donor. High excitation rate pumping of several 100 kW/cm^3 to several MW/cm^3 is necessary to efficiently produce high gain. Pulsed electron beam pumping and self sustained discharge have been employed.

A variety of e-beam pumped RGH lasers have been successfully tried. A few arrangements are shown schematically in Fig. 3.44.

To initiate a volumetrically uniform avalanche discharge in the 2-4 atmosphere rare gas halogen mixture, preionization of the high pressure mixture, prior to the initiation of the main discharge, is necessary. Spatial uniformity of the preionization in the rare gas halogen mixture is the most important issue. Its uniformity perpen-

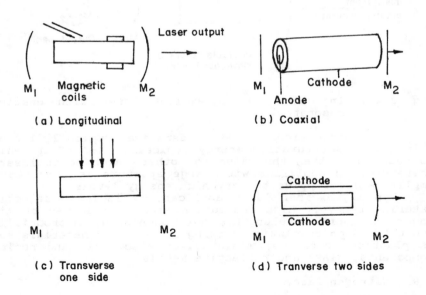

Fig. 3.44 Four major layouts for a beam pumped RGH laser.

dicular to the discharge electric field is of importance. A variety of preionization technologies have been developed, the simplest and most convenient being UV photo preionization using photo electron emission process. The UV photons are generated by the use of a pin-arc discharge and a dielectric surface discharge, both of which are induced in the laser gas mixture. UV preionization via pin-spark discharge is widely used in commercial systems. For large scale devices X-ray preionization is preferred.

Because of the high gain and wide gain bandwidth of RGH excimer lasers, high Fresnel number stable resonators, which are commonly used to efficiently extract the available laser energy, produce a laser output with high spatial divergence. A low divergence laser beam from RGH excimer lasers can be obtained by using a confocal unstable resonator with a large magnification or by simply using a low-Fresnel number stable resonator. A low divergence, narrow bandwidth excimer laser can be obtained by injection locking (Fig. 3.45).

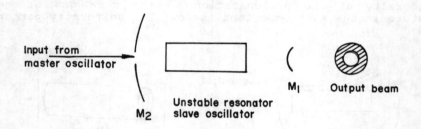

Fig. 3.45. Injection locked excimer laser with unstable resonator.

The storage times of RGH lasers are typically 1-2 ns. This provides saturation energy typically 1-2 mJ/cm^2 which is much lower than the value for other high power lasers. Therefore, an amplifier with large aperture is required to amplify a laser pulse to very high energy levels.

Typical excimer laser applications include: selective material removal, high resolution etching (because of high power at short wavelength), micromachining, microdrilling and ultra high resolution marking in delicate materials such as plastics, ceramics, miniaturized electronic and medical components, glass and reflective metals.

3.9. Nitrogen Laser

Though the molecular nitrogen laser emitting at 331 nm has been superseded by more versatile and powerful excimer lasers in the UV range, the N_2 laser is interesting as an example of its own kind.

The energy level diagram for N_2 is shown in Fig. 3.46. A, B and C states have different multiplicities (a triplet) than does the ground state ($X^1\Sigma_g^+$). The lowest $A^3\Sigma_u^+$ state is metastable and lifetime of the B state is longer than that of the C state. Yet the laser action is obtained on the C→B transition at 0.33 μm. The reason for this performance against lifetime odds is the favourable excitation route from X to C allowed by the Frank Condon principle. Frank Condon principle states that all electronic processes such as electronic transitions and electron molecule inelastic collisions must take place along a vertical path (r

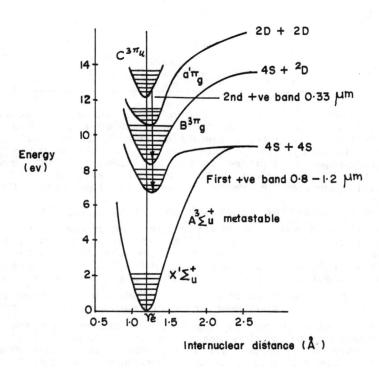

Fig. 3.46. Energy level diagram for N_2 showing common laser transitions.

constant). This is explained classically by considering that bound electrons can readjust their orbits instantaneously compared to the more sluggish vibrational motion of the nuclei. Most of the excitation in a nitrogen discharge would be from the v=0 vibrational level in the $X^1\Sigma_g^+$ state. Since the position r_e of the C state is close to that of the ground state, electron impact excitation will proceed along that path rather than to the B or A states (Table 3.9).

Table 3.9 Energy levels and their lifetimes in N_2 molecule

State	r_e(Å)	τ
$X^1\Sigma_g^+$	1.094	∞
$A^3\Sigma_u^+$	1.293	s
$B^3\pi_g$	1.2123	10 µs
$C^3\pi_u$	1.148	40 ns

In practical terms, nitrogen laser action is achieved by switching very high voltages (20 to 40 kV) with a very fast risetime (<10 ns). A specific configuration called the Blumlein discharge is devised for electrical pumping. The fast discharge is required to sustain the electron temperature at a high enough value to take advantage of the favourable excitation route allowed by the Frank Condon principle.

The N_2 laser has very high gain (typically 50 dB/m) so that multi pass resonant cavities are not needed. Only one mirror is sufficient to direct the energy in one direction. However this radiation does not have the true laser like qualities of high spectral purity and small divergence. The N_2 laser output has been used mostly for pumping dye lasers and as such it has been considered only as a glorified flash lamp. In view of the emergence of excimer lasers the N_2 laser has lost much of its significance except in situations where toxicity of mixing gases and more demanding pumping requirements of excimer lasers are not welcome and simplicity of N_2 laser fabrication is a desirable feature.

3.10. Metal Vapour Lasers

Lasing action in a number of metals has been observed

on their neutral and ionized atomic transitions in both CW and pulsed modes. These transitions span over a large part of the spectrum from IR to UV. In metal vapour lasers the laser action occurs in the atomic or molecular vapour phase of the species at relatively low pressures. These lasers, however, have problems associated with vapours, namely the need to vaporize a solid or liquid into gaseous state either before or during the excitation and the lasing process. Controlling the condensed vapours and corrosive effect of hot metal atoms and ions, present additional problems. The two most well known types of metal vapour lasers are the helium-cadmium ion laser and the pulsed copper vapour laser.

The CW metal ion lasers, typically represented by the He-Cd laser, have been operated on transitions of the singly ionized metal vapour species (Table 3.10).

Table 3.10 Metal vapour lasers

Lasing metal	Wavelength (nm)
Cu	510.6, 578.2
Mn	550, 534
Pb	723
Au	628
He—Cd	441.6, 533.7
He—Cu	259.9, 780.8
He—Ag	800, 840, 318, 408
He—Au	282, 292
He—Zn	602.1, 491.2

Tens of milliwatts to nearly a watt CW powers at wavelengths in the near UV, visible and near IR are obtained. For all these lasers, the active medium is prepared by running a CW discharge in a rare gas (He or Ne) in which relevant metal vapour is admitted. Typical operating pressures are about a few to a few tens of torr of the buffer gas and a small fraction of a torr for the active metal; the latter is produced thermally by sputtering or dissociation of the vapours of a suitable compound of the metal. The main excitation processes in these lasers are through charge transfer or collision between metal atoms and the ionized or excited inert gas atoms, as for example

$$He^+ + M \rightarrow He + (M^+)^* + \Delta E_1$$

$$He^* + M \rightarrow He + (M^+)^{**} + \Delta E_2 \qquad (3.30)$$

where M represents the ground state metal atom, $(M^+)^*$,

$(M^+)^{**}$ are the ionized metal atoms in different excited states, ΔE_1 and ΔE_2 are energy defects in the reactions. Usually $\Delta E_1 \sim 1$ eV, but ΔE_2 can be large. Majority of these systems need operating temperatures more than 400°C and even much higher (>1000°C) in the case of Cu, Ag etc. Devices employing volatile metal compounds can be operated with much more modest currents for laser generation.

3.10.1. *He-Cd Laser*

The He-Cd laser operates primarily in the blue region at 441.6 nm, in the UV at 325 nm, and is useful for printing and lithography. It produces CW power nearly 5-100 mW in the blue and 1-25 mW in UV. The Cd vapour is obtained by heating Cd metal in reservoir located near the He discharge. The Cd vapour diffuses into the excited helium gas, where it is ionized and moves toward the negative potential of the cathode thereby distributing the metal relatively uniformly in the discharge region to produce a uniform gain. The laser levels are excited by collisions with helium metastable atoms, by electron collisions and by photoionization resulting from radiating helium atoms.

3.10.2. *Hg-Br Laser*

This is a representative of the class of metal halide excimer lasers [3.54]. High power pulses with energies up to 0.1 J have been obtained in the blue green on the B→X excimer transitions of the (HgX) molecule. Both mercurous chloride and mercurous bromide excimers have been made to operate successfully. The starting material in all these cases is a mercury dihalide (HgX_2) vapour at a pressure of ~ 1 torr which is dissociated either optically or electrically to form the mercury monohalide in the excited state. The ArF excimer laser at 193 nm has been used for optical dissociation, whereas fast electrical discharges have been used for electrically dissociating the dihalides. Alternatively, mercury and halogen-donor mixtures can be electron beam pumped for the formation of HgX excimer.

$$HgX_2 + h\nu \rightarrow HgX^* + X$$

$$(HgX)^* \rightarrow HgX + h\nu_L \tag{3.31}$$

or

$$Hg + X^* \rightarrow HgX^* \rightarrow HgX + h\nu_L \tag{3.32}$$

The ground state monohalide molecules (HgX) will in turn recombine with free halogens formed as dissociation products to regenerate dihalides (HgX+X→HgX$_2$). This process both vacates the ground state of HgX for B→X transition as well as replenishes the consumed dihalide molecules, thereby making the process a cyclic one. However the recombination process HgX+X→HgX$_2$ is a slow one and lasing action is one of self terminating type due to the ground state bottlenecking.

In the typical discharge pumped lasers, HgX$_2$ vapours are mixed with N$_2$ and He (or Ne) in the ratio of 1:70:700 at about one atmosphere of total pressure. The presence of N$_2$ increases the output by an order of magnitude. Since the upper state lifetimes of these excimer transitions (B→X) are also of the order of a few tens of nanosecond, the rate of energy deposition into the gas mixture should be very fast. This requires discharge stabilized transverse excitation (TE) at high temperature (150°C for HgX$_2$ vapours). High temperature TE discharge presents its own problems. Output pulses over 100mJ with energy conversion efficiency ~ 0.1% have been obtained from HgBr laser.

3.10.3. *Copper Vapour Laser*

Of all the metal vapour lasers, copper vapour laser has gained importance in view of its application for isotope enrichment, large screen optical imaging and pumping of dye lasers. The gold vapour laser that is similar to the copper laser emits at 624 nm and is used for cancer phototherapy.

Copper vapour laser [3.55] operates in the green at 510.5 nm and in the yellow at 578.2 nm. It produces (at nearly 2% efficiency) short laser pulses (10-20 ns duration) of 1mJ energy at repetition rate of up to 20,000 pps yielding average powers of upto 20 W. The size is similar to that of an excimer laser. Commercial versions are designed to heat the metallic copper up to 1600°C in order to provide enough copper vapour to produce laser action. The relevant energy level diagram for this laser is shown in Fig. 3.47.

The lasing transitions are between the resonance levels of the neutral atom ($^2P_{1/2,3/2}$) and the low lying metastable states ($^2D_{5/2,3/2}$). The lifetime of upper laser state is ~ 10 ns while that of the lower laser states is a few hundred microsecond. The laser action is thus self terminating. Since the electron impact excitation cross-section from the ground state to resonance levels are in general high, the upper levels are preferentially populated; thus creating large transient population inversion on these transitions. Very large transient gains (~ 100 to 300 dB/m) have been reported on several of the neutral atom lasing transitions. Also, depending on the ground state population, self radiation trapping on resonance lines effectively lengthens the

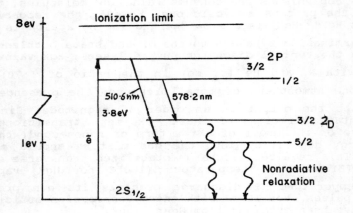

Fig. 3.47. Energy level diagram for copper vapour laser.

life time of the upper laser level by orders of magnitude, thereby enabling substantial energies to be deposited on inverted transition. Therefore, high output power pulses (several kW at ~50 ns duration) can be generated in the copper vapour system.

Essentially two distinct approaches are made in generating the required metal atom densities in the laser discharges. In the commonly used approach, the metal is heated to a sufficiently high temperature in a plasma tube made of an electrically nonconducting refractory material (usually alumina tubes), with typical dimensions of one meter in length by several centimeters in diameter. Tubular electrodes of tungsten, tantalum or other refractory metals are attached concentrically at the ends of the alumina tube. The plasma tube is then heated to the required temperature (up to 1600°C) in a split furnace to produce the metal atom densities up to 10^{15}-10^{16}/cm^3. A rare gas (He or Ne) up to several torr of pressure is also filled in the discharge as a buffer gas to run the discharge. Using suitable heat shielding to the plasma tube, the discharge produced heat at high repetition rates is sufficient to operate the alumina tube at the required temperature. This eliminates the need for a separate furnace for heating. Such self heating permits overall efficiency to exceed 1 percent in practical devices.

In the second approach, used for modest powers, the metal halide is used instead of the pure metal as active material. Usually mono dihalides of the metal with low melting and boiling point are preferred. The vapours of these metal halides are discharge dissociated in a preceding electrical pulse and laser excited in the later pulse. At high enough repetition rates every electrical pulse can act both as a dissociation pulse and excitation pulse. This scheme works at lower temperatures (600 to 800°C). Again, in this arrangement also, the heat produced by the discharge can be utilized for heating the laser medium to improve the overall efficiency of the laser. Further, with transverse excitation, there is possibility of large volumetric scaling for very high powers, higher efficiency and homogeneous field distribution in the discharge volume, leading to high quality of beam.

3.10.4. Plasma Recombination Laser

Electron-ion collisional recombination processes in an expanding plasma leaves the resulting neutral atoms in one of their very highly excited states near their ionization continuum. Since effectively all such atoms contribute to the inverted population on any downward transition, very powerful laser action, with large gain, are possible on these transitions. The possibility of obtaining the laser action in a recombining plasma was first suggested in 1964 [3.56] and first experimental observation of recombination laser was reported in 1973. Since then, there have been many reports on the population inversion in a recombining plasma from IR through X-ray region in neutral, singly, and multiply ionized species. In recombination lasers, the population of the levels are determined by the collisional radiative processes, involving recombination of free electrons with ions followed by collisional and radiative deexcitation of the captured electrons. Sufficient population inversion can be established whenever the free electrons cool in a time short compared to the relaxation time of the ground state of atoms or ions involved. Recombination lasers have been demonstrated in Mg, Cd, Al, Ca, Cu, Zn, Ag, In, Sn, Pb and Bi. Transitions from 289 nm to 1.95 micron have been made to lase with high repetition rates (up to 1 kHz) producing output powers of 1 to 10 W in a pulse of a few microsecond duration.

The excitation mechanism of a plasma recombination laser can be explained with the help of Fig. 3.48.

Ions of an active element are created in some charge state by feeding some energy into the medium. Input energy may be from an electrical discharge, an arc, or a laser beam or an electron beam. Under recombination condition, the laser action takes place between two groups of closely spaced levels, the gaps between such groups being considerably greater than the separations between the levels within a group. To achieve large population inversion and hence the

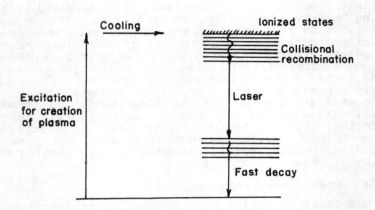

Fig. 3.48. Schematic energy level diagram for a plasma recombination laser.

gain, the following conditions are necessary
(a) fast decay of the lower laser level,
(b) low electron temperature, hence fast cooling of plasma,
(c) large plasma density.

The required plasma generation can be achieved in two ways. In one arrangement known as SPER laser (segmented plasma excitation recombination laser) a series of segmented arcs are produced between a row of strip electrode pairs formed on an insulating substrate [3.57]. Short electrical arcs are run between these pairs to generate ion plasma by sputtering and ionization. The ion density required is 10^{15}-10^{16}/cm^3 for singly ionized species and even more for highly stripped ions. Expansion and cooling of this highly dense plasma is achieved either by adiabatic expansion into a vacuum or (preferably) into a background gas to induce the recombination process.

Alternatively, high density plasma with highly stripped ions can be produced by focussing a pulse of light from a high power laser onto the surface of the metal of interest. The main advantage of the laser produced plasma is that very highly ionized states of atoms can be easily produced leading to lasing transition in VUV and XUV. However, with this technique, gain region cannot be made long enough to achieve required amplification. Lack of homogeneity of the laser medium is also a serious disadvantage when good quality laser beams are needed.

An experimental arrangement for obtaining laser osci-

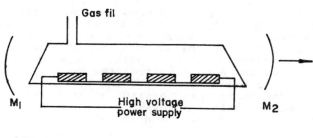

Fig. 3.49. Experimental arrangement for a sealed tube plasma recombination laser [3.58].

llation in Cd II is shown in Fig. 3.49 [3.58]. A low pressure helium gas is used in a sealed discharge tube in which there are eleven 1 mm×2 mm×10 mm long Cd strips placed end to end on a 12 cm long glass substrate in such a way as to leave a 1 mm gap between the adjacent strips. The electrode arrangement is then sealed in a glass tube of 40 mm diameter having Brewster windows. Two extreme end strips are used for connection to a high voltage power supply. A high voltage, high current pulse (400 A, 3-6 μs duration

Table 3.11 Laser transitions in Cd I and Cd II in a plasma recombination laser.

	Transition	Wavelength
Cd I	$6p^3P_2 - 6s^3S_1$	1.4 μm
	$6p^3P_1 - 6s^3S_1$	1.43 μm
	$4f^3F_4 - 5d^3D_3$	1.65 μm
Cd II	$4f^2F_{5/2} - 5d^2D_{3/2}$	533.7 nm
	$4f^2F_{7/2} - 5d^2D_{3/2}$	537.8 nm

obtained from the discharge of a capacitor) is used to create a high density Cd plasma in each gap. The laser resonator is formed by using 2 m radius of curvature dielectric mirrors, 75 cm apart. Laser action was obtained at 2 mm parallel to the row of cadmium strips and the best result was obtained when the optical axis of the resonator was parallel to and 6 to 7 mm from the Cd strips. The output was sensitive to the alignment of the resonator axis and the pressure of the helium gas (maximum at 10 torr).

The laser transitions investigated in Cd I and Cd II in the after glow region are shown in Table 3.11.

3.11. Far Infrared (FIR) Lasers

Molecular emission at wavelengths in excess of 10 μm are dominated by pure rotational transitions. Such transitions can be excited thermally, in glow discharges or via optical pumping. The latter technique has proved to be quite versatile and useful in view of the availability of the CO_2 laser as a line tunable pump. The principle of FIR generation through optical pumping lies in creating large non-equilibrium population differences in atomic or molecular systems. This population difference comes through resonant or near resonant optical excitations and various decay mechanisms. With conventional incoherent optical sources, this technique of optical excitation cannot meet the requirement efficiency-wise since the radiation, being incoherent, does not provide an exact match to narrow absorption lines in gases. But the narrowband pump sources, like CO_2 which are well developed and sufficiently stable, can meet the requirement adequately. A number of molecules (over 50) have been pumped optically with the CO_2 or other mid IR lasers. Emissions occur in the form of discrete lines at wavelengths extending to the millimeter wave region. Some of these are shown in Table 3.12 for illustration [3.59].

Optical pumping of selected rotational levels in a molecule such as NH_3 or CH_3OH leads to the establishment of a population inversion between the excited level and one or more rotational levels at lower energy.

Take the case of a molecule whose rotational states corresponding to the two vibrational states V_1 and V_2 are shown in Fig. 3.50 [3.44]. Let us say that the optical pumping raises the molecule from rotational state R_2 to R_3'. Since the thermal population in the upper vibrational state is very small, an inversion of population between R_3' and R_2' is readily established. Laser action can then be obtained on the rotational transition between these levels provided that the molecule has a permanent dipole moment.

Table 3.12 CW Transitions in the FIR and mm wave region.

Molecule	Pump	Laser wavelength (μm)
CH_3OH	9.3R(10)	97
CH_3OD	9.3R(8)	306
CH_3CN	10.6P(20)	372
HCOOH	9.3R(18)	420
CH_3F	9.6P(20)	496
CH_2CF_2	10.6P(14)	554
CH_2CF_2	10.6P(24)	662
HCOOH	9.3R(20)	744
CH_2CF_2	10.6P(22)	890
CH_3F	9.6P(32)	1222
CH_3CHF_2	10P(22)	2042.5

Sometimes, laser action on the cascade transition ($R_2' \rightarrow R_1'$) and laser action in the lower vibrational state due to depletion of the population from the starting level ($R_3 \rightarrow R_2$) are also observed.

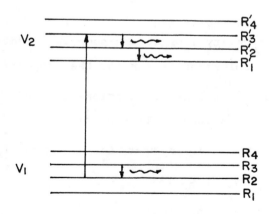

Fig. 3.50. FIR generation through optical pumping.

Optical pumping by laser for achieving population inversion is better than the electrical discharge. This is because the energy difference between the two rotational levels being very small (<kT), highly monochromatic radiation is needed for pumping. This is provided by the narrow-band, high intensity pump source like CO_2 laser. There are many IR laser lines available for pumping with a good number of absorption lines for a given molecule so that the probability of spectral coincidence is reasonably high.

The efficiency of converting pump power P_p into FIR power P_{FIR} can be expressed by [3.60]

$$P_{FIR}/V = (1+g_j/g_k)^{-1}(P_p/V)(\nu_{FIR}/\nu_p)(T/T+A)(\Gamma L/\Gamma L+a_p)$$
$$\cdot [1-(h\nu_{FIR}/kT)f_j(\tau_v/\tau_{\Delta j})], \qquad (3.33)$$

where

P_{FIR}/V = the volumetric FIR power,

g_j, g_k = the degeneracies of the upper and lower FIR levels,

ν_p = the pump frequency,

T = the mirror transmission,

A = the cavity loss at FIR frequency,

Γ = the absorption coefficient at the pump wavelength,

L = the cavity length,

a_p = the cavity loss at the pump wavelength,

f_j = the Boltzmann factor for the upper rotational laser state,

τ_v = the vibrational relaxation time and

$\tau_{\Delta j}$ = the rotational collision time.

From Eq. 3.33, it is clear that for positive P_{FIR},

$$(h\nu_{FIR}/kT)f_j(\tau_v/\tau_{\Delta j}) < 1$$

This is the positive gain condition.

A number of structures for generating millimeter and FIR radiation have been reported (Fig. 3.51). It is impor-

SPECIFIC LASER SYSTEMS

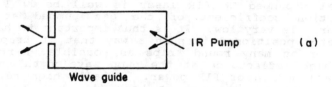

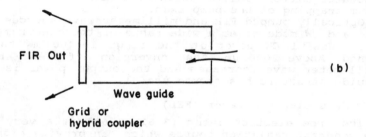

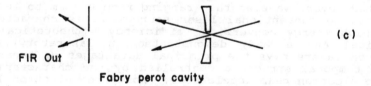

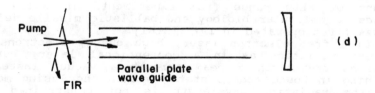

Fig. 3.51. Commonly used FIR laser resonators.

tant that the cavity should offer minimum losses at the pump and the generated wavelength. The operating pressure of an optically pumped CW FIR laser is well below 1 torr. The absorption coefficient of the gas pumped at such low pressures is very low. It is thus important to have the FIR resonator positioned in such a way that it traps the pump beam for as many round trips as possible. Mirrors should show high reflectance at the pump wavelength and transmit only a fraction of FIR power. Usually high reflecting Au coated mirrors are used at both the pump and FIR wavelengths. The pump beam is usually introduced into the resonator by focussing it through a small hole at the centre of a resonator mirror. Mirror curvatures are chosen to produce maximum trapping of the pump beam.

Optically pumped FIR and millimeter wave sources in the pulsed and CW mode cover a wide range of the spectrum (Table 3.12). Useful CW power in the range 10-400 mW has been obtained. Above 2000 μm, the conversion efficiency from IR to millimeter wave decreases and the output power is low. A waveguide structure has proved useful.

3.12. Free Electron Laser (FEL)

The free electron laser [3.61, 3.62] is a very adaptable coherent radiation source which can provide high power output in the wavelength region spanning virtually the entire electromagnetic spectrum. Working systems have operated over wavelengths ranging from 0.3 μm to 1000 μm. Compared to conventional laser sources it is characterized by a high energy conversion efficiency (theoretical 65%, practical 40% already demonstrated in laboratory). Free electron lasers have the principal advantages of tunability and a temporal structure controlled by the characteristics of the electron beam accelerator. Most free electron lasers are built around available electron accelerators, though recently some specific designs have been taken up for development starting from dedicated accelerators for FEL. Applications of FELs range from experiments in solid state physics to molecular biology and ballistic missile defence. FEL was first operated in 1977 [3.61].

In a free electron laser, high energy electrons emit coherent radiation as in a conventional laser, but the electrons travel in a beam through a vacuum instead of remaining in bound atomic states within the lasing medium. Thus the radiation wavelength is not constrained by a particular transition between two discrete energy levels. In quantum mechanical terms the electrons radiate by transitions between energy levels in the continuum and therefore, radiation is possible over a much larger range of frequencies than is found in a conventional laser.

FEL radiation is produced by an interaction among three elements, (i) electron beam, (ii) an electromagnetic wave travelling in the same direction as the electron and (iii) an undulatory magnetic field produced by an assembly of magnets known as Wiggler (Fig. 3.52). The wiggler magnetic field acts on the electrons in such a way that they acquire an undulatory motion. It is a known fact that an accelerated electron radiates. The moving accelerated charge may be considered in simplistic terms, to interact with its own field established at an earlier time when the electron was at a different position along its path. Because of the finite velocity of light, the near field of a charge does not respond instantaneously to the position of the charge, and thus some of the electrons' kinetic energy is lost as radiation. The acceleration associated with the undulatory trajectory of electron makes radiation possible and the electrons lose energy to the electromagnetic wave. As with any laser, gain comes from stimulated emission and the classic term of electromagnetics corresponding to it is E.j, the rate of interchange of energy between the field and the current. This must be negative for an amplifier, signifying that the electrons give energy to the field. For an FEL

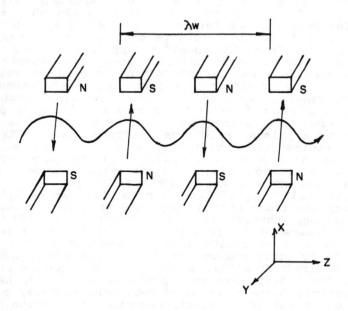

Fig. 3.52. Electron trajectory in a Wiggler.

therefore one must (i) generate a relativistic beam with as much current (electrons) as possible and (ii) periodically accelerate these electrons as they translate along the Z axis.

If the current is high enough, the strength of the Wiggler is strong enough, and feedback is provided, the coherent output at free space wavelength λ_0 is given by

$$\lambda_0 \sim \lambda\omega/2\gamma^2, \qquad (3.34)$$

where $\gamma = (1-v^2/c^2)^{-1/2}$,

$(\gamma-1)m_0c^2$=kinetic energy of the relativistic electrons, m_0=rest mass of the electron and $\lambda\omega$ is the mechanical period of the Wiggler. The wavelength therefore is directly proportional to the Wiggler period and inversely proportional to the square of the streaming energy of electron. This results in a broad tunability that permits the free electron laser to operate across virtually the entire electromagnetic spectrum.

Typical FEL efficiencies range up to approximately 12%, however, significant enhancements are possible when either the Wiggler amplitude or period are systematically tapered. The optimal configuration and type of accelerator used in a FEL design depends on the specific application and issues such as the electron beam quality, energy and current are important consideration in determining both the wavelength and power. Each accelerator type is suited to the production of limited range of wavelengths. The temporal structure of the output radiation from an FEL corresponds to that of the electron beam. Thus, either pulsed or CW electron beam will give rise to a pulsed or CW free electron laser. FELs have been constructed using virtually every type of electron source, including storage rings, rf linear accelerators, microtrons, induction linacs, electrostatic accelerators, pulse line accelerators and modulators.

Overall wall plug efficiency of an FEL can be enhanced by recovery and reuse of the spent electron beam subsequent to its passage through the FEL. Further, an additional source of focussing is required to confine the electrons against the self repulsive forces generated by the beam. This can be accomplished by the use of additional magnetic fields generated by either solenoid or quadrupole current windings. The Wiggler has been produced in planar, helical and cylindrical forms by means of permanent magnets, current carrying coils and hybrid electromagnets with ferrite cores.

The applications of FEL include its use for heating a magnetically confined plasma for controlled thermonuclear fusion. The frequencies of interest are the harmonics of the electron cyclotron frequency and range from about 280 GHz through 560 GHz. Another proposed application for future use against missile attack would need pulses of visible or near IR at an average power of about 10-100 MW over a duration of approximately 1 s. Depending on laser efficiency and target hardness, a collection of ground based FELs would require somewhere between 400 MW and 20 GW of power for several minutes which is yet to be achieved.

3.13. Harmonic Generation of Laser Radiation through Nonlinear Processes

So far we have discussed the generation of laser radiation directly by excitation through various pumping mechanisms in solid, liquid and gaseous media. Some of the laser sources described are tunable and cover a wide range of wavelengths. However, other techniques for extension of wavelength region are available through nonlinear processes. Nonlinear optics by itself is a wide field and covers a number of areas such as frequency conversion, phase conjugation, self focussing etc. Discussion on all aspects of nonlinear phenomena is outside the scope of this book. It is intended here to merely describe the second order effects in nonlinear crystals and third order processes in gaseous media as relevant to extension of frequency region of the primary laser sources. Here again we do not wish to go into nonlinear processes leading to stimulated Raman and Brillouin processes as they deserve a more detailed treatment which is not possible in the present volume. The purpose of including the frequency conversion processes, using nonlinear effects in crystals and gaseous media, is to describe the generation of second and third harmonics of input laser as well as of tunable UV and IR in the frequency range not directly accessible from primary laser sources. This is commonly employed in commercial laser systems for extension of usable wavelength range. The parametric oscillators providing tunable laser radiation and parametric processes for sum and difference frequency generation, are not covered as such since the subject deserves a more detailed treatment than is possible in this discussion. Some excellent texts and reviews on the subject provide details on various nonlinear processes [3.63-3.67].

Frequency conversion essentially involves generation of radiation at wavelengths other than the ones that are contained in the incident radiation (Fig. 3.53). Pump radiation consisting of one or more optical waves at one or more frequencies is incident on a nonlinear medium and interacts with it to generate the new wave. Depending on the interaction, the generated wave can be at a lower or higher frequency than the waves in the pump radiation. In some situations, the generated wave can be an amplified version of one of the incident waves.

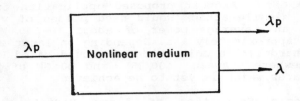

Fig. 3.53. Typical frequency conversion scheme using nonlinear process. Input radiation at pump wavelength λ_p is supplied at one or more incident frequencies and the wave at wavelength λ is generated in the nonlinear interaction.

Frequency conversion interactions can be either parametric processes or stimulated scattering interactions. In parametric interactions (giving sum, difference and harmonic of incident frequencies) energy is conserved among various optical waves. Increase in energy of generated wave is offset by a corresponding decrease in energy of the incident waves through Manley Rowe relations.

The radiation generated through nonlinear frequency conversion processes has all of the special properties usually associated with laser radiation, such as spatial and temporal coherence, collimation and narrow bandwidth. The generated wave can be tunable if the pump radiation is tunable.

Parametric frequency conversion interactions occur through the various higher order terms of the relation

$$P = \varepsilon_0 \chi^{(1)} E + \varepsilon_0 \chi^{(2)} E.E + \varepsilon_0 \chi^{(3)} E.E.E + \ldots \quad (3.35)$$

P is the induced polarization, E is the electric field of the optical wave, ε_0 the dielectric constant of vacuum and $\chi^{(1)}$, $\chi^{(2)}$ and $\chi^{(3)}$ are the nonlinear susceptibilities which are tensors and relate the components of the nonlinear polarization vector to various components of the optical field vectors. For the second order polarization,

$$P_i = \varepsilon_0 \chi^{(2)}_{ijk} E_j E_k \quad (3.36)$$

where i,j,k refer to the spatial directions x,y,z. The susceptibility tensor χ has the symmetry properties of the nonlinear medium.

The most commonly observed processes have interactions

of second and third order, although higher order interactions have also been reported. Parametric interactions are characterized by a growth rate for the intensity of the generated wave that depends on the product of intensities of the incident waves. The interactions are strongly dependent on difference in wave vectors between the nonlinear polarization and the optical fields.

3.13.1. Second Order Effects in Nonlinear Crystals Generation of Second Harmonics

Second order parametric processes include second harmonic generation (SHG), three wave sum and difference mixing, parametric down conversion, parametric oscillation and optical rectification. Because of symmetry restrictions, the even order electric dipole susceptibilities are zero in materials with inversion symmetry. Second order nonlinear interactions are thus observed mostly in certain classes of crystals that lack a centre of inversion. The nonlinear optical polarization for the second order processes can be written as

$$P_{NL}^{(2)} = 2\varepsilon_0 dE^2, \qquad (3.37)$$

where d is nonlinear optical susceptibility related to $\chi^{(2)}$ and E is the total electric field.

For SHG, the intensity is given by

$$I(2\omega) = [8\pi^2 d^2/n_2 n_1^2 c\varepsilon_0 \lambda_1^2] I_0^2(\omega) \mathrm{sinc}^2 \Delta kL/2, \qquad (3.38)$$

where I_0 is the pump intensity at fundamental frequency ω and $I(2\omega)$ is the intensity at second harmonic frequency 2ω, sincx=sinx/x, L is the crystal length and

$$\Delta k = k(2\omega) - 2k_1(\omega) \qquad (3.39)$$

k(2w) is the wave vector at frequency 2ω and $k_1(\omega)$ the wave vector at pump frequency ω. For maximum SHG power, $\Delta k= 0$, which determines the phase matching condition

$$n(2\omega) = 2n(\omega). \qquad (3.40)$$

If $\Delta k \neq 0$ (the phase mismatch condition), the harmonic wave gradually gets out of phase with the polarization as it propagates through the crystal. As a result, the harmonic intensity oscillates with distance, with the harmonic wave first taking energy from the incident wave and then, after a phase change of π relative to the nonlinear polarization, returning it to the incident wave. The shortest distance at which the maximum conversion occurs is termed the coherence length and is given by

$$L_{coh} = \pi/\Delta k. \tag{3.41}$$

Under phase matched condition $\Delta k=0$, the harmonic intensity grows as the square of the crystal length and for large L can even deplete the pump wave. The wave vector mismatch Δk occurs because of the natural dispersion in the refractive index that is present in all materials. For the non-centrosymmetric crystals used for SHG, the most common method of phase matching involves use of the birefringence of the crystal to offset the natural dispersion in the refractive indices. The value of wave vector mismatch can be adjusted by varying the directions of propagation of the various waves relative to the optic axis, or by varying the temperature of the crystal, for a fixed direction of propagation. The angle at which $\Delta k=0$ is called the phase matching angle. For a birefringent crystal, the refractive index, at the phase matching angle ϑ from the optic axis, is given by

$$[n_{2\omega}(\vartheta)]^{-2} = (n^o_{2\omega})^{-2}\cos^2\vartheta + (n^e_{2\omega})^{-2}\sin^2\vartheta \tag{3.42}$$

As ϑ varies from 0 to $90°$, $n_{2\omega}(\vartheta)$ varies from $(n^o_{2\omega})$ to $(n^e_{2\omega})$. o and e refer to ordinary and extraordinary polarization of the wave. The phase matching angle is obtained by the relation

$$n_{2\omega}(\vartheta) = n^o_\omega. \tag{3.43}$$

SHG has been obtained by using crystals like ADP, KDP, AD*P, KD*P and a host of other nonlinear crystals in different wavelength bands. The choice of particular crystal depends on the transmission range (it must be transparent to both pump and second harmonic), reasonable value of the nonlinear coefficient d, sufficient birefringence (n^o-n^e) for phase matching and high value of the damage threshold for pump intensity. SHG has been used to generate light at wavelength ranging from 217 nm (doubling of dye laser radiation at 434 nm) to 5.3 μm (by doubling of 10.6 μm from CO_2 laser). SHG is commonly used with high power pulsed lasers e.g. Nd:YAG, Nd Glass, ruby or CO_2 and tunable dye lasers. Very good efficiency of conversion (up to 75%) has been obtained.

Recently an interesting concept, a self frequency

doubled laser, has appeared in which the active medium of laser is also nonlinear, therefore it generates second harmonic internally without the need for an external nonlinear crystal. $LiNbO_3$ doped with 5% MgO has shown promise. Another group has reported flash lamp pumping of Nd-MgO-$LiNbO_3$ at room temperature giving green wavelength of frequency doubled YAG.

3.13.2. Third Order Nonlinear Processes in Gaseous Media Generation of Tunable UV and IR

The nonlinear electrical susceptibility of many gases has been used for four wave mixing process.

$$\omega_4 = 2\omega_1 \pm \omega_3 \text{ and}$$

$$\omega_4 = \omega_1 + \omega_2 \pm \omega_3,$$

where ω_1, ω_2 and ω_3 are the angular frequencies of input radiations and ω_4 is the generated radiation (Fig. 3.54). The process is commonly used to produce radiation in wavelength ranges that are inaccessible by other means i.e. VUV and mid IR. In particular, the process has been used to generate tunable radiation over most of the vacuum UV from 100 to 200 nm.

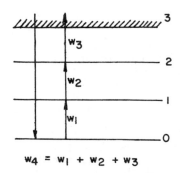

$w_4 = w_1 + w_2 + w_3$

Fig. 3.54. Energy levels and transitions involved in four wave mixing to generate radiation of frequency ω_n

The third-order susceptibility can be written as

$$\chi^{(3)}(\omega_4 = \omega_1+\omega_2+\omega_3) = \frac{(3e^4/4\hbar^3)\,\mu_{01}\mu_{02}\mu_{03}}{(\Omega_{30}-\omega_1-\omega_2-\omega_3)(\Omega_{20}-\omega_1-\omega_2)(\Omega_{10}-\omega_1)},$$

where e is electron charge, μ_{ij} are dipole matrix elements and Ω_{i0}, $i=1,2,3$ denote frequency difference between real states and ground state. $\chi^{(3)}$ has resonances when $\Omega_{20}-\omega_1=0$ and so forth. However, those involving one or three photon processes are accompanied by strong absorption in the non-linear medium. Hence, tunable radiation at frequency $\omega_4 = 2\omega_1 \pm \omega_2$ is usually obtained by tuning $2\omega_1$ to resonate with a two photon transition, preferably chosen such that $2\omega_1+\omega_2$ resonates with an auto ionizing state above the ionization limit of the absorber.

Four wave interactions can be used to generate tunable radiation in resonantly enhanced processes, thereby increasing the efficiency of the process. The pump frequency at ω_1, or the combination $\omega_1+\omega_2$, is adjusted to match a suitable two photon resonance, while the other pump frequency at ω_3 is varied, producing the tunable radiation. The difference frequency processes

$$\omega_4 = 2\omega_1-\omega_3 \text{ and}$$
$$\omega_4 = \omega_1 \pm \omega_2-\omega_3$$

have also been used to generate tunable radiation in the infrared by using pump radiation from visible and near infrared lasers. Resonances between Raman active molecular vibrations and rotations and the difference frequency combination $\omega_1-\omega_3$ can also occur. When the four wave mixing process $2\omega_1-\omega_3$ or $\omega_1+\omega_2-\omega_3$ is used with these resonances it is termed coherent anti Stokes Raman Scattering (CARS).

There are many atomic vapours (Ar, Xe, Kr, Hg, Zn, Mg, Sr etc.) which satisfy the requirement of accessible two photon states, together with a strong autoionization resonance at high energy. Experimentally, the beam from a fixed frequency dye laser (ω_1) is combined with that of a tunable dye laser in a Glan prism to form a collinear beam that enters a cell containing the nonlinear vapour. When $2\omega_1$ is tuned to a two photon resonance in the vapour, a beam at $\omega_4=3\omega_1$ can be generated. However, when the laser emitting at ω_2 is tuned into resonance or near resonance with an autoionizing level, then an enhancement of coherent output at $\omega_4=2\omega_1+\omega_2$ is observed. Phase matching of output and input beams is required for optimum results.

Stimulated scattering processes are also nonlinear interactions which provide frequency conversion. In this process an incident wave at frequency ω_{in} is converted to a scattered wave at a different frequency ω_s. The difference in photon energy between the incident and scattered frequencies is taken up or supplied by the nonlinear medium, which undergoes a transition between two of its internal energy levels. The incident (laser) and scattered (Stokes) frequencies are related by

$$\omega_s = \omega_L - \omega_0$$

where ω_L and ω_s are the frequencies of the laser and Stokes waves and ω_0 is the frequency of the internal energy level of the medium. Similarly the laser and anti Stokes frequencies are related by

$$\omega_{as} = \omega_L + \omega_0.$$

Various types of stimulated scattering processes are possible, each involving a different type of internal excitation. Some of these processes are stimulated Raman scattering (SRS), stimulated Brillouin scattering (SBS) and stimulated Rayleigh scattering.

References

3.1 A.L. Schawlow and C.H. Townes , *Phys. Rev.* **112** (1958) 1940.

3.2 T.H. Maiman, *Nature* **187** (1960) 493.

3.3 Z.J. Kiss and R.J. Pressley, *Proc. IEEE* **54** (1966) 1236.

3.4 J.E. Geusic, H.M Marcos and L.G. Van Uitert, *Appl. Phys. Lett.* **4** (1964) 182.

3.5 C.G. Young, *Proc. IEEE* **57** (1969) 1267.

3.6 E. Snitzer, *Phys. Rev. Lett.* **7** (1961) 444.

3.7 R.D. Maurer, *Appl. Opt.* **2** (1963) 87.
     P.B. Mauer, *Appl. Opt.* **3** (1964) 153.

3.8 E Snitzer and C.G. Young, in *Lasers Vol.2*, ed. A.K. Levine (Marcel Dekker,1968).

3.9 O.K. Deutschbein and C.C. Pautrat, *IEEE Jr. Quant. Electron.* **QE4** (1968) 48.

3.10 R.A. Miller and N.F. Borelli, *Appl. Opt.* **6** (1967) 164.

3.11 D. Olness, *J. Appl. Phys.* **39** (1968) 6.

3.12 G.T. Linford, R.A. Saroyan, J.B. Treholme and M.J. Weber, *Proc. IEEE conf.Laser Engg. and Appl. CLEA* (Washington DC., 1977).

3.13 P.F. Moulton, *IEEE J. Quantum. Electron.* **QE21** (1985) 1582.

3.14 R.W. Nichols and W.K. Ng, *Proc. SPIE* (1986) 610.

3.15 T.Y. Fan and R.L. Byer, *IEEE J. Quantum. Electron.* **QE24** (1988) 895.

3.16 L.F. Mollenauer, in *Laser Handbook* Vol 4, ed. M. Stitch (North Holland, Amsterdam 1985) 143.

3.17 K.R. German and C.R. Pollock, *Opt. Lett.* **12** (1987) 474.

3.18 E.T. Geogiou, T.J. Carrig, C.R. Pollock, *Opt. Lett.* **13** (1988) 987.

3.19 R.N. Hall and G.E. Fenner, J.D. Kingsley, T.J. Soltys, R.O. Carlson, *Phys. Rev. Lett.* **9** (1962) 366.

3.20 G Burns and M.I. Nathan, *Proc. IEEE* **52** (1964) 770. M.I. Nathan, *Proc. IEEE* **54** (1966) 1276.

3.21 J. Hayashi, M.B. Panish and P.W. Foy, *IEEE J. Quantum. Electron.* **QE5** (1969)

3.22 J.J. Hsieh, J.A. Rossi and J.P. Donnelly, *Appl. Phys. Lett.* **28** (1976) 709.

3.23 T. P. Lee, *Proc. IEEE* **79** (1991) 253.

3.24 *Semiconductors and Semimetals*, ed. R. Dingle (Academic Press, 1987) **Vol 24**.

3.25 Y. Arakawa and A. Yariv, *IEEE J. Quantum. Electron.* **QE21** (1985) 1666.

3.26 H. Kogelnik and C.V. Shank, *Appl. Phys. Lett.* **18** (1971) 152.

3.27 H.C. Cassey Jr. and S. Somekh, M Illgems, *Appl. Phys. Lett.* **27** (1975) 142.

3.28 F.K. Reinhart, R.A. Logan and C.V. Shank, *Appl. Phys. Lett.* **27** (1975) 45.

3.29 K. Utaka and K. Kobayashi, Y. Suematsu, *IEEE J.*

Quantum. Electron. QE17 (1981) 651.

3.30 O. Mikami, Japan J. Appl. Phys. 20 (1981) L488.

Y. Suematsu, A. Arai and K. Kishino, IEEE J. Lightwave Techhnol. LT-1 (1983) 161.

3.31 H. Kogelnik and C.V. Shank, J. Appl. Phys. 43 (1972) 2328.

3.32 P.P. Sorokin and J.R. Lankard, IBM J. Res. Develop. 10 (1966) 162.

3.33 F.P. Schafer and W. Schmidt, J. Volze, Appl. Phys. Lett. 9 (1966) 306.

3.34 B.H. Soffer and B.B. McFarland, Appl. Phys. Lett. 10 (1967) 266.

3.35 B.B. Snavely and F.P. Schafer, Phys. Lett. 28A (1969) 728.

B.B. Snavely, Proc. IEEE 57 (1969) 1374.

3.36 O.G. Peterson, S.A. Tuccio and B.B. Snavely, Appl. Phys. Lett. 17 (1970) 245.

3.37 K.H. Drexhage, in Dye Lasers, ed. F.P. Schafer, (Springer Verlag 1973).

3.38 T.W. Hänsch, Appl. Opt. 11 (1972) 895.

3.39 P.P. Sorokin, J.R. Lankard, E.C. Hammond and V.L. Moruzzi, IBM J. Res. Develop. 11 (1967) 130.

3.40 J.B. Marling, D.W. Gregg and S.J. Thomas, IEEE J. Quantum Electron. QE6 (1970) 570.

3.41 R. Pappalardo, H. Samelson and A. Lempicki, IEEE J. Quantum. Electron. QE6 (1970) 716.

3.42 D.C. Hanna, V.V. Rampal and R.C. Smith, Opt. Commun. 8 (1973) 151; IEEE J. Quantum Electron. QE10 (1974) 461.

3.43 R.K. Tyagi, V.V. Rampal and G.C. Bhar, Infrared Phys. 31 (1991) 319.

3.44 R.K. Tyagi, V.V. Rampal and G.C. Bhar, Infrared Phys. 26 (1986) 29.

3.45 A. Javan, W.R. Bennett Jr. and D.R. Herriot, Phys. Rev. Lett. 6 (1961) 106.

3.46 A.D. White and J.D. Rigden, Proc IRE 50 (1962) 2366.

3.47 W.B. Bridges, Appl. Phys. Lett. 4 (1964) 128;

W.B. Bridges, A.N. Chester, A.S. Halstead and J.V. Parker, Proc. IEEE 59 (1971) 724.

3.48 C.K.N. Patel, Phys. Rev. Lett. 12 (1964) 588; Phys. Rev. 136 (1964) A1187.

3.49   A.J. Beaulieu, *Appl. Phys. Lett.* **16** (1970) 504.

3.50   J. Gilbert, *Bull. Amer. Phys. Soc.* **15** sec.**11** (1970) 808;
       K.A. Laurie and M.M. Hale, *IEEE J. Quantum. Electron.* **QE6** (1970) 530.

3.51   N.G. Basov and A.N. Oraevskii, *Sov. Phys. JETP* **17** (1963) 1171;
       V.K. Konyukhov and A.M. Prokhorov, *JETP Lett.* **3** (1966) 286.

3.52   J.V. Kasper and G.C. Pimentel, *Phys. Rev. Lett.* **14** (1965) 352.

3.53   C.A. Brau, in *Excimer Lasers*, ed. C.K. Rhodes, (Springer Verlag, 1979).

3.54   M. Jakob, *IEEE J. Quantum Electron.* **QE15** (1979) 579.

3.55   J.A. Piper, *IEEE J. Quatum Electron.* **QE14** (1978) 405.

3.56   L.I. Gudzenko and L.A. Shelepin, *Sov. Phys. JETP* **18** (1964) 998.

3.57   W.T. Silfvast, L.H. Szeto and O.R. Wood II, *Appl. Phys. Lett.* **36** (1980) 615; *ibid* **39** (1981) 212.

3.58   R.K. Thareja and A.K. Khare, *Opt. Lett.* **12** (1987) 28.

3.59   D.T. Hodges, *Infrared Phys.* **18** (1978) 375; K.J. Button, *Infrared and millimeter waves vol 17* (Academic Press 1983).

3.60   N. Bloembergen, *IEEE J. Quantum. Electron.* **QE20** (1984) 556.

3.61   D.A.G. Deacon, L.R. Elias, J.M.J. Madey, G.J. Ramian, H.A. Schwettman and T.I. Smith, *Phys. Rev. Lett.* **38** (1977) 892.

3.62   T.C. Marshall, *Free Electron Lasers* (McMillan, NewYork 1985).

3.63   N. Bloembergen, *Nonlinear Optics* (Benjamin, 1965).

3.64   S.E Harris, *Proc. IEEE* **57** (1969) 2096.

3.65   *Laser Handbook* eds. F.T. Arecchi, E.O. SchulzDubois, (North Holland, 1972).

3.66   Y.R. Shen, *The Principles of Nonlinear Optics* (Wiley Interscience, 1984).

3.67   D.C. Hanna, M.A. Yuratitch and D. Cotter, *Nonlinear Optics of Free Atoms and Molecules* (Springer Verlag, 1979).

CHAPTER 4

HOLOGRAPHY : PRINCIPLES AND TECHNIQUES

4.1. Introduction

In order to characterize a light wave completely we need to know both its amplitude and phase. The use of a lens to form an image of an object (two or three dimensional) onto a light sensitive film is a well known technique in photography. It is easy to record amplitude by photographic means in which variations in amplitude are converted into variations of opacity of the photographic emulsion. In such techniques, the phase of the impinging wave, i.e. the information concerning the relative lengths of the optical path to different parts of the object, is not recorded as the photographic emulsion is a square law detector and records only the amplitude. In photography, one is thus concerned only with the brightness or irradiance distribution (square of the amplitude) of the image.

In holography, the aim is to record complete wave field (both amplitude and phase) as it is intercepted by a recording medium. The recording plane may not be even an image plane. The scattered or reflected light by the object is intercepted by the recording medium and recorded completely in spite of the fact that the detector is insensitive to the phase differences among the various parts of the optical field.

In 1948 Denis Gabor [4.1-4.3] gave an ingenious solution to the problem of recording of phase information by means of a background wave, which converts phase differences into intensity differences and can thus be recorded by the photographic emulsion. Gabor introduced a two step lensless imaging process known as wavefront reconstruction technique or holography (Greek word *holos* means whole, complete), in which one intermediate recording, the interference between the object field and the background wave (known as reference wave) is formed on a photographic material. The record known as a hologram (whole record) captures the complete wave scattered by the object which can be released at any desired later time by illuminating the hologram with an appropriate beam of light. The reconstructed wave then propagates as if it had never been interrupted.

Gabor's original aim was to improve the image quality of the then electron microscope, which suffered from spherical aberration of the electron lenses. The electron lenses are magnetic fields which cannot be precisely controlled, whereas the spherical aberration of optical lenses can be corrected accurately. Gabor's solution was to dispensed with the electron lenses and record the scattered field of the illuminated object and then generate the field with optical waves.

Thus, in holography the interference between the light reflected and scattered by the object, called the object (signal) beam and an unaltered coherent wavefront, called the reference beam is created and recorded on a photographic emulsion. If the amplitude of the signal beam remains constant and the angle between the beams increases, the fringes will become finer. On the other hand, if the phase relation between the two interfering beams remain constant but the amplitude of the object beam changes, the contrast of the fringes will change. Thus the complex object information gets coded in the form of complicated fringe pattern.

The work on wavefront reconstruction technique was continued by Rogers [4.4], El-Sum and Kirkpatrick [4.5], Baez [4.6] and Lohmann [4.7]. However, the results obtained were very poor and the interest soon faded out. Holography did not attract much attention until the advent of a laser which produced a strong source of high spatial and temporal coherence making it feasible to record objects with considerable depth and size. The revolution in holography began in 1962 with the work of Leith and Upatnieks [4.8-4.11] who suggested modifications in Gabor's original technique.

Figure 4.1 shows the generation of interference fringes when two plane waves fall on a photographic film at a certain angle and produce straight fringes. In the reconstruction process, when the developed film (hologram) is illuminated by coherent light, cylindrical waves propagate from the tranparent areas of the hologram. The constructive interference of these waves produces plane waves, called the diffraction orders. Figure 4.1 shows the zero and first order waves only. The hologram is thus able to exactly duplicate the original wavefront, both its amplitude and phase.

Figures 4.2 shows the generation of modulated interference fringes produced when a plane wave and a wave generated by a point of an object interfere. In this case, when the hologram is illuminated by coherent light, one of the first order waves will be diverging, producing what is called the virtual image of the original object wave. The other will be converging, producing the real image of the original object.

We have so far discussed the simple case of inter-

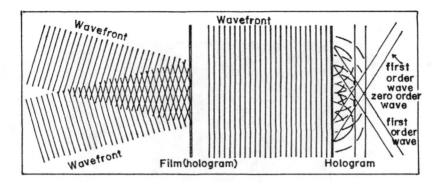

Fig. 4.1. Recording and reconstruction of a hologram of a plane wave.

ference between two plane beams and a plane and spherical beam. The wavefront from a three dimensional object, however, is neither true plane nor true spherical. The object can be considered to be made up of a large number of point sources distributed in a three dimensional space. Each point of the object will interfere with the reference and produce fringes as shown in Fig. 4.2. The fringe patterns generated by different points will be varying in orienta-

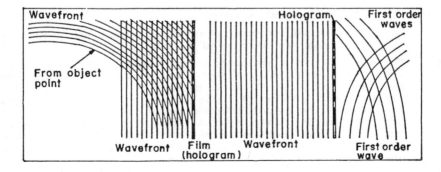

Fig. 4.2. Recording and reconstruction of a hologram of a spherical wave.

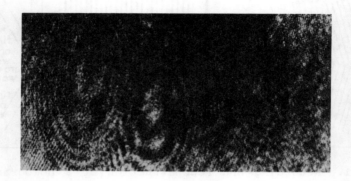

Fig. 4.3. A hologram.

tion, contrast and spacing. The hologram is thus a hodgepodge of whorls (Fig 4.3) and shows the object only when it is illuminated by the reference wave.

Gabor could show the applicability of this new process of wavefront recording by using a mercury discharge lamp and taking collinear object and reference beams. The original in-line technique of Gabor produces both virtual and real images on the same axis, thus an observer focussing on one image, always sees it accompanied by the out-of-focus twin image.

4.2. Characteristics of a Hologram

Holograms have unique characteristics not found in photography. These are discussed below:

(1) The light from a reconstructed image reaching to the observer's eye is the same as that would come from the original object. As a matter of fact, there is no visual test which can distinguish between the real object and the reconstructed image of the object. One can see in the holographic image the depth, parallax and different perspectives available in the actual object scene. Figure 4.4 shows two views of an object scene from a hologram.

(2) The hologram of a diffuse object can be reconstructed even by a small portion of it. In other words, if a hologram breaks into pieces, each piece can reproduce the entire image. However, as the hologram size

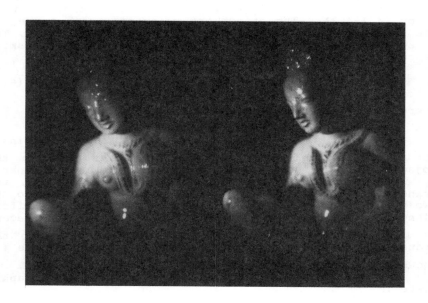

Fig. 4.4. Two views of an object reconstructed from a hologram.

reduces, a loss of image perspective, resolution and brightness result in the reconstructed image. In a particular type of hologram, Fourier transform hologram, very small size hologram can be recorded without the loss of resolution.

(3) A contact print of a hologram will still reconstruct a positive image indistinguishable from the image produced by the original.

(4) Two images, usually a real (pseudoscopic) and a virtual (orthoscopic) can be reconstructed from a hologram.

(5) A cylindrical hologram provides a 360° view of the object i.e. the observer can move round the hologram and see the entire object from all its sides.

(6) More than one independent scenes can be stored in the same photographic plate which can be viewed one at a time, without any cross-talk.

4.3. In-line Holography: Gabor Holography

We shall now consider the process of recording a hologram using an in-line setup similar to that used by Gabor [4.1-4.3]. The object and reference beams travel in the same direction and the angle between them is small. The object is assumed to be highly transmitting with transmittance

$$t(x_o, y_o) = t_o + \Delta t\,(x_o, y_o), \quad (4.1)$$

where $|\Delta t| \ll t_o$.

In Eq. 4.1 t_o is the high average transmission giving a strong uniform reference wave and Δt gives rise to the object wave. When such an object transparency is illuminated by a uniform parallel beam of light of amplitude A, the transmitted light consists of two parts corresponding to the two terms in Eq. 4.1. The first part is an intense uniform plane wave of amplitude R passed by the term t_o. The second is a weak scattered wave of amplitude $O(x,y)$ due to the transmittance $\Delta t(x_o, y_o)$ such that $|O(x,y)| \ll |R|$. Figure 4.5 shows the optical system for recording a Gabor hologram.

The intensity of light incident at the photographic plate is

$$\begin{aligned} I(x,y) &= |R + O(x,y)|^2 \\ &= R^2 + |O(x,y)|^2 + RO(x,y) + RO^*(x,y). \end{aligned} \quad (4.2)$$

A positive transparency is made of this recording. The amplitude transmittance of this transparency is assumed to be linearly proportional to the original intensity and is given by

$$t_A(x,y) = t_b + \beta[|O(x,y)|^2 + RO(x,y) + RO^*(x,y)], \quad (4.3)$$

where $t_b = \beta R^2$ and β is a constant of proportionality depending on the photographic material and processing conditions.

To reconstruct the image, this transparency is repositioned at the recording position and illuminated again with a plane wave of uniform amplitude R_o as shown in Fig. 4.6. The resultant transmitted field amplitude is given by

$$R_o t_A(x,y) = R_o t_b + R_o \beta[|O(x,y)|^2 + RO(x,y) + RO^*(x,y)] \quad (4.4)$$

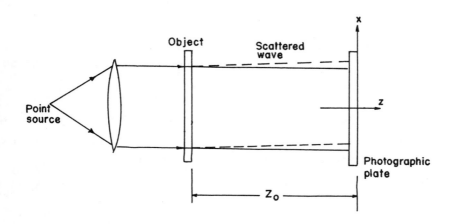

Fig. 4.5. Optical system for recording a Gabor hologram.

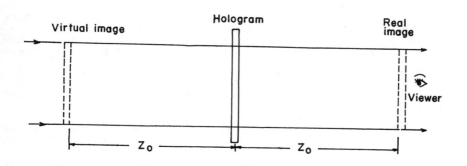

Fig. 4.6. Reconstruction of a Gabor hologram.

From Eq. 4.4 we see that the transmitted amplitude consists of four distinct components which have the following meaning.

$R_o t_b$ Plane wave passing directly through the hologram uniformly attenuated.

$R_o\beta|O(x,y)|^2$ This term is negligibly small in comparison to other terms since $|\Delta t|<<t_o$.

$R_o\beta RO(x,y)$ Except for a simple multiplicative constant, this term represents the wavefront similar to the original scattered wavefront from the object. Hence it gives rise to a reconstructed image of the object located at a distance z_o behind the hologram. It is a *virtual image*.

$R_o\beta RO^*(x,y)$ This corresponds to a field which is proportional to $O^*(x,y)$, the complex conjugate of the object amplitude. This is a *real image* of the object formed at the same distance z_o in front of the hologram.

Thus a Gabor hologram gives simultaneously virtual and real images of the object centred on the hologram axis. These twin images are separated by a distance $2z_o$. We see that while it is possible to reconstruct an image of the original object with the in-line system, this image is always accompanied by a coherent background and a conjugate image. The presence of these poses the serious difficulty of the holographic imaging technique in the form originally proposed by Gabor. When an observer focusses one image, it is always accompanied by the out of focus twin image degrading the quality of the reconstructed image.

Another serious limitation of the Gabor hologram is the need for the object such that $R_o\beta|O(x,y)|^2$ becomes negligible i.e. the object must have a high average transmittance. For objects of low transmittance this term becomes very large and may entirely obliterate the weaker images. Thus it is possible to form images of opaque letters on a transparent background, but not *vice versa*.

Finally it may be noted that if we use the negative of the hologram recording the reconstructed image will also be a negative.

In spite of the limitations of in-line holography, it has potential applications in particle size measurements [4.12-4.14].

4.4. Off-axis Holography: Leith-Upatnieks Holography

Leith and Upatnieks in 1962 demonstrated a technique for recording and reconstruction of a hologram which made it possible to separate the twin images [4.8-4.11]. This opened the way to record holograms of two or three dimensional objects without the drawbacks of in-line holography.

In this technique a separate coherent reference wave, which we shall assume to be a collimated beam of uniform intensity, is allowed to fall on the hologram plate during the recording process, at an offset angle θ to the beam from the object as shown in Fig. 4.7.

We can express the amplitude due to the object at any point (x,y) on the plate as

$$O(x,y) = |O(x,y)|\exp[-j\phi(x,y)], \qquad (4.5)$$

while that due to reference beam is

$$R(x,y) = R\exp[-j2\pi\alpha y], \qquad (4.6)$$

where $\alpha = (\sin\theta)/\lambda$ is the spatial frequency.

The resultant intensity at the photographic plate is

$$\begin{aligned}I(x,y) &= |R(x,y) + O(x,y)|^2 \\ &= |R(x,y)|^2 + |O(x,y)|^2 \\ &\quad + R|O(x,y)|\exp[-j\phi(x,y)]\exp(j2\pi\alpha y) \\ &\quad + R|O(x,y)|\exp[j\phi(x,y)]\exp(-j2\pi\alpha y) \\ &= |R|^2 + |O(x,y)|^2 + 2RO(x,y)\cos[2\pi\alpha y - \phi(x,y)]\end{aligned}$$

$$(4.7)$$

As can be seen from Eq. 4.7 the amplitude and phase of the object beam are encoded respectively as amplitude and phase modulation of a set of interference fringes equivalent to a spatial carrier wave with a spatial frequency.

The exposed plate is developed by normal photographic procedures. If the amplitude transmittance $t_A(x,y)$ of the plate after development is proportional to the exposure, then

$$\begin{aligned}t_A(x,y) = t_b &+ \beta\{|O(x,y)|^2 \\ &+ R|O(x,y)|\exp[-j\phi(x,y)]\exp(j2\pi\alpha y) \\ &+ R|O(x,y)|\exp[j\phi(x,y)]\exp(-j2\pi\alpha y)\}, \qquad (4.8)\end{aligned}$$

where β is the slope of t_A versus exposure response characteristic of the photographic material and t_b is a constant background transmittance.

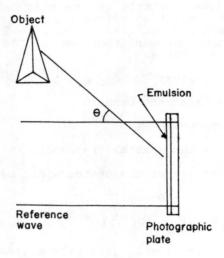

Fig. 4.7. Recording of an off-axis hologram.

To reconstruct the image, the hologram is illuminated, for example by a plane wave of uniform amplitude B. The amplitude $U(x,y)$ of the transmitted wave is

$$U(x,y) = U_1(x,y) + U_2(x,y) + U_3(x,y) + U_4(x,y), \qquad (4.9)$$

where

$$U_1(x,y) = Bt_b,$$
$$U_2(x,y) = B\beta|O(x,y)|^2,$$
$$U_3(x,y) = B\beta R|O(x,y)|\exp[-j\phi(x,y)]\exp(j2\pi\alpha y),$$
$$U_4(x,y) = B\beta R|O(x,y)|\exp[j\phi(x,y)]\exp(-j2\pi\alpha y). \qquad (4.10)$$

The output, thus, consists of the following four terms as shown in Fig 4.8.

(a) The first term $U_1(x,y)$ is merely a plane wave of uniform amplitude travelling along the axis, or in other words attenuated direct beam.

(b) The second term $U_2(x,y)$ corresponds to a halo around

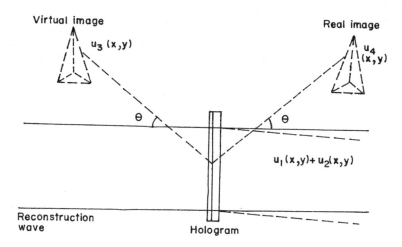

Fig. 4.8. Reconstruction of an off-axis hologram.

the direct beam. If the spatial frequency content of the object is not too large, the angular spread due to the second term is small.

(c) The third term $U_3(x,y)$ is proportional to object wavefront $O(x,y)$ multiplied by a linear exponential factor. Proportionality indicates the generation of a virtual image of the object at the same distance from the hologram. The linear exponential factor $\exp(j2\pi\alpha y)$ indicates that this image is deflected off the axis at an angle θ.

(d) The fourth term $U_4(x,y)$ gives rise to the conjugate image, identical to the object wavefront except that it is of opposite curvature. This results in a real image formed at the same distance from the hologram in front of it. The linear exponential factor $\exp(-j2\pi\alpha y)$ indicates that this image is deflected off the axis at an angle θ in the opposite direction.

Thus we are able to generate virtual and real images angularly separated from each other and from the direct beam also.

The Angle θ

In order to estimate the minimum angle θ or the minimum

262 LASERS AND HOLOGRAPHY

carrier frequency α so that the zero order terms which appear on axis do not interfere with the image terms, we consider the Fourier spectra of various terms.

The one dimensional Fourier transform of the object has a bandwidth of 2B as shown in Fig 4.9. The Fourier transforms of various terms in the hologram reconstruction are

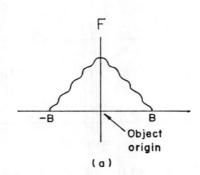

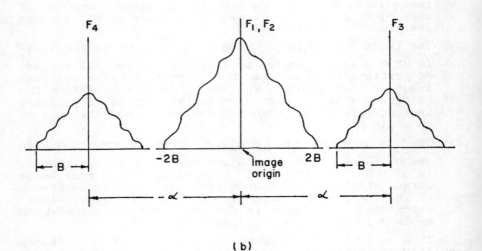

Fig. 4.9. One dimensional Fourier transform of (a) the object and (b) the hologram.

$$\mathcal{F}[t_b] = t_b \delta(u,v) = F_1,$$

$$\mathcal{F}[\beta|O(x,y)|^2] = \beta F_o^*(u,v) F_o(u,v) = F_2,$$

$$\mathcal{F}[\beta R|O(x,y)| \exp[-j\phi(x,y)] \exp(j2\pi\alpha y) = \beta R F_o(u,v-\alpha) = F_3,$$

$$\mathcal{F}[\beta R|O(x,y)| \exp[j\phi(x,y)] \exp(-j2\pi\alpha y) = \beta R F_o(-u,-v-\alpha) = F_4,$$

(4.11)

where $\mathcal{F}[|O(x,y)|] = F_o(u,v)$ is the Fourier transform (spectrum) of the object.

We see that F_1 is simply a δ function at the origin of the Fourier plane. F_2 is the autocorrelation function of F_o and extends twice the extent of the object spectrum. $|F_3|$ and $|F_4|$ are proportional to F_o but are displaced from the centre frequency by $(0,\alpha)$ and $(0,-\alpha)$ respectively. From Fig. 4.9 we see that F_3 and F_4 will not interfere with F_1 and F_2 if the carrier frequency

$$\alpha \geq 3B, \qquad (4.12)$$

where B is the highest spatial frequency of the object. Therefore the minimum angle θ_{min} is given by

$$\sin\theta_{min} \geq 3B\lambda$$

$$\theta_{min} \geq \sin^{-1}(3B\lambda). \qquad (4.13)$$

Eq. 4.13 specifies the condition necessary to guarantee that the two images will be separated in space and will not interfere with each other.

4.5. Holographic Imaging Equations

Let us consider an object point at a distance R_o from the centre of the hologram as shown [4.15] in Fig. 4.10. The object point (R_o, α_o, β_o) scatters spherical wave of complex amplitude U_o at a point (x,y) on the hologram

$$U_o = a_o \exp[-jk_o(r_o + S_o)]. \qquad (4.14)$$

Here $k_o = 2\pi/\lambda_o$ is the wave number of the laser light, λ_o is the wavelength of light used for hologram recording, r_o is

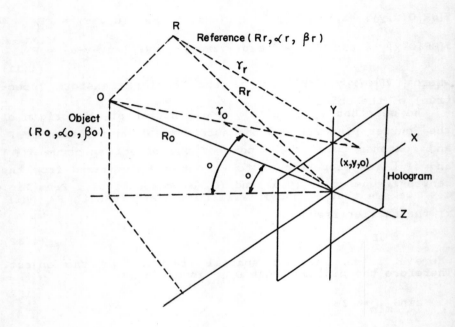

Fig. 4.10. Coordinate system used to study holographic imaging equations.

the distance from the object point $t_o(x,y)$ and s_o is the distance of the object point from the reference point. The distance r_o can be expressed as

$$r_o = [z_o^2 + (x - x_o)^2 + (y - y_o)^2]^{1/2}$$
$$= (x^2 + y^2 - 2xx_o - 2yy_o + R_o^2)^{1/2}. \qquad (4.15)$$

r_o can be expanded in a binomial series about R_o if we assume that R_o^2 is large compared to the remaining terms in the above expression.

$$r_o = R_o + \frac{x^2 + y^2}{2R_o} - \frac{xx_o + yy_o}{R_o} + \cdots . \qquad (4.16)$$

The complex amplitudes of the reference wave (R_r, α_r, β_r) and the reconstruction wave are similarly given by

$$U_r = a_r \exp[-jk_r r_r]$$
$$U_c = a_c \exp[-jk_c r_c]. \qquad (4.17)$$

For coherence $k_r = k_o$.

The complex amplitude of the hologram exposing wave is

$$U_o + U_r = a_o \exp[-jk_r(r_o + S_o)] + a_r \exp(-jk_r r_r). \qquad (4.18)$$

After the plate is developed, the image forming wave is given by

$$U_{virtual} = U_c U_r^* U_o$$

$$= a_1 \exp(-jk_c r_c) \exp[-jk_r(r_o + S_o - r_r)]. \qquad (4.19)$$

Eq. 4.19 represents the complex amplitude of the primary image i.e. the virtual image of the original object point. For a sufficiently thin hologram or for sufficiently small angles $(\alpha_o - \alpha_r)$ and $(\beta_o - \beta_r)$, a conjugate image i.e. the real image is also observed whose complex amplitude is given by

$$U_{real} = a_1 \exp(-jk_c r_c) \exp[jk_r(r_o + S_o - r_r)]. \qquad (4.20)$$

Eqs. 4.19 and 4.20 are similar to spherical waves originating from an image point and can be written in the form

$$U_i = a_i \exp(-jk_c r_i). \qquad (4.21)$$

The image is formed with the light of wavelength λ_c. The location of the image point can be determined by equating the exponent of Eq. 4.21 with the exponent of the Eq. 4.19.

$$k_c r_i = k_c r_c + k_r(r_o + S_o - r_r). \tag{4.22}$$

Substituting the values of r_i, r_c and r_o from Eq. 4.16 and similar expressions for r_c and r_i, we get

$$k_c\left(R_i + \frac{x^2+y^2}{2R_i} - \frac{xx_i + yy_i}{R_i} + ----\right)$$

$$= k_c\left(R_c + \frac{x^2+y^2}{2R_c} - \frac{xx_c + yy_c}{R_c} + ----\right)$$

$$+ k_r\left(R_o + S_o - R_r + \frac{x^2+y^2}{2R_o} - \frac{x^2+y^2}{2R_r}\right.$$

$$\left. - \frac{xx_o + yy_o}{R_o} + \frac{xx_r + yy_r}{R_r} + ---\right). \tag{4.23}$$

Termination of Eq. 4.16 to the third term would result in an approximation which is less stringent than the paraxial approximation.

Equating the coefficients of like powers of x and y appearing on left and right hand sides in Eq. 4.23 yields

$$k_c\left(\frac{1}{R_i}\right) = k_c\left(\frac{1}{R_c}\right) \pm k_r\left(\frac{1}{R_o}\right) \mp k_r\left(\frac{1}{R_r}\right), \tag{4.24}$$

$$k_c\left(\frac{x_i}{R_i}\right) = k_c\left(\frac{x_c}{R_c}\right) \pm k_r\left(\frac{x_o}{R_o}\right) \mp k_r\left(\frac{x_r}{R_r}\right), \tag{4.25}$$

$$k_c\left(\frac{y_i}{R_i}\right) = k_c\left(\frac{y_c}{R_c}\right) \pm k_r\left(\frac{y_o}{R_o}\right) \mp k_r\left(\frac{y_r}{R_r}\right), \tag{4.26}$$

where plus sign refers to the primary image and the minus sign to the conjugate image. But

$$\frac{k_r}{k_c} = \frac{\lambda_c}{\lambda_r} = \mu \text{ (say)},$$

$$\frac{x_i}{R_i} = \sin\alpha_i,$$

$$\frac{y_i}{R_i} = \cos\alpha_i \cdot \sin\beta_i. \tag{4.27}$$

Similar expressions can be found for other terms of Eqs. 4.25 and 4.26.

Eqs. 4.25 and 4.26 become with the help of Eq. 4.27

$$\frac{1}{R_i} = \frac{1}{R_c} \pm \mu\left(\frac{1}{R_o} - \frac{1}{R_r}\right), \tag{4.28}$$

$$\sin\alpha_i = \sin\alpha_c \pm \mu(\sin\alpha_o - \sin\alpha_r), \tag{4.29}$$

$$\cos\alpha_i \sin\beta_i = \cos\alpha_c \sin\beta_c \\ \pm \mu(\cos\alpha_o \sin\beta_o - \cos\alpha_r \sin\beta_r). \tag{4.30}$$

Eqs. 4.28 - 4.30 give the location (R_i, α_i, β_i) of the holographic image of the object point (R_o, α_o, β_o).

4.6. Image Magnification

Lateral Magnification

The lateral magnification of the image is given by

$$M_{lat} = \frac{dx_i}{dx_o} = \frac{dy_i}{dy_o}. \tag{4.31}$$

Eqs. 4.24 - 4.26 give

$$M_{lat} = \left[1 - \frac{R_o}{R_r} \mp \frac{R_o}{\mu R_c}\right]. \tag{4.32}$$

Unit magnification ($M_{lat}=1$) is obtained when $\lambda_o=\lambda_c$ and $R_c=R_r$ for virtual image, and when $\lambda_o=\lambda_c$ and $R_c=-R_r$ for real image. M is also equal to 1 when $R_r=R_c=\infty$ when both the reference and the reconstruction beams are collimated.

Longitudinal Magnification

The longitudinal magnification is given by

$$M_{long} = \pm \frac{1}{\mu} M_{lat}^2. \tag{4.33}$$

Angular Magnification

When the observer's eyes are located in the hologram plane, the angular magnification is given by

$$M_{ang} = \mu \qquad (4.34)$$

for primary and conjugate images.

4.7. Hologram Aberrations

Holograms suffer from aberrations caused by a change in the wavelength from construction to reconstruction and also by a mismatch (eg. angular, spherical etc.) in the reference and reconstruction beams. Both the chromatic and nonchromatic aberrations are quite important even when only small deviation from the recording geometry are present in the reconstruction geometry. The condition that will eliminate all the aberrations simultaneously is to duplicate exactly one construction beam in the reconstruction process, provided that there is no shrinkage in the emulsion during processing.

The third, fifth and seventh order aberrations appear due to the imperfect match between the third, fifth and seventh order phase terms of the equations representing electromagnetic field variations in the holographic image (Eqs. 4.28-4.30) and the corresponding phase terms of the true spherical wave [4.16].

Following Champagne [4.17], Champagne and Massey [4.18], Latta [4.19, 4.20], Meier [4.15] and Mehta *et al.* [4.21] the various aberrations may be written as follows.

Third Order Aberrations

$$\Delta_3 = \frac{1}{\lambda_c}\left[-\frac{1}{8}(x^2+y^2)S + \frac{1}{2}(x^2+y^2)(xC_x + yC_y)\right.$$
$$\left. - \frac{1}{2}(x^2 A_x + y^2 A_y + xy A_x A_y)\right], \qquad (4.35)$$

where x,y are the hologram coordinates where the aberrations are to be evaluated, S, C and A denote the spherical aberration, coma and astigmatism, respectively.

Fifth Order Aberrations

$$\Delta_5 = \frac{1}{\lambda_c}\left[\frac{1}{16}(x^2+y^2)^3 A - \frac{3}{8}(x^2+y^2)^2(xB+yC)\right.$$
$$\left. + \frac{3}{4}(x^2+y^2)(x^2 D + y^2 E + 2xyF)\right.$$

$$- \frac{1}{2} (x^3G + 3x^2yH + 3xy^2I + y^3J) \Big]. \tag{4.36}$$

Various coefficients A to J have been given by Latta [4.19, 4.20].

Seventh Order Aberrations

$$\Delta_7 = \frac{1}{\lambda_c} \Big[- \frac{5}{128} (x^2 + y^2)^4 K + \frac{5}{6} (x^2 + y^2)^3 (xL + yM)$$

$$- \frac{15}{16} (x^2+y^2)^2 (x^2N + y^2O + 2xyP)$$

$$+ \frac{5}{4} (x^2 + y^2) (x^3Q + 3x^2yR + 3xy^2T + y^3U)$$

$$- \frac{5}{8} (x^4V - 4x^3yW + 6x^2y^2X - 4xy^3Y + y^4Z) \Big]. \tag{4.37}$$

Various coefficients A to Z in Eq. 4.37 have given by Mehta et al. [4.21, 4.22]. The coefficients of third, fifth and seventh order aberrations are of similar nature and may be expressed in a general form.

The spherical aberration disappears when $R_r = R_o$ because in that case the curvature of both the interfering beams is the same, hence the phase difference between them changes linearly across the field. This results in straight fringes which cannot introduce spherical aberration. Coma disappears when $R_r = R_o$ and $R_c = \pm R_o$. Astigmatism vanishes when $R_r = R_o$ and $R_o = \mu R_o$. Thus the condition for coma and astigmatism to disappear simultaneously is $\mu = 1$.

A comparison of third, fifth and seventh order aberrations for a holographic optical element with the deviation in the reconstruction angle for various values of f/no. is made in Fig. 4.11. It is observed that the magnitudes of the aberrations decrease rapidly as the value of the f/no. increases. When the holograms are to be used as optical elements, higher order aberrations should also be considered along with the third order to achieve ideal imagery as the aberrations are quite large even when only small deviations from the recording geometry are present in the reconstruction step [4.21, 4.22].

4.8. Orthoscopic and Pseudoscopic Images

A hologram reconstructs two images, one real and the other virtual which are exact replicas of the object. However, the two images differ in appearance to the observer. The virtual image is produced at the same position as the

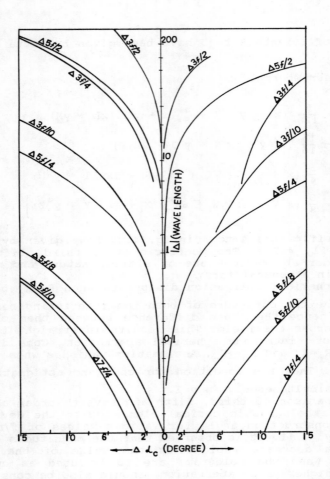

Fig. 4.11. Third, fifth and seventh order aberrations for a holographic optical element with the deviation in the reconstruction angle for various values of f-numbers [4.21].

object and has the same appearance of depth and the parallax as the original three dimensional object. The virtual image appears as if the observer is viewing the original object through a window defined by the size of the hologram. This image is known as orthoscopic image.

The real image is also formed at the same distance from the hologram, but in front of it. In the real image, however the scene depth is inverted. This can be understood with the help of Fig. 4.12. This is due to the fact that the corresponding points on the two images (virtual and real) are located at the same distances from the hologram. The real image is known as pseudoscopic image [4.10] and does not give a pleasing sensation as we do not come across objects with inverted depths in normal life. Such images were not possible to be formed by other optical techniques

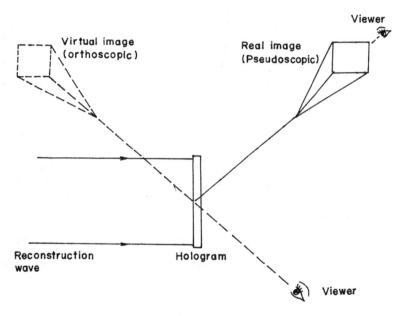

Fig. 4.12. Orthoscopic and pseudoscopic images from a hologram.

till recently. It is now possible to produce such conjugate wavefronts in real-time by optical phase conjugation techniques. Such wavefronts have potential applications in correcting the effects of distorting media on optical imageries [4.23].

A hologram recorded by a lens or a concave mirror [Fig. 4.13] produces an orthoscopic real image of the object.

It is also possible to produce an orthoscopic real image of an object by recording two holograms in succession [4.24]. In the first step, a primary hologram is recorded by a collimated reference beam. This hologram, when recons-

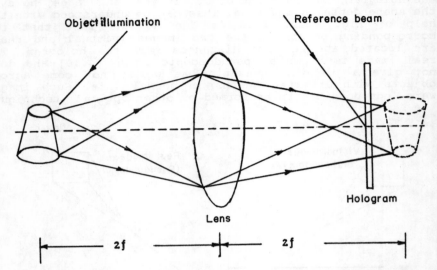

Fig. 4.13. Optical system for the recording of a hologram using a convex lens which reconstructs an orthoscopic real image.

tructed by a collimated beam produces an orthoscopic virtual image and a pseudoscopic real image with unit magnification. The final hologram is recorded using the real image of the primary hologram as the object beam. When this hologram is reconstructed it produces a pseudoscopic virtual image and an orthoscopic real image (Fig. 4.14).

4.9. Classification of Holograms

Holograms may be classified in a number of different ways depending on their thickness, method of recording, method of reconstruction etc.

4.9.1. Amplitude and Phase Holograms

A hologram may be of an absorption type which produces a change in the amplitude of the reconstruction beam. The phase type hologram produces phase changes in the reconstruction beam due to a variation in the refractive index or thickness of the medium. The absorption and phase type holograms correspond respectively to the amplitude modulation (AM) and phase modulation (PM) of a temporal signal in the communication theory. Phase holograms have the advantage over amplitude holograms of no energy dissipation within the

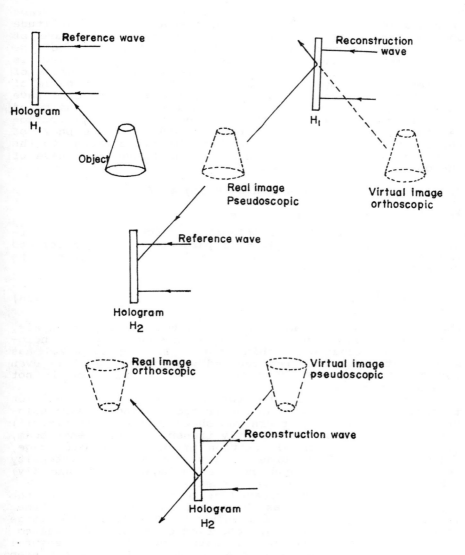

Fig. 4.14. Optical system for the recording of a hologram in two steps which reconstructs an orthoscopic real image and a pseudoscopic virtual image [after 4.24].

hologram medium and higher diffraction efficiency. Holograms recorded in photographic emulsions change both the amplitude and the phase of the illuminating wave. In the case of reflection holograms (section 4.9.3), at the maximum of the energy distribution the density of silver deposited is related to the amplitude of the object wave. The shape of the recorded fringe planes depend on the relative phase of the interfering beams. Consequently the reconstructed wave is reflected from the hologram according to the density of the silver deposited with the amplitude variation proportional to the amplitude of the object. Similarly the phase of the reconstruction wave is modulated in proportional to the phase of the object wave. Thus both amplitude and phase of the object wave are reproduced.

4.9.2. Classification based on Hologram Thickness

Thin Holograms or Plane Holograms

Holograms may be thin (plane) or thick (volume). According to Klein and Cook [4.25] a parameter Q can be defined for distinguishing between a thin and a thick hologram. The parameter Q is defined as

$$Q = \frac{2\pi\lambda t}{nd^2}, \qquad (4.38)$$

where t is the hologram thickness, λ the wavelength in air, n the refractive index and d the fringe spacing. The hologram may be treated as thin if Q < 1. Kasper [4.26] has shown that thick hologram coupled wave theory applies even for Q values of the order of 1. Thus Q criterion is not always sufficient.

A hologram may also be regarded as thin if its emulsion thickness is much less than the fringe spacings. Such holograms produce several orders, as shown in Fig. 4.15 viz. (i) zero order which is the directly transmitted reference beam, (ii) the first order diffraction producing virtual image, (iii) the minus first order diffraction equal in intensity to the first order producing the conjugate image and (iv) higher orders of decreasing intensity.

A hologram may be treated as a plane hologram if the conjugate image is at least half as bright as the true order is diffracted even for small values of Q, whereas for large modulation higher orders are observed even for large values of Q. In some materials like lithium niobate crystal several diffracted orders are possible to obtain [4.27-4.29] even for large values of Q. In general, there is no clear cut regime for plane and volume holograms. There is a gradual transition in the properties of the two holograms as the medium thickness and the angle between the object and reference beams are increased. However, it has been shown [4.27,

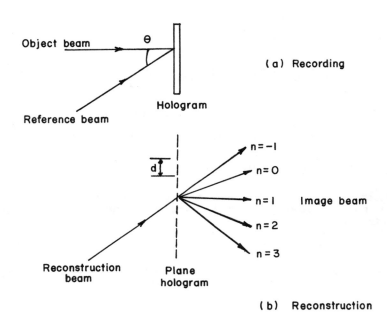

Fig. 4.15. Recording (a) and reconstruction (b) of a thin (plane) hologram. The hologram produces several diffraction orders.

4.28] that for small levels of modulation only one order is diffracted even for small values of θ, whereas for large modulation higher orders are observed even for large values of Q. In some materials like lithium niobate crystal several diffracted orders are possible to obtain [4.27-4.29] for larger values of Q.

In order to calculate the peak diffraction efficiency that may be obtained from a thin amplitude holographic grating, we assume that the amplitude transmittance is proportional to the intensity in the interference pattern. The amplitude transmittance of the grating is then given by

$$|t_A(x)| = t_o + t_1 \cos\left(\frac{2\pi}{d}\right)x, \tag{4.39}$$

where t_o is the average amplitude transmittance of the grating, t_1 is the amplitude transmittance of the spatial variation of $|t_A(x)|$ and d is the spacing of the fringes. The values of $|t_A(x)|$ lies between 0 and 1. The diffracted amplitude will be maximum when

$$|t_A(x)| = \frac{1}{2} + \frac{1}{2}\cos\left(\frac{2\pi x}{d}\right)$$

$$= \frac{1}{2} + \frac{1}{4}\exp\left[\frac{j2\pi x}{d}\right] + \frac{1}{4}\exp\left[-\frac{j2\pi x}{d}\right]. \tag{4.40}$$

The maximum amplitude of each of the diffracted orders is one fourth of the amplitude of the reconstruction wave. The peak diffraction efficiency η_{max} is therefore

$$\eta_{max} = (1/16) = 0.0625. \tag{4.41}$$

Similarly we can calculate the peak diffraction efficiency of a thin phase grating by writing the complex amplitude transmittance as

$$t_A(x) = \exp[-j\phi(x)]. \tag{4.42}$$

The phase shift produced is proportional to the intensity in the interference pattern, hence

$$t_A(x) = \exp\left\{-j\left[\phi_0 + \phi_1\cos\left(\frac{2\pi x}{d}\right)\right]\right\}.$$

$$= \exp(-j\phi_0)\exp\left[-j\phi_1\cos\left(\frac{2\pi x}{d}\right)\right]. \tag{4.43}$$

In Eq. 4.43 the constant phase term $\exp(-j\phi_0)$ may be neglected. Then

$$t_A(x) = \exp\left[-j\phi_1\cos\left(\frac{2\pi x}{d}\right)\right],$$

$$= \sum_{n=-\infty}^{\infty} j^n J_n(\phi_1)\exp\left[jn\frac{2\pi x}{d}\right], \tag{4.44}$$

where J_n is the Bessel function of the first kind of order n. The amplitude of light diffracted in the first order is proportional to $J_1(\phi_1)$. The value of $J_1(\phi_1)$ first increases with phase modulation and then decreases. Thus the diffraction efficiency of a thin phase grating is given by [4.30]

$$\eta = J_1^2(\phi)$$

and
$$\eta_{max} = 0.339. \qquad (4.45)$$

Volume Holograms

A volume (thick) hologram may be regarded as a superposition of three dimensional gratings recorded in the depth of the emulsion each satisfying the Bragg law given by

$$2\Lambda \sin \theta = p\lambda, \qquad (4.46)$$

where θ is the angle of the incident light beam with the grating planes, p the order of diffraction, Λ the spacing between grating planes and λ is the wavelength of the diffracted light. The grating planes in a volume hologram produce maximum change in refractive index and/or absorption index. A consequence of Bragg condition is that the volume hologram reconstructs the virtual image at the original position of the object if the reconstruction beam exactly coincides with the reference beam. However, the conjugate image and higher order diffractions are absent as shown in Fig. 4.16. Thus there is a unique direction of

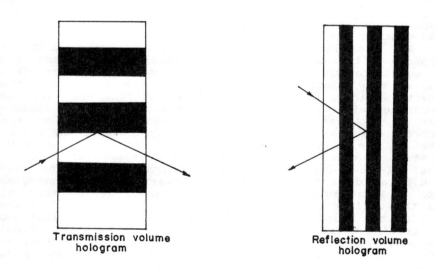

Fig. 4.16. Reconstruction from volume transmission and volume reflection holograms.

reconstruction beam for reconstructing the image. If the illuminating beam deviates from the Bragg angle, the brightness of the image is reduced considerably. For example, a 15 μm thick hologram will reduce the image brightness by 10 dB if the angle of hologram illuminating beam deviates from Bragg angle by $\pm 5°$. In the case of 1500 μm thick hologram, the image brightness is reduced by more than 10 dB if the deviation is ± 2 min of arc.

This property can be used to record a large number of objects in a single hologram with different reference beams either in direction of incidence or in wavelength. The images can be reconstructed without cross-talk by using reconstruction beams with the same direction and wavelength as the corresponding reference beams (section 4.14.3).

It may be mentioned that if the interference pattern is not recorded in depth, the hologram will be a thin hologram even if the emulsion is thick. It is the use of the volume of the emulsion which allows to produce only one image at a time.

Russell [4.31] has analyzed the characteristics of generalized volume gratings. Kogelnik [4.30] has used a coupled wave approach to analyze the diffraction of light by volume gratings. When a thick hologram is illuminated by a beam of light, the amplitude of the diffracted wave increases progressively decreasing the amplitude of the incidence wave as they propagate through the grating. Using coupled wave theory the relations for the efficiencies for a volume transmission phase grating and a volume transmission amplitude grating can be found.

In a volume hologram, the interference patterns recorded in the material may have effective widths less than or equal to the thickness of the material. Kogelnik's coupled wave theory is based on the assumption that the effective widths of the interference patterns are large in comparison with the thickness of the material. This assumption is, however, no longer valid when the size of the recorded hologram is of the order of magnitude of the thickness of the material as in the case of photorefractive crystals. Dubious *et al.* [4.32] have developed an integral model of the diffraction process which predicts the same results as does the coupled wave theory when the hologram size is large as compared to the thickness of the material and the diffraction efficiency is less than 25%. When the size of the hologram is smaller and when the recording and the reading wavelengths are different, a new spatial effect has been predicted which explains the experimental observations [4.33] in photorefractive crystals. The theory needs generalization to involve larger diffraction efficiency.

Volume Transmission Phase Holograms

In this case the efficiency is given by

$$\eta = \sin^2\xi, \quad (4.47)$$

where $\xi = \dfrac{\pi n_1 t}{\lambda \cos\theta}$.

In Eq. 4.47 n_1 is the variation of the refractive index of the medium, t is the emulsion thickness and θ is the Bragg angle. The diffraction efficiency increases as the modulation, ξ increases and becomes maximum at $\xi=\pi/2$. At this point $\eta=1.00$ i.e. all the incident energy is coupled to the diffracted beam. For values of $\xi>1$, energy couples back into the incident wave and the efficiency drops. Thus

$$\eta_{max} = 1.00.$$

For a fixed value of the modulation parameter ξ the efficiency is sensitive to the angular and wavelength deviations $\Delta\theta$ and $\Delta\lambda$ from the Bragg condition and is given by [4.30]

$$\eta = \dfrac{\sin^2(\xi^2 + \chi^2)^{1/2}}{(1 + \chi^2/\xi^2)}, \quad (4.48)$$

where

$$\chi = \Delta\theta \left(\dfrac{2\pi}{\Lambda}\right)\left(\dfrac{t}{2}\right)$$

$$= -\Delta\lambda \left(\dfrac{2\pi}{\Lambda}\right)^2 t/(8\pi n_o \cos\theta_o). \quad (4.49)$$

Figure 4.17 shows [4.30] the variation of the normalized diffraction efficiency of a lossless transmission phase grating as a function of angular and wavelength deviation (i.e. the parameter χ) from the Bragg condition for three values of the modulation parameter $\xi=\pi/4$, $\pi/2$ and $3\pi/4$. It is observed that for a fixed value of ξ the efficiency drops drastically if there is a slight deviation from Bragg condition.

Volume Transmission Amplitude Holograms

In this case the efficiency is given by

$$\eta = \exp[-2D_o \sinh^2(D_1/2)], \quad (4.50)$$

where

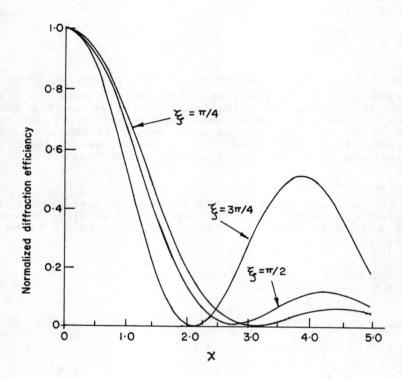

Fig. 4.17. Normalized diffraction efficiency as a function of χ for different values of the modulation parameter ξ for a volume transmission hologram [4.30].

$$D_o = \frac{\alpha_o t_o}{\cos\theta_o},$$

$$D_1 = \frac{\alpha_1 t}{\cos\theta_o}. \tag{4.51}$$

Here α_o is absorption constant of the grating and α_1 is the variation in the absorption constant. D_1 is a measure of the amplitude of the modulation and $D_1/D_o = \alpha_1/\alpha_o$ is the relative depth of modulation. The maximum diffraction efficiency is obtained when $\alpha_1 = \alpha_o$

$$\eta_{max} = (\frac{1}{27}) = 0.037. \qquad (4.52)$$

4.9.3. Classification based on Direction of Reconstructed Image

Holograms may be transmission type or reflection type. In the former type, the image beam is transmitted through the hologram while in the later, the image is reflected. In a transmission hologram the fringes tend to run across the emulsion where as in the reflection hologram the fringes are parallel to the emulsion (i.e. the substrate). The holograms discussed in section 4.9.2 are of transmission type.

A relief hologram on a transparent substrate can reconstruct either as a transmission or a reflection hologram [4.34]. The ratio of the hologram efficiencies for the two types of reconstructions is $\eta_{trans}/\eta_{ref} = n$ the refractive index of the medium. If the hologram surface relief is metallic, the hologram efficiency in reflection can greatly exceed the efficiency in transmission. A reflection hologram can also reconstruct in transmission [4.35].

Figure 4.18 shows the interference between waves from two point sources X and Y which consists of a family of hyperboloids. Depending on the position of the recording material, a transmission or a reflection hologram is recorded. If a photographic plate is exposed at the position A and B, a transmission hologram will be produced. A reflection hologram is produced if the hologram is recorded at the position C.

Reflection Holograms

Reflection holograms are recorded by a coherent light but can be reconstructed either by a coherent light or by a white light illumination. These holograms are known as white light holograms. The recording is in the depth of the emulsion of the hologram. The method is similar to Lippmann method of colour photography and was first introduced by the then Soviet Scientist Denisyuk [4.36].

The reflection hologram is recorded by taking the object and reference beams from opposite side of the recording medium (Fig. 4.19a,c). These beams interfere within the volume of the medium and produce stationary waves. At points where the wavefronts cross each other, constructive interference will result. At these points the intensity will be higher than the surrounding area. When the recording medium is developed, the silver density will be higher at these points. The relatively high silver density will result in partially reflecting planes. The developed medium contains nearly parallel stratified layers of metallic silver,

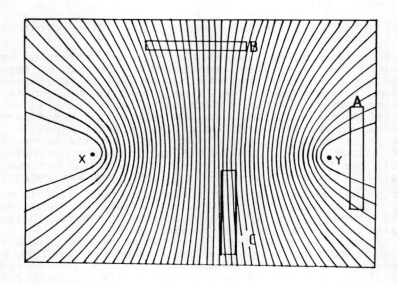

Fig. 4.18. Interference between waves from two point sources X and Y.

similar to a Bragg lattice, with the silver layers acting as Bragg planes (Fig. 4.19c). The spacing Λ between successive silver planes is given by the Bragg condition

$$2\Lambda \sin(\theta/2) = \lambda, \qquad (4.53)$$

where $\theta/2$ is the angle at which the two beams impinge upon the emulsion with respect to the surface normal and Λ is the grating spacing. When the hologram is illuminated with a white light source, only the wavelength that satisfies the Bragg condition will be reflected while the rests are either transmitted or absorbed by the emulsion (Fig. 4.19d).

Figure 4.19 shows the recording and reconstruction of a reflection hologram. When the fringes run exactly or closely parallel to the surface, the hologram acts as a multilayer dielectric coated interference filter and results in a narrow band reflector. When $\theta=180°$, the mirror planes (Bragg) are spaced by $\lambda/2$ within the emulsion. For very narrow band reflectivities, the emulsion must be thick

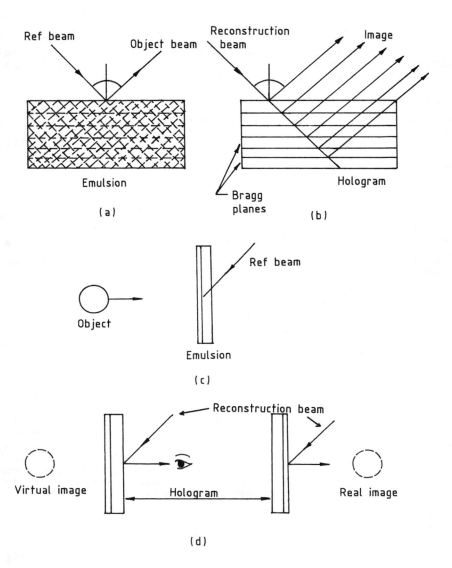

Fig. 4.19. Recording (a,c) and reconstruction (b,d) of a reflection hologram.

enough to produce a large number of such mirror planes. The commercially available recording medium can produce about 15 to 30 reflecting planes which are sufficient for a white light reconstruction.

From the coupled wave theory the relations for the efficiencies for a volume reflection phase grating and a volume reflection amplitude grating can be found [4.30].

Volume Reflection Phase Hologram

In this case the efficiency is given by

$$\eta = \tanh^2 \xi, \tag{4.54}$$

where

$$\xi = \frac{\pi n_1 t}{\lambda \cos\theta}. \tag{4.55}$$

The diffraction efficiency increases asymptotically as the value of ξ increases, $\eta_{max} = 1.00$. The efficiency is sensitive to angular and wavelength deviations $\Delta\theta$ and $\Delta\lambda$ from the Bragg condition and is given by [4.30]

$$\eta = [1 + (1-\chi^2/\xi^2)/\sinh^2(\xi^2-\chi^2)^{1/2}]^{-1}, \tag{4.56}$$

where $\chi = \Delta\theta \left(\frac{2\pi n_o t}{\lambda}\right) \sin\theta_o$,

$$= \left(\frac{\Delta\lambda}{\lambda}\right)\left(\frac{2\pi n_o t}{\lambda}\right)\cos\theta_o. \tag{4.57}$$

Figure 4.20 shows [4.30] the variation of the normalized diffraction efficiency of a lossless reflection phase grating as a function of angular and wavelength deviation (i.e. the parameter χ) from the Bragg condition for the modulation parameter $\xi = \pi/4$, $\pi/2$ and $3\pi/4$. It is observed that for a fixed value of ξ, the efficiency drops drastically if there is a slight deviation from Bragg condition.

Volume Reflection Amplitude Hologram

In this case the efficiency is given by

$$\eta = [2 + 3^{1/2} \coth(3^{1/2}\xi)]^{-2}, \tag{4.58}$$

$$\eta_{max} = [2 + 3^{1/2}]^{-2} = 0.072. \tag{4.59}$$

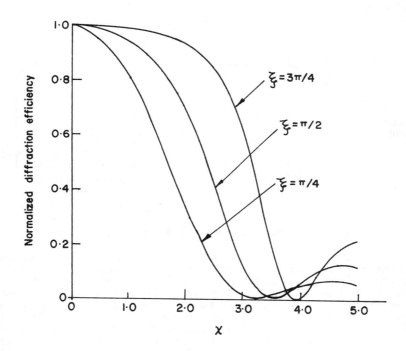

Fig. 4.20. Normalized diffraction efficiency as a function of χ for different values of the modulation parameters ξ for a volume reflection hologram [4.30].

4.9.4. Classification According to Recording Arrangement

Holograms may be classified according to the type of diffracted field associated with the object and the distance between the object and the recording plane. The hologram may be recorded with a plane, diverging or converging reference beam.

Fresnel Hologram

The hologram is recorded by interfering the Fresnel diffraction wave from the object with a plane reference wave i.e. the object is at some finite distance from the recording medium and the reference source is at infinity. With an

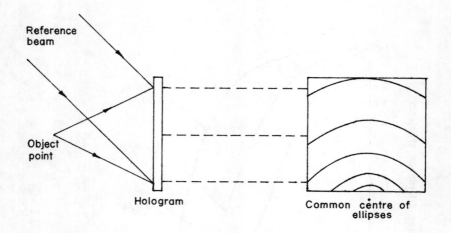

Fig. 4.21. Recording of a Fresnel hologram.

off-axis reference wave a point on the object produces interference maxima which lie on paraboloids of revolution with the object point as common focus. The intersection of these with the plane of the recording medium produces a system of ellipses similar to a zone plate pattern. The fringe spacing decreases with increasing distance from the axis of the paraboloids (Fig. 4.21). The hologram is reconstructed by a plane beam. The focussed image has a unit magnification. In such a hologram, the image moves with the motion of the hologram. A detailed mathematical treatment is given by DeVelis and Reynolds [4.37].

Fraunhofer Hologram

The hologram is recorded by interfering the Fraunhofer diffraction pattern of the object formed in the far field with a plane reference wave. The Fraunhofer holography with collinear reference wave was developed for the application of particle size analysis [4.13].

The object may be placed in the focal plane of a lens placed between the object and the recording medium. The interference pattern for an object point with the plane reference beam is a system of parallel planes. The intersection of this with a plane of the recording medium is a line grating with sinusoidal intensity variation (Fig. 4.22). The spacing and orientation of this grating structure is uniquely correlated with the orientation of the object relative to the reference beam. The hologram is recons-

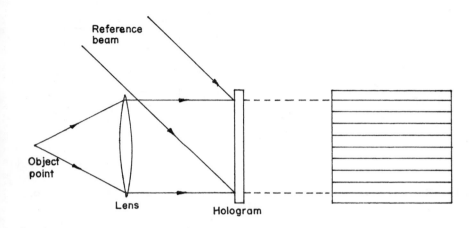

Fig. 4.22. Recording of a Fraunhofer hologram.

tructed with a plane reference beam. The image does not move with the motion of the hologram.

For particle size analysis the hologram is recorded such that the particle distance from the hologram plate must satisfy the far field diffraction pattern condition

$$z_o \gg \frac{(x_o^2 + y_o^2)}{\lambda}, \qquad (4.60)$$

where x_o and y_o define the size of the particle. For point scatterers such as aerosol particles, z_o of the order of a few millimeters ensures far field condition [4.13].

Lensless Fourier Transform Hologram

Holograms having properties similar to those of Fourier transform holograms which are recorded by lenses can be made without the use of lenses by a method first proposed by Stroke et al. [4.38]. A lensless Fourier transform hologram is produced when the object and reference source are placed equidistant from the recording medium. A single point on the object will produce hyperboloids of revolution with the object and reference points as common foci (Fig. 4.23).

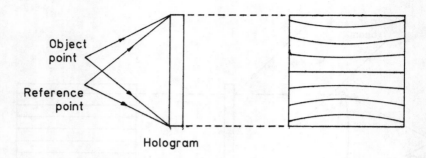

Fig. 4.23. Recording of a lensless Fourier transform hologram of a point object.

Since the object and reference points are equidistant from the recording plane, the sphericities of the two spherical waves at the recording plane tend to cancel producing a plane grating type of fringes. It may be noted that the lensless Fourier hologram is a modification of Fraunhofer and Fourier holograms with the important feature that it yields image immobility without the use of a lens.

Consider the general case described in Fig 4.24. Let the reference point source be situated on the axis at a distance z from the recording medium. Its complex amplitude $R(x)$ is given by

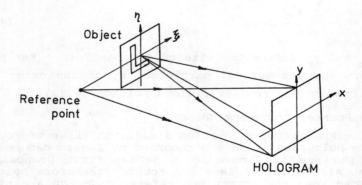

Fig. 4.24. Recording of a lensless Fourier transform hologram of an object.

$$R(x) = R\exp(\tfrac{j\pi}{\lambda z} x^2). \tag{4.61}$$

The object is situated at a distance ξ_o from the centre. Each point of the object sends a spherical wave with a complex amplitude

$$O\exp[\tfrac{j\pi}{\lambda z}(x-\xi)^2]. \tag{4.62}$$

For a small object and recording on a large photographic plate

$$(x-\xi)^2 \simeq x^2 - 2x\xi.$$

The total object field produced by the object is

$$O(x) = \int_{\xi_1}^{\xi_2} O\exp[\tfrac{j\pi}{\lambda z}(x-\xi)^2]d\xi,$$

$$= \int_{\xi_1}^{\xi_2} O\exp[\tfrac{j\pi}{\lambda z}x^2]\exp[-\tfrac{j\pi}{\lambda z}2x\xi]d\xi. \tag{4.63}$$

In Eq. 4.63 a multiplicative phase factor has been omitted.
The intensity of the interference pattern formed by the object and reference waves as recorded in the hologram is

$$I(x) = |R(x) + O(x)|^2,$$

$$= |R(x)|^2 + |O(x)|^2 + R^*(x)O(x) + R(x)O^*(x). \tag{4.64}$$

The terms of interest are the last two terms. We assume that the photographic plate is developed linearly to yield a transparency with amplitude transmittance proportional to the exposure. The image forming term is thus

$$R^*(x)O(x) = R\exp[-\tfrac{j\pi}{\lambda z}x^2]O\int_{\xi_1}^{\xi_2}\exp[\tfrac{j\pi}{\lambda z}x^2]\exp[-\tfrac{j\pi}{\lambda z}2x\xi]d\xi,$$

$$= R\int_{\xi_1}^{\xi_2} O\exp[-\tfrac{j\pi}{\lambda z}2x\xi]d\xi. \tag{4.65}$$

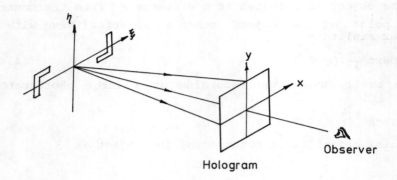

Fig. 4.25. Reconstruction of virtual images from a lensless Fourier transform hologram.

Similarly the fourth term in Eq. 4.64 will produce a term which is complex conjugate of Eq. 4.65

$$R(x)O^*(x) = R \int_{\xi_1}^{\xi_2} O \exp[\frac{j\pi}{\lambda z}2x\xi]d\xi. \qquad (4.66)$$

Eqs. 4.65 and 4.66 are the Fourier transforms of the object and its complex conjugate, respectively without a multiplicative phase factor. To reconstruct the virtual images, the hologram is illuminated by a point source as shown in Fig 4.25. Two images are obtained which are situated on either side of the point source. The real images are obtained when the hologram is illuminated by a plane beam of light and a lens is used to take the Fourier transform of the diffracted wave. The hologram may be placed either before or after the lens (Fig. 4.26). Two symmetric and separated off-axis images are obtained, one centred at ξ_o', the one at $-\xi_o'$. The zero order terms will be focussed on the axis. The direct image appears inverted and centred at ξ_o and the conjugate image is upright and centred at $-\xi_o$. It may be noted that with the reconstruction schemes shown, two virtual or two real images are obtained.

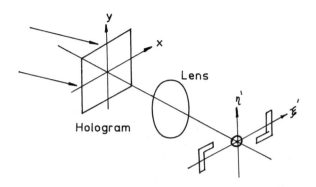

Fig. 4.26. Reconstruction of real images from a lensless Fourier transform hologram.

Fourier Transform Hologram

A Fourier transform (FT) hologram is a special case of Fraunhofer hologram and is equivalent to a lensless Fourier transform hologram. The hologram records the interference of Fourier transforms of both the object and the reference point source [4.18]. The reference point source and the object are located in the front focal plane of a Fourier transform lens and the photographic plate is placed at its back focal plane (Fig. 4.27). Each point of the object produces a parallel beam of light which interfere with the parallel off-axis reference beam. Thus both object and reference sources are at infinity with respect to the hologram. This hologram is also known as Fourier-Fraunhofer hologram. This arrangement of recording Fourier transform hologram produces images without any multiplicative phase factor and hence the location of the reconstructed image becomes invariant under a lateral translation of the hologram. If in the reconstruction step (Fig. 4.28) the focal length of the lens is f_2 then both the direct and the conjugate images formed are magnified by f_2/f_1 where f_1 is the focal length of the lens used in the recording step.

The Fourier transform of a continuous tone object has very large amplitude of the low order frequencies as compared to higher frequencies. Consequently, the lower frequencies are overexposed due to limited dynamic range of the recording materials. This affects the diffraction efficiency. The situation can be improved by slightly

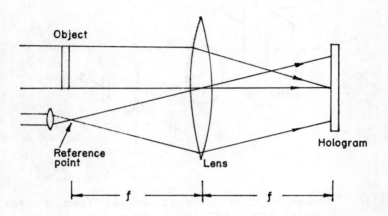

Fig. 4.27. Recording of a Fourier transform hologram.

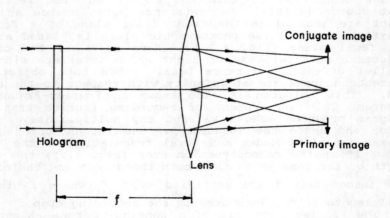

Fig. 4.28. Reconstruction of a Fourier transform hologram.

defocussing the Fourier spectrum i.e. recording away from the focal plane. The defocussing should be large enough to reduce the dynamic range while retaining the advantages of Fourier hologram. Generally a defocussing by 10% of the focal length of the Fourier transform lens may be used.

Fourier hologram is of fundamental practical importance due to its unique features: (i) FT holograms produce aberration free images, (ii) the location of the output image is invariant under a lateral translation of the hologram and (iii) it allows a higher density of recorded information than a Fresnel hologram. High capacity information storage and character and pattern recognition are the classical application areas for FT holography.

Image Hologram

The image hologram is recorded by imaging the object by a lens, concave mirror or primary hologram on the recording medium and interfering it with a coherent reference beam [4.39] as shown in Fig. 4.29. The projected image of the object can straddle the recording plate so that in the reconstruction one half of the image appears to be in front

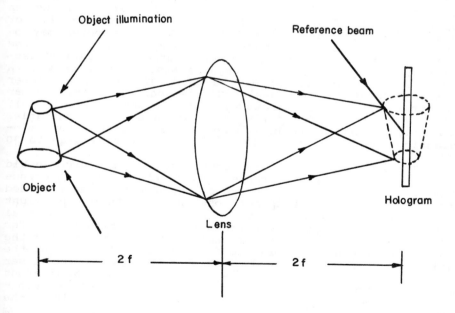

Fig. 4.29. Optical system for recording of an image hologram

of the plate while the other half behind it. This type of hologram can be reconstructed with an incoherent source of light. The viewing zone of the reconstructed image is limited by the aperture of the imaging system. This, however, results in increased image luminance as all of the diffracted light is restricted to the smaller viewing zone. When viewed directly, the image is visible with low contrast with a fine grating structure.

4.10. Practical Holography

This section considers the requirements for recording a good hologram. We have seen that recording of a hologram involves interference phenomenon. Therefore, all the precautions are needed to create and record high contrast interference fringes.

4.10.1. Laser

The heart of any holographic system is the laser. Although the early work in holography was done with ordinary sources of light, with the use of lasers the holograms of large 3D objects can be recorded. He-Ne laser is the most commonly used light source due to its low cost, high beam quality and long life. For most of the applications an optical power of 2 to 35 mW at 633 nm is sufficient. For small objects, a He-Ne laser with 2mW output power may be used. Longer exposure times can record holograms of large size objects but that would need perfect vibration isolation. For very large size objects argon ion laser with a power of 5 W (all lines) may be used. The argon ion laser which lases at several wavelengths between 476 to 514 nm can be tuned to a particular wavelength. Krypton-ion (647 nm), He-Cd (442 nm) and ruby (694 nm) lasers are also suitable. Recently, diode lasers have also become available for holographic applications at 740 nm with output powers up to 30 mW.

The spatial and temporal coherence of the source should be high. The total difference between object and reference beam paths plus the total depth of the object being recorded must not exceed the temporal coherence length of the light source, and the reference beam must have high spatial coherence. The temporal coherence is the limiting factor for the depth of the object to be recorded. Also the greater the coherence length, the more the flexibility in setting up the holographic system. Depending on the length of the laser cavity, a He-Ne laser can have a coherence length of about 20-100 cm. The shorter cavity lengths provide better temporal coherence but their output power is also low. The coherence length of argon ion lasers is only about 1 cm which can be extended up to 10 m by incorporating an optical etalon in the laser cavity, but at the expense of output power. Similarly, the temporal coherence length of a Ruby laser is very small and as such it is not suitable for most of the holographic work unless it is extended by intracavity etalons. Recently diode lasers with coherence length of several metres have been commercialized. The spatial coherence of the reference beam can be ensured by focussing the beam and passing through a fine pinhole.

The spatial and temporal coherence of a light source can be measured by interferometers (section 1.19). A very quick and easy way of the measurement of coherence length is by using holographically generated phase conjugated wavefront [4.40]. The schematic diagram is shown in Fig 4.30. Initially the object beam path is made approximately equal to the reference beam path and the reconstructed conjugate beam is observed on the screen. Then the path length of the object beam is continuously increased by moving the mirror M_2 till the conjugate object beam disappears on the screen.

The path difference between object beam and reference beam under this condition gives a direct measure of the temporal coherence length of the laser. In this method, one has to see only the presence of a laser point on the screen rather than to observe interference fringes as in Michelson method. This method is also applicable to laser sources other than in the visible wavelength region, i.e. IR and UV lasers. In such cases, the hologram recording/writing is done at the desired wavelength and the reconstruction beam R_1 used from

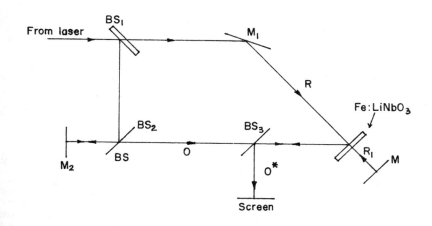

Fig. 4.30. Optical system for the measurement of coherence length of a laser [4.40].

a He-Ne laser so that on the screen a red point will be observed [4.39]. However, the recording crystal should be sensitive to the desired wavelength application.

4.10.2. Reference-to-Object Intensity Ratio

In order to obtain bright reconstructed images from a hologram it is necessary to optimize the interference fringe contrast recorded in the hologram which depends on the exposure time and the ratio of the reference beam intensity to that of the object beam. The intensity at any point due to the interference between the reference and object beams is

$$I = |R + O|^2$$
$$= |R|^2 + |O|^2 + RO^* + R^*O$$
$$= I_1 + I_2 + 2(I_1 I_2)^{1/2} \cos(\phi_1 - \phi_2), \quad (4.6?)$$

where ϕ_1 and ϕ_2 are the phases of the object and reference beam. The contrast of the interference pattern is measured by the fringe visibility V which is given by

$$V = \frac{I_{max} - I_{min}}{I_{max} + I_{min}}, \quad (4.68)$$

where I_{min} and I_{max} are the minimum and maximum values of the intensity. From Eq. 4.67 the value of V becomes

$$V = \frac{2(I_1 I_2)^{1/2}}{(I_1 + I_2)},$$
$$= \frac{2\rho^{1/2}}{1 + \rho}, \quad (4.69)$$

where $\rho = I_2/I_1$ is the reference-to-object beam intensity ratio. The fringe visibility becomes maximum when $\rho=1$ i.e. when reference and object beam amplitudes are equal. The exposure time should be such that amplitude transmittance of the finished hologram is about 0.5 (neutral density of about 0.6). This ensures the recorded density of the processed film on the linear region of the exposure versus amplitude transmittance curve of the film.

As noted above the high contrast fringes are produced when the reference-to-object beam intensity ratio is unity.

However, as most 3D objects have poor reflectivities, the reference beam intensity together with object beam intensity may be too less and require very long exposure times. This requires very high order of stability of the optical system. Highly reflecting points on the object may lead to nonlinear recording. To overcome these difficulties, the reference beam intensity may be increased to make ρ equal to 3 to 5 so as to produce a density of 0.6-0.8 on the photographic plate within desired exposure time. This will produce an amplitude hologram of high diffraction efficiency [4.41]. For bleaching the hologram, the density before bleaching should be [4.40, 4.41] greater than 1.5 and the intensity ratio may be [4.42] as high as 10. A general rule of thumb is that no point of the object should be more intense than the reference point source. Painting of the object by a diffuse reflecting paint may help to increase the overall object reflectivity. The beam balance ratio can be adjusted by using a variable beam splitter, a neutral density filter in either beam or polarizers in the setup.

4.10.3. Angle between Reference and Object Beams

The contrast of the fringes also depends on the angle between the interfering beams. The angle between the beams decides the spacing between the fringes. The larger the angle the finer are the fringes. The recording materials have a certain resolution. The angle between the beams should be adjusted such that the frequency of the interference fringes lies within the resolution limit of the material. However, there is a minimum angle that will just separate the image of the object from the reconstruction beam.

4.10.4. Polarization of Light Beams

For the interference phenomenon to occur, the interfering beams must have the same polarization. In Eq. 4.67 it was assumed that the two waves were polarized with their electric vectors parallel. If the two electric vectors make an angle ψ with each other the intensity at any point in the interference pattern becomes

$$I = I_1 + I_2 + 2(I_1 I_2)^{1/2} \cos(\phi_1 - \phi_2) \cos\psi \qquad (4.70)$$

and the visibility of the fringes becomes

$$V = \frac{2\rho^{1/2} \cos\psi}{(1 + \rho)}. \qquad (4.71)$$

Thus the fringe visibility decreases as the value of ψ increases from 0 and drops to zero when $\psi = \pi/2$, i.e. light waves polarized linearly but at right angle to each other will not interfere. The correct polarization requirements can be ensured if the two beams are polarized with the electric vector normal to the plane of the optical table. Although most of the lasers produce linearly polarized beams, polarization changes can occur as a result of reflections from dielectric coatings (beam splitter) and metallic surfaces (mirrors, object). It is always advisable to keep the total number of reflections in the object path and the reference path the same. The polarization problem can be rectified by changing the surface properties of the object or by rotating the plane of polarization of the beams. The laser beam can be made circularly polarized before it is split for forming object and reference beams (section 4.16).

4.10.5. Vibration Isolation Table

Another important equipment for making a high quality hologram is the vibration isolation table which is necessary because of the requirement that the object and reference beam paths must not vary more than a fraction of a wavelength ($\lambda/8$) during exposure. This can be achieved to a fair degree by mounting all the optical components rigidly on a platform such as a heavy block of granite which in turn is placed on a sand bath or over a pile of news papers or rubber mattresses. This system would not work if the environmental vibrations are severe.

Pneumatic isolation is the most effective way of protection against vibration. The table platform can be supported on inflated automobile inner tubes. Such a system has resonant frequencies around 10-20 Hz. System resonances of less than 1 Hz are achievable by supporting the optical table on air-filled legs. The compressed air provides a soft spring to the system.

Electronic vibration isolation system uses active feedback to remove the effects of vibrations. These systems isolate not only the vibrations in the ground but also those generated by equipment mounted on the table top and also vibrations produced due to air flow. The isolation is not based on weak floating spring but is provided by a set of electromechanical servo circuits which actively cancel the vibration even due to any time varying forces.

The table top is usually made up of a honeycomb core sandwiched between two faces of skin material. The honeycomb core makes the table top a perfect rigid body. The skin material should have a large Young's modulus, light in weight and a low coefficient of thermal expansion. The top surface of the table is the working surface. A magnetic surface with lots of tapped holes is the most convenient mounting surface.

The commercially available table systems have an arrangement for self leveling the table top so that the equipment can be added or removed on the top without disturbing the level [4.43]. In addition to this, the table top is made up of acoustically damped honeycomb material for further reduction in vibration sensitivity.

For longer exposure durations any residual disturbances can be eliminated by a close loop feedback system in which any motion of the interference fringes are sensed by a photodetector which in turn controls the position of a small mirror in the reference path mounted on a piezoelectric transducer (Fig.4.31). This automatically restores the path difference between the two beams to its original value [4.44, 4.45].

4.10.6. Optical Components and Mounts

The optical components should be of high quality. All mirrors should be front surface coated. All optical mounts should be stable. The optical mounts of adjustable height with provision for rotation and tilt are preferred than fixed height mounts. The optical components should have

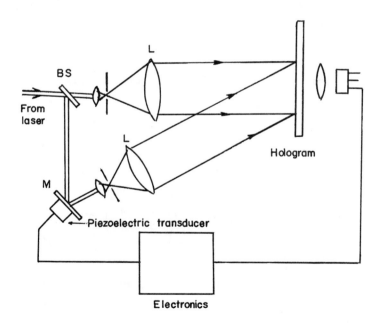

Fig. 4.31. Schematic diagram of a fringe stabilization system [after 4.44].

magnetic mounts. The object to be recorded should also be firmly mounted.

Lens-pinhole Spatial Filter

Holograms made without spatial filter produce poor quality images. For quantitative measurements and display holograms, spatial filters are essential. A spatial filter cleans the laser beam from cosmetic defects which may be due to dust particles, multiple reflections and scattering.

Spatial filters remove random fluctuations from the intensity profile of a laser beam. This greatly improves the resolution of holographic images. Fig. 4.32 shows the spatial filtering action of a pinhole on an input laser beam which has intensity variations due to scattering by optical defects and particles in air [4.43]. When a Gaussian beam is focussed by a positive lens of focal length F, the image at the focal plane will contain the scattered noise information around the image on the optical axis. A pinhole centred on the axis can block the unwanted noise while passing more than 99% of the laser energy. The optimum diameter of the pinhole D_{opt} is given by

$$D_{opt} = \frac{F\lambda}{a}, \qquad (4.72)$$

where a is the laser beam radius where its intensity falls

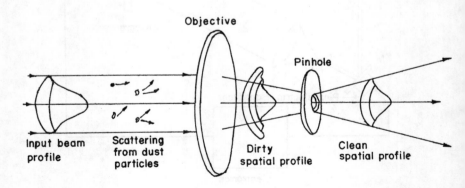

Fig. 4.32. Spatial filtering action of a pinhole [4.43].

to $1/e^2$ value.

For a 10X objective and input laser beam diameter of 5.5 mm, a pinhole of 25 μm is used. Similarly, for 60X objective and input laser beam diameter of 3.5 mm, a pinhole of 5 μm is required to filter out the undesired noise pattern. For accurate spatial filtering, micrometers are used for objective focus and pinhole placement. The objective is moved in z-axis to precisely position the focus on the pinhole which can be moved in x-y direction.

The quality of the pinhole may be checked by examining its Airy diffraction pattern which must show perfectly circular and uniform rings. If needed, the pinholes can be cleaned by ultrasonic cleaning or by audio tape head cleaner or carbon tetrachloride.

4.10.7. Hologram Recording Geometries

Transmission Holograms

Figure 4.33 shows a simple schematic diagram for recording of a transmission hologram. The laser beam height is adjusted by a beam steerer which is a combination of two

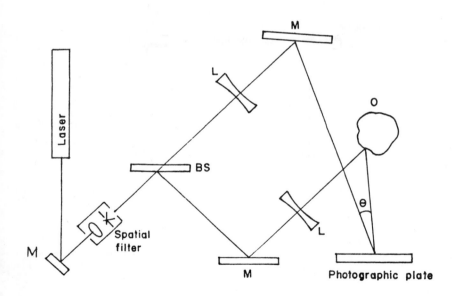

Fig. 4.33. Schematic diagram for recording of a transmission hologram.

mirrors. The laser beam is split by a beam splitter. One part of the beam (say reflected part), after expansion illuminates the object whose hologram is being recorded. The laser beam is expanded by a lens-pinhole spatial filter

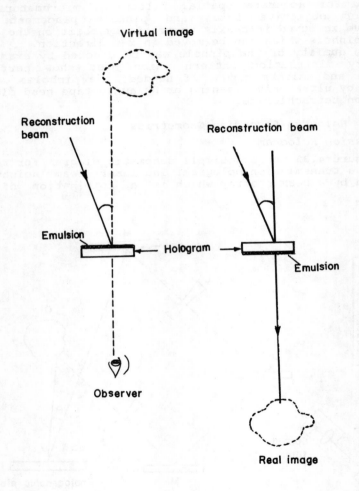

Fig. 4.34. Reconstruction of virtual and real images from a transmission hologram.

combination. The scattered beam from the object falls on a photographic plate (emulsion side towards light beams), which also receives the directly transmitted part of the laser light. The two beams can be further expanded by negative lenses. These two beams, one via object (known as the object beam) and the other directly from the source (known as the reference beam), interfere and produce a fine fringe pattern which exposes the photographic plate. The exposure time is adjusted so as to produce an optical density of 0.6-0.8 after photographic processing.

To reconstruct the image, the developed hologram is repositioned at its recording position and illuminated by the reference beam which produces the virtual image of the object as shown in Fig 4.34. When the hologram is illuminated from the glass side a real image is produced which can directly be recorded by a photographic plate placed at the focussed image.

Figure 4.35a shows a very simple arrangement for recording a transmission hologram. This method requires only one component, either a spatial filter or simply a microscopic objective. However, the ratio between the object and reference beam intensities cannot be adjusted and the angle between the two beams is also restricted. As the number of components in the setup is very small, this geometry does not demand high order of stability. A mirror can be used in the setup to adjust the angle between the beams (Fig.4.35b).

Opaque objects made up of plaster of Paris or solid plastic make the best subjects for display holograms. Metallic and dull objects may be painted by a thin layer of diffuse reflecting paint. For better object illumination, more than one object beams may be needed.

If the reference point source and the object are at equal distances from the hologram, then a lensless Fourier transform hologram is recorded producing image immobility. If the reference point source is near the object the resolution of the image is maximum but its field-of-view is restricted. On the other hand, if the reference point source is moved to infinity (i.e. plane reference wave), the field-of-view is maximized but the resolution of the image is reduced. A compromise between the field-of-view and resolution is obtained [4.46, 4.47] if the reference point source is suitably positioned between the object and infinity. If the reference beam is not derived from a point source (spatially incoherent), the resolution of the image is reduced by the size of the source.

Reflection Holograms

The recording of a reflection hologram requires that the object be incident on the opposite side of the photographic plate from the reference. This can be readily accomplished [4.10] by illuminating the object through the photogra-

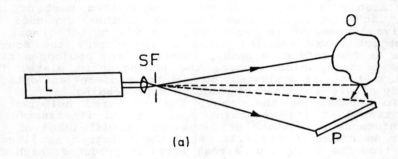

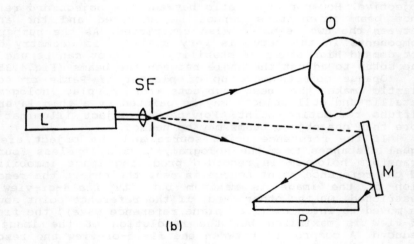

Fig. 4.35 Simple arrangements for the recording of a transmission hologram using only (a) a spatial filter or (b) a spatial filter and a mirror.
L-laser, SF-spatial filter, P-photographic plate, M-mirror.

phic plate as shown in Fig. 4.36a. The collimated laser beam which is incident on the plate serves as the reference beam. Obviously the object must be highly reflective otherwise the

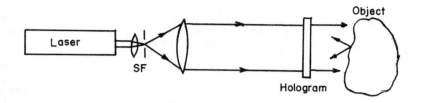

(a)

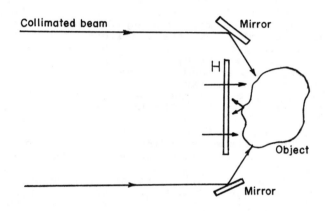

(b)

Fig. 4.36. Simple arrangements for recording a reflection hologram using (a) a collimated beam and (b) additional beams for object illumination.

reference-to-object beam intensity ratio becomes much higher than 1. Object beam intensity can be improved by illuminating the object by additional beams from sides (Fig. 4.36b), keeping all the path length differences within the coherence length of the source.

If the photographic plate is attached to the object, then the effect of vibration will be minimum as there can be no differential motion between the object and the plate. It is pointed out that the stability requirement is much critical for reflection holography than that for transmission holography because in a reflection hologram, the fringes spaced by half a wavelength are recorded within the volume of the hologram and a slight vibration will result in a complete failure. Even an improper plate holder might cause the emulsion layer to vibrate like a thin membrane. Thermal and humidity shocks to the emulsion may ruin the hologram.

The colour of the reconstructed image depends on the fringe spacing established during recording. When the hologram is illuminated with a white light source, the image formed is expected to be of the same colour as the laser wavelength by which the hologram has been recorded if the angle of illumination is the same as the angle of the reference wave used in the recording. If the illuminating angle is different from the Bragg angle for the recording wavelength, the image in different colour is observed.

The colour of the image may be shifted to lower wavelength due to emulsion shrinkage as a result of photographic processing which reduces the spacing between the fringes. Thus a hologram recorded with the red light of a He-Ne laser will reconstruct in green or yellow. The shrinkage can be minimized to a larger extent by not fixing the hologram, but then the hologram becomes unstable against light. The shifted colour of the hologram can be restored by reswelling the hologram [4.48], say by soaking the hologram in a bath of triethanolamine or D-sorbitol. The thickness of the emulsion may be deliberately varied between exposure and viewing in order to obtain reconstruction wavelengths that are very different from that of the recording laser (section 4.13.5).

The sharpness of the image depends on the size of the reconstruction source. Best results are obtained by a point light source such as sun and short halogen bulbs. Objects recorded as close as possible to the emulsion are reconstructed sharply. The portions of the object within the plane of the emulsion will be sharp even when reconstructed with an extended source of light such as a milky bulb or a fluorescent light. The object can be positioned very close to the emulsion by projecting the image of the object by a lens. A reconstruction source at a large distance from the hologram also produces a sharp image.

4.10.8. Hologram of a Moving Object

Since holography involves basically the recording of interference fringes, it is important that the total change in path length of the two beams at the recording plane must be less than $\lambda/2$. If a path difference of $\lambda/2$ occurs at the recording plane it will shift the fringe position by one half of one fringe spacing making the recorded fringe modulation to zero. Thus the geometry of the optical arrangement for recording the hologram becomes extremely important for the successful recording of a moving object.

In evolving a practicable geometry for recording the

hologram of an object in motion, it is kept in view that there is no restriction on the motion of the object during the exposure, but it should be such that the path difference between the interfering beams remains constant [4.49, 4.50]. The constancy property of an ellipse is quite beneficial for evolving a practical geometry for motion holography [4.49, 4.50]. According to this property all paths similar to a, b, and c as shown in Fig. 4.37 are constant and equal to twice the semi major axis value.

Abramson [4.51] has shown the construction of a holo-diagram which consists of a set of spheroids with their foci at A and B as shown in Fig. 4.38. The laser source is placed at one of the foci A and the recording plate at the other focus B. The circles, called the k lines, indicate the relative fringe sensitivity that can be achieved when the object surface (C) moves normal to the ellipse. Suppose the object moves along the line PP', tangent to the surface of an ellipse. Thus, even if the object moves, the path length of the object beam remains equal to twice the semi major axis. This is true because as the object moves from point Q to the right, it intercepts the successive ellipses (Figure 4.39) at points P_1, P_2 etc. In other words, as the object moves, the original ellipse can be positioned to grow while the foci separation remains constant. The total linear motion

$$\Delta x = (\Delta d/2)^{1/2} a^{3/2}/b = vt, \qquad (4.73)$$

where Δd is the change in the optical path length of the

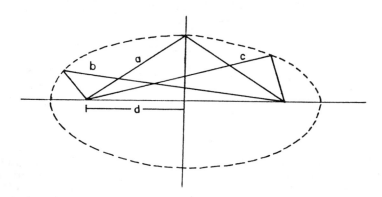

Fig. 4.37. Constancy property of an ellipse [4.49].

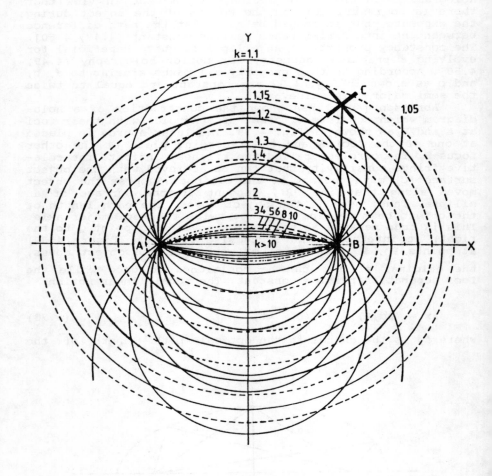

Fig. 4.38. Holodiagram [4.51].

object, a and b are the values of the semi major and semi minor axes of the original ellipse and v is the distance travelled by the object in time t. But since $\Delta d < \lambda/2$, the ellipse can be constructed for the allowed Δx. Kurtz and Perry [4.52] have experimentally verified this method for holographic recording of objects in motion.

When the object is moving with a random velocity vector

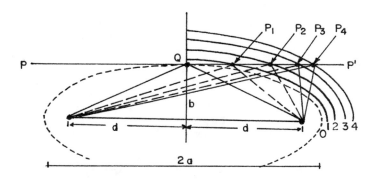

Fig. 4.39. Ellipses with constant foci separation 2d. The object moves along PP' [4.49].

a composite elliptical arrangement may be constructed [4.49] as shown in Fig 4.40. The two ellipses record two orthogonal velocity components thereby permitting the object to move with a random velocity in the plane of two ellipses. In order to record a completely random velocity vector, a composite system consisting of three mutually orthogonal ellipses may be employed. The recording plate is placed at the common foci of the ellipses [4.49, 4.52]. If the available laser power is inadequate, one could use three laser sources, one for each ellipse system.

Smigielski et al. [4.53] have experimentally verified the application of ellipse method for a speed of 2000 m/s. Recently Parker [4.54] has used a pulse laser to record the hologram of hypervelocity armour piercing shells. The hologram captured the jet of copper travelling at a speed of 6500 m/s, a fastest moving event that has ever been captured in a hologram.

4.10.9. Efficiency of a Hologram

The image luminance from a hologram depends on its diffraction efficiency. The diffraction efficiency of a hologram is defined as the ratio of the power diffracted into the image to that incident on the hologram. Thus

$$\eta = I_1/I, \qquad (4.74)$$

where I is the power of the light illuminating the hologram and I_1 is the power diffracted in the image beam. Sometimes,

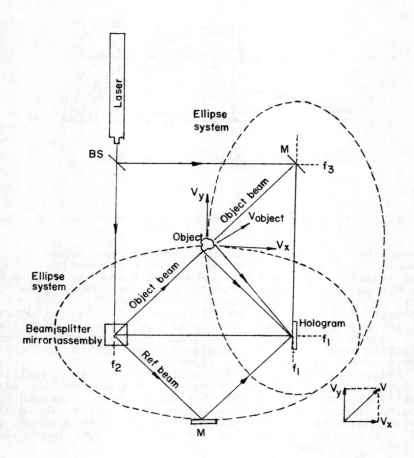

Fig. 4.40. A compact elliptical arrangement for recording a hologram of an object in motion [4.49].

the diffraction efficiency is defined as

$$\eta = I_1/(I_1 + I_0), \qquad (4.75)$$

where I_0 is the power of the zero order beam. This definition neglects the light loss due to the front surface reflection, light diffracted into other orders and scattering and absorption by the medium. The diffraction efficiency of a hologram depends on the recording medium, recording and reconstruction geometry and the visibility of the carrier fringes. The maximum efficiency is obtained where the slope of the amplitude transmittance versus log E curve is steepest. It varies as the square of the visibility of the fringes. The diffraction efficiency can also be expressed in terms of the modulation of the intensity distribution $V(f)$ i.e. fringe visibility and the modulation transfer function $M(f)$ as [4.55]

$$\eta = (1/4)[\beta<E>V(f)M(f)]^2, \qquad (4.76)$$

where $\beta<E>$ is the slope of the t_A-log E curve at the average exposure $<E>$. According to Lin [4.55], the plots of η versus $V(f)$ with $<E>$ as a parameter and plots of η versus $<E>$ with $V(f)$ as a parameter should be straight lines for ideal recording material. These curves help in determining the exposure needed to obtain maximum efficiency.

Table 4.1 gives the values of the maximum theoretical diffraction efficiencies that can be obtained with different types of holographic gratings.

Table 4.1 Maximum theoretical diffraction efficiency η_{max} (in %) for holographic gratings.

Transmission hologram				Reflection hologram	
Thin		Thick			
Amplitude	Phase	Amplitude	Phase	Amplitude	Phase
6.25	33.9	3.7	100	7.2	100

4.10.10. Refractive Index Modulation

In the phase modulating media, the diffraction efficiency depends on refractive index modulation (Δn) of the material. The Δn values may be calculated by measuring the diffraction efficiency η and grating thickness and using Kogelnik's coupled wave theory [4.30].

The refractive index modulation for a lossless unslanted transmission grating is obtained from Eq. 4.47 and is given by

$$\Delta n = \frac{\lambda_a \cos\theta_m \sin^{-1}(\eta)^{1/2}}{\pi t}, \qquad (4.77)$$

where λ_a is the probe wavelength in air, t is the grating thickness and θ_m is the incidence angle in the medium. The value of θ_m can be calculated using Snell's and Bragg's laws.

$$\sin\theta_m = \sin\theta_a / n_m \tag{4.78}$$

and

$$\sin\theta_m = \lambda_a / (2 n_m \Lambda), \tag{4.79}$$

where θ_a is the angle of incidence in air and n_m is the average refractive index of the medium. Material thickness is taken as the grating thickness α which can be measured using an electronic comparator or a surface profile monitoring system.

The Δn for unslanted reflection grating is obtained from Eq. 4.54 and is given by

$$\Delta n = \frac{\lambda_o \cos\theta \, \tanh^{-1}(\eta)^{1/2}}{\pi t}, \tag{4.80}$$

where λ_o is the Bragg wavelength. The value of η is experimentally measured.

In some of the materials like dichromated gelatin and photopolymers very high degree of refractive index modulation is achieved, hence these materials record very high efficiency holograms.

4.10.11. Signal-to-Noise Ratio (SNR)

Due to the imperfections in the recording medium and coherent illumination for reconstruction of a hologram, holographic images are corrupted by noise. The light randomly scattered from the hologram is superimposed on the reconstructed image. The nonlinearity of the recording medium (section 4.11.2) produces higher order images which also add to the noise. If these images are angularly separated from the primary image, then they are not considered in the signal-to-noise (SNR) measurements.

The SNR may be defined as the ratio of the average of the light power forming the image signal and the average light power of the noise signal and the average light power of the noise. Mathematically, it is given by [4.56]

$$SNR = I_i/\sigma = \frac{(I_i/<I_n>)}{(1+2I_i/<I_n>)^{1/2}}, \tag{4.81}$$

where I_i is deterministic image irradiance, $<I_n>$ is the average scattered irradiance and σ is the standard deviation of the total irradiance. The scattered flux is dependent on the sensitivity of the film, mean transmittance, type of developer, development time and density of the developed film. The scattered flux is higher for emulsions of higher sensitivity. It increases with the increase in development time. The scattering also varies inversely on the fourth power of the wavelength. While the glass substrate does not scatter significantly, the acetate and ester film substrates may scatter as much light as from the emulsion [4.57].

To measure the SNR of a holographic image, a hologram of a point source at infinity is recorded and the ratio of the intensity of the reconstructed image point and that of the scattered flux around it is measured by a detector. Fig. 4.41a shows the measurement of the light power due to reconstructed beam and noise while the measurement at a large distance (Fig. 4.41b) gives the light power due to reconstructed beam only.

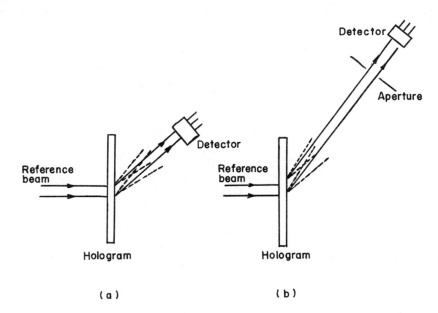

Fig. 4.41. Measurement of light power due to (a) reconstructed beam and noise, and (b) reconstructed beam only. The dotted lines show scattered noise.

The SNR of a diffuse object may be determined by recording a hologram of a back illuminated diffuse object [Fig. 4.42]. The SNR is equal to the ratio of the average light intensity in the reconstructed image of the object to the average intensity of the scattered light in the black square at the center. The SNR can be plotted against different recording parameters such as reference-to-object beam angle θ or spatial frequency (sinθ)/λ, the object distance from the hologram, exposure, etc.

For holographic recording of two plane beam interference, the diffraction efficiency maximizes at beam intensity ratio ρ equal to 1. But in the case of recording of a diffuser, the intermodulation term OO* limits the modulation for the image term. Hence in such a case the maximum diffraction efficiency has been observed [4.58] at ρ=2 rather than ρ=1. The efficiency lowers down by 25% with ρ=7.5. Therefore a balance between efficiency and background flux has to be maintained. For a maximum SNR of 20 dB, an efficiency of the order of 1% is obtained both for amplitude and phase holograms. Phase holograms can offer much higher efficiencies in silver halide emulsion with some sacrifice of SNR.

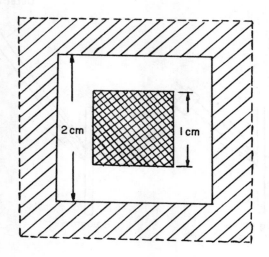

Fig. 4.42. Diffuse object for the measurement of SNR.

4.11. Holographic Recording Materials

A hologram may be recorded in a medium as a variation of absorption or phase or both. The recording material must respond to incident light pattern causing a change in its optical properties. The complex amplitude transmittance t_A, of a recording material can be written as

$$t_A = \exp(-\alpha T)\exp[(-j2\pi nT)/\lambda] = |t|\exp(-j\phi), \quad (4.82)$$

where α is the absorption constant, T is thickness and n is the refractive index of the material.

In the absorption or amplitude modulating materials, the absorption constant changes as a result of exposure, while the thickness or the refractive index changes due to the exposure in phase modulating materials. In the phase modulating materials there is no absorption of light and all the incident light is available for image formation, while the incident light is significantly absorbed in an amplitude modulating medium. Thus a phase material can produce a higher efficiency than an amplitude material. Also in phase modulating media the amount of phase modulation can be made as large as desired [4.59]. The resolution capability of a recording material depends on its modulation transfer function. The nonlinear effects of the recording material are minimized for obtaining high quality holographic images.

4.11.1. Modulation Transfer Function

A recording material has a limited frequency response. The range of frequencies over which the material shows a significant response varies widely from material to material depending on the emulsion thickness, grain size and processing conditions.

The modulation transfer gives the ability of the emulsion to record the spatial frequency of a test object. In other words, it defines the resolution capability of that emulsion.

Modulation transfer may be determined experimentally by exposing the material with a sinusoidal flux of increasing frequency. It is defined as

$$M = \frac{I_{max} - I_{min}}{I_{max} + I_{min}}, \quad (4.83)$$

$$= \frac{C-1}{C-1}, \quad (4.84)$$

where $C = I_{max}/I_{min}$ is the contrast. Figure 4.43 shows the modulation of sinusoidal irradiance. It is observed that the modulation is the ratio of amplitude A to the mean value B. The modulation value can be between zero and one. The modulation given by Eq. 4.83 is not faithfully transferred in a recording material due to light scattering. The actually

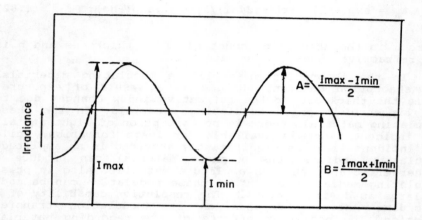

Fig. 4.43. Modulation of sinusoidal irradiance.

recorded modulation will be lower than the input modulation and this difference increases at higher spatial frequency.

For any spatial frequency the ratio of the modulation in the effective exposure to the input modulation is known as the modulation transfer function (MTF). If M_{in} is the modulation of the true exposure and M_{eff} is the effective exposure distribution, the MTF of the material is given as

$$\text{MTF} = \frac{M_{eff}}{M_{in}} \qquad (4.85)$$

4.11.2. Nonlinear Recording

In a linear recording, the transfer of exposure to amplitude transmittance corresponds exactly to the irradiance distribution to which the emulsion is exposed. The irradiance distribution for hologram exposure consists of a sum of four terms

$$I = RR^* + OO^* + RO^* + OR^*. \qquad (4.86)$$

The linear transfer from exposure to amplitude transmittance is described by the sum of the above four terms, one of which is the desired reconstructed wavefield ORR^*.

If the transfer from exposure to amplitude transmittance is nonlinear, higher order terms appear. Kozma [4.60] and Friesem and Zelenka [4.61] investigated the effects of nonlinear recording for simple objects. The amplitude transmittance of the recording material can be represented by a polynomial [4.62] of the form

$$t_A = t_o + \beta_1 E + \beta_2 E^2 + \beta_3 E^3 + \ldots$$

$$\begin{aligned} = t_o &+ \beta_1 [RR^* + OO^* + R^*O + RO^*] \\ &+ \beta_2 [RR^* + OO^* + R^*O + RO^*]^2 \\ &+ \beta_3 [RR^* + OO^* + R^*O + RO^*]^3. \end{aligned} \quad (4.87)$$

When the hologram is reconstructed by the wave R which may be plane wave of unit amplitude, the amplitude of the wave transmitted by the hologram is

$$\begin{aligned} U &= \text{linear terms} \\ &+ \beta_2[(OO^*)^2 + O^2 + O^{*2} + 2O^2O^* + 2O^{*2}O] \\ &= \text{linear terms} + U_1 + U_2 + U_3 + U_4, \end{aligned} \quad (4.88)$$

neglecting the higher order terms. Let us consider the terms other than linear terms in Eq. 4.88.

$U_1 = \beta_2(OO^*)^2$ This corresponds to OO^* of the linear recording and produces a halo around the directly transmitted beam with a spatial frequency spectrum that has twice the width of the spectrum of OO^*.

$U_2 = \beta_2 O^2$ This is higher order image term diffracted at an angle approximately twice that of the object wave and curvature having twice that of the object wave.

$U_3 = \beta_2 O^{*2}$ This is higher order conjugate wave.

$U_4 = \beta_2[2O^2O^* + 2OO^{*2}]$ These are intermodulation terms producing a noise halo about the true image.

The effects of nonlinearity are pronounced in phase holograms. The complex amplitude transmittance in a phase hologram is given by

$$t_A(x,y) = \exp[-j\phi] = 1 - j\phi - (1/2)\phi^2 + (1/6)j\phi^3 + \ldots \quad (4.89)$$

For obtaining high diffraction efficiency, the phase

modulation is increased, in which case the higher order terms in Eq. 4.89 cannot be neglected.

The nonlinearity effects are dominant when the exposure produced by the reference beam is comparable to that produced by the object. The effect of nonlinearity is to produce false images and halos around the images due to multiple autoconvolution and autocorrelations of the desired reconstructed field [4.62, 4.63]. A linear t_A-E relationship will ensure linear recording for an amplitude hologram. However, it is not possible to entirely eliminate the effects of nonlinearity in high efficiency holograms by lowering the beam intensity ratio.

4.11.3. Silver Halide Emulsion

The photographic emulsions are the most convenient, versatile and commonly used materials for recording holograms. These are available in a wide range of spectral sensitivities. The emulsion contains silver halide microcrystals, dispersed in gelatin deposited on a glass substrate or on a plastic film. The average size of the grains in the holographic emulsion is about 0.08-0.03 μm. In conventional emulsion, the size of silver halide crystals is of the order of 1 μm which is larger than the wavelength of light. Hence according to Mie theory, these grains scatter strongly. To decrease the scattering the crystals must be smaller than the wavelength of light so that Rayleigh theory becomes applicable. In this case, the scattered radiation goes with the sixth power of the grain size [4.64]. The emulsions with smaller grains are 'slower' than those with larger grains. The emulsion thickness range from 5 μm to 15 μm. They can be used for recording thin or volume holograms of amplitude or phase type. These materials have excellent shelf life.

Response

The behaviour of a photographic emulsion is described by the characteristic curve, also called as Hurter and Driffield (H&D) curve or D-log E curve. Fig. 4.44 shows a typical characteristic curve for a negative film. The characteristic curve shows the variation of density with log exposure. The log scale is a logical consequence of the logarithmic response of the human eye. The density D is defined as

$$D = \log (1/T)$$
$$= \log (I_o/I), \qquad (4.90)$$

where T is the transmittance, I_o is incident flux (influx)

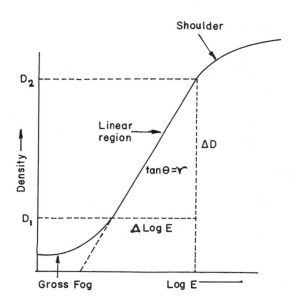

Fig. 4.44. Hurter and Driffield curve for a photographic emulsion.

and I is the transmitted flux (efflux).

The exposure E is equal to the energy per unit area incident at each point on the emulsion. If the emulsion is exposed for time t by the incident light of intensity I,

$$E = It. \qquad (4.91)$$

The H&D curve can be segmented into three parts: toe, straight line portion and shoulder. Some emulsions show a fourth segment called as solarization. In the straight line portion the output is a linear function of the input. The toe and shoulder portions correspond to exposure ranges for nonlinear behaviour. When the exposure is below a certain level, the density does not depend on exposure and has a minimum value known as gross fog. The actual shape of the curve depends on the nature of the emulsion, type of developer, time of development and the conditions during development.

The slope of the straight line portion of the characteristic curve is called gamma (γ),

$$\gamma = \Delta D/\Delta \log E = \tan\theta. \tag{4.92}$$

The low-contrast and high-contrast films are distinguished by the low and high values of γ, respectively (Fig. 4.45).

On the straight line portion, the difference in densities at two points is proportional to the difference in corresponding exposures.

$$D_2 - D_1 = \gamma(\log E_2 - \log E_1). \tag{4.93}$$

Putting the values of D_1 and D_2

$$\log(T_1/T_2) = \log(E_2/E_1)^\gamma. \tag{4.94}$$

For holography the curve between amplitude transmittance t_A and exposure is more meaningful. The amplitude transmittance is equal to the square root of transmittance.

$$t_A = T^{1/2}, \tag{4.95}$$

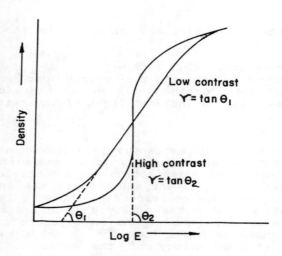

Fig. 4.45. H&D curves for a low contrast and a high contrast film.

$$T = t_A^2. \tag{4.96}$$

The curve for amplitude transmittance against exposure for a typical holographic emulsion is shown in Fig. 4.46. For linear recording

$$t_{A2} - t_{A1} \propto E_2 - E_1. \tag{4.97}$$

Equation 4.94 with the help of Eq. 4.96 becomes

$$\log (t_{A1}/t_{A2}) = \log (E_2/E_1)^{\gamma/2}. \tag{4.98}$$

Equation 4.98 shows that unless $\gamma=+2$ or -2 linearity is not achieved. However, in practice for holographic emulsions linearity is achieved even for emulsions with γ different

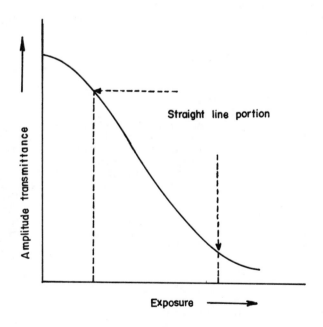

Fig. 4.46. Amplitude transmittance versus exposure curve

than 2. In such cases any point on the characteristic curve where the instantaneous gradient is 2 will satisfy the requirement.

Latent Image Formation

The mechanism of latent image formation in a silver halide emulsion with regard to the recording of a hologram has been described by Johnson et al.[4.65]. During exposure, the absorption of a photon by a grain of the emulsion can free an electron from a surface localized halide ion. This electron can move through the crystal lattice of the grain to combine with a surface silver ion. The reactions are

$$Br^- + h\nu \longrightarrow Br + e^-,$$
$$Ag^+ + e^- \longrightarrow Ag.$$

The silver atom has a lifetime of about two seconds and dissociates back to an atom. If, however, another photon is available producing another electron, a larger silver speck is formed. This silver speck does not contain enough atoms to be developable. At least four silver atoms are needed for development.

$$Ag + Ag^+ + e^- \longrightarrow Ag_2$$
$$Ag_2 + Ag^+ + e^- \longrightarrow Ag_3$$
$$Ag_3 + Ag^+ + e^- \longrightarrow Ag_4$$

The silver speck is now stable and developable. This is known as a latent image. The latent image can be converted into a hologram by development and fixing. During development process each exposed grain is entirely reduced to silver by catalytic action. The effect of development is about 10^6 times that of the effect of exposure. The unexposed grains remain without any change. The development process is an amplification process.

If the photon energy is not sufficient to produce developable silver atoms, an initial bias exposure can be given to the emulsion to bring it to the threshold value [4.66].

Sensitivity and Speed

Intrinsic sensitivity of silver halide is in the ultraviolet and blue region of the spectrum. Dyes which absorb in the longer wavelength region are adsorbed to the silver halide crystals, so that they can transfer the absorbed energy to the crystals for the latent image formation. Thus the silver halide emulsion can be sensitized to record holograms at different laser wavelengths in blue, green and red region. The spectral sensitization can be extended to near infrared region up to about 1.3 μm. But due to their MTF and granularity characteristics, they are unsuitable for recording holograms with 1.06 μm laser radiation from a Nd:YAG laser.

The speed of the emulsion can be increased by hypersensitization by bathing. The hypersensitization process removes the excess bromide ions and/or increases the relative concentration of free silver ions in the emulsion. While soaking the photographic plate in a water bath for a few second can increase the speed of the film, bathing the plate in a solution of ammonia yields holographic plates with significant increase in sensitometric speed. Table 4.2 shows the procedure for hypersensitization [4.67].

Table 4.2 Procedure for hypersensitization of a holographic silver halide emulsion (Temperature of baths : 20°C) [4.67].

1.	Soak in 25% ammonia (spec. density 0.91)	3 min
	Ammonia 10 ml	
	Distilled Water 1000 ml	
2.	Wash in distilled water + wetting agent	5 min
3.	Dry at room temp.	

Material

The holographic silver halide materials are manufactured by Eastman Kodak company, Agfa-Gevaert company and Ilford company, besides Russian materials. The materials are available on glass substrates or plastic films. Since the interference fringes to be recorded are usually of the order of magnitude of the wavelength of light, a very high resolving power is needed. A high speed is also desirable to allow short exposures. However, high speed and high resolving power are incompatible properties. The choice between high speed and high resolution material can be made depending on the nature of the object. For recording of a high quality reflection hologram, silver halide material with highest possible resolving power is used. Russian Lippmann emulsions have grains of the order of 0.01 μm to 0.025 μm in size.

Table 4.3 compares the important characteristics of different photographic materials for holography. The amplitude transmittance versus exposure curves for a number of silver halide materials are shown in Fig. 4.47, while spectral sensitivity curves are given in Fig. 4.48.

Table 4.3 Silver halide holographic recording materials

Type	Sensitivity	Emulsion thickness (μm)	Grain size (nm)	Exposure ($\mu J/cm^2$)	Resolution limit (lines/mm)
Eastman Kodak					
649F	Panchromatic	15	58	50	>3000
120	Red	6	50	20	>3000
131	Red	9	70	0.3	1250
125	Blue, green	7	65	0.05	1250
Agfa-Gevaert					
8E75HD	Red	7	35	10	5000
10E75	Red	7	88	0.5	3000
8E56HD	Blue, green	7	37	25	5000
10E56	Blue, green	7	90	1	3000
Ilford					
HotecR	Red	>15		5	
SP695T	Blue, green	>15			

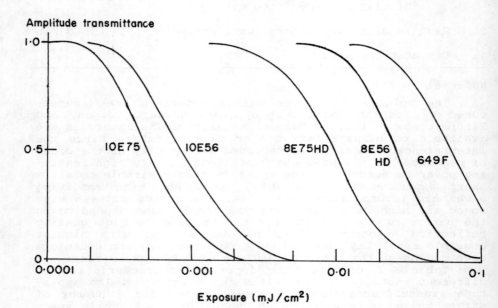

Fig. 4.47. Amplitude transmittance versus exposure curves for silver halide holographic emulsions.

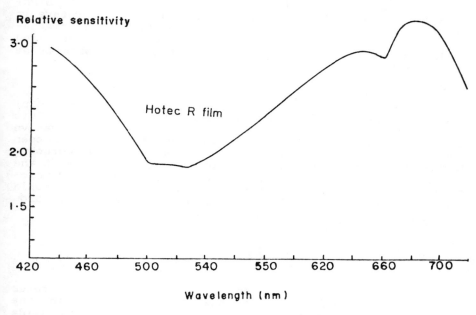
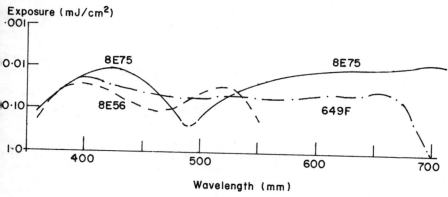

Fig. 4.48. Spectral sensitivity curves for silver halide holographic emulsions.

Ilford HotecR emulsion contains a neutral material which washes out during chemical process thereby reducing the layer thickness. This is known as Built-in-Swell (BIPS) factor. The emulsion has a BIPS factor of 8.8%, therefore it becomes unnecessary to pre-swell the emulsion with triethanolamine if a gold coloured reflection hologram is required with a He-Ne exposure. If the exposure is made with pulse ruby laser, then the replay will be at 633 nm.

Development

Kodak developer D-19 (or Agfa-Gevaert developer G3p) has been the most popular since beginning. This developer has been optimized for black and white negative development. Different types of developers have since been evolved through a process of trial and error for holography. During development process the exposed silver halides in the emulsion are chemically reduced to metallic silver. Developing agent is usually a benzene molecule containing ammonia (NH_3), hydroxyl (OH), methyl (CH_3) etc. groups. For hydroquinone developer the reduction can be summarized as

$$2Ag^+X^- + HOC_6H_4OH \rightarrow 2Ag^o + O=C_6H_4=O + 2HX$$

An alkali is added in the developer which accelerates the reduction and consumes the acid produced in the reaction. The developer will reduce the silver halide crystals only when there is a reaction latent image.

The unexposed silver halide crystals remain in emulsion after development. These are still photosensitive and limit the life of the developed emulsion. They can be removed by fixing with sodium thiosulphate (hypo) which forms a number of water-soluble silvery complexes along with a few water-insoluble complexes, for example

$$AgX + 2Na_2S_2O_3 + 2H_2O \rightarrow Na_3[Ag(S_2O_3)_2] \text{ (soluble)}.$$

Table 4.4 gives chemical composition of D-19, GP61 and Agfa 80 developers which can be used for the development of transmission holograms. It is better to use two developing agents rather than one as their activity together is greater than individual one. Most developing agents work efficiently only in alkaline solution with pH>7. The addition of Na_2CO_3 maintains a high amount of OH^- which neutralizes the H^+ produced in the reduction process and keeps the pH at desired level. The developing agent together with an alkali may be sufficient to develop an emulsion but the solution will lose its reducing power on reaction with atmospheric oxygen. Therefore a preservative like sodium sulphite is added in large quantity. A restrainer is also added to prevent the developer from attacking unexposed crystals. KBr or an organic antifoggent may be used as a restrainer. The restrainer hinders the production of development fog.

A chemical developer contains five principal ingredients viz.

(a) the reducing agent such as amidol, glycin, hydroquinone, metol, paraaminophenol, pyro and pyrocatechol
(b) the accelerator, usually an alkali,
(c) the preservative and stain preventer,
(d) the restrainer and
(e) water as solvent.

Without an accelerator the reducing agent would be ineffective as a developer. The preservative and stain preventers are sulphites, bisulphites and metasulphites. With amidol and metol, sodium sulphite plays a dual role as an accelerator and stain preventer. Neutral salt such as sodium sulphate in concentrated solutions check the swelling of gelatin.

Table 4.4. Chemical composition of some developers for transmission holograms.

Composition	Kodak D-19	Agfa 80	GP61
Water	750 ml	750 ml	700 ml
Metol	2 g	2.5 g	6 g
Hydroquinone	8 g	10 g	7 g
Phenidone			0.8 g
Sodium sulphite	90 g	100 g	30 g
Sodium carbonate	50 g		60 g
Potassium carbonate		60 g	
Potassium bromide	5 g	4 g	2 g
Sodium EDTA			1 g
Add Water to make	1 l	1 l	1 l
Development time	5 min	4 min	2 min

The developer due to "direct" or "chemical" development converts grains of silver halide to filaments or worms of black silver [4.68]. A significant amount of silver can be added to the grain by "solution-physical" development by transferring the silver atoms from neighbouring undeveloped micro crystals. Silver halide solvents like sulphite, thiosulphate or thiocyanate etch the microcrystal surfaces and produce mobile silver complexes. These diffuse to nearby already developed silver grain and produce spherical particles. The increase of silver in each grain can lead to substantial increase in the diffraction efficiency of the holograms if bleached. However, in the development process a certain amount of silver is precipitated at randomly located chunks. This gives a milky appearance to the bleached hologram. The random precipitation is pronounced in weak silver solvents such as sodium sulphite and weak in strong solvents such as ammonium thiocyanate. The strong solvents, if added in a minimum required quantity in the developer, will produce low noise holograms.

The requirements of development of reflection holograms are different than those of transmission holograms. This is so because the emulsion shrinkage has little effect on the fringe spacing in a transmission hologram. But in a reflecttion hologram because the fringe planes are parallel to the emulsion layer, shrinkage can reduce the fringe spacing and can rotate the fringe planes in slanted gratings. This will affect the efficiency and peak reflectivity of the hologram. Thus, in a reflection hologram it is desirable to minimize shrinkage of emulsion during processing. Table 4.5 gives composition of a few developers for reflection holograms.

The drawbacks of solution-physical development can be removed if silver halide solvents are not used in the developers. Such developers are known as direct, chemical, surface or nonsolvent developers. Ascorbic acid is a developing agent with photographically inert oxidation products and has been used without any sulphite protection in MAA-3 developer [4.69]. A more active developer is a PAAP developer which is vulnerable to aerial oxidation [4.68]. These developers produce very clean holograms with high efficiency and low scatter noise. Addition of 0.05g ammonium thiocyanate to PAAP developer reduces the scatter noise considerably. Holograms developed in catechol and bleached in p-benzoquinone bleach gives high efficiencies [4.70].

Russian developers have been formulated to produce colloidal silver (particle diameters lie in the range of the order of 0.01-0.05 μm) with a brown colour. Such developers permit the formation of holograms using only development without any fixing and bleaching stages. They are capable of creating both amplitude and phase modulation simultaneously. Since brown silver has lower opacity than black silver, very high efficiencies ($\approx$ 55%) can be achieved by colloidal silver development [4.71]. The efficiency depends on the size of the developed silver particles which in turn depends on the exposure and processing.

The composition of Russian developer is given in Table

4.6. The developer contains ammonium thiosulphate, a silver halide solvent which helps to create the required conditions for the colloid formation.

Table 4.5 Chemical composition of some developers for reflection holograms.

Composition	MAA-3	PAAP	Composition	CWC-1	CWC-2
Water	500 ml		Catechol	10 g	10 g
Metol	2.5 g		Sodium sulphite	10 g	5 g
Phenidone		0.5 g	Sodium carbonate	30 g	
Ascorbic acid	10.0 g	18.0 g	Ascorbic acid	5 g	
Sodium carbonate	55.6 g		Water	1 l	1 l
Sodium hydroxide		12.0 g			
Sodium phosphate	8	28.4 g			
Add Water to make	1.0 l	1.0 l			
Development time	4 min	4 min		2 min	2 min

Table 4.6 Composition of Russian developer GP-8.

Phenidone	0.026 g
Hydroquinone	0.65 g
Sodium hydroxide	1.38 g
Ammonium thiosulphate	3.12 g
Water	1.00 l
Development time	6.00 min

Tanning developers based on pyrogallol produce high modulation required to achieve highly efficient reflection holograms. The composition of a pyro developer, known as pyrochrome developer is given in Table 4.7. Addition of 200 g/l K_2SO_3 in pyrogallol developer increases efficiency as well as SNR [4.73].

Table 4.7 Composition of pyro developer [4.72].

Part A		Part B	
Pyrogallol	10 g	Sodium carbonate	60 g
Water	1 l	Water	1 l

Mix equal parts of A and B just before use.

Tray life	10 min
Development time	4 min

Bleaching Process

The silver halide materials usually record an absorption hologram which can be converted into a phase hologram by bleaching process. This can be done by modulating the thickness or refractive index in accordance with the absorption constant. A bleached hologram gives very high diffraction efficiency.

In direct bleaching process the developed and fixed silver in the photographic plate is converted into transparent silver salt using strong oxidizing agents like ferric chloride, mercuric chloride, cupric bromide, potassium dichromate and potassium ferricyanide. The bleached holograms usually have both thickness and refractive index modulation. The holograms bleached in these oxidizing agents produce low frequency intermodulation noise [4.74] and light scattering [4.75]. The emulsion shrinkage effects are also pronounced in these bleaches.

The dielectric silver compounds formed by bleaching agent are light sensitive. Hence extended exposure to light darkens the hologram due to the formation of photolytical silver. This effect is known as print out effect. In order to increase the stability of bleached holograms to light, desensitizing dyes may be added to the bleached emulsion [4.66, 4.76]. The photosensitive dyes may be replaced by non photosensitive dyes. To obtain further durability, the hologram is dipped into a bromine solution and is sealed between glass plates in a nitrogen gas atmosphere [4.76]. Figure 4.49 shows the behaviour of a hologram against bright illumination. The diffraction efficiency first increases under the exposure of 320 and 650 W halogen lamps which may be attributed to the emulsion expansion with heat from infrared rays. However, emulsion starts to crack after 500 hours. If a 200 W halogen lamp with infrared cut filter is used to illuminate the hologram, high diffraction efficiency can be obtained even after 1000 hours of exposure [4.76].

Amongst the direct bleaches mentioned above, the one producing silver chloride shows the least diffraction effi-

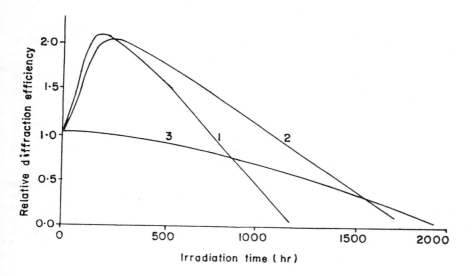

Fig. 4.49. Hologram durability against bright illumination. The light source power for curve 1 is 650 W without IR cut filter, for curve 2, 320 W without IR cut filter and for curve 3, 200 W with IR cut filter [4.76].

ciency but also the least scattered flux and highest SNR. The bleach formula is given below [4.77].

Water	800 ml
Copper sulphate crystals	100 g
Sodium chloride	100 g
Sulphuric acid	25 ml
Add Water to make	1000 ml

The diffraction efficiency can be improved by soaking the bleached holograms in a solution of either 20 g/l AgBr or 20 g/l AgI for 2 min. The conversion of AgCl to AgBr or AgI increases the diffraction efficiency and also the scattered flux. The mercuric chloride bleach (20 g/l) produces the higher diffraction efficiency and scattered flux and the bleached hologram shows fast print out effects.

An alternative to wet bleach process is the bromine vapour bleach [4.78] which produces clean and bright holograms with very high efficiency. This bleach, however, attacks the emulsion on prolonged bleaching. Holograms bleached in bromine vapour have been observed to deteriorate even after bleaching is completed. Bromine water bath is a good substitute for a bromine vapour bleach. Bromine water bath is prepared by taking about 5 ml cold bromine in a stoppered bottle, pouring cold water in it and keeping the bottle for a day. The water becomes orange as the bromine slowly dissolves in it. The hologram is bleached in this water. The holograms bleached in bromine are less sensitive to light if kept dry.

In the reversal bleach process the intermodulation noise and light scattering are suppressed by keeping a distribution of those silver halide microcrystals that were present in the unexposed emulsion. In this process, the hologram is not fixed and the developed silver grains are dissolved away leaving undeveloped silver halide crystals which are smaller than the developed silver grains. The small size of silver grains produce less scattering. The development in a tanning developer results in a reduction in intermodulation noise. Tanning developer and tanning bleach produce local tanning which minimizes emulsion shrinkage [4.70]. The components of tanning bleaches are given in Table 4.8.

Table 4.8 Composition of tanning bleaches GP 432, GP 433 and PBQ-2.

	GP 432	GP 433	PBQ-2
Potassium iodide		30 g	
Potassium bromide	50 g		50 g
Boric acid	15 g	3 g	
Citric acid			15 g
Water	1 l	1 l	1 l
p-benzoquinone*	2 g	2 g	2 g
Tray life 4 hr			

* add just before use.

The bleaches based on p-benzoquinone are extremely toxic. A substitute of PBQ-2 bleach is the ferric sulphate bleach which yields the same result [4.79]. Its composition is given below.

Water	700 ml
Ferric sulphate	30 g
EDTA-sodium salt	30 g
Potassium bromide	30 g
Sulphuric acid (conc.)	10 ml
Add Water to make	1000 ml

Another reversal bleach based on potassium dichromate is given in Table 4.9.

Table 4.9 Reversal bleach for high efficiency stable holograms [4.80].

Solution A

Potassium dichromate	8 g
Sulphuric acid conc.	10 ml
Water	1000 ml

Solution B

Potassium iodide	2 g
Water	1000 ml

Working solution

A : B : Water = 1 : 1 : 8

In potassium dichromate bleach, usually potassium bromide has been used. However, the use of iodine makes the hologram insensitive to ambient light. Experiments on Agfa 8E75HD and Ilford SP673 films have shown [4.81] that an increase in the amount of potassium dichromate in the bleach will increase the diffraction efficiency, while a low potassium dichromate concentration will minimize the noise in the hologram. A low sulphuric acid concentration will reduce the swelling of the emulsion.

Earlier it was thought that the diffraction efficiency of a bleached hologram would be better if higher be the refractive index of the silver salt. Van Renesse and Bouts [4.82], however, showed that polarizability is the important factor. The Lorentz-Lorenz equation [4.74, 4.82] gives the local index of refraction

$$n^2 = \frac{1 + (8\pi/3)\sum_i \alpha_i N_i}{1 - (4\pi/3)\sum_i \alpha_i N_i}, \qquad (4.99)$$

where N_i is the partial molecular concentration and α_i is the electrical polarizability of substance i contributing to the index of refraction. If the Lorentz formula is applied to bleached photographic layers, a simple relation for the phase shift in the emulsion is obtained [4.82].

$$\Delta\phi/\Delta D = (8.7/a) \cdot kq\alpha, \qquad (4.100)$$

where $k = 2\pi/\lambda$, q is the number of molecules of silver compound from one atom of silver, D is the photometric density before bleaching, a is related to the density and mass of silver and α is the polarizability of the compound embedded in the gelatin.

Equations 4.99 and 4.100 show that the diffraction efficiency of the bleached hologram is proportional to the square of the polarizability of the compound produced by the bleaching. Table 4.10 compares the data of Eq. 4.100 for some silver compounds produced by bleaching. It is clear that $AgHgCl_2$ gives much higher efficiency than AgCl, even though it has a lower refractive index.

Table 4.10 Relative diffraction efficiency for bleached holograms [4.82].

Silver compound	n	q	$\alpha(10^{-30} m^3)$	$\eta(rel)$
AgCl	2.07	1	5.3	1.0
AgBr	2.25	1	6.6	1.6
$Ag_4Fe(CN)_6$	1.56	1/4	35.9	2.9
AgI	2.20	1	9.2	3.0
$AgHgCl_2$	1.82	1	12.4	5.5

The bleached holograms can be redeveloped [4.83] into a structure partially composed of colloidal silver. For this the bleached hologram is exposed to saturation by incoherent light and developed in ascorbic acid solution of a pH of 3. This produces holograms with high efficiency and low noise typical of a colloidal developer.

In a series of papers, recently Kumar and Singh [4.84] have reported the behaviour of different developers, bleaches and their combinations on various silver halide emulsions.

Processing Steps

Technical literature describes a number of processing steps which produce highest possible diffraction efficiency and lowest possible noise. Table 4.11 gives a process for making transmission holograms. Distilled water should be used for making different baths. It is essential to use recommended temperatures of different baths to obtain optimum results. The final film or plate is required to be dried in a vertical position and in a dust free room until the emulsion is completely dry. Irregular drying or remaining water drops may cause stains on the holograms.

Reflection holograms require very careful processing. Use of demineralized water increases efficiency. The temper

Table 4.11 Transmission hologram processing (all solutions at 20±2 °C).

1.	Develop	GP 61	2 min
		or Kodak D-19	5 min
		or Neofin blue (≤ 18 °C)*	5 min
2.	Wash in water		5 min
3.	Fix	G321 fixer	5 min
		or Kodak rapid fixer	5 min
		or G334 fixer	5 min
4.	Wash in running water		5 min
5.	Bleach	ferricyanide bleach	
		or ferric nitrate bleach,* continue for 1 min after plate is cleared.	
6.	Wash in running water		15 min
7.	Wash in	Water + photoflo	2 min
		or methanol	2 min
		or Drysonal	2 min
8.	Dry vertically		

---

* [4.85]

ature control of processing baths are essential. With a suitable combination of developer and bleach images in different colours can be obtained [4.80]. Table 4.12 gives steps for processing of a reflection hologram [4.80]. The exposure should be such that a density between 1.5 and 2.5 is obtained when the plate is developed for the prescribed time.

It has been observed that by varying the pre-bleach density of the plate and development time, fine tuning of the colours can be obtained.

Drying is an important step in reflection hologram processing. Forced drying should never be used. The plate must not be turned around in the course of the drying process. Irregular drying will degrade the quality.

In order to provide a background to the bleached reflection hologram an absorbing paint can be sprayed on the back of the glass plate. A black tape can also be laminated on the glass side of the hologram. To chemically blacken the hologram, the dry glass plate can be placed in a bath of sodium borohydride (0.5 g) in 65/35 mixture of methanol and water. After the hologram becomes dark it is thoroughly washed and left to dry.

Table 4.12 Reflection hologram processing [Exposure with He-Ne laser (633 nm), all baths at 20±2 °C].

For red colour image

1. Develop in pyrochrome developer — 4 min
2. Wash in running water — 5 min
3. Bleach in PBQ-2 bleach, continue for 1 min after plate is cleared.
4. Wash in running water — 15 min
5. Wash in distilled water + wetting agent — 2 min
6. Stand plate upright and leave to dry

For yellow/gold colour image

Process as for red colour image. The bleaching bath, however, should be replaced by potassium dichromate bleach.

For green colour image

Process as for red colour image. The developer should be replaced by Kodak D-19 and the bleaching bath by potassium dichromate bleach.

If an amplitude hologram is baked at 400°C over an alcohol burner or a heating wire until the hologram turns into orange-yellow colour, on cooling to room temperature it retains orange-yellow colour. The hologram becomes more transparent and behaves as a phase hologram exhibiting high efficiency [4.86]. During heating, relatively large silver particles are split into smaller ones and spread in the medium making its refractive index higher. A diffraction efficiency of more than 60% has been achieved in this manner by Guo and Cai [4.86] in their preliminary experiments.

In applications such as real-time holographic interferometry, the photographic process can be speeded up by using a monobath which does not require separate steps of development and fixing [4.88]. The process is a combination of chemical and physical developments. The formula for monobath bleach for Agfa 10E75 plates is given in Table 4.13 [4.81]

Archival Storage of Holograms

Holograms recorded in silver halide emulsion may over a period of time suffer [4.87, 4.88] from one or more of the deterioration listed in Table 4.14. The effect of these deterioration will lead to distortions in fringe spacing, grain size, shape and loss of fringe contrast. This will degrade the quality of the hologram as a display. The holograms intended for measurements over a long period of time also require proper storage.

Table 4.13 Monobath formula for Agfa 10E75 plates [4.81].

Solution A

Metol	3.5 g
Sodium sulphite (anhydrous)	50 g
Hydroquinone	15 g
Potassium alum	5 g
Sodium thiosulphate	50 g
Distilled water	800 ml

Solution B

Sodium hydroxide	10 g
Distilled water	200 ml

Mix 4 parts of A with 1 part of B before use. Process plates for 2 min at 20± 0.5°C.

Table 4.14 Causes of deterioration in holograms recorded in silver halide emulsion [4.87].

Effect	Cause
Darkening of bleached plates	Print out silver in silver halides
Overall yellow stain	Residual complexes (thiosulphate, silver thiosulphate complexes)
Microspots	Oxidizing atmospheric gases (oxygen + ammonia, hydrogen sulfide, SO_2, ozone, nitric oxides etc.), oxidising gases from containers (plastic, wood fresh paint, varnish etc.). Peroxy radicals (from car exhausts)
Local or general damage	Bacterial or fungal attack on gelatin by acidic atmospheric gases
Dimensional changes	High humidity and temperature

Brown and Jacobson [4.87] have recommended proper storage of holograms to minimize the problems of image deterioration. These recommendations are given in Table 4.15.

Table 4.15 Recommendations for hologram storage to minimize the problems of image deterioration [4.87].

---

1. Final washing should be proper to ensure minimal retention of processing chemicals.
2. Avoid storage containers made of wood, chip board and PVC. Enameled steel, stainless steel or anodized aluminium are preferred.
3. Temperatures below $20^{\circ}C$ and humidities within 25-50% are recommended. Also avoid too many humidity and temperasure cycles within short period of time.
4. If possible, avoid areas known to contain oxidizing gases and freshly painted areas.
5. Display holograms using as low light intensity as possible for adequate image brightness.

4.11.4. Hardened Dichromated Gelatin (DCG)

Dichromated gelatin (DCG) is currently the best holographic material for volume phase holograms capable of producing very high diffraction efficiency and low scattering noise. Unfortunately because of poor self life of DCG plate, it cannot be commercialized.

The dichromated gelatin has the following unique characteristics which make it an ideal material for hologram recording.

a) DCG has resolution capability extending beyond 5000 lines/mm. Its response is uniform over a broad range of spatial frequencies from 100 lines/mm to 5000 lines/mm.
(b) The refractive index modulation capacity of DCG is very high.
(c) DCG has low absorption over a wide range of wavelengths.
(d) DCG can give reconstructions even without development.
(e) DCG hologram can be redeveloped to get any desired refractive index modulation and peak diffraction wavelength.
(f) The thickness of DCG can be increased or decreased by controlling the exposure and processing conditions. This property is unique because other materials normally shrink after development.
(g) DCG hologram has high SNR.

Hologram Formation

Dichromated gelatin was first used hologram recording material by Shankoff [4.89] who found that very large index modulation can be obtained in relatively thin films. When a gelatin film, sensitized with ammonium dichromate, is exposed to actinic radiation, the hexavalent chromium ion Cr^{6+} is photoinduced [4.90] to trivalent Cr^{3+} either directly

$$Cr_2O^{2-} + 14H^+ + 6e \longrightarrow 2Cr^{3+} + 7H_2O$$

or indirectly

$$Cr_2O^{2-} + H_2O + 2H^+ \longrightarrow H_2CrO_4$$

$$2H_2CrO_4 + 6H^+ + 3e \longrightarrow 2Cr^{3+} + 5H_2O + 3O.$$

The trivalent chromium forms cross-link bonds between carboxylate groups of gelatin chains (Fig.4.50). This cross-link differentially hardens the gelatin of the exposed and relatively unexposed regions. The hardness differential forms the latent image of the exposure pattern.

Recording of holograms in DCG involves (a) preparation of gelatin layer (b) sensitization, (c) exposure and (d) development. The hologram may be either relief or volume type, determined by the initial bias hardness of the DCG [4.59, 4.91]. The unhardened (soft) gelatin forms a relief

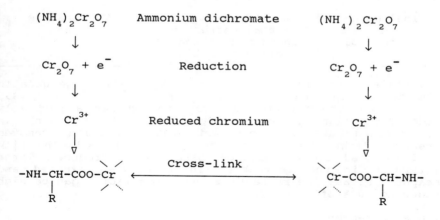

Fig. 4.50 Cross-link mechanism

hologram like a negative resist. Table 4.16 gives a generalized processing procedure for hardened DCG holograms [4.59].

Table 4.16 Generalized processing procedure for hardened DCG hologram [4.59].

Step	Function	Key parameter
1. Preprocessing	Prehardening Prevention of crystallization (amorphous film) anealing and dry	Hardener concentration Temp.>25°C
2. Sensitization	Ammonium dichromate Solution Dry	 Relative humidity
3. Exposure	Creation of hardness differential between exposed and unexposed regions	
4. Initial development with water	Removal of residual chemicals, creation of initial refractive index modulation by swelling and transferring gelatin	pH and temp. of water
5. Final development with isopropanol	Amplification of initial refractive index modulation by rapid dehydration	Temperature

Phase holograms are obtained by processing the gelatin film to obtain large refractive index modulation [4.89, 4.92, 4.93]. The interference pattern exposes a partially hardened DCG film and increases the number of cross-links in the exposed areas. Subsequent water washing removes unreacted dichromate leaving the film unmolested. The film is dehydrated in an alcohol bath which produces large difference in the refractive index between exposed and unexposed areas. The refractive index modulation can be high producing 100% efficiency.

Film Preparation

A widely used method of preparing a DCG film is to

dissolve out the silver halide in a silver halide photographic plate by soaking the unexposed plate in fixer. Kodak 649F spectroscopic plates give good results. However, since these plates are not manufactured for recording of DCG holograms, they require lengthy procedures for the recording of high quality holograms. Table 4.17 gives the processing sequence for the preparation of DCG plates from Kodak 649F plates [4.59]. The concentration of ammonium dichromate used in the sensitization of the gelatin controls the exposure time and the Bragg angle shift. The amount of hardener in the fixer controls the diffraction efficiency and scattering noise. The plate may be dried [4.94] by placing it on a preheated hot plate. If crystallization on the plate is observed, the temperature of the plate should be lowered.

Table 4.17 Preparation of DCG plates from Kodak 649F plates (All steps are at 20°C) [4.59].

Step	Time
Lighting : Room light	
1. Soak in fixer without hardener	10 min
2. Wash in running water (start at temperature 20°C and raise slowly to 33°C, then lower slowly to 20°C).	5 min
3. Soak in Kodak fixer with 3.25% hardener	10 min
4. Wash in running water	10 min
5. Rinse in distilled water	5 min
6. Rinse in photoflo solution	30 s
7. Remove the excess water and dry the plate in horizontal position	Over night
Lighting : Red	
8. Soak in 5% ammonium dichromate solution with 0.5% Photoflo	5 min
9. Wipe glass side and dry the plate in horizontal position	Over night

It is also possible to coat glass plates with gelatin films. A 10 μm film can be made by mixing 1 g ammonium dichromate with 3 g of gelatin and 25 g of water. It is stirred together, heated, filtered and coated on a 8x10 inch glass plate. The coating can be done by spin method (75 rpm) or doctor's blade method.

The plates should be hardened initially so that the film will not dissolve in water during processing. It should be soft enough so that photochemical reaction produces a significant difference in hardness. The degree of prehardening is a critical parameter for obtaining good results. If the hardening is too high, the refractive index modulation decreases. If it is too low, the hologram appears milky. If the hardening is done with formaldehyde or H_2PtCl_6 $6H_2O$, an increase in sensitivity by a factor of 10 is obtained [4.95]. Both of these materials form cross-links with amino groups and not with carboxyl groups.

If the plate is stored in dark, the dark reaction takes place due to which Cr^{6+} ion reduces to Cr^{3+} changing its colour from yellowish to reddish brown. The dark reaction reduces the self-life of the DCG layers. It also increases the hardness of the film which reduces the refractive index modulation capacity of DCG. The rate of dark reaction is low when temperature and relative humidity are low, and the pH is high [4.59]. At relative humidity lower than 40% the shelf-life of DCG can be several weeks. The plates store well in a refrigerator for at least a year [4.96].

The sensitized plates should be kept under relative humidity below 40%. The sensitized emulsion can be placed in a liquid gate containing xylene. This would avoid the need for keeping the relative humidity of the exposure laboratory below 40%.

DCG is most sensitive in hot moist environment [4.96]. At 441 nm less than 10 mJ/cm^2 is required. The spectral sensitivity of DCG is very poor at 633 nm. The sensitivity to longer wavelength can be increased by sensitizing the DCG with dyes like triphenylmethane [4.97] and methylene blue (MB) [4.98]. However, the sensitivity of such plates is an order of magnitude lower than for blue light. To improve the sensitivity of MB sensitized DCG, "external electron donor" compounds are added [4.99, 4.100]. These are tertiary amines with the three bonds of the nitrogen atom attached to groups other than hydrogen. The donor compounds are ethylenediamine tetra-acetic acid (EDTA), triethanolamine (TEA) and N,N'-dimethylformamide (DMF). The advantage of addition of these compounds to MB sensitized DCG is rather limited [4.101]. Addition of tetramethylguanidine as electron donor to MB sensitized DCG gives improved results [4.102]. The high alkalinity of the system allows the unexposed films to have an effective life-time of weeks at room temperature.

Development

For developing the exposed DCG plate, it is soaked in a 0.5% solution of ammonium dichromate and then in a hardener bath. Next step is to swell the film in a bath of warm water. The unhardened areas swell to a larger extent than the hardened areas. The last step is the dehydration of the film. This is the most crucial step which may result in a very high efficiency hologram or in a very poor hologram. Table 4.18 gives the processing steps. If the film is properly prepared and processed efficiencies approaching 100% can be achieved (Fig. 4.51). The most important step is the 100% isopropyl alcohol bath. The plate is placed with emulsion side up in a tray and alcohol is poured onto the plate. Rapid agitation is continued for the entire 3 min ensuring rapid and complete dehydration of the gelatin.

Table 4.18 Processing of DCG plates [4.59].

Step	Direction	Time
1.	Agitate in 0.5% ammonium dichromate solution	5 min
2.	Agitate in rapid fixer without hardener (with hardener for 649 F plates)	5 min
3.	Wash in running water	10 min
4.	Rinse in Photoflo 200 solution	30 s
5.	Agitate in distilled water	2 min
6.	Dehydrate in a 50:50 solution of distilled water and isopropanol	3 min
7.	Dehydrate in 100% isopropanol	3 min
8.	Free-air dry	1 hr
9.	Bake over a hot plate	2 hr

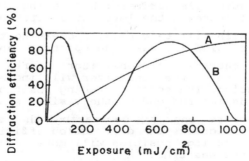

Fig. 4.51. Diffraction efficiency of a DCG hologram as a function of exposure. Curve A- inside a water gate, curve B- in air [4.59].

Optimization

Dichromated gelatin is a very flexible and versatile holographic material. By suitable control of various parameters in the film preparation, exposure and development process, a wide range of holographic properties can be achieved.

The holographic properties in DCG is affected by several parameters as shown in Table 4.19.

Table 4.19 Parameters affecting recording in DCG.

Process	Critical parameters
Film preparation	Film thickness Film hardness Concentration of sensitizer
Exposure	Energy Wavelength Geometry Direction of incidence of reference beam
Development	Temperature of baths Dehydration
Storage	Relative humidity Temperature

The effect of various parameters on the hologram behaviour is discussed below [4.103].

(a) Gelatin film hardness: Decreasing the film hardness will increase the level of scattering.

(b) Gelatin film thickness: Decreasing the film thickness will increase the spectral bandwidth.

(c) Concentration of sensitizer: Decreasing the concentration of the sensitizer will increase the fringe spacing. The spectral tuning position will move towards the red end of the visible spectrum.

(d) Exposure energy: Decreasing the exposure energy will decrease the diffraction efficiency. The spectral tuning position will move towards the red end of the spectrum.

(e) Exposure wavelength: Decreasing the exposure wavelength will shift the spectral tuning position towards the blue end of the spectrum.

(f) Exposure geometry: Decreasing the angle of incidence decreases the fringe spacing. The spectral tuning position will move towards the blue end of the spectrum.

(g) Temperature of the developer bath: Decreasing the process temperature will decrease the diffraction efficiency and the spectral bandwidth.

The factors which influence the reconstruction wavelength are listed below in order of decreasing importance [4.104]:

1. Concentration of dichromate sensitizer
2. Exposure wavelength
3. Exposure energy
4. Initial hardness
5. Processing temperature
6. pH of baths in processing
7. Concentration of water/alcohol baths
8. Processing rate

Depending on the temperatures and times spent in each bath, a variety of effects can be obtained [4.96]. Following points are useful.

(a) Cold baths produce better uniformity and lower noise.

(b) Hot baths can yield very high index modulation but often with increased noise.

(c) Processing and reprocessing can vary the index modulation from nearly nothing to a very high value.

(d) Repeated processing can produce variations of colours and bandwidth. A red shifted hologram can be converted into a blue shifted hologram. Also a 100 nm bandwidth HOE can be shifted to a 20 nm bandwidth HOE by reprocessing.

The direction of shift can be controlled by the ratio of water and alcohol in the bath preceding 100% alcohol. A 30% water-70% alcohol bath will give different results than a 70% water-30% alcohol bath. The amount of shift can be varied by the time in this bath. For example, a red shifted broad band hologram can be converted into a blue shifted narrow band hologram by agitating in 50% water-50% alcohol mixture for 30 s and then agitating in 100% hot alcohol for 30 s [4.96].

The recording in dichromated gelatin is highly affected by the relative humidity in the laboratory. If the sensitized plate is accidentally exposed to relative humidity higher than 40% before, during or after the exposure, it may result in complete failure.

For making HOEs in DCG requires uniformity of holograms to be precisely controlled over the full aperture of the hologram. This needs very tight control over the temperature and relative humidity of the environment at every stage of the process. The uniformity of the fringe structure through the entire gelatin thickness is also very important as this affects the shape of the main diffraction peak and sideband structure. The unwanted sidebands may sometimes be suppressed by producing controlled non uniform fringe structure.

Gelatin grabs moisture easily from the air even from glue and most plastics. The processed hologram must be properly sealed to protect it from environmental changes. It has been observed [4.105] that microwave (2.45 GHz) drying increases the life-time of DCG hologram without affecting the wavelength of reconstruction. The best way is to laminate between glass with enough gelatin removed around the edges to form a 'O' ring seal with a UV curable resin. Spray sealants are not recommended. Steps for sealing DCG holograms are [4.92, 4.94]:

1. Bake hologram at $90^\circ C$ for 2 hours or place hologram in vacuum chamber for 2 hours.
2. Mix 5 ml of cement and one drop catalyst.
3. Place cement in vacuum to remove air bubbles.
4. Cement glass plates together.
5. Cure the cement.

A procedure which requires less number of steps and produces high efficiency holograms even if the relative humidity in the laboratory is more than 40% is given in Table 4.20 [4.106]. The important steps are 10 min soaking in hot water at $63^\circ C$ before sensitization and 10 min baking at $70^\circ C$ after sensitization. These steps ensure adequate hardness of the emulsion so that even higher humidity has little effect on the hologram recording. The technique has produced diffraction efficiencies as high as 95%. Jeong and Song [4.92] have recently shown that if a non-hardening fixer (steps 1 and 9) is used in the process listed in Table 4.20, a diffraction efficiency approaching 100% can be obtained.

Mechanism of Recording

Several theories have been put forward to account for large index changes in DCG. According to the Curran and Shankoff model [4.107], the index differential comes from the gelatin/air interface. During development gelatin is removed from the film surface and cracks are formed at the fringe boundaries. As the film thickness increases the cracks extend from the surface to the substrate. The crack size increases with increase in the exposure. According to

Table 4.20 Simplified procedure for the fabrication of a DCG hologram from Kodak 649F plates [4.106].

	Step	Time
	Preprocessing	
1.	Soak in Kodak fixer with 3.25% hardener	10 min
2.	Wash in running water	10 min
3.	Soak in hot water at 63°C	10 min
	Sensitization	
4.	Soak in 10% ammonium dichromate solution with 1% photoflo	5 min
5.	Wipe glass side of the plate	10 min
6.	Bake at 70°C	10 min
7.	Exposure	
	Development	
8.	Soak in 0.5% ammonium dichromate solution	5 min
9.	Soak in fixer with hardener	5 min
10.	Wash in running water	5 min
11.	Dehydrate in 50:50 solution of distilled water and isopropanol	3 min
12.	Dehydrate in 100% isopropanol	3 min
13.	Bake at 70°C	10 min

this model, the grating structure can be represented as a square-wave function of attenuating refractive index. The existence of cracks is confirmed by the fact that immersing a DCG grating in a index matching liquid would reduce the diffraction efficiency. The efficiency will be restored when the liquid is evaporated. However, this model cannot explain the effect of relative humidity on the diffraction efficiency.

Meyerhofer [4.91] postulated that during dehydration the index modulation is caused due to the formation of chromium/alcohol bonds. Although it may be possible to bind alcohol to reduced chromium sites, it cannot explain the observation that the holographic image is destroyed with only a small amount of water present in the alcohol bath. Further, if a high efficiency DCG hologram is washed in water, liquid chromatography would reveal no alcohol, where-

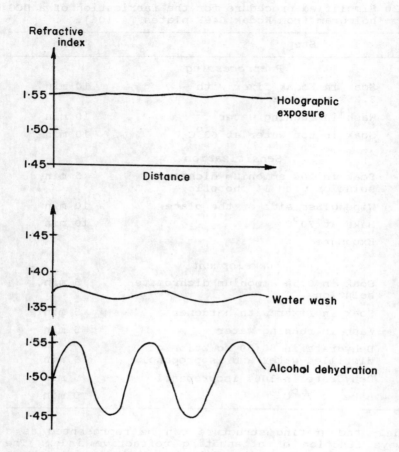

Fig. 4.52. Refractive index changes for low exposure energies in a DCG hologram.

as alcohol is released only by grating of medium efficiency; e.g. for gratings with η= 70 and 99%, the released alcohol concentration has been found to be 1.7 and 0.05 wt. percent respectively [4.97].

The most satisfactory model as proposed by Case [4.108, 4.109] postulates the formation of a very large number of small voids, with dimensions much smaller than the wavelength of light in soft gelatin areas. Lack of higher order diffraction at low exposure energies suggests a sinusoidal index modulation distribution. Figure 4.52 shows the refractive index changes during hologram processing for low exposure energies. The holographic exposure produces a slight index modulation. In the water development, the overall index drops, but the index modulation remains low. During alcohol dehydration the index modulation increases dramatically to very high values.

Grube [4.97] has proposed a molecular model that postulates the sequential formation of a chemical small void and large crack. According to this model the DCG image is formed by gelatin coagulation in the fringe boundary. The dry film swells in water and returns to its original condition on drying. If, however the film is alcohol dehydrated, alcohol enters the unexposed areas more easily and voids are formed as the gelatin strands curl, kink and becomes mutually attractive. If the precipitation is violent enough, the small voids grow to become physical cracks. The whole mechanism of image formation can be visualized as: low index modulation caused only by reduced chromium cross-links, medium index modulation caused by small voids during development and finally large index modulation appearing as a result of void joining into physical cracks.

The recording mechanism has also been explained by isopropanol-gelatin gel formation [4.110, 4.111].

Dichromated gelatin is currently the best material for recording holographic optical elements. DCG applications were initially based on the 10 to 25 μm film thickness range. In recent years, the DCG technology has advanced to films ranging in thickness from 4 to several hundred micrometers [4.112].

Table 4.21 gives some of the applications of DCG of different film thicknesses.

Table 4.21 Applications of dichromated gelatin holograms [4.112]

Film thickness (μm)	Bandwidth (FWHM) (nm)	Applications
4	35	Wide band, increased FOV head-up displays
20	10-15	Current head-up displays based on P43 phosphors
50	10-30	Multiplexers (700-950 nm), high reflection filters
150-300	2-4	Position tunable multiplexers (1250-1650 nm)

Self-Developing Dichromated Gelatin

Relative humidity present during the exposure of dichromated gelatin material plays an important role in controlling the efficiency of the hologram [4.113-4.115]. The latent image induced in DCG is transformed to a relief or phase information under the action of water and alcohol. The introduction of alcohol increases the retention of water molecules in DCG. This implies that the exposure of DCG material under conditions of increased humidity will result in 'self-development' of the material [4.116, 4.117]. If, therefore, water vapour is applied during the exposure by an atomizer, an efficiency of 7.5% can be achieved in real-time [4.118]. If, however, the hologram is recorded on a wet DCG plate (surface layer dried only for 5 minutes at 40°C), a higher efficiency of 20% may be realized which decreases when the hologram dries up [4.119]. Addition of glycerol in DCG layers which acts as a good plasticizer will retain unbound water molecules in DCG. Sherstyuk et al. [4.119] have achieved a real-time efficiency of 30% using a glycerol concentration of 96% with respect to dry gelatin weight.

Silver Halide Sensitized Gelatin

It is basically a silver halide material which is processed to obtain highly efficient volume phase holograms in hardened gelatin. It thus combines the sensitivity of photographic materials and high efficiency, low scattering and high light stability features of DCG. It is 10^3 times more sensitive than DCG.

The technique was originally described by Pennington et al.[4.120] and reviewed by Chang and Winick [4.121] and Graver et al. [4.122]. The emulsion is exposed, developed and bleached. Due to the action of bleach, the developed silver is oxidized to Ag^+ and Cr^{6+} ion is reduced to Cr^{3+}. The Cr^{3+} ion is linked to the gelatin chains near the oxidized silver grains producing a differential hardening between exposed and unexposed areas of the emulsion [4.123]. The emulsion is then fixed to remove excess silver halides.

The efficiency depends on the original gelatin hardness. The hardness can be adjusted by presoaking in hot water [4.124]. Angell [4.123] obtained good results without preprocessing (Table 4.22). The silver halide sensitized gelatin holograms are deteriorated by humidity, hence require proper sealing as DCG holograms. The main drawback of this material is its poor spatial frequency response as compared to DCG [4.124, 4.125]. Figure 4.53 shows the plot of diffraction efficiency of a silver halide DCG hologram as a function of spatial frequency [4.125]. It is observed that the material works efficiently only in the range between 600 and 1200 lines/mm. At higher spatial frequency, the diffraction efficiency is limited due to the granular distribution of silver halide crystals in the emulsion [4.125].

Table 4.22 Processing of silver halide sensitized gelatin holograms for Agfa 8E75HD plates [4.123].

1.	Develop with Kodak D-19	5 min
	or PAAP developer	4 min
2.	Wash in running water	1 min
3.	Bleach in modified R-10 bleach, continue for 30 s after plate is cleared.	
4.	Wash in water	1 min
5.	Soak in fixer	2 min
6.	Wash in water	10 min
7.	Dehydrate in 50% isopropanol	3 min
8.	Dehydrate in 90% isopropanol	3 min
9.	Dehydrate in 100% isopropanol	3 min
10.	Dry in vacuum chamber	

Modified R-10 Bleach

Solution A		Solution B	
Water	500 ml	KBr	92 g
Ammonium dichromate	20 g	Water	1000 ml
Sulphuric acid conc.	14 ml		
Add water to make	1000 ml		

Mix one part of A with 10 parts of water, then add thirty parts of B.

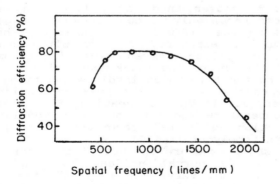

Fig. 4.53. Diffraction efficiency of a silver halide DCG hologram versus spatial frequency [4.125].

4.11.5. Photopolymers

Photopolymer is a material for recording volume phase holograms. The film consists of a dye, initiator, acrylic monomers and polymeric film forming binder. Coatings can be formed from solvent by spin coating method or by doctor knife method. Spectral sensitivity can be adjusted to include blue, green and red wavelengths or combination of these by varying the type of dye sensitizer. The speed of the film varies with composition and range from 10 to 100 mJ/cm^2. The advantage of these materials over DCG is that it has high durability against environmental changes.

Photopolymer is not available commercially. Polaroid and Du Pont companies have allowed laboratory evaluations. Du Pont material is available on Mylar support and protected by a thin Mylar cover sheet. To use the film, the cover sheet is removed and the film is mounted on a glass plate by laminating the tacky photopolymer coating on the glass surface. The Mylar film support protects the coating during handling, exposure and processing operations. After exposure the film is cured for about 60 s with ultraviolet light (5 mW/cm^2). Weak reconstruction can be obtained during exposure. The efficiency increases by ultraviolet curing. Very high efficiencies are obtained by heating the film at $100^{o}C$ in a forced-air convection oven for an hour. Processed holograms are insensitive to humidity and temperature.

During exposure, monomer polymerization starts and proceeds rapidly in regions of bright interference fringes [4.126-4.129]. As a result the monomer is converted into polymer and fresh monomer diffuses in from neighbouring dark fringes. This is responsible for refractive index modulation. The conversion of monomer into polymer is associated with a decrease in volume (shrinkage) which causes the diffusion of unreacted monomer into the polymerizing region. During this process the highly viscous composition gels and hardens, suppressing the diffusion and makes the hologram stable toward further exposure. Uniform ultraviolet (or white light) exposure polymerizes any remaining monomer and permanently fixes the image. The index modulation is greatly enhanced by heating the film which involves additional diffusion of the components [4.129]. The peak reflection wavelength changes very little due to heating. Fig. 4.54 shows the effect of temperature and time on index modulation in plane beam reflection holograms recorded at 514 μm in a typical Du Pont photopolymer [4.130].

Figure 4.55 shows exposure characteristics for Du Pont HRF-352 films [4.129] for normal incidence holographic gratings recorded at 514 nm. Very high reflectivities with a narrow bandwidth are obtained in polymer films. Table 4.23 summarizes the characteristics like thickness (t), speed, exposure wavelength (λ_e), peak reconstruction wavelength

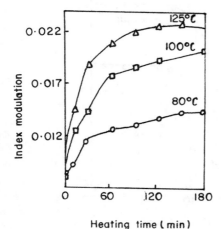

Heating time (min)

Fig. 4.54. Effect of temperature and time on index modulation in plane beam reflection holograms in a typical Du Pont photopolymer [4.130].

Table 4.23 Characteristic parameters for reflection holographic gratings (mirrors) recorded in Du Pont films [4.129].

Type	t (μm)	λ_e (nm)	Speed (mJ/cm^2)	λ_p (nm)	Δλ (nm)	η %	Δn
HRF-150	34	514	60	502	8	88.7	0.008
HRF-352	25	514	20	507	13	99.94	0.029
HRF-410	25	514	100	512	10	98.22	0.018
HRF-410	26	633	200	624	15	99.17	0.024
HRF-410	25	647	400	636	13	99.11	0.019
HRF-700	18	514	15	502	28	99.99	0.068
HRF-150#	38	514	100			99	0.0076

For transmission grating, efficiency measured at 632.8 nm.

(λ_p), bandwidth (FWHM, Δλ), diffraction efficiency and index modulation (Δn) of reflection holographic gratings (mirrors) recorded in Du Pont polymers [4.129].

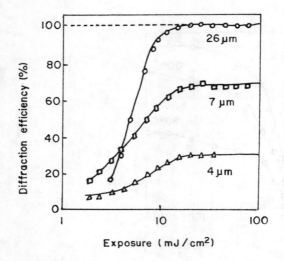

Fig. 4.55. Diffraction efficiency of non-slanted reflection gratings recorded in Du Pont HRF-352 films of three thicknesses (heat processed) [4.129].

The characteristics of photopolymer holograms can be altered by immersing them in organic liquids that swell the coating. The hologram playback wavelength shifts to longer wavelength and the bandwidth also increases. Most of the solvents like ketones, esters, chlorocarbons, xylene, pyridine etc. give good results but the shift is not permanent and the hologram playback wavelength slowly shifts back to its original value as solvent evaporates. The refractive index of the solvent does not affect the diffraction efficiency of the hologram which suggests that microscopic pores, cracks or voids do not play any significant role in forming the hologram, otherwise effect of index matching would have reduced the efficiency.

Russians have developed a material "Reoksan" which uses polymer as a host but the recording is not based on photopolymerization [4.131, 4.132]. Reoksan is an acronym for recording oxidizable material with anthracene. It contains a PMMA matrix, an anthracene derivative as active material and a sensitizer. The material has a resolution capability exceeding 6000 lines/mm. Its sensitivity is in 0.45-0.60 μm range and requires 0.5-1 J/cm^2 for exposure. The material is first saturated with oxygen. During exposure, a photo-oxidation reaction takes place which converts highly polarizable anthracene into anthracene peroxide. This induces phase modulation in real-time. The material is fixed by storing it in the dark. A diffraction efficiency of 80% for holograms of diffuse objects has been demonstrated in this material [4.133].

Calixto [4.134] has reviewed the work on photopolymers carried out prior to 1987.

Materials which are based on polyvinyl carbazole (PVCz), can produce [4.129, 4.130] large refractive index modulation. Holograms recorded in (PVCz) have shown no degradation at 70°C, 95% R.H. over hundreds of hours without any protecting cover plates. The material consists of PVCz as base polymer, an initiator and a sensitizer. Thioflavine-T [4.135] and carbon tetraiodide [4.136] can be used as sensitizer and camphorquinone as reactive initiator. The materials are dissolved in an organic solvent (chloroform) and spin coated to a thickness of about 2 µm. Figure 4.56 shows the spatial frequency response of a PVCz film, while Fig. 4.57 gives the exposure characteristics.

On exposure the sensitizer generates radicals which cause cross-linking reaction between the polymers. The cross-linking results in the difference of the degree of polymerization. Thus the interference pattern is written in the material as differential polymerization. After exposure the material is developed by swelling in toluene and shrinking in pentane.

A very high refractive index modulation is achieved even in a film of 2 µm. This gives a larger angular tolerance to the hologram. Figure 4.58 compares the change in diffraction efficiencies of a PVCz hologram and a silver

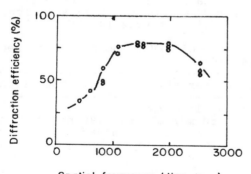

Fig. 4.56. Diffraction efficiency versus spatial frequency of a typical PVCz material [4.135].

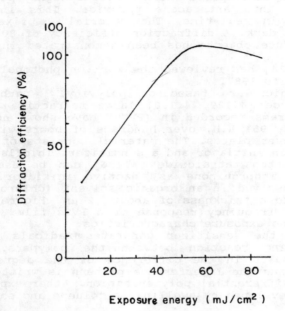

Fig. 4.57. Diffraction efficiency as a function of exposure for a typical PVCz material [4.136].

halide hologram with a change in Bragg angle [4.135]. The holograms in PVCz can give a higher efficiency over a larger angular change (11°) than the silver halide emulsion. Table 4.24 gives the characteristics of PVCz [4.136].

Table 4.24 Characteristics of PVCz holographic recording material [4.136]

	Characteristics	Condition
1. Resolution	>3500 lines/mm	
2. Sensitivity	10-50 mJ/cm^2	at 488 nm for $\eta=35\%$
3. Recording wavelength	<560 nm	$\eta=96\%$
4. Maximum transmittance	93%	
5. Environmental behaviour	unchanged	720 hr without any protecting glass plate at 70°C and 95% R.H.

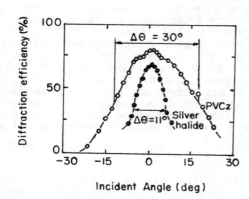

Fig. 4.58. Diffraction efficiency versus incident angle of a PVCz hologram and a silver halide hologram [4.135].

Dye-polymer systems have also high environmental stability. Couture et al. [4.137] have proposed fluorescein dye/polyvinyl alcohol (PVA) and eosin dye/PVA system as high efficiency materials for hologram recording. Table 4.25 gives method for preparation of dye sensitized dichromated

Table 4.25 Procedure for making dye sensitized DC-PVA holograms [4.137].

Solution A

Mix PVA 7% by weight in demineralized water at 80°C.

Solution B

Mix ammonium dichromate 10% by weight in water.

1. Mix solution A and B in ratio of 6:1.
2. Add dye.
3. Make a film about 30 μm thick and dry it in 20-25% R.H. and 25°C for 24 hours.
4. Expose
5. Development

   (a) Develop in 20% water and 80% ethanol for 10 s.
   (b) Soak in 100% ethanol for 3 min.
   (c) Dry with hot air.

---

polyvinyl alcohol (DC-PVA) film for recording of holograms. The dye sensitization improves the diffraction efficiency and speed of recording. A diffraction efficiency of 58% has been achieved in both the dye systems.

Introduction of azo dyes in PVA matrix produces colour films which can be used to record polarization volume holograms in real-time [4.138, 4.139]. The material is erasable and can be reused for many thousand write/read/erase cycles. The samples (15 μm thick) can be prepared by pouring 1.1 ml of a mixture including dye (chrysoidine, Mordant Yellow 3R or methyl orange) dissolved in water in a PVA solution over a leveled microscopic slide. The sample can be used after 12 hours. The material has a uniform frequency response from 500 to 3300 lines/mm. With an exposure of 300 mJ/cm^2, polarization volume hologram with an efficiency of 0.27% can be produced in real-time.

4.11.6. Photoresists

Photoresists are organic materials for producing thin relief phase holograms. Holograms in photoresist are ideal for making nickel masters which are used for embossing replicas.

Photoresists are of two types, negative or positive working. In negative working photoresist the unexposed areas are dissolved in the development, while in the positive working photoresist, the exposed areas are dissolved away in the developer. The negative working photoresist requires a larger exposure usually through the plate so that the exposed photoresist adheres to the substrate during development.

The most widely used photoresist is Shipley Microposit 1350 which is a positive working resist. The sensitivity of this resist is maximum in the ultraviolet. It has adequate sensitivity at 458 nm of argon ion laser and 442 nm of He-Cd laser. However, the sensitivity at 488 nm is poor (600 mJ/cm^2). Its physical and optical properties are given in Table 4.26. The material produces a maximum efficiency at 1500 lines/mm.

The material is coated on a glass substrate by spin method to a layer 1 to 2 μm thick. It is then baked at 75°C for 15 min. On exposure (E) to light, the layer thickness changes. The change in layer thickness Δd is determined by the exposure constant α_o for the photoresist and on the rates of removal (etching) of absorbed molecules (r_1) and unabsorbed molecules during development. Δd is expressed as [4.140]

$$\Delta d = T(r_1 - \Delta r \exp(-\alpha_o E)], \qquad (4.101)$$

where $\Delta r = (r_1 - r_2)$ and T is the exposure time. Thus the

Table 4.26 Properties of Shipley photoresist

Physical properties	
Type of solution	solvent
Appearance- liquid	clear, amber red
solid	pale yellow
Solids content	19.5%
Specific gravity	1.0-1.1 (20°C)
Viscosity	5 centipoises (20°C)
Flash point	46°C
Optical properties	
Useful sensitivity range	340-450 nm
Index of refraction	1.66 - 1.70

change in layer thickness is a nonlinear function of exposure.

To eliminate the effects of material nonlinearity, it is desirable to have a linear depth versus exposure characteristic over a wide exposure range. When the exposed plate is developed in the conventional developer Shipley AZ 1350, the resulting hologram exhibits nonlinear effects. If, however, an alkaline developer, such as Shipley AZ 303 is used, the exposed portions are etched at a faster rate than the unexposed portions. The etch rate can be controlled by diluting the AZ 303 developer with four parts distilled water. Fig. 4.59 shows the thickness change as a function of exposure for two developers [4.140, 4.141].

In the development of photoresist, many physiochemical parameters such as temperature, developer concentration, agitation and degree of solvation complicate the process control. Sthel et al. [4.142] have measured the index modulations during the holographic exposure of sinusoidal pattern and observed a lowering of the recorded index modulation with the decrease in the period.

Ichimura and Ohe [4.143] have reported a visible light sensitive positive working resist which is based on poly-(methacrylate). It requires only 40-60 mJ/cm^2 at 488 nm [Fig 4.60]. A copolymer hexafluorophosphate and sensitizer (2-benzoyl-3-(p-dimethyl aminophenyl)-2-propene-nitrile) are dissolved in diglyme and spin coated on a glass plate. The coated plate is baked at 80°C for 10 min. The exposed plate is again baked at 110°C for 10 min (post baking). The plate is developed in (1-butanol-methanol). An efficiency of 23% at a spatial frequency of 800 lines/mm has been achieved [4.143].

Photoresist materials based on a derivative of polystrene, poly(p-t-butoxycarbonulyxostyrene) (t-BOC) has been shown [4.144] to have sensitivity below 4mJ/cm^2 at 442 nm and resolution up to 500 lines/mm. The material is dissolved

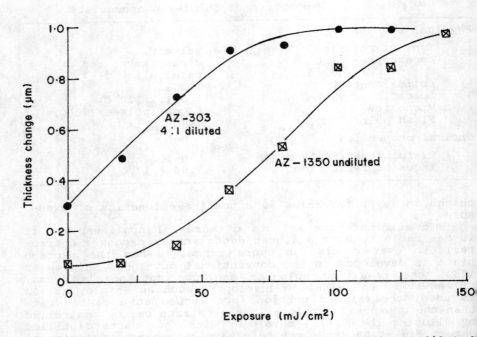

Fig. 4.59. Thickness changes versus exposure for undiluted and diluted developer [4.140, 4.141].

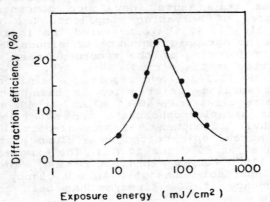

Fig. 4.60. Diffraction efficiency of a holographic grating recorded at 488 nm in a poly(methacrylate) based positive resist [4.143].

in propylene glycol methyl ether acetate and spin coated on a substrate to achieve a film of 1 to 3 µm thick and up to 20 cm in diameter. The films are stable at room temperature. During exposure the absorbed energy is transferred to the onium salt. The excited onium salt generates acidic decomposition products which catalyzes the thermolysis of the t-BOC group. The reaction is very fast at 100°C. The photochemically altered polymer poly-(p-hydroxy- styrene) has a higher refractive index and a smaller thickness than the original material. Thus the material produces surface relief holograms. This material requires much lower exposure compared to other photoresist materials. The material has been used to make holographic elements for photoablation using high power pulsed lasers [4.144].

4.11.7. Photothermoplastics

Thermoplastic is a material for producing surface relief thin phase holograms. The thermoplastic material repeatedly softens and hardens when heated and cooled. The material has a multilayer structure on a substrate of glass or film [4.145]. It consists of three thin layers: doped tin or indium oxide (a transparent conductor), polyvinyl carbazole sensitized with trinitro-9-fluorenone (photoconductive organic polymer) and staybelite Ester 10 (thermoplastic substance, a resin about 0.7 µm thick). The material has high sensitivity over the whole visible spectrum. Figures 4.61 and 4.62 respectively show the variation of diffraction efficiency with exposure and spatial frequency for a typical thermoplastic material [4.146].

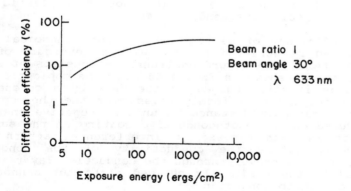

Fig. 4.61. Variation of diffraction efficiency of a hologram recorded in photothermoplastic with exposure energy [4.146].

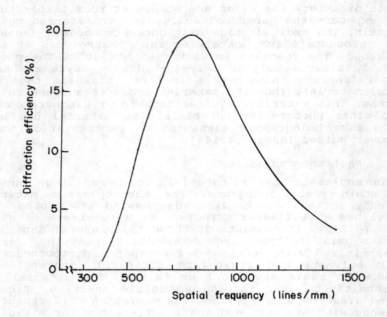

Fig. 4.62. Variation of diffraction efficiency of a hologram recorded in photothermoplastic with spatial frequency [4.146].

An efficiency of as high as 34% for sinusoidal phase gratings can be achieved [4.147] with an exposure of less than 100 mJ/cm^2. The recording technique consists of a number of steps as shown in Fig. 4.63. The thermoplastic resin is first positively charged in the dark by a corona discharge device which uniformly moves over the thermoplastic plate at a constant distance. A uniform negative charge is thus induced on the photoconductive coating on the surface. The plate is then exposed to interference pattern which periodically alters the conductivity of the conductive layer. Electrons travel through the conductive layer and are attracted to the positively charged plate but cannot pass through the photo conductor.

The electrostatic field is further increased by recharging the surface (after blocking the light) once again by the corona discharge as in the first step. The charge pattern creates a spatially periodic static electric field of

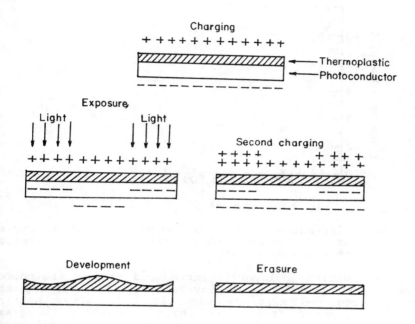

Fig. 4.63. Hologram recording steps in a photothermoplastic material.

-10^6 V/cm. The film is developed by passing a current through the conductive layer, heating the plate and softening the thermoplastic film. The softened film deforms under the static electric field, becoming thicker in the unexposed areas and thinner at the illuminated areas (where the field is higher). As the film cools to the room temperature, the thickness variation gets frozen in.

The complete charging, exposing and developing cycle takes less than a minute. The plate can be erased by flooding it with light and passing a current pulse through the conductive layer which heats up the thermoplastic layer and resoftens it. This smooths out the deformations in the film and the previously recorded hologram is erased. Before the next exposure, the plate is blasted with compressed air or dry nitrogen to cool the thermoplastic layer to the room temperature.

The thermoplastic holograms are usually associated with frost which scatters light. The frost is reduced substanti-

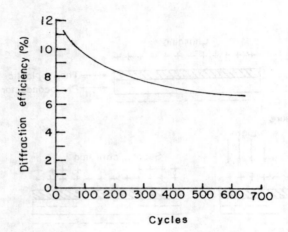

Fig. 4.64. Diffraction efficiency versus number of cycles of hologram recordings in a typical thermoplastic material.

ally if an insulating layer is introduced between the photo conductor and thermoplastic layers [4.148]. Applying the corona charge and exposure simultaneously after heating would also produce frost free, high resolution holograms [4.146]. Photothermoplastics can be recycled several hundred times without any appreciable deterioration. Fig 4.64 shows the results of diffraction efficiency measurements for 700 cycles of hologram recordings in a typical thermoplastic material for exposure at 633 nm and 30° angle between the two beams [4.146]. The material is optically inert when not charged, so there is no degradation from exposure to heat and actinic radiation. The stability of the developed thermoplastic hologram is excellent. The main drawback of the material is the requirement of a complex apparatus for controlled charging and development. The material is ideal for holographic nondestructive testing of materials.

4.11.8. *Photochromic Materials*

Photochromics are real-time recyclable materials. They require no processing for development and can be erased and reused. The holograms can be readout during or immediately after the recording. This property is useful for holographic interferometry. There is no inherent resolution limit since they are essentially grain free and operate on atomic or molecular scale. Their storage capacity is high since the storage process occurs throughout the volume of the material. This property is useful for storing a large number of holograms in the same volume [4.149].

Photochromic materials change their transmission spectrum in response to exposure of light of appropriate wavelength. They, in general, become dark under the action of shortwave visible or ultraviolet radiation. They are bleached by exposure to longwave visible or infrared radiation. This reversibility of the colour change distinguishes these materials from other photosensitive materials [4.150].

Both organic [4.151] and inorganic [4.152, 4.153] materials have been studied for photochromism. The photochromic behaviour in inorganic crystalline materials can be explained with the help of Fig. 4.65. The photochromism involves photo reversible charge transfer between two types of electron traps (colour centres). The two species of traps have characteristic absorption bands corresponding to their photo excitation into the conduction band. At thermal equilibrium, the trapped electrons occupy the lower energy A traps. The material is said to be in *unswitched state* when the absorption spectrum of the material exhibits band-A only.

A transformation from band-A to band-B can be achieved by exposing the material in band-A. The electrons in band-A are excited into the conduction band and are captured by B-traps. This modifies the absorption spectrum which now contains only band-B. Under such a condition the material is said to have gone into the *switched stste*. The transformation from switched state to unswitched state can be achieved by illuminating the material with light of appropriate wavelength corresponding to band-B or raising the temperature of the crystal. The transformation from switched state to unswitched state corresponds to the recording step, while the reverse transformation constitutes the erase step.

Holograms are recorded in photochromic materials generally by selective optical erasing or bleaching of material which has been darkened by uniform exposure of light. When the material is exposed to an interference pattern of erase light, the transmission increases at the bright portion of the pattern due to bleaching effect. This creates an absorption hologram. To erase the hologram, it is illuminated by the switching light which darkens the crystal uniformly.

A large number of photochromic materials have been studied, some of which are CaF_2:La,Na [4.154], $SrTiO_3$:Ni,Mo [4.155] and alkali halide crystals [4.153, 4.156, 4.157]. In most of the materials, the switching requires exposure to radiation near 400 nm. The optical erase is carried out by exposing the material to visible light.

In spite of several advantages, the photochromic materials have not been popular for hologram recording. This is because their sensitivity, efficiency and storage time are low. The reconstruction beam usually degrades the stored information. In alkali halide crystals the playback can be

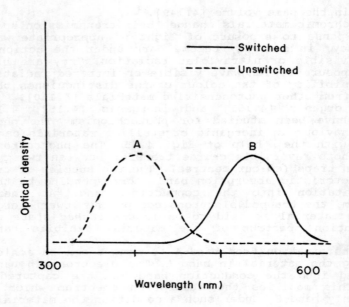

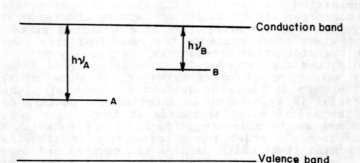

Fig. 4.65. Photochromic behaviour in inorganic crystalline materials [4.149].

nondestructive if the temperature of the material is lowered after recording.

Photodichroics

Photodichroism can also be used to create switched and

unswitched states in suitable crystals. This happens through selective alignment of anisotropic absorption centres induced by exposure to linearly polarized light [4.158]. The material selectively absorbs light of a certain polarization only. Na-doped KCl crystal is a good example of photodichroism [4.159]. It has been shown that photodichroism can produce better writing efficiency than photochromism [4.160].

4.11.9. Photorefractive Crystals (Electro-Optic Materials)

Photorefractive crystals are excellent for recording volume phase holograms in real-time. These materials have excellent resolution, efficiency, storage capacity, sensitivity and reversibility. The application of these materials for hologram recording was first considered by Chen et al. [4.161], who suggested that "optical damage" effect [4.162] in these crystals could be exploited to record a thick hologram.

Materials

Lithium niobate was the first ferroelectric material used for hologram recording [4.161]. Since then several perovskites and paraelectric crystals have been used as recyclable holographic materials. The important crystals are lithium niobate [$LiNbO_3$], lithium tantalate [$LiTaO_3$], barium titanate [$BaTiO_3$], potassium tantalate niobate KTN, a mixture of $KTaO_3$ and $KNbO_3$ [$KTa_{0.65}Nb_{0.35}O_3$], barium sodium niobate (SBN) [$NaBa_{0.65}Na_{0.35})_3$, bismuth silicon oxide (BSO) [$Bi_{12}SiO_{20}$], bismuth germanium oxide (BGO) ($Bi_{12}GeO_{20}$], GaAs, InP and PLZT ceramics. The crystals are grown using Czochralski method. The grown crystals are cut and polished. The optic axis (c-axis) of the crystal is kept parallel to the plane of polarization of the hologram recording beams.

Mechanism of Recording

The photorefractive materials contain localized centres with trapped electrons that can be excited into the conduction band by the action of light [4.163]. These materials have dark conductivity which allows charges to freeze in place. The dark storage time varies from crystal to crystal. For example SBN has a dark storage time of a few seconds, while for lithium niobate it is a few years.

The magnitude of the refractive index change Δn under the action of light is determined by the material's effective Pockels coefficient r_{eff}

$$\Delta n = -(1/2)(n^3 r_{eff} E), \qquad (4.102)$$

where E is the light-induced electrostatic field. $BaTiO_3$ has a large Pockels coefficient, while GaAs and BSO have small Pockels coefficient.

When the material is exposed to an interference pattern the electric charges from interference maxima drift and/or diffuse and are (trapped) collected at the interference minima (Fig. 4.66). The space charge pattern creates a strong spatially periodic field (10^3 V/cm). This field

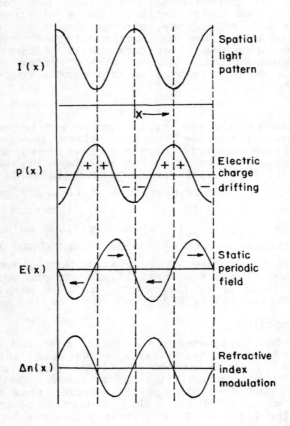

Fig. 4.66. Mechanism of photorefractive effect [4.164].

deforms the crystal by the Pockels effect and causes a refractive index modulation producing a hologram as shown [4.164].

Materials which lack inversion symmetry exhibits a strong photorefractive effect. The change in the index of refraction is independent of the total light intensity. Thus even weak beams can induce strong optical nonlinearities in photorefractive crystals. In crystals like $BaTiO_3$ and SBN though the Pockels coefficient is large, the speed of charge migration is slow, while opposite is the case for BSO, GaAs and InP. $LiNbO_3$ on the other hand exhibits weak Pockels effect and slow speed of charge migration while KTN is strong and fast.

The undoped crystals have an impurity of Fe which introduces traps for hologram recording [4.165]. Crystals are usually doped with iron impurity. In hologram recording process, the Fe^{3+} ions are empty traps and Fe^{2+} ion are occupied traps. Their relative concentrations determine the holographic behaviour of the crystal. When $LiNbO_3$ is doped with Mn, Fe or Cu, it produces large increase in diffraction efficiency and recording sensitivity [4.166]. The Fe doped $LiNbO_3$ gives diffraction efficiencies approaching 100% and shows a 500 fold increase in sensitivity over undoped crystal [4.167]. Table 4.28 gives the dark storage time and energy density required for 1% efficiency and other useful parameters in some of the photorefractive crystals [4.168].

Table 4.28 Some parameters for photorefractive crystals.

Crystal	Refractive index n	Index modulation Δn	Linear Electro optic Tensor (pm /V)	λ (nm)	Exposure Energy (mJ/cm^2)	Storage Time
Fe:LiNbO$_3$	2.259	10	31	633	200	1 yr
Fe:LiTaO$_3$	2.227	10	31	488	11	1 yr
Ce: SBN					1.5	1 m
KTN	2.350		400	488	0.05	7 m
BSO	2.540		5	620	0.3	2 hr
BGO	2.550		3.4	510		
GaAs	3.500		1.2	1060		
BaTiO$_3$	2.365		80	546		15 hr

The spatial resolution is limited only by the distance between traps. The distance between traps is of the order of 100 nm, so very high spatial frequencies can be recorded. However, due to statistical fluctuations of the trapped electrons scattering noise may be severe at smaller grating spacings. The speed and diffraction efficiency can be enhanced by the application of electric field across the crystal. The holograms can be recorded sequentially at different angles.

Fixing and Erasure

The recorded hologram is immediately visible. It can be erased by exposure to a uniform beam of light which releases the trapped electrons. These released electrons redistribute evenly throughout the volume. This erases the hologram as the crystal returns back to its original condition having a uniform distribution of trapped electrons. It is thus clear that continued readout of a hologram will erase it. This is a undesirable feature of photorefractive crystals.

The hologram in these crystals can be fixed by converting the charge patterns to the patterns of ions which are not sensitive to light [4.166]. The hologram in $Fe:LiNbO_3$ can be fixed via thermally activated ionic conductivity [4.169]. The $LiNbO_3$ crystal is heated during or after storage to a temperature above $100°C$. The resulting ionic pattern is frozen upon cooling to the room temperature. The fixing process neutralizes the space charge field but the ionic pattern is not erased. Hence the hologram can be readout without erasure. The write/erase behaviour of $LiNbO_3$ is shown [4.166] in Fig 4.67. A high efficiency fixing is obtained in heavily doped crystals [4.170].

The fixed holograms in $LiNbO_3$ can be erased [4.170] by heating them to the fixing temperature and exposing them to uniform light.

$BaTiO_3$ and SBN have low coercive fields. These crystals require respectively an electrical field of 1.1 kV/cm and 970 V/cm to reverse the internal polarization [4.171]. After recording the hologram, the application of the field will produce a spatial pattern of domain reversal. The hologram can then be readout without erasure. To erase the hologram the crystal is poled with a field to remove all domains. If the hologram is recorded in SBN crystal heated above the Curie temperature ($\approx 50°C$) at which the crystal switches from a ferroelectric to a paraelectric state, the field pattern of the hologram induces polarization domains as it cools to room temperature [4.172]. This results in an optically and thermally stable hologram. In some materials like SBN the high level of intensities during recording fixes the holo-

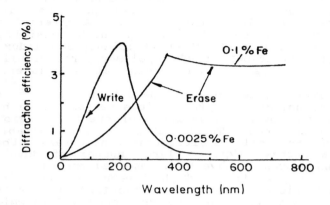

Fig. 4.67. The write/erase behaviour of LiNbO$_3$ with different Fe concentrations [4.166].

gram due to reversal of polarization or heating effect [4.173].

It is possible to record a hologram with 100% diffraction efficiency in a 1 cm thick crystal of LiNbO$_3$. It is also possible to record 1000 holograms with usable levels of diffraction efficiency in the reconstructed image. An efficiency of 95% has been achieved in BGO crystal [4.174].

Holographic recording in photorefractive crystals has opened up a new branch of modern optics, known as *dynamic holography* or *real-time holography* in which the read-write processes are performed simultaneously [4.161, 4.164]. The volume and the real-time write-erase capability of the material allow the interference of incident light beam with its own diffracted beam within the material. The writing beams can also be diffracted by the recorded hologram. This, known as 'self diffraction', can change the wave amplitude, phase, polarization and frequency. The phase shift between the incident fringe pattern and the recorded hologram causes a dynamic redistribution of the interference field [4.164].

Das and Singh [4.175] have recently reviewed the work on dynamic holography, nicely covering the important developments, current status and suggestions for future research directions.

4.11.10. *Summary of Recording Materials*

In the preceding sections several hologram recording materials have been considered. While some of these materials like dichromated gelatin can be prepared in the laboratory, the others like photopolymers have shown potential for commercialization. No single material possesses all the requirements of a holographic material. A material is yet to be discovered which will have the high sensitivity of silver halides, high diffraction efficiency and index modulation capability of DCG and photopolymers, recyclability of photorefractive crystals, and useful at all laser wavelengths.

The silver halide materials have been the most popular choice of the holographers for obvious reasons of high exposure sensitivity over a wide range of spectral regions and high resolving power. These materials are suitable for transmission as well as reflection holograms, both of amplitude and phase type. A large number of developers, bleaches and processes have been reported for silver halide materials. Most of the work has been on improving the diffraction efficiency and SNR, particularly for reflection holograms. Recently, Ilford has developed a film which can be a true panchromatic film useful for recording a multicolour hologram. The introduction of BIPS layer in red sensitive Ilford emulsion is an advantage for making master reflection holograms for replication.

The recording sensitivity of DCG has been extended to red wavelength making it possible to record multicolour reflection holograms. This material is very difficult to handle and the recorded holograms are sensitive to environmental conditions, yet this is the only material suitable for very high efficiency and low noise holograms. Photopolymers are expected to replace DCG as these are also capable of producing large index modulation and high diffraction efficiencies and are free from the disadvantages of DCG. Photopolymers do not require lengthy controlled processing techniques and can be processed *in situ*. The photopolymer holograms are insensitive to environmental changes.

Photoresists are suitable for producing surface relief holograms for making masters needed for replication by embossing techniques. This material is most sensitive to ultraviolet-blue light only. Efforts have recently been made to make photoresist material sensitive at red wavelength.

Photorefractive crystals are very promising materials for real-time holography. They can be recycled. Photothermoplastics can also be recycled several hundred times and are most suitable for holographic interferometry.

Table 4.29 shows the useful parameters of different hologram recording materials. Various values given are those reported for commercial and experimental materials. The diffraction efficiencies for a number of materials have been compared in Fig. 4.68 for plane beam holograms at unity beam ratio [4.59].

Several other recording materials have also been used

Table 4.29 Characteristics of recording materials for holography

Material	Usable thickness (μm)	Sensitivity (nm)	Exposure (mJ/cm^2)	Resolution (lines/mm)	Efficiency (%)	Type of hologram *
Silver halide	5-16	400-700	5×10^{-2}	1000-10000	75	A, P
DCG	0.5-100	250-700	5-300	10000	9	P
Photopolymers	3-100	350-700	1-1000	5000	99	P
Photoresists	1	250-500 633	10-100	3000	35-95	P
Photothermoplastics	0.3-1.2	350-650	10^{-2}	1500	30-90	P
Photochromic	>10^3	300-700	10-100	>5000	2	A
Photorefractive crystals	>10^3	350-700	10-1000	1500-10000	20-95	P

* A - Amplitude, P - Phase

for hologram recording. These include magneto-optic materials, metal films, elastomer devices, liquid crystal photoconductor devices, As_2S_3 etc.

4.11.11. Health Hazards of Hologram Processing Chemicals

Most of the processing chemicals pose health hazards when inhaled, swallowed or when come in contact with body. While some chemicals like ferric nitrate produce minor skin irritation, others like bromine vapour can lead to mental deterioration upon prolonged exposure. In many cases the damage is irreversible when working with these materials. Rubber gloves, rubber apron, full face shield respirator and ventilation should be used. A good exhaust system should be used so that the air should enter, pass around the worker, and chemicals and then exit out. A fume hood is essential. The hands should be washed well with soap and hot water before leaving the processing laboratory. All work surfaces should be wet wiped frequently. In many cases immediate flushing with water helps. If a holographer has a casual

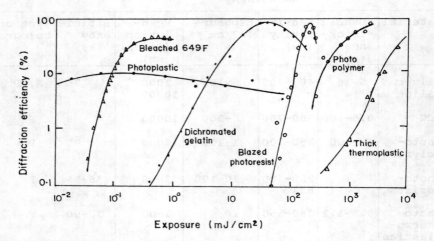

Fig. 4.68. Diffraction efficiency versus exposure curves for plane beam holograms recorded in different recording materials [4.59].

attitude towards chemical handling, it can be dangerous, even deadly. Table 4.30 gives the possible hazards of different chemicals used for holography [4.176, 4.177].

Table 4.30 Health hazards of hologram processing chemicals [4.176, 4.177].

Chemical	Appearance	Symptoms
Acetic acid CH_3COOH	Clear liquid with pungent odour	Irritating, conjuctivities
Ammonium chloride NH_4Cl	White crystals	Fumes on inhalation can cause nausea, vomiting and kidney damage.
Ammonium dichromate $(NH_4)_2Cr_2O_7$	Yellow needles	Ulcers, sores of the skin, perforation of nasal septum, kidney damage.
Ammonium thiosulphate $(NH_4)S_2O_2$	White crystals	Possible carcinogen, small quantity ingested may cause permanent injury.

Table 4.30 contd.

Chemical	Appearance	Symptoms
Ammonium thiocyanate NH_4CNS	Colourless solid	Skin eruptions, izziness, nausea and disturbance of the nervous system.
Ascorbic acid $C_2HO_3:C_4H_7O_3$ (Vitamin C)	White crystals	Not hazardous
Boric acid H_3BO_3	White crystals	Not highly toxic, quantity exceeding 15 g if ingested is highly toxic.
Bromine B_2	Crystals or dark red liquid	Irritating to eyes and respiratory tract, pulmonary edema (fluid in lungs), depression, psychosis and mental deterioration.
Ethanol C_2H_5OH	Colourless liquid	Central nervous system depressant.
Ferric nitrate $Fe(NO_3)_3 6H_2O$	Crystals	Local irritant on skin.
Glycerine $C_3H_8O_3$	Colourless or pale yellow liquid	An irritant if inhaled as a mist.
Hydrogen peroxide H_2O_2	Clear liquid	Blistering of skin, irritation to body tissues.
Hydroquinone $C_6H_4I_{14}(OH)_2$		Dermatitis, ingestion of 1 g causes nausea and death.
Iodine I_2	Violet black	Vapour irritating to lungs, ingestion of small amounts may cause permanent injury.
Isopropanol C_3H_8O	Clear liquid	Local irritant, corneal burns and eye damage, blurred vision, nasal and respiratory irritation, dizziness, kidney damage, unconsciousness.

Table 4.30 contd.

Chemical	Appearance	Symptoms
Mercuric chloride $HgCl_2$	White crystals	Slow poison, dryness of the throat and mouth, tremors and psychic disturbances, memory loss, insomnia, lack of confidence, depression, colitis gum and mouth effects.
Methanol CH_3OH	Clear liquid	Cumulative poison, toxic to nervous system and to optic nerves and retina, permanent damage to visual system may lead to blindness.
Metol P-methyl aminophenol sulphate $(CH_3NHC_6H_4OH)_2 \cdot H_2SO_4$	Colourless	Causes allergic skin reaction.
Naptha (Index matching fluid)	Clear liquid	Irritating to skin, headache, drowsiness and possibly coma.
Perchloro-ethylene CCl_2CCl_2	Colourless liquid	Kidney and liver damage, irritating to gastro-intestinal tract, nausea, diarrhea, injurious to eyes and mucous membrane.
Parabenzo-quinone (Quinone) (PBQ)	Characteristic irritating odour	Damage to skin, cell death, disturbances of vision, conjuctivitis, loss of vision.
Phenidone $C_9H_{10}N_2O$	Crystals	Low toxicity, no skin irritation
Potassium aluminium sulphate $KAl(SO_4)_2$	White powder	May cause dermatitis
Potassium bromide KBr	Colourless	Acne-type skin eruptions, irritating to eyes and mucous membranes.

Table 4.30 contd.

Chemical	Appearance	Symptoms
Potassium carbonate K_2CO_3	Colourless	Strong caustic, irritant.
Potassium ferrocynide $KFe(CN)_6 3H_2O$	Yellow crystals	Low toxicity
Potassium hydroxide KOH	White	Highly toxic
Potassium iodide KI	White granules	Skin rashes, headache, irritation, mucous membranes, pimples, boils.
Potassium permanganate $KMnO_4$	Dark purple crystals	Strong irritant, highly toxic if ingested or inhaled.
Pyrocatechol (Phenol)		Dermatitis, digestive disturbances, vomiting, difficulty in swallowing, excessive salivation, headache, dizziness, mental disturbances.
Pyrogallol (Pyrogallic acid) $C_6H_3(OH)_3$	White crystals	Readily absorbed through skin permanent injury or death resulting from short exposure to small quantities, can cause vomiting, diarrhea, convulsions, circulatory collapse, kidney and liver damage, destruction of red blood cell renal and heptic damage.
Sodium carbonate Na_2CO_3	White powder	Concentrated solutions in contact with skin or eyes causes local necrosis (death of a part of a body).
Sodium hydroxide NaOH	Clear liquid	Upper respiratory tract and lung tissue damage.
Sodium phosphate	Translucent crystals	A mild irritant

Table 4.30 contd.

Chemical	Appearance	Symptoms
Sequestrene (Na-EDTA) $C_{10}H_{14}O_8Na_2 \cdot 2H_2O$	White crystals	Kidney damage
Sodium thiosulphate $Na_2S_2O_3H_2O$	Colourless crystals	Low toxicity
Sulphuric acid H_2SO_4	Colourless liquid	Severe burns and rapid destruction of tissues, inflammation of the upper respiratory tract leading to chronic bronchitis.
TEA	Clear liquid	Liver and kidney damage.
Xylene $C_6H_4(CH_3)_2$	Colourless liquid	Irritating to the skin and upper respiratory tract, common air contaminant

4.12. Display Holography

Holographic displays have been a subject of interest right from the beginning when the first laser hologram of a three dimensional object was recorded. The striking three dimensional properties of holographic images make holograms ideal for displays. Holography can produce spectacular displays for scientific, educational, medical, artistic and commercial purposes. Holographic displays would avoid the risk of theft of art objects. The same object can be displayed at several places at the same time, thus permitting much wider exposure of rare items.

To day holographic studios exist in several countries for display holography. These holographic studios are equipped for recording of a wide range of objects and compositions including people and animals. A laser display hologram of the full size statue of Venus de Milo (height 2.18 m) has been made on a 1.0×1.5 m^2 size photographic plate by Tribillon and Fournier of Besancon, France. The holographic studio of State Optical Institute, Leningrad has recorded holograms of art work from Hermitage collection and museum articles. It has also made holographic portraits of Soviet Scientists. The studio has created the facilities for recording of 120×80 m^2 monochrome reflection and 60×40 cm^2 size colour reflection holograms. A mobile hologram recording unit is under fabrication at this studio to record holograms of sizes up to 1.5x1.5 m² directly in museums, exhibitions and other places of interests [4.178].

4.12.1. Requirement of a Display Hologram

An ideal display hologram should meet the following requirements [4.179].

White Light Viewable Image

The holograms should be preferably viewed with a white light source. Finite size of the source produces blur in the image. A typical white light source is the sun. It is equivalent to the case of reconstructing a hologram with a light source of 9 mm diameter from a distance of one metre. The image blur is around 0.6 mm for an image point at 10 cm from the hologram. In other light sources, the apparent size of the source can be reduced.

Focussed Image

The image should appear to be well focussed, when the hologram is viewed by a white light source. The image is blurred due to the finite size of the source and due to its spectral bandwidth. The blurring increases with the increase in the distance of the image point from the hologram. The average distance of the image from the hologram can be minimized if the image straddles the hologram plane.

Wide Field-of-View

The field-of-view available should be wide to allow a large parallax. This also allows simultaneous viewing by several observers. The vertical field-of-view may be restricted (as in a rainbow hologram) but the horizontal field must be large.

High Resolution

The hologram should produce an image with excellent sharpness.

Low Image Aberrations

The image must be aberration free and without distortions. The image should not move as the observer moves horizontally or vertically in the field. Sometimes it becomes difficult to fuse the views seen by two eyes in the brain which makes the viewing uncomfortable.

Low Background Noise

Background noise reduces the contrast of the image. Hence the processing chemistry should not introduce scattering noise. Interference between light scattered by different parts of the body also introduces background noise (flare light).

True Colour

The hologram should produce the image with true colour. For monochrome object, while a display with only one colour may be satisfactory for many purposes, the addition of false colours may improve the display.

Ambience

A hologram must display the image with proper contrast in high ambient light brightness. The image should be acceptably bright.

Hologram Durability

The hologram medium should be resistant to humidity, temperature, heat, abrasion, intense light, etc.

Some of the techniques developed for holographic displays are discussed below.

4.12.2. $360°$ Holograms

A hologram reconstructs only that portion of the object whose scattered light is received by the photographic plate during recording. The front, sides and back of the object can be recorded on three, four or more photographic plates [4.180, 4.181]. Such holograms give $360°$ view of the object. A hologram recorded on a cylinder of film which surrounds the object gives complete $360°$ view of the object [4.182-4.185].

Figure 4.69 shows a simple set up for making a $360°$ hologram [4.183]. The object is placed in a glass cylinder at its centre. A strip of photographic film is fixed to the inner surface of the cylinder with the emulsion side facing the object. The laser beam is expanded by a powerful microscope objective. The central portion of the laser beam illuminates the object and the remaining portion which falls on the film acts as the reference beam. After development the film is put back in its original position and illuminated again with the laser beam. The object, of course, is removed. The image is viewed from all sides of the cylindrical film.

A $360°$ hologram on a flat plate can be recorded by multiple exposure technique [4.186]. For this, the object is placed on a turn table and its normal off-axis hologram is recorded on a photographic plate which is exposed through a vertical slit placed in front of it (Fig. 4.70). The object is rotated by small angles about a vertical axis and a series of holograms are recorded successively by moving the slit in opposite direction about the horizontal axis by the width of the slit. Thus a $360°$ view of the object can be recorded on a flat photographic plate. When this hologram is reconstructed, a single image of the object is formed. When the observer moves his head from side to side, the

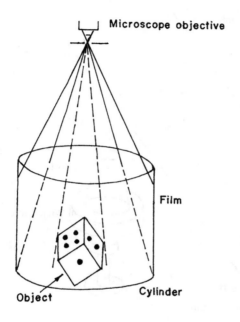

Fig. 4.69. Optical system for recording a 360° hologram [4.183].

image appears to rotate about the vertical axis.

Another technique which can be used to produce a 360° view flat hologram utilizes both the sides of the photographic plate [4.187]. Such a double-sided hologram reconstructs the front of the object from one side and its back from the other side of the plate. If properly recorded, the hologram preserves the correct perspective of the object. Figure 4.71 shows the steps involved in the double-sided reflection hologram [4.187]. First a transmission hologram of the object (front side) is recorded as hologram H_1 using the reference beam R_1. The back side is recorded as a reflection hologram (H_2) by using the reference beam R_2. Then the hologram H_1 is illuminated with the conjugate of R_1

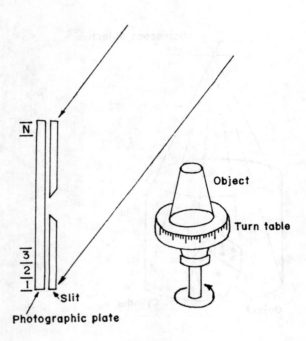

Fig. 4.70. Recording of a 360° hologram on a flat holographic plate [4.186].

to produce the real image of the object at its original place. Now plate H_2 is given another exposure to record this real image (front side of object) by using the reference beam R_3. The hologram H_2 after development is reconstructed by beams R_2 and R_4 to produce front and back side of the object from the two sides of the hologram (Fig. 4.71).

Figures 4.72 and 4.73 show arrangements of optical components for recording of a double-sided hologram which involves minimum changes from the recording of primary hologram to the final hologram [4.188]. Laser beams O_1 and O_2 illuminate the front and the back side of the object from a vertical inclination of 45°. Hologram H_1 is recorded by the beams 1 and 2, while the beams 3 and 4 record the hologram

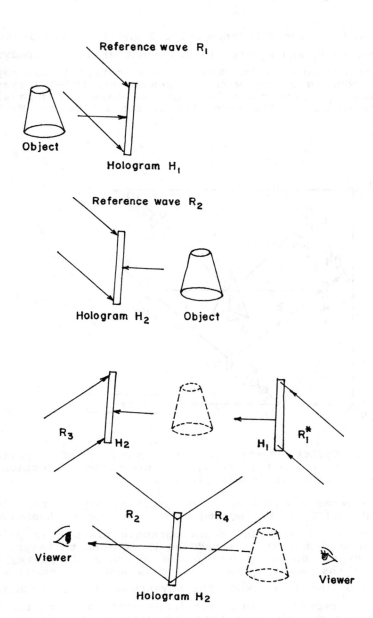

Fig. 4.71. Recording of a double-sided reflection hologram [4.187].

H_2 (Fig. 4.72). The reference beams 2 and 4 are collimated. The holograms H_1 and H_2 are recorded either in photopolymer or in dichromated gelatin. Both of these materials are capable of producing images without processing. PVC-dye system also gives excellent recordings in real-time . Figure 4.73 shows the setup for the recording of final hologram H. The

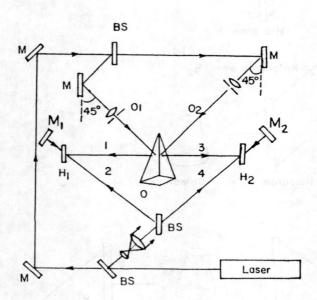

Fig. 4.72. Optical system for the recording of a primary hologram for making a double-sided reflection or rainbow hologram [4.188].

reference beams 2 and 4 are conjugated simply by two plane mirrors M_1 and M_2 by reflecting these beams onto themselves. The object is removed and a photographic plate H is placed in its place such that it straddles the composite image.

Two exposures are given to the photographic plate, the first with the beams O_2 and 1^* blocked and the second with the beams O_1 and 3^* blocked. The advantage of this technique is that the recording setup remains almost the same for the master (primary) and the transfer (final) holograms. The same setup can be used for the recording of reflection and the rainbow holograms.

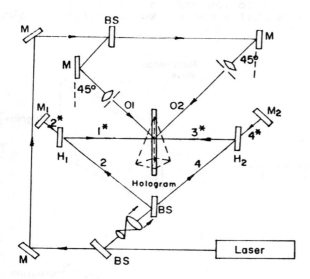

Fig. 4.73. Optical system for the recording of the final double-sided hologram.

The techniques of George [4.187] and Mehta [4.188] require two reference beams to produce front and back side of the object from the two sides of the hologram. Aggarwal and Kaura [4.189] have recorded a full view rainbow-reflection hologram which requires a single reference beam to produce both the sides of the object. This method is described in the next section.

4.12.3. Rainbow Hologram

For displays holograms it is desirable to use a white light source for hologram illumination. When a transmission hologram is reconstructed with a non-monochromatic source, the image is smeared due to dispersion property of the hologram. Benton in 1969 invented a special type of hologram called rainbow hologram in which parallax is eliminated in vertical direction to reduce the coherence requirements [4.190]. The technique utilizes the full advantage of placing the image very close to the plane of the hologram. As there is no vertical parallax, colour smearing is minimized. The absence of parallax in vertical direction does not affect the display as complete depth perception is preserved in the horizontal direction and the viewer normally moves his head in a horizontal direction to get different perspectives of the image.

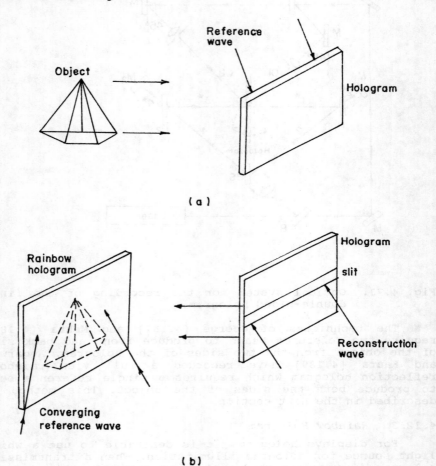

Fig. 4.74. Optical system for the recording of a rainbow hologram.

A rainbow hologram is made in two steps [4.190]. First an ordinary hologram H_1 of the object is recorded. The object is placed close to the recording medium so that in the final display hologram, different perspectives of the object may be obtained. The final hologram is recorded by a spherical or plane wavefront as the reference beam (Fig. 4.74).

In the second stage of the recording the hologram H_1 is masked by a narrow horizontal slit and is reconstructed by the conjugate of the reference beam to obtain the conjugate image (real image). A second hologram H_2 is made of this real image by using a converging reference beam inclined in the vertical plane. The real image of hologram H_1 is placed very close to the photographic plate for making the hologram H_2.

When the hologram H_2 is illuminated with a diverging reference beam, a real image of the slit is produced. The image of the object is viewed through this slit pupil. When the hologram is reconstructed with a white light source, the slit image is dispersed in the vertical plane (Fig. 4.75). As the viewer moves vertically, the colour of the image changes but do not smear. Moving in the same slit horizontally, different perspectives of the object in a single colour are observed. Since all the diffracted light is

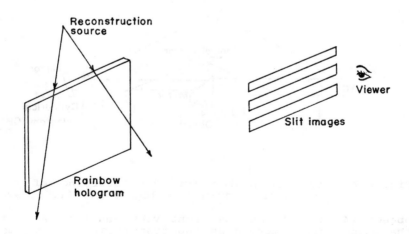

Fig. 4.75. Reconstruction of a rainbow hologram.

contained in a narrow slit, a very bright image is produced even by a weak light source.

Several variations of the Benton's original technique have been proposed to widen the viewing zone. Murata and coworkers [4.191, 4.192] have made a cylindrical rainbow hologram (CRH) using a large cylindrical primary hologram. The recording technique requires the use of conical and cylindrical mirrors. The conical mirror reflects the incident laser beam to the cylindrical film CH (Fig. 4.76). Figure 4.77 shows the side view of the recording arrangement for the cylindrical hologram, while the reconstruction setup is shown in Fig. 4.78. For recording a cylindrical hologram of 10 cm diameter, the diameter of the primary hologram should be more than 50 cm.

A glass rod can be used [4.193] for recording large size rainbow hologram. The glass rod is used to generate a narrow line of light which serves as reference beam for recording and reconstruction of the primary hologram. Rainbow holograms of size up to 1 m×1 m can be recorded by this technique.

Figure 4.79 shows the technique for recording a full view rainbow-reflection hologram which requires only a single illumination beam to produce both the sides of the

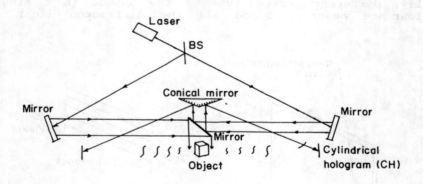

Fig. 4.76. Optical system for the recording of a large cylindrical primary hologram [4.191, 4.192].

object [4.189]. First the front view and the back view of the object are recorded as two transmission holograms. The real images are produced from these master holograms illuminated through a horizontal slit and recorded as rainbow reflection holograms on a single photographic plate on its

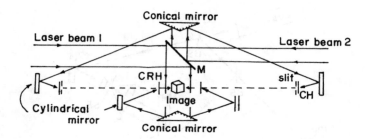

Fig. 4.77. Optical system for the recording of a cylindrical rainbow hologram [4.191, 4.192].

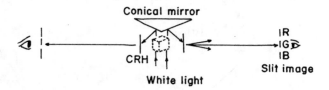

Fig. 4.78. Reconstruction setup for a cylindrical rainbow hologram [4.191, 4.192].

both sides by reference beam R_1 and R_2. Each side of the hologram reconstructs only one view of the object (Fig.4.79b).

It is possible to use a small size collimator (100 mm diameter) to synthesize a large size collimated beam to record a large (200 mm diameter) rainbow hologram [4.194]. The recording plate is divided into two parts and exposed separately. The upper portion is exposed by reference beam R_2 (100 mm diameter) at an angle θ_2 with RP. Since R_1 and R_2 form different angles with the photographic plate it is essential to tilt the RP by an angle $\alpha = \theta_1 - \theta_2$ before the second exposure (Fig 4.80).

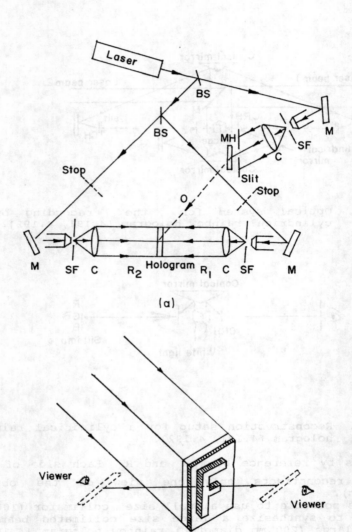

Fig. 4.79. Optical system for the recording (a) and reconstruction (b) of a full view rainbow-reflection hologram [4.189].

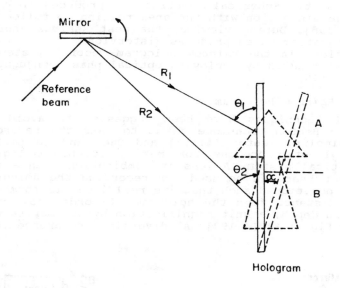

Fig. 4.80. Synthesis of a large size collimated beam from a small size collimator [4.194].

Aberrations Control

For recording of a rainbow hologram it is essential that the primary hologram (H_1) projects an undistorted real image into space which is achieved by illuminating the holograms with the phase conjugate of the reference beam. The conjugate of a diverging reference beam is a converging beam which needs a large aperture convex lens. The conjugate of a plane beam is a plane beam propagating in reverse direction. However, large aperture collimating lenses are needed to produce a large aperture plane reference beam. A diverging beam at a distance of 8-10 m may be approximated as a plane beam. However, perfect phase conjugation is difficult. The rainbow hologram process usually involves two conjugations, during the recording and viewing steps. Aberrations commonly arise in both the cases and affects the perception of the 3D image.

Coma present in the primary hologram image causes "bowing" of the final image which produces a vertical misalignment between left and right eyes during viewing. Proper choice of the primary hologram tip angle (i.e. object beam normal to the photographic plate) will minimize coma.

Astigmatism is overcome by the depth-of-focus of the narrow horizontal slit. Spherical aberration produces barrel-bending image distortion with the apex rolling to follow the observer [4.195]. During viewing, the most serious aberration is astigmatism which produces distortion.

Aberrations in the rainbow hologram recording step is completely eliminated by employing optical phase conjugation technique [4.196].

Single-step Rainbow Hologram

Several techniques have been suggested to avoid two steps of the Benton technique so as to make the recording procedure simpler. Benton [4.197] and Chen and Yu [4.198] used a single-step process for making rainbow holograms using an optical system. A lens in combination with a slit in the object beam can be used for recording the hologram. The slit is positioned such that its real image is formed at a suitable distance from the hologram. In order to avoid distortion in depth, a unit magnification system may be used as shown in Fig. 4.81 [4.199]. A diverging reference beam

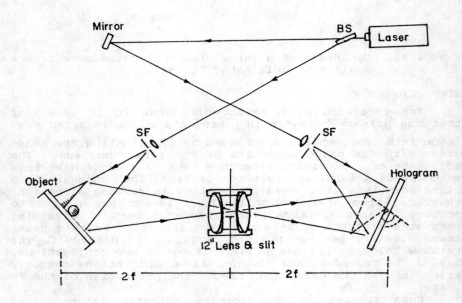

Fig. 4.81. Optical system for the recording of a single-step rainbow hologram using a lens [4.199].

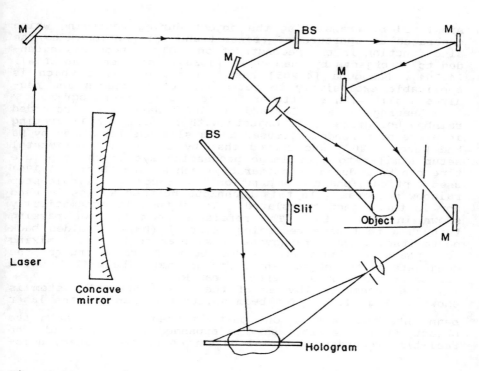

Fig. 4.82. Optical system for the recording of a single-step rainbow hologram using a concave mirror [after 4.200].

can be used in this technique. The linear fringe structure is obtained when the distance of lens to hologram is kept equal to that between reference point source to the hologram. This ensures equal curvatures of the interfering beams producing linear fringes with constant frequency.

The field-of-view is limited by the aperture of the lens. Cheaper optical systems such as a large concave mirror of small focal length may be used [4.200]. Figure 4.82 shows an optical system for recording a rainbow hologram in a single-step using a concave mirror. The object is placed on the axis of the mirror at its centre of curvature. A beam splitter BS_1 allows to form an image with unit magnification on a photographic plate where a diverging reference beam is incident.

Grover et al. [4.201, 4.202] proposed a technique for single-step rainbow holography for 2D objects which does not require a narrow slit in the object wave. The slit can be simulated by translating the object during recording which generates an aperture of transmittance proportional to the sinc function in the reconstruction. The method was extended to 3D objects by Shan et al.[4.203] and Joenathan et al. [4.204]. Bahuguna [4.205] has proposed a method which is applicable exclusively to retroreflective objects and requires a slit with a diffuser for recording the hologram.

Beauregard and Lessard [4.206] have also recorded rainbow holograms of 3D objects with no slit. A 1D grating in place of a random diffuser and a slit has been used by Da and Wang [4.207] who showed that by using an experimental setup similar to a 4f image processing system the exposure time can be reduced considerably. Mehta and Bhan [4.196] used optical phase conjugation to record a single-step rainbow hologram. In this technique neither a lens is used nor is the object translated. The method is schematically shown in Fig. 4.83. The reference beam 1 after passing through a real time recording material (DH) is folded back on to itself. The arrangement is similar to four wave mixing for phase conjugation. The beam 2 reconstructs the conjugate of the object and forms an image (beam 4) O^* which can be recorded by a reference beam R.

The schematic diagram of the actual optical system is shown in Fig. 4.84. The beam splitter BS_1 splits the laser beam into two parts. One part is used to illuminate the object, while the other part is expanded and collimated. The real-time recording material DH serves as the primary holo-

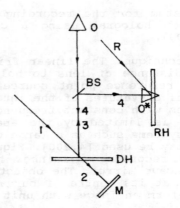

Fig. 4.83. Principle of recording a single-step rainbow hologram using optical phase conjugation [4.196].

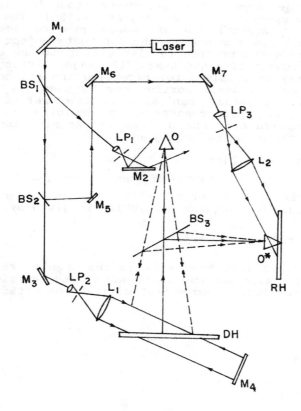

Fig. 4.84. Optical system for the recording of a single-step rainbow hologram using optical phase conjugation [4.196].

gram of the Benton two step process while its vertical dimension acts as a slit. The phase conjugated object beam is separated by a pellicle beam splitter BS_3 to form the image on the photographic emulsion for making the rainbow hologram RH.

The field-of-view is restricted by the horizontal dimension of the primary hologram DH. A very large strip of dichromated gelatin or photopolymer may be used for the hologram DH. Alternatively, a small real-time material (photorefractive crystal) may be used, which can be translated in one direction to simulate a large strip hologram. In such a case the rainbow hologram will be an incoherent superposition of a large number of subholograms producing different horizontal perspectives of the object [4.196]. The use of photorefractive crystal for DH would provide additional flexibilities besides producing the conjugate object beam in real-time. Since the photorefractive effect can be obtained even with microwatt of light beams, the major portion of the laser light may be utilized for recording the rainbow hologram RH and only a small portion may be used for DH. The photorefractive crystals can also produce reflectivities greater than one. This property may be exploited in manipulating the amplitude of the O beam for keeping the proper beam balance ratio for recording RH.

Resolution and Image Blur

If R_i is the distance of the image point from the hologram, R the distance of the slit from the hologram and b the slit width, then angular resolution is given by

$$bR_i/R(R + R_i) = \lambda/b. \qquad (4.103)$$

In Eq. 4.103 left hand side is the angular resolution due to the slit size and the light source and the right hand side is the resolution because of diffraction effects. From Eq. 4.103

$$b^2 = (\lambda/R_i)[R(R + R_i)]. \qquad (4.104)$$

If $R_i << R$ then

$$b = R(\lambda/R_i)^{1/2}. \qquad (4.105)$$

The size of the slit is critical for achieving optimum image quality from white light holograms. It has been observed that if the slit is made narrower than 3 mm, the diffraction effects take place and the image becomes speckled and if it is made wider than 5 mm, the image becomes blurred. The image brightness L_i is proportional to [4.208]

$$L_i \propto [(R_i + R)/R_i]^2. \qquad (4.106)$$

Thus the image will be brighter if it is at a larger distance from the hologram.

It may be pointed out that though the reconstructed

images appear quite sharp, there is some blurring of the images due to the finite source size and the dispersion of the hologram. While viewing the rainbow hologram the finite diameter of the eye allows only a narrow range of wavelengths to form the image, thus in many circumstances, the image appears sharp [4.209]. The wavelength spread ($\delta\lambda$) in the image and the corresponding angular blur ($\Delta\theta$) as seen by the eye due to the angular dispersion of the rainbow hologram are given by [4.210, 4.211]

$$\delta\lambda = \left(\frac{b+D}{R}\right)\left(\frac{d\lambda}{d\theta}\right) \tag{4.107}$$

$$\Delta\theta = \left(R_i \Delta\lambda\right)\left(\frac{d\theta}{d\lambda}\right), \tag{4.108}$$

where D is the aperture of the eye pupil, $\Delta\lambda$ the bandwidth of the reconstruction source, b/R and D/R the respective angular subtenses of the slit and the eye pupil as measured from the hologram.

The dispersion of the hologram grating, (dθ/dλ) is given by

$$d\theta/d\lambda \simeq (\sin\theta)/\lambda, \tag{4.109}$$

where θ is the reference beam angle. Eqs. 4.108 and 4.109 give

$$\delta\lambda = \left(\frac{b+D}{R}\right)\left(\frac{\lambda}{\sin\theta}\right) \tag{4.110}$$

and

$$\Delta\theta = (R_i/R)(\Delta\lambda/\lambda)\sin\theta.$$

$$\simeq \left(\frac{b+D}{R^2}\right)R_i$$

Eqs. 4.109 and 4.110 are useful in finding out the effect of various parameters on the image quality. The image blur depends on the distances of the slit and the image from the hologram. The image blur can be reduced by placing the image close to the hologram. But for image points away from the hologram the blurring can be reduced by increasing the slit (primary hologram) distance from the final hologram and by decreasing the slit width. The slit width however cannot be reduced too much because the diffraction effects will degrade the image in that case.

The wavelength spread in the image as observed by the eye is a function of the angular subtenses of the slit and the eye pupil measured from the hologram. It decreases by increasing the hologram-to-slit distance as in the case of angular blur. If we select various parameters as $\lambda=600$ nm, $R=600$ mm, $R_i=60$ mm, $D=3$ mm, $b=3$ mm, $\theta=45^\circ$, we get an angular blur of 1 mR and a wavelength spread of 8.5 nm. These values are low and suggest that all image points at distance of ± 60mm from the hologram plane will be sharp.

4.12.4. Holographic Stereogram

For making holographic stereograms, photographs of objects rather than the real objects are used. The two dimensional photographs are recorded in the form of a composite hologram such that the image appears to be three dimensional. The main advantage of this technique is that the images can be magnified or demagnified and it makes holography of outdoor scenes possible.

The holographic stereogram invented by Cross is recorded in two steps [4.212, 4.213]. In the first step, the subject is revolved on a turn table and a series of photographs are taken by a movie camera. If T is the period of revolution and t the time between two consecutive frames then 360(t/T) degree is the field recorded by each frame.

In the second step, each frame is recorded [4.214] as a strip hologram by the optical system shown in Fig. 4.85a. Each frame is projected through a cylindrical lens to make a line image which is recorded as a strip hologram. A sequence of these holograms are recorded on a film. The processed film is shaped in the form of a cylinder and illuminated by a white light source situated on the axis of the cylinder. A three dimensional image is perceived from all sides as one moves round the cylinder (Fig. 4.85b). The image appears to be on the axis of the cylinder. Although each frame contains a 2D information yet the image perceived is 3D because each eye sees a different perspective of the object.

The distortions in the image of the holographic stereogram should be eliminated to produce acceptable images [4.215-4.218]. In order to minimize the image distortion [4.219], the distance D of the camera to the centre of rotation of the subject should be

$$D = \frac{d}{m}, \qquad (4.112)$$

where d is the diameter of the cylindrical hologram and m (m<1) is the image magnification.

The height-to-width distortion in the image points that are away from the hologram can be eliminated by the concept of 'Ultra Gram' [4.220], in which with suitable predistortion of the component images a flexible hologram format is realized by placing the intended viewing zone and the centre

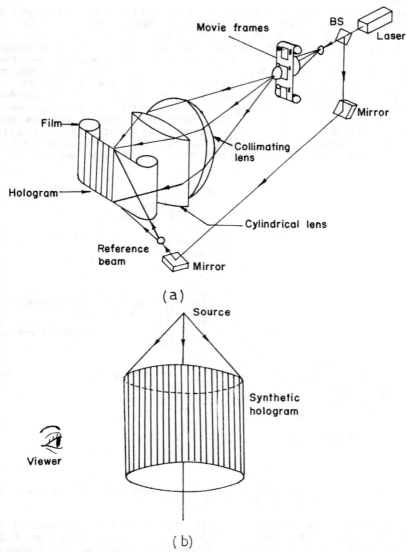

Fig. 4.85. Optical system for the recording (a) and reconstruction (b) of a holographic stereogram [4.214].

of the image at arbitrary locations with respect to the final hologram.

An interesting variation of the cylindrical stereogram has been demonstrated by Okada and Tsujiuchi [4.221] in which a conical hologram is recorded to reconstruct the image floating above the surface of the cone. Such a hologram gives strong three dimensional impression to the viewer. For making the hologram, a series of photographs of the object are taken from the upper inclined direction at small angular intervals. These photographs are then recorded as rainbow or Lippmann hologram. Each frame of transparency is imaged on a directional diffuser and the holograms are recorded through a narrow fan shaped aperture (Fig. 4.86). After each exposure, the film is turned for next exposure such that all exposures combined give a wide fan shape.

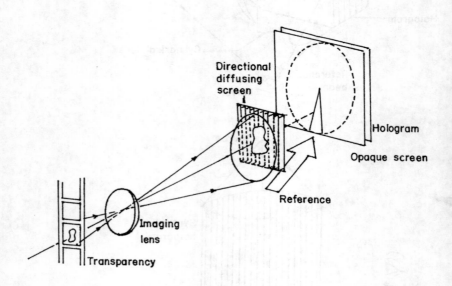

Fig. 4.86. Optical system for the recording of a conical holographic stereogram [4.221].

For reconstructing the image, the hologram is shaped as a cone and placed to make the apex downward as shown in Fig. 4.87. A real image floating above the hologram is reconstructed with an illumination from the bottom of the hologram.

Reflection holographic stereograms have been used for

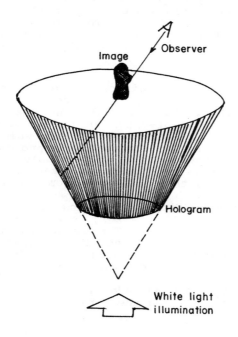

Fig. 4.87. Reconstruction of a conical holographic stereogram [4.221].

recording of cultural properties [4.222] and medical images [4.223] and for producing 3D hard copies of computer processed objects [4.224].

Animation has also been included in holographic stereograms [4.225]. However, this introduces confusion in the perceived location of the moving object, because each eye sees the scene as recorded in the animated stereogram at different instants of time. If the motion of the object is large, then the brain will be unable to locate the object position.

The Cross holographic stereogram combines sequence of views separated by a few degrees as narrow strip holograms. This hologram does not show any vertical parallax. In a multiplane-multiplex hologram [4.226-4.229], photographs of serial sections of an object are recorded as planar holograms, each of which is reconstructed at a different distance from the hologram plane. When the multiplane-multiplex hologram is played back the serial sections are "stacked" so that all sections are seen simultaneously, each at the appropriate depth [4.228]. This hologram has both vertical and horizontal parallaxes and is more suited for the study of biological data [4.226-4.229].

4.12.5. Viewing Zone

The viewing zone of the rainbow hologram and holographic stereogram is limited by the size of the primary (master) hologram, the width of the slit and the size of the final hologram. The angle-of-view can be increased by increasing the width of the primary hologram which requires wide aperture collimators. An alternative approach may be to reduce the distance between the master hologram and the final hologram. But this results in height-to-width distortion in the image parts that are away from the final hologram [4.230].

Another approach to get wide viewing zone is to record an 'alcove hologram' [4.231, 4.232] which can provide a viewing zone as large as $180°$. The recording material is curved into a hemicylinder in front of or behind the image. The image is projected from the hologram surface at a large distance which makes the hologram almost disappear from view [4.231-4.233].

The viewing zone ϕ can be made wide by selecting proper widths of the primary (w_p) and the final hologram (w_f). The viewing angle that can be obtained is given by [4.234]

$$\tan\phi = [(k-1)/(k+1)]\tan\theta, \qquad (4.113)$$

where θ is the reference beam angle and $k = w_p/w_f$. The values of ϕ calculated for different values of k are given in Table 4.31. From table it is clear that a wide viewing zone is produced by increasing the relative width of the primary hologram and taking a larger reference beam angle for the final hologram.

Table 4.31. Values of viewing angle ϕ for different values of k [4.234].

k	2.0	4.0	8.0	16.0
ϕ ($\theta = 45°$)	18.4	31.0	37.8	41.4
ϕ ($\theta = 56°$)	26.3	41.7	49.0	52.6

4.12.6. Change of Size

Holograms of reduced size objects such as living persons, outdoor scenes etc. will open new application areas of display holography. Such holograms may find place on newspaper front pages or in magazines. They can be used in holographic movies where small size holograms may be projected by telescope systems to real life size. Holographic stereogram is one technique of recording reduced size objects.

Direct magnification or reduction of 3D images in a hologram are associated by unequal longitudinal (M_{long}) and lateral magnifications (M_{lat}). They can become equal only by a manipulation of reconstruction-to-construction wavelength ratio (μ). The size of image can also be changed by changing the curvature of the reference beam, but the image suffers from non uniform magnification. The points nearer to the hologram suffer smaller amount of reduction. Moreover, it requires a large size master to obtain wide viewable panoramic reduced size hologram. A change in wavelength and in fringe spacing may reduce the image but grating reduction is not easy.

Fargion [4.235] has designed telescope and zoom systems which produce undistorted 3D magnified or demagnified images. In a two lens system of focal lengths f_1 and f_2 separated by a distance d, the longitudinal and lateral magnifications are

$$M_{long} = -f_1^2 f_2^2 /[R_i(d - f_1 - f_2) - f_1(d - f_2)] \qquad (4.114)$$

and

$$M_{lat} = f_1 f_2 /[R_i(d - f_1 - f_2) - f_1(d - f_2)], \qquad (4.115)$$

where R_i is the distance of image point from the hologram. Both M_{long} and M_{lat} become independent of R_i if

$$d = f_1 + f_2. \qquad (4.116)$$

This is the well known telescope system with

$$M_{long} = -\frac{f_2^2}{f_1^2} = -M_{lat}^2. \qquad (4.117)$$

The minus sign shows space reversal (up side down, left

side right) of the image.
The following combination of lenses are possible [4.235].

Galileo Telescope

(a) $f_1 > 0$, $f_2 > 0$; $f_1 > f_2$ reduced inverted image

(b) $f_1 > 0$, $f_2 > 0$; $f_2 > f_1$ enlarged inverted image

Newton Telescope

(a) $f_1 < 0$, $f_2 > 0$; $|f_1| > f_2$ reduced erect virtual image

(b) $f_1 < 0$, $f_2 > 0$; $|f_1| < f_2$ enlarged erect virtual image

To obtain a large panoramic reduced size 3D image, a large diameter lens (first lens) with a focal length larger than the object size is needed. Large size plastic lenses or Fresnel lenses may be used. The reduced image may be directly recorded as a transmission hologram or a reflection hologram. It is also possible to use the optical system during the image transfer from a master to rainbow hologram. A three lens zoom system may also be used [4.235] for any demagnification. Mehta et al. [4.236] have used lens systems to record transmission holograms which throw the images in space several metres away from the hologram.

4.12.7. Dispersion Compensation

When a transmission hologram is reconstructed by a white light source, the image is dispersed due to the recorded grating structure. The finite source size and its bandwidth produce blur in the reconstructed image (Fig.4.88). The longitudinal image blur is greater than the transverse image blur in magnitude. The image hologram of an object of limited depth produces an acceptable achromatic image. De Bitteto [4.237] suggested a powerful technique for dispersion compensation which uses a compensation holographic plane grating with fringe spacing equal to the average fringe spacing of the hologram. Figure 4.88 shows the technique to produce an achromatic image [4.237, 4.238]. The dispersion of the hologram is cancelled due to the opposite dispersion produced by the grating. The directly transmitted 0 order beam can be blocked by a light shield in the form of a venetian blind placed between the hologram and the grating [4.239]. If the grating and hologram do not have any wavelength selectivity, then the image will be achromatic. If the carrier frequency is large, this arrangement will produce distorted image. A distortion free image can be obtained by predispersing the illuminating source such that after dispersion by the hologram all wavelengths are dispersed in the same direction.

In a reflection hologram, the wavelength selectivity can be improved by using a thick recording medium. The

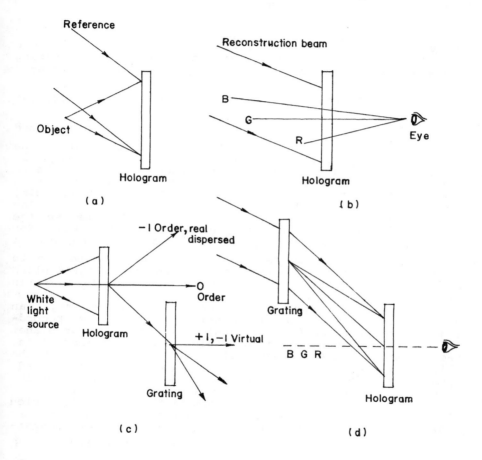

Fig. 4.88. Wavelength dispersion compensation of a transmission hologram using a grating [4.237, 4.238]. (a) Hologram recording, (b) hologram reconstruction with a white light source, (c) and (d) dispersion compensation.

thickness of a commonly used photographic emulsion is 6 μm and the bandwidth of the reconstructed image may be 15-20 nm. A thick emulsion like DCG or photopolymer would reduce this bandwidth but the uniform development along the entire thickness becomes difficult as the thickness increases.

The estimate of the dispersion from a reflection hologram is obtained by the grating equation

$$\Lambda(\sin\theta_1 - \sin\theta_2) = \lambda, \qquad (4.118)$$

where λ is the mean replay wavelength. The dispersion can be obtained [4.240] by differentiating Eq. 4.118, assuming the angle θ_2, the reconstruction angle to be typically $45°$.

$$d\theta_1/d\lambda = 1/\Lambda\cos\theta_1 = (\sin\theta_1 - \sin\theta_2)/\lambda\cos\theta_2. \qquad (4.119)$$

For $\theta_1 = 0°$ and $\theta_2 = 45°$ and $\lambda = 515$ nm, the dispersion produced by the hologram is 1.8×10^{-3} deg/nm. This dispersion in the image is acceptable. One difficulty with the image reconstruction at $45°$ is that the viewer cannot come very close to the hologram as he begins to obstruct the reconstruction source. If one tries to increase the reference beam angle, the dispersion in the image becomes unacceptable. The solution as given by Bazargen [4.240] is to observe the image with the observer at an angle to the hologram. For example, for $\theta_1 = 15°$ and $\theta_2 = 45°$, the dispersion produced is -1.2×10^{-3} deg/nm, which is acceptable. The blurring due to dispersion can also be reduced by tilting the hologram plane by $15°$ to the vertical plane during recording as shown in Fig. 4.89.

The angular dispersion of the reconstructed image can be minimized by recording a conformal hologram, i.e. recording the interference fringes parallel to the hologram plane. In this case a sharp image is obtained and the colour of the reconstructed image is independent of the direction of the illuminating beam from the hologram normal. The main disadvantage is that the image overlaps with the specular reflection of the illuminating beam from the hologram surface.

Another method [4.241] of compensating the dispersion of a reflection hologram is to use a correcting hologram as shown in Fig. 4.90. H_1 hologram which is a reflection grating directs the illuminating beam to the actual hologram H_2 at proper angle. The reconstruction from H_1 and H_2 compensates the dispersion of each other. This arrangement produces a very sharp image as compared to similar arrangement for a transmission hologram (Fig 4.88).

It may be mentioned that the image resolution for a display better than the resolution of the eye has no advant-

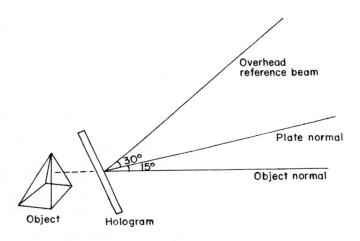

Fig. 4.89. Blurring reduction geometry for recording a reflection hologram.

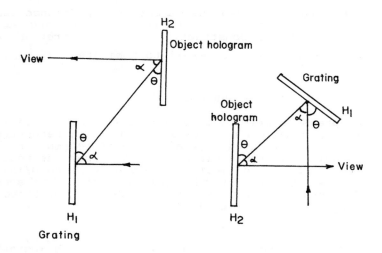

Fig. 4.90. Wavelength dispersion compensation of a reflection hologram [4.241].

age. The resolution of the eye is about 0.5 mrad, which means that the eye can resolve a separation of 0.5 mm at a viewing distance of 1 m. A hologram producing this much resolution is suitable for display purposes.

4.12.8. Sources for Reconstruction

In the early years of development of holography, it was necessary to use a source of laser light at the same wavelength and angle as used during the recording. Techniques were evolved to record the hologram with appropriate compensation introduced for any desired change in reconstruction wavelength and angle. Mercury arc lamps with a filter at desired wavelength can be used for reconstruction of a transmission hologram [4.242]. Rainbow holograms and holographic stereograms can be viewed by using an ordinary vertical filament light bulb. Reflection holograms are best viewed by halogen lamps. Grating-hologram combinations described in section 4.12.7 can be used for reproducing dispersion free images.

The source requirement can be estimated from the equation for image blur [4.243] in the direction of the illumination for central image points.

$$\theta_{blur}^2 = (\Delta\alpha \cos\theta)^2 + [(\Delta\lambda/\lambda_c)\sin\theta]^2, \qquad (4.120)$$

where θ is the angle between the object and reference beam, $\Delta\alpha$ is the angular width of the (square) source, and $\Delta\alpha$ is the bandwidth of the reconstruction source centred at λ_c.

For $\theta=45°$ and acceptable blur equal to 2 minutes,

$\Delta\theta \leq 2.0$ min,

$\Delta\lambda \leq 0.32$ nm at $\lambda_c = 550$ nm.

A high-pressure mercury short-arc lamp with a diameter of 5 mm, filtered with a narrow band filter can reconstruct an acceptable image located at a distance of 100 mm from the hologram illuminated from a distance of 1 m from the source. If the depth of the image limited within $D/(N-1)$ to $D/(N+1)$, where D is the distance of the observer, then the source size and its spectral width can be increased by a factor N without increasing the image blur [4.243].

4.12.9. Hologram Display Systems

There are numerous requirements that a holographic display system must satisfy in order to be used in environments other than a dark laboratory. The display is required to be used in a variety of ambient lighting conditions, therefore the images should be much brighter than normally being displayed in dimly lit rooms. A hologram display system must contain an efficient hologram, an illumination source and optical system. The display must be bright and the viewing zone must be wide. The illumination source should not be visible to the viewer.

Figure 4.91 shows a display system that guarantees optimum illumination for a hybrid system consisting of a

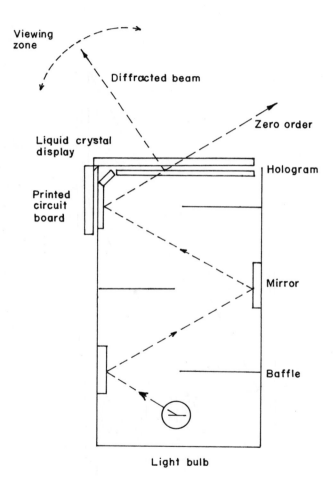

Fig. 4.91. Hologram display system for a transmission hologram [4.244].

hologram and a liquid crystal display containing electronically activated icons which has been designed by Andrews et al. [4.244] to function as a user interface to a business machine. The system combines the versatility of liquid crystal displays with the 3D perception and look around possible by the holograms [4.245, 4.246].

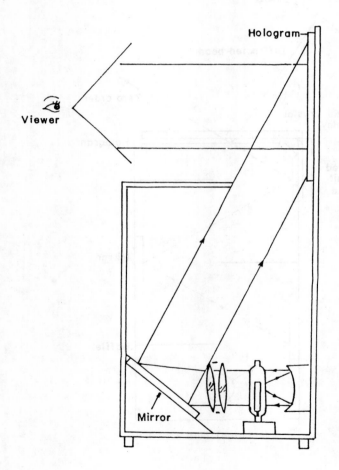

Fig. 4.92. Hologram display system for a reflection hologram [4.247].

The reflection holograms are best displayed by fixing them on a wall and the illuminating source on the ceiling at a proper angle. A single illuminating bulb can be used to display a number of reflection holograms. In some places where it is not possible to fix the source at the ceiling, the display boxes can be constructed which contain bulb, optics and hologram as shown [4.247, 4.248] in Fig. 4.92. Figure 4.93 shows a photograph of the display systems.

For display of transmission holograms, the display system can be based on dispersion compensation [4.249] which can produce sharp, full parallax and achromatic image. The rainbow hologram does not require dispersion compensation. They can be reconstructed by a source at an angle of $45°$ from above and at a distance of about 2 m. This requires the hologram to be hung. In order to mount the hologram directly onto a wall, a mirror can be placed behind the hologram. The light passes through the hologram twice from source to viewer's eyes. In such a display the background light level severely degrades the holographic image.

In order to provide a deep black background, a prism/ mirror reflection mount behind the hologram may be used [4.250, 4.251]. To make the display thin, the mirror/prism geometry can be reduced in scale as a multifaceted Fresnel-type optics as shown in Fig. 4.94.

Fig. 4.93. A hologram display system.

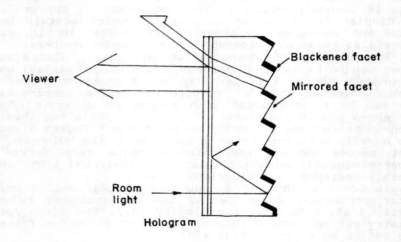

Fig. 4.94. Prism-mirror reflection backing for display of a transmission hologram [4.251].

If a edge-lit hologram [4.251] is recorded, then the source can be fixed at the lower edge of the hologram (section 4.14.6). This makes the display very compact. A compact 1 inch thick display system can be made consisting of three parts: a light guide, a rainbow hologram and a light source [4.252]. The light guide is a pair of glass sheets placed parallel with an air gap between them. the rainbow hologram is laminated to one of the glass sheets. The light source is placed at one end of the light guide.The hologram needed for this display system is recorded by using free space reference beam rather than a guided reference beam, thus avoiding the requirement of glass block and index matching liquid. The technique offers the possibility of making more compact and self contained commercial holographic displays.

4.12.10. Holographic 3D Printer

Three dimensional information is useful for medical, engineering and scientific fields. In medical field, doctors take the help of 3D data produced by x-ray, computed tomography and magnetic resonance imaging for diagnosis purposes. 3D computer-aided design (CAD) systems assist the design of industrial structures. The 3D processing systems display the processed 3D data on a cathode ray tube, which with the help of stereoscopic display systems [4.253] produce 3D images with or without glasses. However, no device can produce a 3D hard copy except that based on a holographic stereogram [4.224, 4.254].

Figure 4.95 shows the schematic diagram of a holographic 3D printer system [4.224]. The optical system consists of a liquid crystal panel and a focussing system in a reflection hologram recording arrangement. The graphic processor receives 3D data from the host computer and calculates the image by perspective projection for each exposure. The calculated image is displayed on the liquid crystal panel. The film is moved to a horizontal or vertical direction after each exposure.

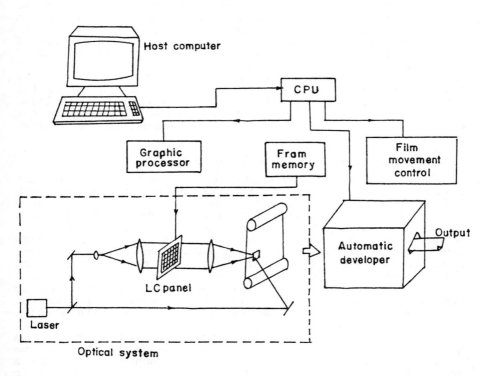

Fig. 4.95. Schematic diagram of a holographic 3D printer system [4.224].

The main advantage of the method is that both the parallaxes (horizontal and vertical) are preserved. The small aperture optical system can be used for even larger size 3D hard copy by simply increasing the number of elemental holograms. The reconstruction can be achieved only by a single spherical lens.

4.12.11. Holographic Television

Transmission of holograms via television has been a natural desire of holographers due to its possible impact in the field of entertainment. The main difficulty has been due to the enormous information content of the holograms which cannot be handled by TV channels. The techniques can be employed to reduce the information content of the hologram. The transmission of holograms via TV was executed [4.255] in 1966 but the actual holographic TV could not be realized in the absence of a suitable recording material.

Recently, Sato et al. [4.256] have used a liquid crystal spatial light modulator for recording and display of TV signals. The holographic interference pattern is transformed into electrical signals and transmitted as TV signal. The fringe pattern is displayed on a liquid crystal device (LCD) as shown in Fig. 4.96. The image is displayed in real-time using a He-Ne laser. Spatial filtering in the reconstruction arrangement can be incorporated to improve the signal-to-noise ratio.

4.12.12. Holographic Cinematography

Holographic cinematography (or cine holography) has not reached a stage of commercial exploitation so far although the demonstration of its principle was made quite early [4.257, 4.258]. The difficulty arises from the requirement of creating and presenting a sequence of holograms of a scene to a large number of viewers. The recording of large size objects and scenes in their natural colours is another problem. Present pulse laser technology permits to record only monochrome holograms of several cubic metres.

The motion of the object also poses problem during recording. The velocity of the object must be low of the order of 1 m/s in the direction of the camera with a pulse duration of 20 ns.

Komar [4.258] in 1977 demonstrated for the first time a holographic movie of a few minutes duration which could be viewed by four persons at a time. The concept based on a projection technique could show the promise of holographic cinematography in the field of entertainment. A wide aperture lens (200 mm diameter) is used to record a series of image holograms on a 70 mm film. The processed holographic images can be projected through an identical lens onto a holographic screen which is a multiplexed holographic opti-

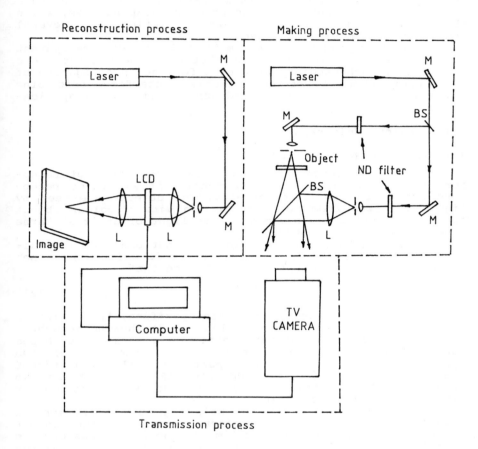

Fig. 4.96. Concept of a holographic TV using a liquid crystal spatial light modulator [4.256].

cal element serving as concave mirrors. The holographic screen forms multiple real images of the projection lens in front of the viewers. Looking through the pupil shows a full size three dimensional image.

Work on holographic cinematography is also being carried out in USA [4.259] and France [4.260]. Smigielski and his coworkers [4.260] have produced a few short duration holographic movies. The interesting ones are *Christiane et les holobulles* and *La belle et la bébé*. The former of 40 s duration shows a young lady who produces soap bubbles, while the later of 80 s duration shows a fiction with several sequences with music and speech.

Cine holography in transmitted light of transparent object does not show the objects but only their shadows are visualized. Such studies are useful for the investigation of air flows, shock waves and plasmas. Interferometric cine holography can be useful for industrial and medical applications.

4.13. Colour Holography

4.13.1. Chromaticity Diagram

Our colour perception is due to three types of receptors having different spectral sensitivities. These receptors are activated differentially. The objects are seen by light reflected from some source. To view a colour object the source must contain a mixture of light of at least three monochromatic wavelengths.

According to CIE (Commission Internationale de l'Eclairage) chromaticity diagram, each colour is defined by a set of chromaticity coordinates (x,y,z) such that

$$x + y + z = 1 \qquad (4.121)$$

If two coordinates x and y are specified, the third is fixed so a 2D map can be drawn (Fig. 4.97) showing monochromatic light of different wavelengths on a horse shoe-shaped curve known as spectrum locus [4.261]. Mixing of two spectral colours in varying proportions produces different hues which lie on the straight line AB joining these two primaries. When a third colour is also used, any colour within the triangle ABC can be obtained.

CIE chromaticity diagram helps in the selection of wavelengths for recording colour holograms. The colour produced in the holographic image should correspond as closely as possible to those in the object. In other words the colour fidelity should be high. The colour reproduced in the holographic image critically depends on the laser wavelengths used for recording the hologram. Only those colour information will be recorded which are reflected at the chosen wavelengths. This seems to imply that a large number of laser wavelengths would be required to reproduce exact colour of the object. However, as shown by Thornton [4.261] only three colours near 450, 540 and 610 nm can render high degree of colour fidelity. This combination of wavelengths also produces a brighter image.

Several combinations of laser wavelengths have been used for recording colour holograms. The wavelength

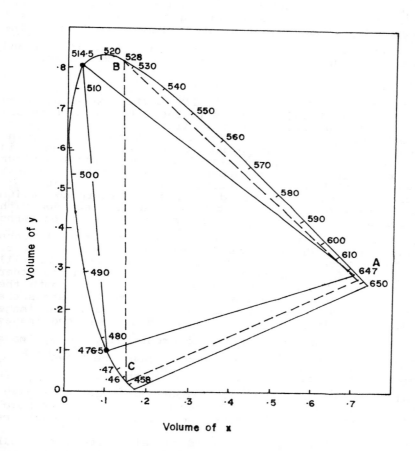

Fig. 4.97. CIE chromaticity diagram showing colours obtainable by using 476.5, 514.5 and 632.8 nm ———; and 458, 528 and 647 nm ---- [after 4.261].

combinations 476, 514 and 633 nm and 488, 514 and 633 nm are available from argon ion and He-Ne lasers. These combinations of colours do not produce yellow, violet, purple and dark blue colours [4.262-4.264]. Hubel and Ward [4.265] have recently used 458, 528 and 647 nm wavelengths which gives a much larger gamut of colour. Bazargan [4.179] has used a single laser system which produces three recording wavelengths 459, 532 and 633 nm in pulsed mode. It is based on a frequency-doubled Nd-YAG laser producing 532 nm. The output of this laser is focussed into a pressurized gas cell (Deuterium D_2) producing stimulated Raman scattering. The emergent beam is accompanied by a Stokes (red 633 nm) and anti Stokes (blue 459 nm). All the beams are collinear.

4.13.2. Recording of Colour Holograms

Colour holograms are basically multiplexed holograms which produce multicolour images [4.266]. They can be recorded with three wavelengths as was first pointed out by Leith and Upatnieks [4.10]. When reconstructed with the recording wavelengths the hologram produces overlapping images in three colours producing a multicolour image. The behaviour of the reconstructed image depends on whether the hologram has been recorded in a thin medium or in a thick medium.

Colour holograms recorded in a thin recording medium suffer from cross-talk. If the hologram is recorded with two wavelengths λ_1 and λ_2, then the two fringe patterns produced will be different, even though same angle between the object and reference waves for each wavelength are kept. When the hologram is illuminated with one wave, it will produce two images, one at its proper location and the other displaced. If the hologram is reconstructed with both the waves, four images are reconstructed. Figure 4.98 shows the reconstructed beams from a two-colour hologram. The image λ_1,λ_2 is the desired two-colour image. The other two images λ_1,λ_2 and λ_2,λ_1 are produced due to cross-talk. λ_1,λ_2 image corresponds to λ_1 wavelength image when the wave of λ_1 illuminates the hologram recorded with the wave of λ_2. In general, a N-beam hologram will produce N^2 images on replay. Of these N^2-N images are undesirable. A three-colour hologram would produce nine images, out of which six would be unwanted.

Several solutions were tried to eliminate cross-talk images. The spatial filtering technique using an aperture as a filter to pass the image waves and to block the unwanted images can be used for two dimensional object. Other techniques are based on carrier multiplexing [4.10], spatial multiplexing [4.267], coded reference beams [4.268] and division of aperture field [4.269]. All these techniques require multiple monochromatic sources for reconstruction. The images have low resolution and low signal-to-noise ratio.

4.13.3. Volume Colour Holograms

Volume holograms effectively eliminate cross-talk

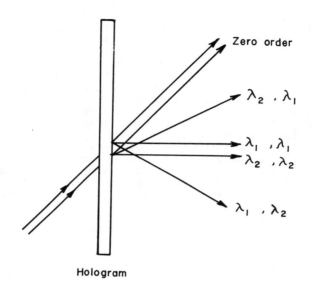

Fig. 4.98. Reconstructed beams from a two-colour hologram.

images utilizing Bragg effect. Both transmission and reflection volume colour holograms can be recorded in thick media. The transmission volume holograms are reconstructed with the laser beams used to record it, while the volume colour reflection holograms are reconstructed with white light due to their inherent wavelength discrimination ability. In general, a colour volume hologram will not produce any cross-talk images if the bandwidth of the constituent gratings is smaller than the difference in any two of the recording wavelengths.

The first two-colour transmission volume hologram of a transparency was recorded by Pennington and Lin [4.270]. The technique was subsequently extended [4.263] to record three-colour transmission volume hologram of a three dimensional object using blue (488 nm) and green (514 nm) light from an argon ion laser and red (633 nm) light from a He-Ne laser.

Multicolour reflection holograms [4.271-4.274] recorded on a single photographic plate like Kodak 649F plate suffer from low diffraction efficiency. This is because the dynamic range of the material is shared between three recordings for three wavelengths. The conventional processing techniques reduce the thickness of the recording material. The shrinkage shifts the colour of the reconstructed image towards shorter wavelength. The emulsion thickness can be restored by soaking the emulsion in an aqueous solution of TEA [4.263, 4.275]. The holograms treated with TEA show rapid print-out effects due to the formation of photolytic silver. A substitute of TEA is D-sorbitol or glycerol [4.276]. A suitable combination of tanning developer and bleach bath can be used to minimize the emulsion shrinkage.

Hariharan [4.276] eliminated the problem of low luminance of reconstructed image from reflection holograms by recording the component holograms for different primary wavelengths on separate photographic plates. The final multicolour hologram is made up by combining the separate holograms. This technique works well because such reflection holograms uniquely diffract only a relatively narrow band of wavelengths. Each component hologram can be separately optimized. Thus the red component can be recorded in a red sensitive emulsion, while green and blue components can be recorded in a green-blue sensitive material. The composite hologram can have an improvement in image luminance by a factor of 2 or more over a three exposure hologram in a single photographic layer [4.276].

A further gain in the image luminance is achieved by restricting the available solid angle of viewing [4.277] as in the case of a rainbow hologram. Figure 4.99 shows a schematic diagram for recording multicolour reflection hologram with high luminance [4.276]. The red component (633 nm) hologram can be recorded on Agfa 8E75HD plate with emulsion side towards the reference beam while the green (514 nm) band blue (488 nm) component holograms can be recorded on Agfa 8E56HD plate with emulsion side facing the mirror. After processing, the two plates are placed with their emulsions in contact.

By suitably modifying the recording geometry of Fig. 4.99 so that both the object beam and reference beam are incident from the same side of the photographic plate, multicolour rainbow and transmission holograms can be recorded [4.200, 4.278-4.280].

Hubel and Ward [4.265] have recorded a muticolour reflection hologram in Ilford films by using 647 nm (red), 528 nm (green) and 458 nm (blue) wavelengths and obtained a large gamut of colours including yellow, purple, dark blue and violet which are not produced by using other combination of wavelengths. The 647 nm component hologram can be recorded in Ilford P673T film and 528 nm and 458 nm component holograms in Ilford SP672T film. The two finished holograms are laminated together with double-sided optically clear adhesive film. The efficiency of red hologram will be higher than the blue/green hologram as in the former case, there is only one fringe structure. The red image hologram should therefore be placed behind the blue/green image hologram.

While efforts have been made to make suitable

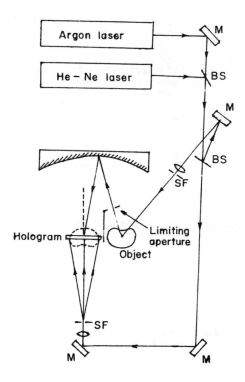

Fig. 4.99. Optical system for the recording of multicolour reflection hologram with high luminance [4.276].

sandwiches of different materials for making true colour holograms, it appears that sandwiching is a stop-gap technique at best [4.281, 4.282]. Recently, Weber *et al.* [4.129] have been successful in recording a multicolour hologram in Du Pont photopolymer HRF705 film. Figure 4.100 shows that the playback wavelengths from a three-colour reflection grating are nearly the same as that used for recording. Thus, this film appears to be a true full colour recording material, therefore is a significant development for the recording of colour holograms with extremely low noise, high diffraction efficiency, true colour reproduction, freedom from colour processing and good storage characteristics.

Dichromated gelatin is another material which can be used for multicolour holograms. DCG films have a weak sensitivity to red light [4.283] due to a charge transfer transition, yet they can not be used for recording multicolour holograms without sensitization. Methylene Blue sensitized dichromated gelatin (MBDCG) has been used for recording colour holograms [4.98, 4.99, 4.284].

The sensitivity and diffraction efficiency of MBDCG plates can be increased by controlling the moisture of the gelatin layer during the exposure [4.284]. This is achieved by storing the plate in a thermohygrostate for 60 min before exposure and exposing the plate in a liquid gate. Table 4.32 gives a complete procedure for preparing a MBDCG plate and

Table 4.32 Preparation and processing of MBDCG plate [4.284]

Step	Direction	Time
	Preparation	
1.	Coat 4% by weight aqueous solution of gelatin on 3 mm glass plate.	
2.	Dry	
3.	Soak in Kodak rapid fixer with hardener	10 min
4.	Dry	
5.	Soak in ammonium dichromate 1.5% by weight, methylene blue 0.02 % by weight	
6.	Dry in atmosphere of ammonia	
	Exposure	
7.	Store in a thermohygrostate	60 min
8.	Expose the plate in a liquid gate	
9.	Soak in ethanol	
	Development	
10.	Wash in water	
11.	Soak in warm water	2 min
12.	Dehydrate in 70% isopropanol	3 min
13.	Dehydrate in 90% isopropanol	1 min
14.	Dehydrate in 100% isopropanol	3 min
15.	Dry at 120°C	

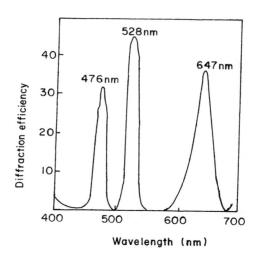

Fig. 4.100. Playback wavelengths from a three colour reflection grating recorded in Du Pont photopolymer HRF-705 film [4.129].

its processing to obtain a high quality multicolour hologram. The exposure energy for 647, 514 and 476 nm wavelengths are 400, 120 and 30 mJ/cm^2 respectively. Mizuno et al. [4.284] have obtained a diffraction efficiency of over 80% for all the three wavelengths as shown in Fig. 4.101.

4.13.4. Recording Geometry

The recording geometries for colour holograms are basically the same for those used for monochrome transmission and reflection holograms. Special considerations are, however, required for the optical components which should be corrected for different wavelengths to be used for colour hologram recording.

The first and most important component is the device for combining different laser beams. A combination of a beam splitter and a mirror (Fig. 4.102a), which reflects one wavelength and transmits the other two can be used. A prism can also be used with the different laser beams incident at appropriate angles. It is quite time consuming to obtain the three collinear laser beams from these devices. Some light power may also go waste out of the device at the beam splitter.

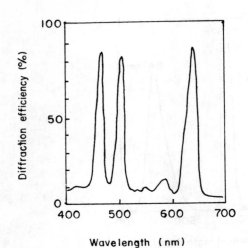

Fig. 4.101. Playback wavelengths from a three colour reflection grating recorded in MBDCG plate [4.284].

Use of optical fibres is a very convenient method of beam ombination [4.281, 4.285]. The output from three lasers are launched in three optical fibres, the other ends of which are fixed close to each other (Fig. 4.102b). Another innovative device used by Jeong and Wesly [4.281] is based on a holographic optical element (HOE). As illustrated in Fig. 4.102c, a hologram multiplexed at blue, green and red wavelengths at appropriate angles produces three beams all at the same angle. Thus, the requirement of alignment is automatically met.

In a monochrome recording set up, the reference-to-object beam intensity ratio may be controlled by incorporating half-wave plates in the beam paths. But half-wave plates for polychromatic beam cannot be used for polarization rotation. Jeong and coworkers [4.281, 4.285] have suggested to use double Fresnel rhombs of angle $54.6°$ and made of glass with index of refraction 1.51 (Fig. 4.103a). A plane polarized beam entering the first rhomb emerges as a circularly polarized, which after passing through the second rhomb becomes plane polarized again, but rotated. Such a device can be used in a two beam multicolour hologram recording setup as shown in Fig. 4.103b.

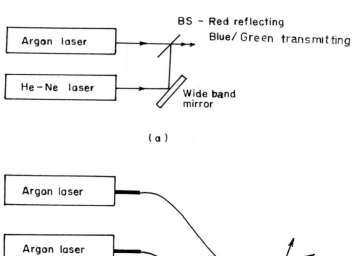

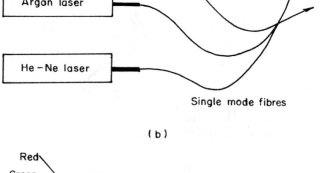

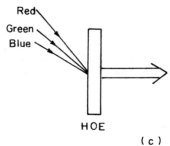

Fig. 4.102. Beam combiners for recording of a colour hologram using (a) a beam splitter and a mirror, (b) singlemode fibres [4.281, 4.285] and (c) a multiplexed HOE [4.281].

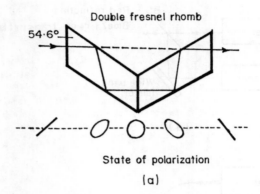

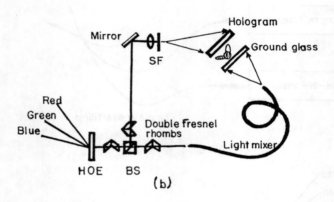

Fig. 4.103. (a) Ray diagram in a double Fresnel rhomb and (b) recording setup for a multicolour hologram [4.281, 4.285].

4.13.5. Pseudocolouring

The thickness of a volume hologram affects the wavelength of maximum diffraction efficiency. The reflection display holograms therefore require special care to keep the emulsion thickness and refractive index the same during exposure and after processing. The earlier reflection holograms developed in Kodak D-19, fixed and bleached were associated with a shrinkage effect in the emulsion leading to a shift of the playback wavelength towards shorter side. Thus a red wavelength exposed hologram would reconstruct images in green colour. A great deal of research has gone into the chemistry of processing baths to keep the thickness of the emulsion with minimal change so that the playback wavelength remains the same as the exposure wavelength.

Blyth [4.286] and Hariharan [4.287] were the first to exploit the thickness of the emulsion as a key parameter for recording multicolour holograms. It is possible to vary the thickness of the emulsion deliberately between exposure and viewing in order to obtain reconstruction wavelengths that are very different from the recording wavelengths. This 'pseudocolour' technique makes multicolour reconstruction possible with only one colour of laser light.

Two approaches have been adopted for controlling the thickness of the emulsion. In the first approach, the shrinkage of the recording material is controlled with a proper combination of developers and bleaches [4.80, 4.288]. Table 4.12 gives three combinations for producing red, yellow/gold and green colour images by exposing the emulsion with red laser light. By varying the exposure energy and developing time combination, it is possible to vary the amount of silver removed which controls the thickness of the final emulsion. However, too much deviation from the optimum exposure/developer times would reduce the diffraction efficiency.

To obtain a gold colour image, a wavelength shift of 114 nm is required when the hologram is recorded with a pulsed ruby laser and a shift of 53 nm when He-Ne laser is used for the recording. These wavelength shifts are possible with proper choice of developers and bleaches [4.80]. Figure 4.104 shows the reconstructed image from a pseudocolour reflection hologram recorded by using a He-Ne laser. It shows a number of pencils in a beaker, with alternate pencils of red colour and remaining of golden yellow colour. The red component hologram was exposed on a Agfa 8E75HD plate with the emulsion side towards the reference beam. The yellow component hologram was then recorded on another Agfa 8E75HD plate with the emulsion side towards the object side. Both the plates were developed and bleached as shown in Table 4.12. After drying, the plates were cemented together with their emulsion sides in contact.

The second approach for the control of the thickness of the emulsion is to pre-swell the emulsion before exposure by soaking the plate in triethanolamine [4.288, 4.290, 4.291]. Table 4.33 gives percentage of TEA to produce the image in a particular colour [4.291]. To obtain a green colour image 12% TEA is used whereas a 28% TEA solution would produce blue colour image. A white colour image (i.e. reconstruction in red, green and blue colours) can also be produced. The results shown in Table 4.33 have been obtained by exposing Agfa 8E75HD plates with 647 nm light and developing the

Fig. 4.104. Reconstructed image from a pseudocolour reflection hologram [4.289].

Table 4.33 Generation of pseudocolours in 8E75HD plates. exposure at 647 nm, developer D-19 and bleach GP 432 [4.291].

Colour	Wavelength (nm)	Δλ (nm)	TEA %
Red	652	23	No soak
Green	532	20	12
Blue	458	20	28
Yellow	648/529	R=23, G=20	R=0, G=12
Cyan	527/452	G=22, B=18	G=12, B=28
Magneta	642/452	R=24, B=18	R=20, B=28
White	640/529/450	R=23, G=20, B=18	R=0, G=12, B=28

plates with D-19 developer and bleached in GP 432 bleach.

Walker and Benton [4.292] have suggested to control the thickness of the emulsion during exposure by immersing the plate in a water-propanol solution. This method can be used in situ.

A different technique for transmission pseudocolour holograms has been used by Bazargan [4.238] who combined a dispersion compensating grating with the hologram. When the grating and the hologram have same spatial frequencies and negligible wavelength selectivity, an achromatic image is produced. However, if a wavelength selectivity is introduced in the hologram by using a recording medium of sufficient thickness, then the grating-hologram combination can produce the image in a single colour. A multicolour image is obtained by superimposing several wavelength selective holograms, each reconstructing at a different wavelength. The average spatial frequency in each case should be equal to that of the grating.

4.14. Special Techniques

4.14.1. Local Reference Beam Hologram

In a local reference beam (LRB) hologram the reference wave is derived from a portion of the object wave. The LRB hologram is useful in the following situations:

(a) When object is very far away and we do not have a laser of sufficient coherence length. In such a situation reference beam path may be made equal to the object beam path either by an optical delay line (by reflecting the laser beam many times between mirrors or prisms) or by placing a small mirror near the object. The more convenient way is to record a LRB hologram.

Since the reference beam originates from the object wave in a LRB hologram, the requirement of temporal coherence of the source can be relaxed.

(b) When the object is to be holographed through a turbulent medium, a portion of the object illuminating wave is focussed on the object [4.293] which acts as an extended reference source in the plane of the object, resulting in a lensless Fourier hologram. Such a speckled reference beam hologram can be combined with a LRB hologram [4.294]. Since object and reference waves travel through the same turbulent medium, the distortions are equal on both the waves. Hence the effects of turbulent medium will be cancelled [4.295]. The resolution of the image depends on the size of the focussed spot acting as reference source. A small mirror can be placed on or near the object which would form the reference wave travelling the same path as the object wave.

(c) When the object is unstable or even slightly moving.

Both transmission or reflection type LRB holograms can

be recorded by appropriate recording geometries [4.296, 4.297]. A transmission geometry is shown in Fig. 4.105. The object wave is split into two parts. One part is focussed on an iris. This beam acts as the reference beam and interferes with the other part of the beam. The quality of the reconstructed image depends on the size of the iris hole. The small size of the iris hole ensures uniform phase over the reference but it results in reduced energy. When the size of the iris hole is large, a small image of the object is produced which acts as the reference. This will result in a reduction in image quality. The iris and lens can be eliminated if an image plane hologram is recorded [4.298].

Local reference beam holograms can be recorded using Fourier transform, lensless Fourier transform, Fresnel and image type geometries. The image resolution and the image field-of-view depend on the location of the reference source. In general, the reference source cannot be a perfect point source, hence the image resolution may be low but larger field-of-view may be obtained. Even though the image quality may be low but it is acceptable for display.

4.14.2. Multiple-Exposure Holography
(Scanning Object Beam Holography)

When the power of the laser available is low and the object is large, it becomes difficult to record good holograms by conventional single-exposure method. An alternate technique is to record a large number of holograms on the same area of the emulsion by dividing the object in a number

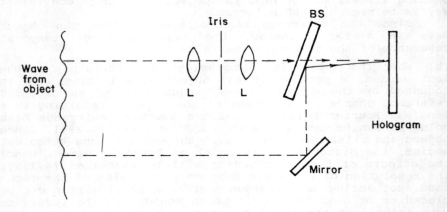

Fig. 4.105. Setup for recording a transmission LRB hologram [4.296, 4.297].

of parts and scanning the whole object by illuminating only one part during each exposure. This will reduce the exposure time to record the hologram of each part. However, the multiple exposures decrease the reconstruction efficiency and the material becomes overexposed if the number of exposures and each exposure time are not adjusted properly. The technique of multiple-exposures can be thought of in terms of pre-or post-exposure, each exposure serving as a bias for the previous or next recording. Under proper recording conditions, it is possible to record a multiple-exposure hologram with the same diffraction efficiency as that of a single-exposure hologram [4.299-4.302].

In the conventional single-exposure, single-object recording, the exposure E to the emulsion is given by

$$E = T\, I_o [1 + \rho + 2\rho^{1/2} \cos\theta], \qquad (4.122)$$

where T is the recording time, I_o is the object beam intensity, ρ is the reference-to-object beam intensity ratio and ϕ is the phase difference between the object and the reference beam.

In the multiple-exposure technique, the object is divided into N parts and a hologram of each part is recorded by N exposures on the same recording emulsion. Let the object beam amplitude of the nth exposure at the recording plane be $O_n \exp(j\phi_o)$ and that due to the reference beam be $R \exp(j\phi_o)$. If T' is the exposure time for each exposure, the total exposure E_n to the emulsion for all the N exposures is given by

$$E_n = T' I'_o [N(1 + \rho') + 2\rho'^{1/2} g(N)], \qquad (4.123)$$

where I'_o and ρ' are the object beam intensity and reference-to-object beam intensity ratio, respectively, and $g(N)$ is known as the interference function given by [4.299]

$$g(N) = \sum_{n=1}^{N} \cos(\phi_r - \phi_n). \qquad (4.124)$$

Equation 4.123 assumes that the recording time for each exposure is the same (T'=tT) and the absolute amplitude of the object beam at the recording plane is the same in each case. Due to multiple-exposures the bias of the hologram is increased and the visibility of the recorded fringes and hence the diffraction efficiency of the hologram is decreased.

In order to retain the characteristics of conventional single-exposure, single-object recording in the multiple-exposure case, the values of the bias and the fringe visibility in both the cases must be equal. This leads to the condition [4.300]

ρ [single-exposure, single-object]
    = $N\rho$ [multiple-exposure, multiple-object].

If all the exposures are of the same duration, the image of the first exposure will be much brighter.

An advantage of scanning object beam hologram is the reduction in what is called flare light due to the on-axis distribution [4.303]. The on-axis distribution arises due to the mutual interference of the object waves which is quite low in the scanning object beam hologram. In between exposures, slight motion of the object may be tolerable. The diffraction efficiency, however, decreases due to increased background exposure.

4.14.3. Multiplexed Hologram

A multiplexed hologram is one in which more than one images are stored. They may be holographed on the surface of the photographic emulsion or within its volume, superimposed or separated, as a permanent record or an erasable one. In holographic memories a large number of signals are recorded holographically on a single photographic plate by a single-exposure or multiple-exposure technique. Multiple-exposure multiplexing has normally been done by either changing the angle between the reference beam and the object beam between the exposures (carrier multiplexing) [4.10, 4.304, 4.305] or by using complementary areas of the emulsion to record different signals (spatial multiplexing) [4.306]. Both of these techniques require separate exposures for each hologram.

In the carrier multiplexing either the reference beam is kept constant or both the reference beam and the object beams are changed between the exposures. If the reference directions are separated sufficiently in angle, many signals can be stored on a single photographic plate and subsequently decoded independently by a beam at the corresponding construction angles.

The main draw back of this technique is that the total power that can be diffracted [4.306, 4.307] into any one of the N reconstructed images is approximately $1/N^2$ of the power that can be diffracted by storing only one information on the same photographic plate and that the signal-to-noise ratio of one image varies as $1/N^2$. This is due to the increased bias level of the hologram resulting in poor contrast of the recorded fringes. Different types of holograms like Fresnel, lensless Fourier transform etc. can be

combined with carrier multiplexing technique [4.308].
The phenomenon of holographic reciprocity law failure is also responsible for the chronological decrease in the efficiency of the sequentially recorded holograms in the same recording material [4.309]. Time elapsed between exposure may be responsible for this behaviour because a prespeck created by an exposure may dissociate into an ion before the arrival of second photon from the next exposure.
Table 4.34 gives the amount of exposure needed in successive exposures to get high efficiencies. The values in the table are normalized to the first exposure. Thus it is clear that the successive exposures should be increased from the previous exposures.

Table 4.34 Exposure values and obtainable efficiency in a four exposure hologram [4.309].

Order of exposure	1	2	3	4
Exposure	1.00	1.11	1.24	1.45
Efficiency	1.00	0.98	1.00	1.12

Collins and Caulfield [4.310] have shown that for recording 100 high efficiency holograms a material with thickness of 1 mm would be required. The maximum number of superimposed holograms N, each with 100% diffraction efficiency that can be recorded in a material of thickness t is given by

$$N = \frac{n(\Delta n)t}{\lambda (n^2 - \sin^2\theta)^{1/2}}, \qquad (4.125)$$

where n is the refractive index of the unexposed material, Δn is the maximum possible change in the refractive index of the material and θ is the angle between the object beam and the reference beam. Figure 4.106 shows the plot of Eq. 4.125 for three different values of Δn. It is observed that with an emulsion of 15 μm thickness, approximately 5 holograms, each with 100% efficiency can be recorded.

In the spatial multiplexing technique, there are two approaches. The recording plate is divided into N separate regions and N different holograms are recorded into each region. In this case the resolution becomes less as aperture size of each hologram is proportional to $1/N^{1/2}$. The other approach is to record each image on a spatially distinct area of the photographic plate, the distinct area occupies, in parts, the full area of the plate. Thus, each information is recorded in a large number of small areas distributed over the full area of the plate. For recording other information complementary areas are used by complementary masks.

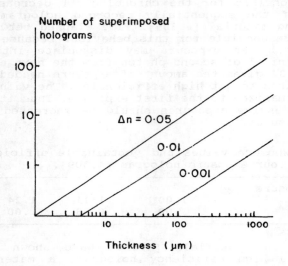

Fig. 4.106. Possible number of superimposed holograms, each with 100% efficiency versus thickness of the recording material for three values of Δn [4.310].

In this technique the image quality is inversely proportional to N, since as N increases the total area available for each information decreases. Moreover, the reconstruction requires the perfect alignment of the masks with the hologram. Different types of holograms like Fresnel, lensless Fourier transform etc. can be combined with spatial multiplexing technique also [4.311].

In the single-exposure multiplexing technique all the signals are generated and recorded simultaneously [4.299]. The interference between signals produces unwanted modulation terms producing flare light extending to the reconstructed image.

Suppose there are N objects to be stored in a single exposure on the same photographic plate. The hologram is produced by the superposition of the reference beam and the total object beam reaching the plate. If the reference wave be

$$E_r = O_o \exp[j(\phi_o - \omega t)] \qquad (4.126)$$

and the total object wave be

$$E_o = \sum_{n=1}^{N} O_n \exp[j(\phi_n - \omega t)], \qquad (4.127)$$

where O_o and O_n are the absolute amplitudes of the reference and the object beams, the total light intensity on the recording emulsion is therefore given by

$$I = O_o^2 + \sum_{n=1}^{N} O_n^2 + \sum\sum 2O_n O_{n'} \cos(\phi_n - \phi_{n'})$$
$$+ \sum_{n=1}^{N} 2O_o O_n \cos(\phi_n - \phi_o). \qquad (4.128)$$

In Eq. 4.128 the third term arises due to interference between the object beams themselves, while the fourth term arises due to the interferences between the object beams and the reference beam. The mutual interference between the object waves produces on-axis spread of light.

When the objects are sequentially recorded in N exposures, the light intensity on the recording emulsion is given by

$$I = \sum_{n=1}^{N} [O_n^2 + O_o^2 + O_o O_n^* + O_o^* O_n]. \qquad (4.129)$$

In the sequential recording the on-axis spread is narrow. Eqs. 4.128 and 4.129 reveal that the modulation and the efficiency of the hologram are different than the case of a single-object, single-exposure recording. However, the recording in all the three cases can be equally good if exposure times and reference-to-object beam intensity ratios are accurately controlled [4.299, 4.302, 4.312].

Several holographic multiplexing techniques have been developed by using deterministic [4.313, 4.314] and random [4.315] encoding phase. A coded reference wave can be used for each object wave. The reference beam is coded by changing its phase by passing the reference beam through a diffuser and moving the diffuser by a short distance between exposures. The objects can be reconstructed without crosstalk by using the corresponding reference wave. In such a way more than 1000 holograms of point sources have been superimposed on a single recording medium [4.13]. A high signal-to-noise ratio (>20 dB) has been obtained.

For multiplexing of photographic transparencies, binary or continuous tone, a convex lens and a diffuser may be used for recording the hologram as suggested by Mehta and Singh [4.316]. The lens is positioned so as to obtain real images when the hologram is reconstructed.

Holographic multiplexing in photorefractive crystals is useful for applications such as neural networks [4.313, 4.317] due to their large storage capacity and fast parallel access to the stored information [4.318, 4.319]. But the information recorded in this material is erased due to subsequent exposures if the hologram is not fixed. This problem has been solved by a recording technique in which each hologram is recorded with multiple short exposures in random order [4.313, 4.317] or in an incremental fashion [4.320]. The individual exposures are extremely short compared to the material's response time. For N recordings, each object and reference pair is sequentially displayed repetitively through all the N recordings. In the incremental method [4.320], the first hologram recorded with an exposure equal to Δt will experience erasure during the next (N-1) exposures while the other holograms are being recorded. However, the effects of recording and erasure become equal after many cycles.

4.14.4. Multifaceted Hologram

A multifaceted hologram performs a redistribution of an input intensity function into desired output intensity pattern [4.321-4.324]. Multifaceted holograms are useful for producing arbitrary laser illumination patterns [4.325], space variant optical elements [4.326], coordinate transformations [4.327] and optical interconnects (Chapter 7). A wedge and ring shaped multifaceted hologram [4.328] is useful for diffraction pattern sampling experiments. An important application of multifaceted holograms is in the fabrication of scanners.

In an image forming type of multifaceted hologram, the hologram is subdivided into many small square areas (known as facets), each of which diffracts light to an arbitrarily prescribed location in the output plane, thus forming an image composed of small square patches of light [4.324]. Each facet of the hologram is a grating structure of specific spatial frequency and orientation. The output plane is located a small distance from the hologram plane. As shown in Fig. 4.107, the image is built up of small patches of light mapped by Fresnel diffraction from different facets.

For recording an imaging multifaceted hologram the facet shape is defined by a small square aperture in Mylar film which is pressed lightly against the holographic film. The reference beam is introduced at an angle, say $35°$ with the normal to the film. The object beam can be introduced

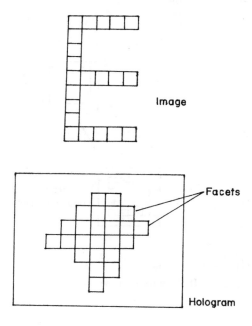

Fig. 4.107. A multifaceted hologram and the reconstructed image [4.324].

at a calculated angle by a pivoting mirror which is computer controlled. The film is translated after each exposure by computer controlled stepping stages to expose different facets.

The hologram facets may be arranged in a particular shape (a letter, a sketch, a word, etc.) so that when these facets diffract light leave dark image of the facets in the zero order. The light which is diffracted from these facets form an image in the diffraction order. The binary images can be converted into grey level images by reconstructing the multifaceted hologram with the required spatial intensity distribution.

Multifaceted holograms can also be used to transform a wavefront into a specific shape. A system of two multifaceted holograms is used as a wavefront transformation for illumination of a hollow-box object [4.329]. The first hologram redistributes the Gaussian input beam into a square pattern. The second hologram corrects [4.329] the phases of various beams for illumination of a hollow-box object (Fig. 4.108).

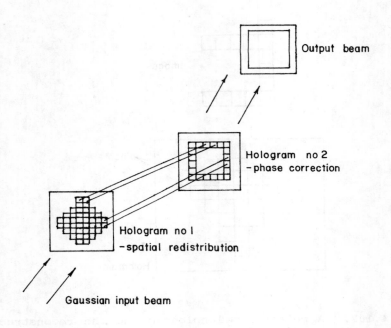

Fig. 4.108. Use of two multifaceted holograms to illuminate a hollow-box object [4.329].

Another example of a multifaceted hologram consists of wedge and ring shaped holograms (Fig. 4.109) that diffract light from the Fourier plane of a lens to an array of photodiodes and is useful for real-time polar coordinate diffraction pattern sampling applications [4.328]. Each facet of the HOE in the Fourier plane receives light from object of a particular spatial frequency and orientation which may be present in any part of the object plane. The detectors may be arranged linearly or angularly. In order to avoid crosstalk between opposite detectors different off-axis angles may be used for the ring facets and the wedge facets so that ± orders do not coincide with each other.

Fig. 4.110 shows the recording arrangement of such a

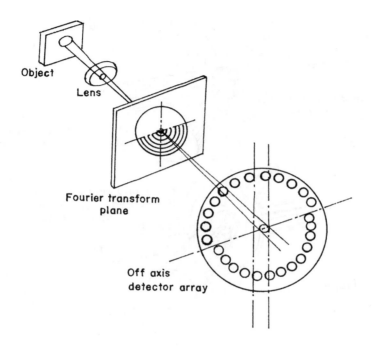

Fig. 4.109. A multifaceted wedge and ring shaped hologram used in a Fourier transform setup [4.328].

hologram [4.328]. The wedge sectors are recorded by using a fixed wedge aperture in contact with the photographic plate. The plate is turned in steps by the wedge angle $\Delta\theta$ between exposures. For recording of the ring facets, ring shaped masks and the recording plate are rotated together by $\Delta\theta$. A different mask is used for each exposure.

It is also possible to generate the wedge-ring HOE by the technique of computer generated holography [4.330].

4.14.5. Pinhole Hologram

Pinhole holography [4.331] utilizes pinhole camera approach for recording the hologram. During recording the object beam passes through a pinhole which plays the role of a lens and produces an image hologram. When reconstructing

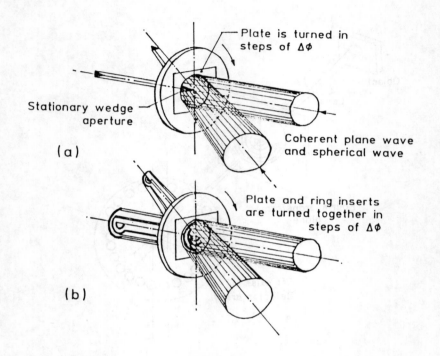

Fig. 4.110. Recording of wedge (a) and ring (b) type multifaceted holograms [4.328].

the hologram, the pinhole can be removed. Pinhole holography suppresses intermodulation noise in the reconstructed image. Spatial filtering operation can be carried out on the stored image of the object by using optical information processing elements in the position of the reconstructed pinhole image.

Fig. 4.111 shows the recording and reconstruction of a pinhole hologram of a 3D object [4.331]. For the recording of a planar object, recording arrangement of Fig. 4.112 can

$$D = 1.8 \sqrt{\lambda d_i}, \qquad (4.130)$$

where D is the diameter of the pinhole and d_i is the distance between the pinhole and the hologram recording plane (Fig. 4.112).

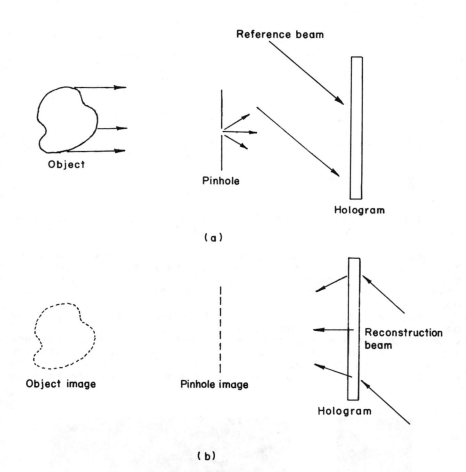

Fig. 4.111. Recording (a) and reconstruction (b) of a pinhole hologram of a 3D object [4.332].

Multiple objects can be sequentially recorded in a single hologram by associating a unique position of the pinhole for each object. The image of interest can be selected by allowing the corresponding reconstructed pinhole image and blocking all other pinhole images. Pinhole holograms thus have the potential of realizing programmable optical interconnects (Chapter 7). Different interconnect

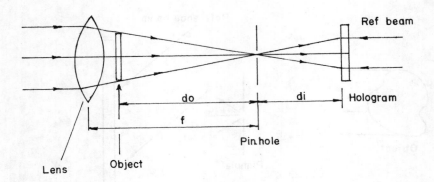

Fig. 4.112. Recording of a pinhole hologram of a mask of an array of points [4.331].

patterns may be recorded, each through a separate pinhole. The desired pattern can be played back by selecting the corresponding pinhole image in the reconstruction. The pinhole array can be formed by a pixelated spatial light modulator.

Figure 4.113a shows a reconstructed image of a pinhole hologram of a mask consisting of an array of 277 points, each 50 μm in diameter with a separation of 250 μm between adjacent points [4.331]. Figure 4.113b shows the replication of the pattern by inserting a grating of 300 lines/mm in the reconstructed pinhole plane. This example shows the promise

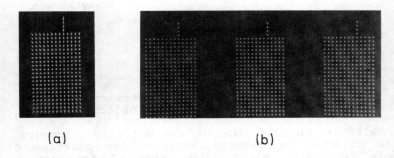

Fig. 4.113. (a) Reconstruction image of a pinhole hologram of an array of points and (b) effect of a grating in the reconstructed pinhole plane [4.331].

of a programmable holographic interconnect with high resolution and low cross-talk.

4.14.6. Edge-lit Hologram

Conventional holograms generally require a considerable distance between the light source and the hologram for satisfactory reconstruction. For 3D displays this is a limitation requiring illuminating sources to be fixed on the room ceiling or on a side wall. Edge-lit holograms can produce self-contained and compact 3D displays without the need for an external source [4.252, 4.333-4.337]. Figure

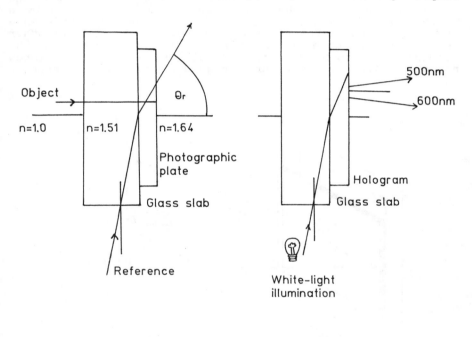

Fig. 4.114. Recording (a) and reconstruction (b) of an edge-lit hologram [4.252].

4.114 shows the recording and reconstruction of an edge-lit hologram. The reference beam is entered from one edge of the substrate. To achieve this a relatively thick glass plate or plastic sheet or slab is optically contacted to the photographic plate by placing an index matching liquid between the slab and the photographic plate. In this way intra-emulsion reference beam angle of up to 70° from the plate perpendicular can be formed. The large reference beam angle produces a very high frequency grating with a large diffraction angle resulting in an increased dispersion. The diffraction angle is given by

$$\frac{d\theta}{d\lambda} = \frac{n}{\lambda} \sin\theta_r. \tag{4.131}$$

This equation gives an angle of 7.4° between 500 and 600 nm output beams when the hologram is illuminated by a white light at angle of 45° in air. But when the reference angle is 65.1° in the emulsion, $d\theta$ becomes 15.5°. In other words, in the case of an edge-lit rainbow hologram, the viewing zone can be more than doubled than a normal rainbow hologram [4.252]. However, Bragg selectivity is stronger

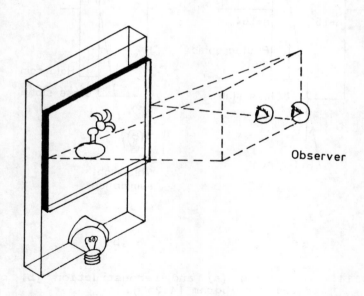

Fig. 4.115. Compact display using an edge-lit hologram [4.252].

than that for ordinary holograms.
It is also possible to take the reference beam as a collimated beam which can be fed using a cylindrical lens from one edge of the substrate. This type of hologram is strictly neither a transmission hologram nor a Lippmann hologram [4.336].
Edge-lit holography is a significant progress in the area of display holography. The reconstruction source can be an integral part of the display system and the hologram can be hung on a wall [4.252] as shown in Fig. 4.115. Besides compact displays, edge-lit holography has potential application in proximity recording of micro-images for photolithography [4.336].

4.15. Hologram Replication

Hologram replication or copying is a complicated operation as compared to copying an ordinary photograph. The hologram basically contains closely spaced fringes of the order of 1000 lines/mm and more and may be distributed in the volume of the emulsion. A high quality copy lens is not sufficient to image such fine details. Two methods are employed for hologram replication which do not need any lenses. These methods are optical interferometric replication technique and mechanical replication technique [4.338]. Vanin [4.339] has reviewed different techniques.

4.15.1. Optical Interferometric Techniques

Non Contact Interferometric Method

In this technique the reconstructed image from the hologram is used as the object wave for recording a new hologram. The method offers flexibility because the reference beam angle and the beam ratio can be adjusted which allows even to record a high efficiency copy hologram from a low efficiency master hologram [4.340]. Another advantage is that a transmission hologram can be copied as a reflection one or as an image hologram [4.339]. Even the copying wavelength can be different then the original recording wavelength for the master. This is useful for making holographic optical elements for near infrared operation or for other spectral region using materials sensitive to visible light [4.341].
Either virtual image or the real image of the master hologram can be used to make the copy hologram (Fig.4.116). When copying is done by using the real image of the original master hologram, the reconstructed real image of the copy hologram is orthoscopic whereas the real image of the master hologram was pseudoscopic. The orthoscopic real image of the copy hologram which is formed towards the observer is quite appealing to the viewer as he can even touch the image.

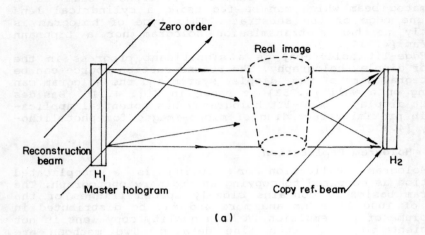

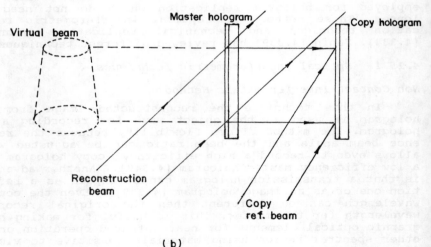

Fig. 4.116. Hologram copying using real image (a) and virtual image (b) from the master hologram.

Contact Interferometric Method

Hologram replication is extremely difficult to make

using contact printing. If the original hologram and the copy emulsion can be brought microscopically close together by vacuum contact, conventional photolithography techniques may be employed to transfer the hologram fine details to the copy emulsion. This technique may succeed for copying a low frequency absorption type master hologram. It may be noted that even though the fringes in the copy hologram have reversed contrast, the image of the copy hologram has normal contrast. This is because the copying process introduces a phase change of $180°$ in the image beam with respect to the reference beam. The contact copying method cannot be used for copying of volume holograms.

Contact interferometric method is suitable for copying both the transmission and reflection types of thin and volume holograms of any spatial frequency. Figure 4.117a shows an arrangement for copying a transmission hologram. The master is placed in front of the copy emulsion and illuminated with the reconstruction beam at proper angle. For best results, the reconstruction beam should duplicate the reference beam of the master hologram H_1. The reconstructed image wave interferes with the undiffracted zero order wave

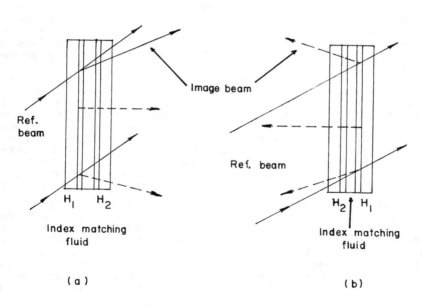

Fig. 4.117. Contact interferometric copying of a transmission (a) and a reflection (b) hologram.

and the interference pattern is recorded in the copy emulsion H_2. The emulsion of the two plates H_1 and H_2 are placed against each other and an index matching fluid such as xylene is placed between them to reduce the unwanted reflections. The efficiency of the master hologram should be 50% to achieve maximum efficiency of the copy hologram. The paths travelled by the two interfering beams are very short, therefore the coherence and stability requirements are greatly relaxed. Broad band thermal or discharge type of sources and filtered sun light can be used [4.342, 4.343]. The reduced coherence of the source increases the signal-to-noise ratio of the copy hologram.

When the master hologram is a thin hologram, it will reconstruct both the virtual and the real images. These two image waves will interfere with the undiffracted zero order wave and will produce two sets of interference patterns of equal amplitude and contrast which are recorded in the copy hologram. In addition to these a weak contrast interference pattern between the two image waves will also be recorded. One should remember that the master hologram has only one interference pattern. The copy hologram on reconstruction thus produces two real and two virtual images. The separation between double images (virtual or real) will be twice the separation between the master and the copy holograms if a plane reference beam is used [4.339]. If the two holograms are placed very close to each other the two virtual images (and also real images) can be superimposed. The two superimposed images add coherently and may produce interference effects.

For copying of reflection holograms, the copy emulsion is placed in front of the master hologram and illuminated with the reconstruction beam (Fig. 4.117b). For best results the curvature, direction and the wavelength of the reconstruction beam should match to the original reference beam. The direct incident beam interferes with the back diffracted image beam similar to Lippmann hologram and the interference pattern is recorded in the copy emulsion. The use of index matching fluid between the emulsions of the two plates H_1 and H_2 greatly reduces the unwanted reflections. The efficiency of the master hologram should be nearly 100% to achieve the maximum efficiency of the copy hologram. Obviously, even by using a phase master hologram the efficiency of the copy hologram can never exceed the master hologram if the same type of emulsion is used for both the holograms. However, if the master is recorded in a silver halide emulsion and the copy is recorded in photopolymer or dichromated gelatin, higher efficiencies in the copy hologram is possible [4.344].

For mass replication of transmission and reflection volume holograms, it is possible to develop an automatic copier based on a pulsed ruby laser which can produce 400000 square inches of holograms per hour on Ilford film [4.345]. The master is held stationary and the copy film is transported frame by frame. The complete roll of the film is exposed, developed, bleached, laminated and die cut.

A low power He-Ne laser can also be used for mass production [4.346, 4.347]. The direct laser beam is scanned across the hologram. This results in a point-wise copy of the hologram. Although the power of the laser is low, yet the direct unexpanded beam intensity is sufficient to expose the copy film. The copy rate can be quite high. The technique has been used [4.346] for copying of multicolour holograms on flexible polyester film base. The multicolour copy is sequentially exposed with each of the contact masters at the correct reference angles. The film is soaked in triethanolamine and dried before the exposure for each master (except for red colour) to copy the hologram by a He-Ne laser beam. CW argon ion laser has also been used along with a step-and-repeat camera to mass produce Lippmann holograms for security applications [4.348].

4.15.2. Mechanical Replication Technique (Embossed Holograms)

Embossed holograms can be produced on PVC, polyester and Mylar sheets. The basic steps involved in making embossed copies are [4.349, 4.350]:

(a) Recording of a surface relief hologram on a positive photoresist.

(b) Deposition of electrically conductive layer on the photoresist hologram.

(c) Electroplating the conductive photoresist hologram to create a metal master

(d) Embossing on plastic using metal master.

The hologram may be first recorded on a silver halide material. It is then copied on a positive photoresist using an argon ion laser or a He-Cd laser. The hologram is made electrically conductive by depositing silver onto it using a spray gun or vacuum deposition. A chemical bath for silvering similar to mirror formation can also be used. Electroless nickel deposition [4.350] is an inexpensive method and is preferred for large format. The hologram is first sensitized in a stannous chloride solution which is followed by dipping in a palladium chloride solution. Electroless nickel deposition then takes place in a heated dip tank. The immersion time, temperature of solutions and the agitation technique are critical.

Nickel sulfamide is used for electroforming bath for making the nickel master. The essential components for an electroplating system are DC rectifier, nickel anodes, filteration pump, solution heater and solution agitation system. The current density, cleanliness, and the pH, temperature and the specific gravity of the bath must be tightly controlled to get metal master free from cracks, pinholes, blisters, bumps and stains.

The metal master is carefully separated from the hologram and rinsed in a solvent to remove any adhered resist. Small holograms can be embossed by a press system. Large size holograms (6"x6") are embossed by a rolling machine. Roll embossing is the fastest method of replication [4.351]. Very large quantities of holograms (millions of copies) can be made by roll embossing. Embossed aluminized polyester holograms are durable in handling.

4.16. Polarization Holography

4.16.1. Depolarization Effects

The basic equations for holographic exposures have been derived assuming that the object and reference beams are similarly polarized. However, in many practical cases the situation is different. The light output of the majority of lasers is linearly polarized. Therefore, the reference beam is always linearly polarized, but polarization of the object beam will depend on the nature of the object surface. Depolarization can occur as a result of reflections from metallic and dielectric surfaces; it is also caused by scattering in some media like aerosol particles eg. rain, fog, dust etc. when the laser beam passes through them. The discrete particles of irregular shape and complex refractive index strongly scatter the light beam in all directions, resulting in a significant depolarization of the laser beam. If the particle number density is high, even spherical particles can introduce some depolarization [4.352].

An extended object surface with large curvatures can cause depolarization of the incident linearly polarized wave to produce the cross-polarized component [4.353]. In the case of a diffuse object, even a polarized beam becomes randomly polarized after reflection from it. The hologram will record only that component of the electric vector of the object wave which is parallel to the electric vector of the reference wave. The remaining portion of the object wave will be adding to the background exposure of the hologram.

The effect of depolarization is to degrade the reconstructed image quality with the possibility of a complete loss of information of those parts of the object from where the scattered beam is depolarized. Depolarization and nonlinearity of the recording medium result in a multiplicative distortion of the reconstructed image [4.354] which depends on the strength of the cross-polarized component of the object beam.

Fig. 4.118 shows [4.355] different polarization components in the scattered light from a laser illuminated object which is partly specularly reflecting, partly diffusing and partly polarization rotating. Figure 4.118a is the image when the object is illuminated by both parallel and orthogonally polarized beams. Figure 4.118b and 4.118c show the two polarized components of the object beam when it is illuminated by a linearly parallel polarized laser beam. The annular ring marked orient appears in all the three images because it is diffusing. The presence of orthogonal component in the scattered beam is due to depolarization.

Let an object be illuminated by a linearly polarized beam. The scattered beam is depolarized and its amplitude is given by $O = O_p + O_n$, where O_p denotes that component of the beam which is polarized parallel to the reference beam and O_n denotes the orthogonal polarization component.

The degree of polarization P can be defined as [4.356]

$$P = I_p/(I_p + I_n),$$

$$= 1/(1 + \delta), \qquad (4.132)$$

where $\delta = I_n/I_p =$ linear depolarization ratio, which is a direct

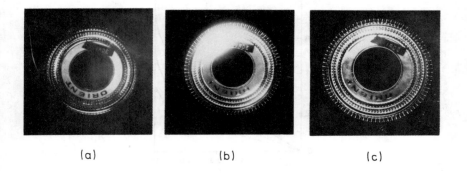

(a) (b) (c)

Fig. 4.118. (a) An object when illuminated by parallel and orthogonally polarized laser beams. (b) and (c) Two polarized components of the object when it is illuminated by a linearly parallel polarized beam [4.355].

measure of the amount of linear depolarization caused by reflection and scattering from the object surface. The efficiency of the hologram in terms of degree of polarization is given by [4.355]

$$\eta = \frac{K\rho}{(1 + \rho)^2 [1 + \{(1 - P)/P\}^{1/2}]} , \qquad (4.133)$$

where ρ is the reference-to-object beam intensity ratio and K is a constant which depends on the transfer characteristics of the recording medium.

The dependence of diffraction efficiency on the degree of polarization is shown [4.355] in Fig. 4.119. It is noted that when P decreases from 1 to 0.9, the efficiency drops by 43%.

4.16.2. Polarization Recording

The effects of depolarization may be reduced by using a rotated polarization reference beam for recording the hologram [4.354, 4.356]. The angle of rotation would assign part of the reference power to the cross-polarized component of

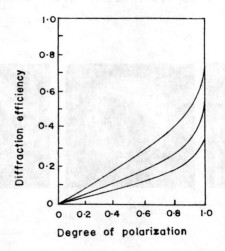

Fig. 4.119. Diffraction efficiency versus degree of polarization [4.355].

the object beam. The angle of rotation will depend on the relative strengths of the two polarized components.
The complete polarization information can be recorded in a volume hologram by using two reference waves with their polarization orthogonal to each other [4.356-4.358]. The two reference beams can be generated by splitting a beam into two and recombining them by mirrors and beam splitters. Before recombination the polarization of one of the beams is rotated through 90° with respect to the other. Depending on the degree of polarization of the object beam, the reference -to-object beam intensity ratio is adjusted for each object beam component. The hologram is an incoherent superposition of the two holograms. Such a hologram produces image which has full properties of the object wave.
In early work the two orthogonally polarized reference beams were used to record thin holograms which produced cross-talk images [4.359, 4.360]. Reference beam encoding can also be used [4.361] but it requires *in situ* development.
Photoelastic stress analysis is an important application of polarization holography.

4.17. Evanescent Wave Holography (Waveguide Holography)

An evanescent wave (also known as inhomogeneous wave or surface wave) is a wave whose planes of constant amplitude and phase cross each other at a finite angle. This is different than a homogeneous plane wave in which planes of constant amplitude and constant phase coincide. Evanescent waves occur during total reflection and diffraction by ultrafine structures [4.362].
If a monochromatic plane wave strikes the boundary of two media with refractive indices n_1 and n_2 ($n_1 > n_2$), the wave will be totally internally reflected if the angle of incidence of the wave in the denser medium is greater than the critical angle. The evanescent wave is created under this condition and propagates along the interface. The wavelength of the evanescent wave λ_e is given by

$$\lambda_e = \frac{\lambda_o}{n_1 \sin\theta} ,\qquad (4.134)$$

where λ_o is the wavelength in vacuum and θ is the angle of incidence. Figure 4.120 shows the creation of evanescent wave by total internal reflection. The penetration of the evanescent waves in rarer medium is damped exponentially. The penetration depth becomes appreciable when the angle of incidence becomes equal to or more than the critical angle.
The penetration depth increases for higher wavelengths of incident radiation. The propagation direction of the

454 LASERS AND HOLOGRAPHY

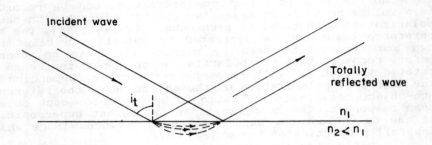

Fig. 4.120. Creation of evanescent wave by total internal reflection.

evanescent wave is influenced by the state of polarization of the radiation. The subject has been reviewed by Bryngdahl [4.363].

In evanescent wave holography (or waveguide holography) evanescent waves take part in the formation and reconstruction of the hologram [4.363-4.368]. During recording either or both the interfering beams may be evanescent.

Although the evanescent waves can be generated by diffraction, evanescent wave holograms are conveniently recorded by producing evanescent waves by total internal reflection technique. Photographic emulsions have refractive index of about 1.62 for a wavelength 633 nm before their processing and about 1.55 after development and fixing. To produce and record evanescent waves in these emulsions, the surrounding medium in which the internal reflection takes place must be higher than that of the emulsion. One way is to submerge the photographic plate in a liquid with high refractive index such as di-iodomethane whose refractive index is 1.73 at 633 nm [4.363]. The photographic emulsion may also be coated on a substrate of higher refractive index [4.364].

Let us consider the interference between a plane homogeneous object wave and an evanescent reference wave as shown in Fig. 4.121. The complex amplitude of the normally incident homogeneous wave is

$$O = O_o \exp(j2\pi z n_e/\lambda_o), \qquad (4.135)$$

where n_e is the refractive index of the emulsion and $\lambda_o/n_e = \lambda_h$ is the wavelength of the homogeneous wave.

The complex amplitude of the evanescent wave is given by

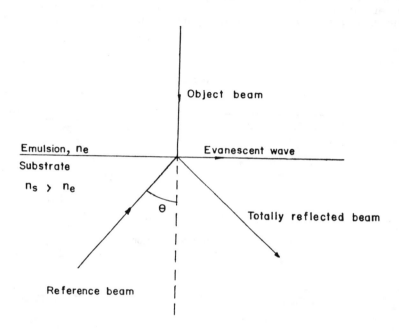

Fig. 4.121 Recording of a hologram using a plane homogenous object wave and an evanescent reference wave.

$$R = R_o \exp(-2\pi n_s z/p) \exp(j2\pi x/\lambda_e), \qquad (4.136)$$

where R_o is the amplitude at the surface and p is the penetration depth of the evanescent wave. The holographic exposure I is given by

$$I = |O + R|^2$$
$$= R_o^2 + O_o^2 \exp(-4\pi n_s z/p)$$
$$+ 2O_o R_o \exp(-2\pi n_s z/p) \cos\{2\pi[(x/\lambda_e) - (zn_e/p)]\}$$
$$\qquad (4.137)$$

The last term in Eq. 4.137 is the interference term

which is rapidly attenuated in the z-direction in the emulsion. The penetration depth of the evanescent wave is of the order of 1 μm, hence the recording may be treated as thin. The fringes have a period equal to the wavelength λ_e of the

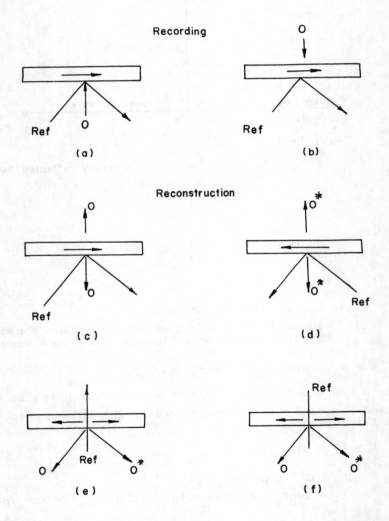

Fig. 4.122. Recording and reconstruction of an evanescent wave hologram [4.363, 4.365].

evanescent wave. A consequence of this is that the change in the direction of homogeneous wave does not change the hologram recording (Fig. 4.122a and 4.122b).

Evanescent wave holograms reconstruct images which are quite different in nature than those reconstructed by conventional holograms [4.363, 4.365]. When the evanescent wave hologram is reconstructed with the evanescent wave, the original homogenous wave is generated. In addition to this wave, an additional mirror image copy is also reconstructed. The two images are symmetrically placed on either side of the hologram (Fig. 4.122c). On reversing the direction of illumination of the evanescent wave, two symmetric complex conjugate waves are reconstructed (Fig. 4.122d). It may be noted that either both the images are original waves or conjugate waves unlike conventional holography in which a thin hologram produces one original wave and the other its conjugate [4.363, 4.366].

We can also record a hologram using both the object and the reference waves as evanescent. The two evanescent waves can interfere at a given angle if they have a common component of polarization.

Let the object and reference beams are represented by

$$O_e = O_o \exp(-2\pi n_s z/p_1) \exp(j2\pi x/\lambda_{e1}) \qquad (4.138)$$

and

$$R_e = R_o \exp(-2\pi n_s z/p_2) \exp(j2\pi x/\lambda_{e2}), \qquad (4.139)$$

where p_1 and p_2 are the penetration depths. The holographic exposure is proportional to

$$|O_e + R_e|^2 = O_o^2(-4\pi n_s z/p_1) + R_o^2 \exp(-4\pi n_s z/p_2)$$
$$+ 2O_o R_o \exp[-2\pi n_s z(1/p_1 + 1/p_2)]$$
$$\cdot \cos[2\pi x(1/\lambda_{e1} - 1/\lambda_{e2})]. \qquad (4.140)$$

The first two terms on the right hand side of Eq. 4.140 are attenuated rapidly in the z direction. The interference term (3rd term) has some penetration depth from the surface. The period of this structure is equal to

$$d = \frac{\lambda_{e1} \lambda_{e2}}{(\lambda_{e2} \pm \lambda_{e1})}. \qquad (4.141)$$

The minus sign applies when both the interfering beams propagate in the same direction and the positive sign applies when they propagate in the opposite direction.

When $\lambda_{e1} = \lambda_{e2}$, the fringe spacing becomes equal to $\lambda_e/2$ if the beams propagate in opposite direction.

$$\text{Fringe spacing} = \lambda_e/2 < \lambda_h/2 \qquad (4.142)$$

Thus it is possible to produce finer fringes (spacing $\lambda_e/2$) with evanescent waves than with homogeneous standing waves (spacing $\lambda_h/2$) of the same frequency.

Evanescent wave holography has some unique features. These are

a) It is possible to record very fine structures, finer than those possible by using homogeneous waves of the same frequency. Thus the object information beyond the conventional resolution limit of optics can be recorded and studied. This is a very useful property which may be exploited for optical lithographic applications. The configuration of Figs. 4.122b and 4.122c are of particular interest for this application as the reference and illumination beams are on the one side of the recording medium while the object and the reconstructed image are on the other side.

(b) The diffraction efficiency of an evanescent wave hologram can be higher than the conventional homogeneous wave hologram. A diffraction efficiency as high as 22.6% has been achieved by Nassenstein [4.364] for an amplitude holographic grating illuminated with a plane polarized beam at an angle of total internal reflection. This is much higher than the theoretical limit of 6.25% for an amplitude homogeneous wave hologram. Theoretical studies on the diffraction efficiency of evanescent wave hologram show [4.369, 4.370] that the diffraction efficiency for p-polarized waves (electric vector in the plane of polarization) exhibits dips which are not present for s-polarized waves (electric vector perpendicular to the plane of incidence). In the later case, the diffraction efficiency exhibits sharp maxima when the angle of incidence of the reconstruction wave or the angle of diffraction is equal to the critical angle.

(c) An evanescent wave hologram can be reconstructed by a white light source [4.363] similar to conventional volume type hologram. Fig 4.123 shows the reconstruction process for a conventional volume hologram and an evanescent wave hologram. In the former case constructive interference occurs for a particular wavelength only, while in the later case the wave propagates in the plane of the thin hologram and reconstructs the image when a condition equivalent to Bragg condition is satisfied. However, there is no effect of

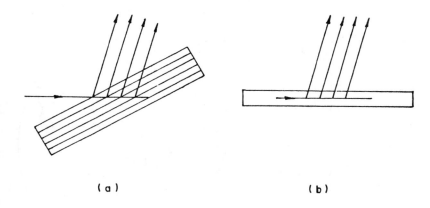

Fig. 4.123. Comparison of a conventional volume hologram (a) with an evanescent wave hologram (b) [4.363].

emulsion shrinkage in the evanescent wave hologram because the hologram is confined to the surface of the emulsion and not in the volume of the emulsion.

An evanescent wave holographic beam splitter has been used for making a compact recording system for wide field-of-view display transmission holograms [4.371]. Such a beam splitter is recorded as shown in Fig 4.124a.

The reference beam is guided in the glass plate with a prism coupler. Due to twin image property of such a hologram, the waveguide hologram beam splitter will produce two beams on reconstruction which can be used for object illumination and as a reference beam as shown in Fig 4.124b. In such a configuration, the object can be placed very close to the recording medium without any difficulty in object illumination, thus field-of-view comparable to the reflection hologram is possible. An achromatic image can be reconstructed with white light.

The waveguide holography has potential applications in information transfer [4.372], optical interconnects [4.373], lithography [4.374] and displays [4.371].

4.17.1. Holographic Lithography

With the scaling down of VLSI chips, the number of components per chip has increased to an astronomical figure of 2^{24}/chip. This has put very stringent requirements of microlithography such as field size, resolution, power re-

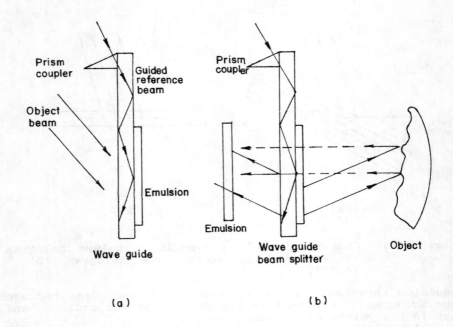

Fig. 4.124. (a) Recording of a waveguide hologram beam splitter [4.371]. (b) Configuration for recording of an off-axis transmission hologram [4.371].

quirements and aberrations. These requirements are very difficult to be met by traditional optical lithography methods. Figure 4.125 shows the narrowest line widths that has been experimentally achieved with different lithography techniques [4.375]. The dotted line shows the possibility of further reducing the line widths. Though X-ray and electron beam lithography can produce line widths smaller than 1 μm, their equipments are very expensive. Besides these, electron beam lithography machines have limited throughput and proximity effects on exposures.

An interesting and very useful application of holography is in lithography. Holographic lithographic process can produce [4.374, 4.375] line widths lower than 0.5 μm using lower laser recording wavelengths and employing evanescent wave illumination. Figure 4.126 shows the principle of holographic lithography with total internal reflection setup [4.374, 4.375]. In the recording process a volume hologram

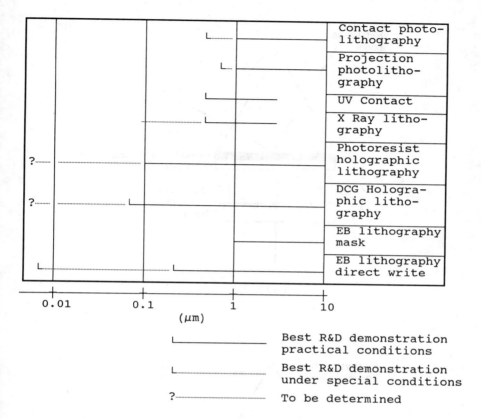

Fig. 4.125. Capabilities of different lithographic techniques [4.375].

in thick photopolymer or dichromated gelatin (20-40 μm thick) is recorded. The volume hologram produces high diffraction efficiency and eliminates the conjugate image in the reconstruction process. In the reconstruction process the hologram is rotated by 180° and a wafer is located in place of the mask for exposure. The separation distance from the hologram should be kept exactly the same for recording and reconstruction process to achieve high resolution.

The holographic lithography machine can provide the following superior features over the existing X-ray or e-beam machines.

(a) <u>High volume</u>: The processing speed for each individual wafer is extremely fast. For example, a 20 cm wafer can be processed in one minute.

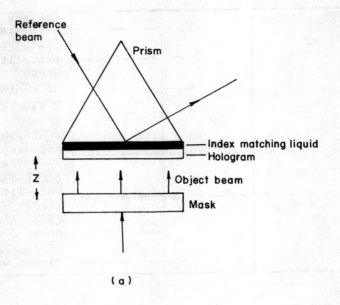

(a)

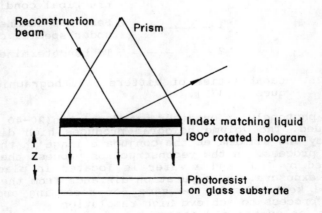

(b)

Fig.4.126. Principle of holographic lithography system [4.374, 4.375].

(b) **High resolution:** The resolution is limited only by the diffraction limit as other aberrations are absent due to the lensless recording and use of the optical wave for both recording and reconstruction.

(c) **Cost effectiveness:** The holographic lithography system is comparatively cheaper as it does not involve any complicated optical system.

Reference

4.1 D. Gabor, *Nature* **161** (1948) 777.

4.2 D. Gabor, *Proc. Roy. Soc. A.* **197** (1949) 454.

4.3 D. Gabor, *Proc. Phys. Soc. (London) B* **64** (1951) 449.

4.4 G.L. Rogers, *Proc. Roy. Soc. (Edinburgh)* **63 A** (1952) 193.

4.5 H.M.A. El-Sum and P. Kirkpatrick, *Phys. Rev.* **85** (1952) 763.

4.6 V.A. Baez, *Am. J. Phys.* **20** (1952) 311.

4.7 A. Lohmann, *Optica Acta* **3** (1956) 97.

4.8 E.N. Leith and J. Upatnieks, *J. Opt. Soc. Am.* **52** (1962) 1123.

4.9 E.N. Leith and J. Upatnieks, *J. Opt. Soc. Am.* **53** (1963) 1377.

4.10 E.N. Leith and J. Upatnieks, *J. Opt. Soc. Am.* **54** (1964) 1295.

4.11 E.N. Leith and J. Upatnieks, *J. Opt. Soc. Am.* **55** (1965) 569.

4.12 G.B. Parrent and B.J. Thompson, *Optica Acta* **11** (1964) 183.

4.13 B.J. Thompson, *J. Phys. E* **7** (1974) 781.

4.14 J. Trolinger, *Opt. Eng.* **14** (1975) 383.

4.15 R.W. Meier, *J. Opt. Soc. Am.* **55** (1965) 987.

4.16 M. Born and E. Wolf, *Principles of Optics* (Pergamon Press, Oxford 1959).

4.17 E.B. Champagne *J. Opt. Soc. Am.* **57** (1967) 51.

4.18 E.B. Champagne and N.G. Massey, *Appl. Opt.* **8** (1969) 1879.

4.19 J.N. Latta, *Appl. Opt.* **10** (1971) 599, 609, 2698.

4.20 J.N. Latta, *Appl. Opt.* **10** (1971) 666.

4.21 P.C. Mehta, K.S.S. Rao and R. Hradaynath, *Appl. Opt.* **21** (1982) 4553.

4.22 P.C. Mehta, *Optica Acta* **21** (1974) 1005.
4.23 R.A. Fisher, *Optical Phase Conjugation* (Academic Press, 1983)
4.24 E.B. Rotz and A.A. Friesem, *Appl. Phys. Lett.* **8** (1966) 146.
4.25 W.R. Klein and B.D. Cook, *IEEE Trans. Sonics & Ultrasonics* **SU-14** (1967) 123.
4.26 F.G. Kaspar, *J. Opt. Soc. Am.* **63** (1973) 37.
4.27 R. Magnusson and T.K. Gaylord, *J. Opt. Soc. Am.* **67** (1977) 1165; **68** (1978) 809.
4.28 B. Benlarbi, D.J. Cook and L. Solymer, *Optica Acta* **27** (1980) 885.
4.29 R. Alferness, *J. Opt. Soc. Am.* **66** (1976) 353.
4.30 H Kogelnik, *Bell Syst. Tech. J.* **48** (1969) 2909.
4.31 P.St. J. Russel, *Phys. Reports* **71** (1981) 209.
4.32 F. Dubois, F. De Schryver and B. Biran, *J. Opt. Soc. Am.* **8A** (1991) 270.
4.33 F. Dubois, F. De Schryver and P. Lemaire, *Topical Meeting on Photorefractive Materials, Effects and Devices* (Aussors, France, Jan 17-19, 1990).
4.34 P.C. Mehta, *Rev. d. Opt.* **6** (1975) 175.
4.35 P.C. Mehta, *J. Opt. (Paris)* **13** (1982) 293.
4.36 Y.N. Denisyuk, *Soviet Phys-Dokl.* **7** (1962) 543.
4.37 J.B. DeVelis and G.O. Reynolds, *Theory and Applications of Holography* (Addison-Wesley, Reading, Massachusetts, 1967).
4.38 G.W. Stroke, D. Brumm and A. Funkhouser, *J. Opt. Soc. Am.* **55** (1965) 1327.
4.39 G.W. Stroke, *Phys. Lett.* **23** (1966) 325.
4.40 C. Bhan, K.S.S. Rao and P.C. Mehta, *Appl. Opt.* **30** (1991) 4282.
4.41 K. Biedermann and K.A. Stetson, *Phot. Sci. Eng.* **13** (1969) 361.
4.42 J. Upatnieks and C. Leonard, *J. Opt. Soc. Am.* **60** (1970) 297.
4.43 Newport Catalog, A-14 (1990).
4.44 D.B. Neumann and H.W. Rose, *App. Opt.* **6** (1967) 1097.
4.45 D.R. MacQuigg, *Appl. Opt.* **16** (1977) 291.
4.46 R.J. Collier, C.B. Burkhardt and L.H. Lin, *Optical*

Holography (Academic Press, New York, 1971).

4.47 W.T. Cathey, Optical Processing and Holography (Wiley Interscience, New York, 1974).

4.48 N. Nishida, Appl. Opt. **14** (1970) 238.

4.49 R.L. Kurtz and L.M. Perry, Appl. Opt. **12** (1973) 2815.

4.50 D.B. Neumann, J. Opt. Soc. Am. **58** (1968) 447.

4.51 N. Abramson, Appl. Opt. **8** (1969) 1235; **9** (1970) 97, 2311; **10** (1971) 2155, 1143.

4.52 R.L. Kurtz and L.M. Perry, Appl. Opt. **11** (1972) 1998.

4.53 P. Smigielski, H. Fagot, A. Stimpfling and J. Schwab, Nouv. Rev. Opt. Appl. **2** (1971) 205.

4.54 R.J. Parker, Proc. SPIE **1358** (1990) 52; 3rd IEE Conf. on Holographic Components and Applications (1991) p. 234.

4.55 L.H. Lin, J. Opt. Soc. Am. **61** (1971) 203.

4.56 J.W. Goodman, J. Opt. Soc. Am. **57** (1967) 493.

4.57 H.M. Smith, Appl. Opt. **11** (1972) 26, 1869.

4.58 F.G. Kaspar and R.L. Lamberts, J. Opt. Soc. Am. **58** (1968) 970.

4.59 B.J. Chang and C.D. Leonard, Appl. Opt. **18** (1979) 2407.

4.60 A. Kozma, Optica Acta **15** (1963) 527; J. Opt. Soc. Am. **56** (1966) 428.

4.61 A.A. Friesem and J.S. Zelenka, Appl. Opt. **6** (1967) 201.

4.62 O. Bryngdahl and A. Lohmann, J. Opt. Soc. Am. **58** (1968) 1325.

4.63 J.W. Goodman and G.R. Knight, J. Opt. Soc. Am. **58** (1968) 1276.

4.64 P. Hariharan, G.S. Kaushik and C.S. Ramanathan, Opt. Commun. **5** (1972) 59.

4.65 K.M. Johnson, M. Armstrong, L. Hesselink and J.W. Goodman, Appl. Opt. **24** (1984) 4467.

4.66 B.J. Chang and N. George, Appl. Opt. **9** (1970) 713.

4.67 K. Biedermann, Appl. Opt. **10** (1971) 584.

4.68 S.A. Benton, in Handbook of Optical Holography, ed. H.J. Caulfield (Academic Press 1979) p. 349.

4.69 T.H. James and W. Vaneslow, PSA Tech. Quart. **2** (1955) 135.

4.70 D.J. Cooke and A.A. Ward, *Appl. Opt.* **23** (1984) 934.
4.71 Yu. N. Denisyuk, *Proc. SPIE* **136** (1977) 365.
4.72 W. Spierings, *Holosphere* **10** (7 & 8) (1981) 1.
4.73 L. Joly, R. Phelan and M. Redzikowski, *Proc. SPIE* **615** (1986) 66.
4.74 J. Upatnieks and C. Leonard, *J. Opt. Soc. Am.* **60** (1970) 297.
4.75 K. Biedermann, in *Holographic Recording Materialls*, ed. H.M. Smith (Springer-Verlag, Berlin 1977) p. 21.
4.76 H. Suzuki and A. Ono, *Proc. SPIE* **1212** (1990) 334.
4.77 H.-G. Weide, E. Lau and S. Dähne, *Z. Wiss. Phot., Photophysik Photochem.* **61** (1967) 145.
4.78 A. Graube, *Appl. Opt.* **13** (1974) 2942.
4.79 M.M. Crenshaw, *Proc. SPIE* **747** (1987) 104.
4.80 P. Hariharan, C.S. Ramanathan and G. S. Kaushik, *Opt. Commun.* **3** (1971) 246; *App. Opt.* **12** (1973) 611.
4.81 R.K. Kostuk, *Proc. SPIE* **1212** (1990) 54.
4.82 R.L. Van Renesse and F.A.J. Bouts, *Optik* **38** (1973) 156.
4.83 N.J. Phillips and W. Ce, *3rd IEE Conf. on Holographic Systems, Components and Applications* (1991) p. 30.
4.84 S. Kumar and K. Singh, *Optik* **86** (1990) 99; *J. Opt. (India)* **19** (1990) 108; *Optik* **88** (1991) 45; *J. Opt. (Paris)* **22** (1991) 22; *Opt. Laser Technol.* **23** (1991) 37; *J. Opt. (India)* **21** (1992) 1.
4.85 N.J. Phillips and D. Porter, *J. Phys. E* **9** (1976) 631.
4.86 C.S. Guo and L.Z. Cai, *Opt. Lett.* **16** (1991) 1777.
4.87 K.C. Brown and R.E. Jacobson, *J. Phot. Sc.* **33** (1985) 177.
4.88 R.E. Jacobson and K.C. Brown, *Proc. SPIE* **1051** (1989) 60.
4.89 T.A. Shankoff, *Appl. Opt.* **7** (1968) 2101.
4.90 J. Kosar, *Light Sensitive Systems* (Wiley, New York 1968) p 74.
4.91 D. Meyerhofer, *RCA Rev.* **33** (1972) 110; *Appl. Opt.* **10** (1971) 416.
4.92 M.H. Jeong And J.B. Song, *Appl. Opt.* **31** (1992) 161.
4.93 L.H. Lin, *Appl. Opt.* **8** (1969) 963.
4.94 C.D. Leonard and B.D. Guenther, *A Cookbook for*

Dichromated gelatin holograms, US Army Missile Res. & Dev. Command, Technica Report T-79-17, Jan. 1979, p. 13.

4.95 M. Mazakova, M. Pancheva, P. Kandilarov and P. Sharlandjiev, *Opt. Quant. Elect.* **14** (1982) 317.

4.96 R.D. Rallison and S. E. Bialkowski, *Proc. SPIE* **1051** (1989) 68.

4.97 A. Grube, *Opt. Commun.* **8** (1973) 251, *Phot. Sc. Eng.* **22** (1978) 37.

4.98 T. Kubota and T. Ose, *Appl. Opt.* **10** (1979) 2538; *Opt. Lett.* **4** (1979) 289.

4.99 T. Kubota, T. Ose, M. Sasaki and K. Honda, *Appl. Opt.* **15** (1976) 556.

4.100 R. Changkakoti, S.S.C. Babu and S.V. Pappu, *Appl. Opt.* **27** (1988) 324.

4.101 R. Changkakoti and S.V. Pappu, *Opt. Laser Technol.* (1989) 4.

4.102 J. Blyth, *Proc. SPIE* **1212** (1990) 190; *Holographic International* **No. 7** (1989) 22.

4.103 H. Owen, *Proc. SPIE* **523** (1985) 296.

4.104 S.P. McGrew, *Proc. SPIE* **215** (1980) 24.

4.105 A.A. Andrade and J.M. Rebordao, *Proc. SPIE* **1051** (1989) 96.

4.106 T.G. Georgekutty and H.K. Liu, *Appl. Opt.* **26** (1987) 372.

4.107 R.K. Curran and T.A. Shankoff, *Appl. Opt.* **9** (1970) 1651.

4.108 S. Case, *Doctoral Dissertation,* University of Michigan (1976).

4.109 S. Case and R. Alferness, *Appl. Opt.* **10** (1976) 41.

4.110 J.R. Magarinos, *Holographic Mirrors, Doctoral Dissertation,* University of Santiago, Spain (1984).

4.111 J.R. Magarinos and D.J. Coleman, *Proc. SPIE* **523** (1985) 203.

4.112 D. Sheel, *Proc. SPIE* **1212** (1990) 2.

4.113 S. Calixto and R.A. Lessard, *Appl. Opt.* **23** (1984) 1989.

4.114 S. Calixto and R.A. Lessard, *Appl. Opt.* **24** (1985) 317.

4.115 J.C. Newell, L. Solymer and A.A. Ward, *Appl. Opt.* **24**

4.115 (1985) 4460.
4.116 A.I. Erko and A.N. Malov, *Fund, Osnovy Optich, Pamyati* Sredy, **11** (1980) 62.
4.117 N.F. Balan, V.V. Kalinkin, N.W. Losevsky and A.N. Malov, in *Materials and Devices for Holographic Recording* ed. V. A. Barachevsky (Leningrad 1986) p.68
4.118 A.K. Aggarwal and S.K. Kaura, in *Holography and Speckle Phenomenon & their Industrial Applications*, ed. R.S. Sirohi (World Scientific 1990) p. 213.
4.119 V.P. Sherstyuk, A.N. Malov, S.M. Maloletov and V.V. Kalinkin, *Proc. SPIE* **1238** (1989) 218.
4.120 K.S. Pennington, J.S. Harper and F.P. Laming, *Appl. Phys. Lett.* **18** (1971) 80.
4.121 B.J. Chang and K. Winick, *Proc. SPIE* **215** (1980) 172.
4.122 W.R. Graver, J.W. Gladden and J.W. Eastes, *Appl. Opt.* **19** (1980) 1529.
4.123 D.K. Angell, *Appl. Opt.* **26** (1987) 4692.
4.124 P.G. Boj, A. Fimia and J.A. Quintana, *Proc. SPIE* **702** (1986) 105.
4.125 J.A. Quintana, P.G. Boj, A. Bonmati, J. Crespo, M. Pardo and C. Pastor, *Proc. SPIE* **1051** (1989) 160 .
4.126 W.S. Colburn and K.A. Haines, *Appl. Opt.* **10** (1971) 1636.
4.127 R.H. Wopschall and T.R. Pampalone, *Appl. Opt.* **11** (1972) 2096.
4.128 B.L. Booth, *Appl. Opt.* **14** (1975) 593.
4.129 A.M. Weber, W.K. Smothers, T.J. Trout and D. Mickish, *Proc. SPIE* **1212** (1990) 30.
4.130 W.K. Smothers, B.M. Monroe, A.M. Weber and D.E.Keys, *Proc. SPIE* **1212** (1990) 20.
4.131 G.I. Lashkov and E.N. Bodunov, *Opt. Spectrosc.* **47** (1979) 625.
4.132 G.I. Lashkov and V.I. Sukhanov, *Opt. Spectrosc.* **44** (1978) 590.
4.133 V.I. Sukhanov and V.L. Korzinin, *Sov. Tech. Phys. Lett.* **8** (1982) 491.
4.134 S. Calixto, *Appl. Opt.* **26** (1987) 3904.
4.135 Y. Yamagishi, T. Ishizuka and T. Yagishita, *Proc. SPIE* **600** (1985) 14.
4.136 K. Matsumoto, T. Kuwayama, M. Matsumoto and N

Taniguchi, *Proc. SPIE* **600** (1985) 9.

4.137 J.J.A. Couture, R.A. Lessard and R. Changkakoti, *Proc. SPIE* **1319** (1990) 281.

4.138 J.J.A. Couture and R.A. Lessard, *Appl. Opt.* **27** (1988) 3368.

4.139 J.J.A. Couture, *Appl. Opt.* **30** (1991) 2858.

4.140 R.A. Bartolini, in *Holographic Recording Materials*, ed. H. M. Smith (Springer-Verlag, 1977) p. 209.

4.142 M.S. Sthel, C.R.A. Lima and L. Cescato, *Appl. Opt.* **30** (1991) 5152.

4.143 K. Ichimura and Y. Ohe, *Proc. SPIE* **1212** (1990) 73.

4.144 F.M. Schellenberg, C.G. Willson, M.D. Levenson, K.M. Sperley and P.J. Brock, *Proc. SPIE* **1051** (1989) 317.

4.145 J.C. Urbach, in *Hoographic Recording Materials* ed, H.M. Smith (Springer-Verlag, 1977) p. 161.

4.146 T.C. Lee, J. Skogen, R. Schulze, E. Bernal G., J.Lin, T.Daehlin and M. Campbell, *Proc. SPIE* **215** (1980) 192.

4.147 T.L. Credelle and F.W. Spong, *RCA Rev.* **33** (1972) 206.

4.148 T.C. Lee, J.W. Lin and D.N. Tufte, *Proc. SPIE* **123** (1977) 74.

4.149 R.C. Duncan, Jr. and D.L. Staebler, in *Holographic Recording Materials* ed. HM Smith, (Springer-Verlag, 1977) p. 133.

4.150 G.M. Brown and W.G. Shaw, *Rev. Pure Appl. Chan* **11** (1961) 2.

4.151 Z.J. Kiss, *IEEE J. Quant. Electron.* **QE-5** (1969) 12.

4.152 Z.J. Kiss, *Phys. Today* **23** (1970) 42.

4.153 S.V. Pappu, *Contemp. Phys.* **26** (1985) 79.

4.154 W. Phillips and R.C. Duncan, Jr., *Metallurg. Trans.* **2** (1971) 769.

4.155 B.W. Faughnan, D.L. Staebler and Z.J. Kiss, *Inorganic Photochromic Materials,* in *Applied Solid State Science,* ed. R. Wolfe **Vol.** 2 (Academic Press, New York 1971).

4.156 S.V. Pappu, *Phys. Edu. (India)* **6** (1989) 104.

4.157 G.E. Scrivener and M.R. Tubbs, *Opt. Commun.* **6** (1972) 242; **10** (1974) 32.

4.158 I. Schneider, *Appl. Opt.* **6** (1967) 2197.

4.159 F. Lanzl, V. Rodor and W. Waidelich, *Appl. Phys. Lett.* **18** (1971) 56.

4.160   D. Casasent and F. Caimi, *Appl. Opt.* **15** (1976) 815.
4.161   F.S. Chen, J.T. LaMacchia and D.B. Frazer, *Appl. Phys. Lett.* **13** (1968) 223.
4.162   A. Ashkin, D.D. Boyd, J.M. Dziedzic, R.G. Smith, A.A. Ballman, H.J. Levinstein and K. Nassau, *Appl. Phys. Lett.* **9** (1966) 72.
4.163   J.J. Amodei and D.L. Staebler, *RCA Rev.* **33** (1972) 71.
4.164   J. Feinberg, *Proc. SPIE* **532** (1985) 119.
4.165   G.E. Peterson, A.M. Glass and T.J. Negran, *Appl. Phys. Lett.* **19** (1971) 130.
4.166   D.L. Staebler, in *Holographic Recording Materials*, ed. H.M. Smith, (Springer-Verlag, Berlin, 1977) p. 101.
4.167   W. Phillips, J.J. Amodei and D.L. Staebler, *RCA Rev.* **33** (1972) 94.
4.168   E. Krätzig and R. Orlowski, *Appl. Phys.* **15** (1978) 133.
4.169   J.J. Amodei and D.L. Staebler, *Appl. Phys. Lett.* **18** (1971) 540.
4.170   D.L. Staebler, W. Burke, W. Phillips and J.J. Amodei, *Appl. Phys. Lett.* **26** (1975) 182.
4.171   F. Micheron and G. Bismuth, *Appl. Phys. Lett.* **20** (1972) 79; **23** (1971) 71.
4.172   F. Micheron and J.C. Trotier, *Ferroelectrics* **8** (1974) 441.
4.173   J.B. Thaxter and M. Kestigian, *Appl. Opt.* **13** (1974) 913.
4.174   J.P. Herriau, D. Rojas, J.P. Huignard, J.M. Sassat and J.C. Launay, *Ferroelectrics* **75** (1987) 271.
4.175   T.K. Das and K. Singh, *J. Opt. (India)* **20** (1991) 1.
4.176   N. Cheung, *Holosphere* **8** (Dec 1979) 6.
4.177   M.M. Crenshaw, *Proc. SPIE* **615** (1986) 92.
4.178   V.Z. Bryskin, V.N. Krylov and D.I. Staselko, *Proc. SPIE* **1238** (1989) 448.
4.179   K. Bazargan, *Proc. SPIE* **1051** (1989) 6.
4.180   T.H. Jeong, P. Rudolf and A. Luckett, *J. Opt. Soc. Am.* **56** (1966) 1263.
4.181   H.M. Chau, *Appl. Opt.* **9** (1970) 1479.
4.182   R. Hioki and T. Suzuki, *Jap. J. Appl. Phys.* **4** (1965) 816.

4.183 T.H. Jeong, *J. Opt. Soc. Am.* **57** (1967) 1396.
4.184 B.A. Stirn, *Am. J. Phys.* **43** (1975) 297.
4.185 J. Upatnieks and J.T. Embach, *Opt. Eng.* **19** (1980) 696.
4.186. M.C. King, *Appl. Opt.* **7** (1968) 1641.
4.187. N. George, *Opt. Commun.* **1** (1970) 457.
4.188 P.C. Mehta, Unpublished Work.
4.189 A.K. Aggarwal and S.K. Kaura, *Opt. Laser Technol.* **19** (1987) 209.
4.190 S.A. Benton, *J. Opt. Soc. Am.* **59** (1969) 1545A.
4.191 K. Murata and K. Kunugi, *Appl. Opt.* **16** (1977) 1978.
4.192 R. Sato and K. Murata, *13th Congress of the International Commission for Optics*, Sapporo, Japan (1984) p. 50.
4.193 Q. Zhimin, X. Yingming, L. Meiyue and C. Xueqang, *Proc. SPIE* **673** (1986) 93.
4.194 A.K. Aggarwal, S.K. Kaura and D.P. Chhachhia, *Proc. SPIE* **1238** (1989) 18.
4.195 S.A. Benton, *13th Congress of the International Commission for Optics*, Sapporo, Japan (1984) p. 392.
4.196 P.C. Mehta and C. Bhan, *Opt. Eng.* **31** (1992) 369.
4.197 S.A. Benton, in *Applications of Holography and Optical Data Processing*, eds. E. Marom, A.A. Friesem and M. Wiener-Avnear, (The Pergamon Press, Oxford, 1977) p. 401.
4.198 H. Chen and F.T.S. Yu, *Opt. Lett.* **2** (1978) 85.
4.199 S. A. Benton, *Opt. Eng.* **19** (1980) 686.
4.200 P. Hariharan, Z.S. Hegedus and W.H. Steel, *Optica Acta* **26** (1979) 289.
4.201 C.P. Grover and H.M. Van Driel, *J. Opt. Soc. Am.* **70** (1980) 335.
4.202 C.P. Grover, R.A. Lessard and R. Tremblay, *Appl. Opt.* **22** (1983) 3300.
4.203 Q. Shan, Q. Chen and H. Chen, *Appl. Opt.* **22** (1983) 3902.
4.204 C. Joenathan, R. Kamala and R.S. Sirohi, *J. Opt. (India)* **17** (1988) 12.
4.205 R.D. Bahuguna, *Opt. Eng.* **27** (1988) 243.
4.206 A. Beauregard and R.A. Lessard, *Appl. Opt.* **23** (1984) 3045.

4.207 X.-Y. Da and Q.-Q. Wang, *Appl. Opt.* **30** (1991) 5143.
4.208 *RCA Electro-Optics Handbook, Tech. Ser.* **EOH-11**, RCA, Harrison, New Jersey (1974) p. 16.
4.209 P.C. Mehta, C. Bhan and P. Lal, *Indian J. Pure Appl. Phys.* **19** (1981) 995.
4.210 J.C. Wyant, *Opt. Lett.* **1** (1977) 130.
4.211 C.B. Brandt, *Appl. Opt.* **8** (1969) 1421.
4.212 M.C. King, *Appl. Opt.* **7** (1968) 1641.
4.213 L. Cross, Unpublished Work.
4.214 L. Huff and R.L. Fusek, *Opt. Eng.* **19** (1980) 691.
4.215 S.A. Benton, *J. Opt. Soc. Am.* **68** (1978) 1440A.
4.216 K. Okada, T. Honda and J. Tsujiuchi, *Opt. Commun.* **41** (1982) 397.
4.217 S.M. Jaffy and K. Dutta, *Proc. SPIE* **367** (1982) 130.
4.218 K. Sato, S. Furukawa, T. Kawabe and H. Katsuma, *Proc. SPIE* **1051** (1989) 177.
4.219 T. Honda, K. Okada and J. Tsujiuchi, *Opt. Commun.* **36** (1981) 11.
4.220 M.W. Halle, S.A. Benton, M.A. King and J.S. Underkoffler, *Proc. SPIE* **1461** (1991) 142.
4.221 K. Okada and J. Tsujiuchi, *Proc. SPIE* **1319** (1990) 296.
4.222 H. Katsuma, *Proc. SPIE* **1212** (1990) 166.
4.223 K. Sato, I. Akiyama, A. Ohimura, K. Fukazawa and M. Gomi, *Proc. SPIE* **1212** (1990) 136.
4.224 M. Yamaguchi, N. Ohyama and T. Honda, *Proc. SPIE* **1212** (1990) 84.
4.225 M.A. Teitel, *Proc. SPIE* **1051** (1989) 204.
4.226 G.-P. Zhou, *Opt. Laser Technol.* (Feb 1985) 23.
4.227 R. Blackie, R. Bagby, L. Wright, S. Dover, J. Drinkwater and S. Hart, *J. Pathol.* **149** (1986) 202A.
4.228. R. Bagby, *Int'l Rev. Cytol.* **105** (1986) 67; *Proc. SPIE* **1212** (1990) 102.
4.229 M. Suzuki, M. Kanaya and T. Saito, *Proc. SPIE* **523** (1988) 38.
4.230 I. Glaser and A. Friesem, *Proc. SPIE* **120** (1977) 150.
4.231 S.A. Benton, *Proc. SPIE* **761** (1987) 53.
4.232 S.A. Benton, *Proc. SPIE* **884** (1988) 106.

4.233 S.A. Benton, *3rd IEE Conf. on Holographic Systems, Components and Applications* (1991) p. 1.

4.234 J.R. Andrews, M.D. Rainsdon and W.E. Haas, *Proc. SPIE* **1051** (1989) 200.

4.235 D. Fargion, *Proc. SPIE* **1238** (1989) 428.

4.236 P.C. Mehta, C. Bhan and L. Joshi, Unpublished Work.

4.237 D.J. De Bitteto, *Appl. Phys. Lett.* **9** (1966) 417.

4.238 K. Bazargan, *Proc. SPIE* **673** (1986) 68.

4.239 C.B. Burckhardt, *Bell Syst. Tech. J.* **45** (1966) 1841.

4.240 K. Bazargan, *3rd IEE Conf. on Holographic Systems, Components and Applications* (1991) p. 82.

4.241 T. Kubota, *Proc. SPIE* **1051** (1989) 12.

4.242 P.C. Mehta, *Phys. News* (1975) 136.

4.243 S.A. Benton, *Opt. Eng.* **14** (1975) 402.

4.244 J.R. Andrews, W.E. Haas, R. Mileski, A. Yesul, J. Havranek, K. Blaksley, G. Johnston, L. Kawacz, G. Muller and M. D. Rainsdon, *Proc. SPIE* **1051** (1989) 156.

4.245 W.E. Haas, J.R. Andrews and M.D. Rainsdon, *Proc. SPIE* **1051** (1989) 171.

4.246 J.R. Andrews, M.D. Rainsdon and W.E. Haas, *Proc. SPIE* **1051** (1989) 200.

4.247 P.C. Mehta, P. Lal, D. Mohan, and R.D. Khajuria, Unpublished Work.

4.248 P.C. Mehta, P. Lal and T.C. Jain, Unpublished Work.

4.249 K. Bazargan, *Proc. SPIE* **523** (1985) 24.

4.250 S.A. Benton, *J. Opt. Soc. Am.* **73** (1983) 933A; *Proc. SPIE* **462** (1984) 2.

4.251 S.A. Benton, S.M. Birner and A. Shirakura, *Proc. SPIE* **1212** (1990) 149.

4.252. Q.Huang, *Holography* **2** (1992) 5.

4.253 G.T. Herman, *Proc. SPIE* **671** (1986) 124.

4.254 T. Honda, M. Yamaguchi, D.-K. Kang, K. Shimura, J. Tsujiuchi and N. Ohyama, *Proc. SPIE* **1051** (1989) 186.

4.255 L.H. Enloe, J.A. Murphy and C.B. Rubinstein, *Bell Syst. Tech. J.* **45** (1966) 335.

4.256 K. Sato, K. Higuchi and H. Katsuma, *3rd IEE Conf. on Holographic Systems, Components and Applications* (1991) p. 20.

4.257   D.J. De Bitteto, *Appl. Opt.* **9** (1970) 498.
4.258   V.G. Komar, *Proc. SPIE* **120** (1977) 123.
4.259   A.J. Decker, *Opt. Lett.* **7** (1982) 122.
4.260   P. Smigielski, H. Fagot and F. Albe, *Proc. SPIE* **673** (1986) 22.
4.261   W A. Thornton, *J. Opt. Soc. Am.* **61** (1971) 1155.
4.262   A.A. Friesem and R. J. Fedorowicz, *Appl. Opt.* **5** (1966) 1085.
4.263   A.A. Friesem and R.J. Fedorowicz, *Appl. Opt.* **6** (1967) 529.
4.264   K. Bazargan, *Proc. SPIE* **391** (1983) 11.
4.265   P.M. Hubel and A.A. Ward, *Proc. SPIE* **1051** (1989) 6.
4.266   P. Hariharan, in *Progress in Optics,* **20** ed. E Wolf (North Holland, Amsterdam, 1983) p. 265.
4.267   R.J. Collier and K.S. Pennington, *Appl. Opt.* **6** (1967) 1091.
4.268   W.T. Cathey, in *Handbook of Holography* (Academic Press, 1979) 199.
4.269   R.A. Lessard, S.C. Som and A. Boivin, *Appl. Opt.* **12** (1973) 2009.
4.270   K.S. Pennington and L.H. Lin, *Appl. Phys. Lett.* **7** (1965) 56.
4.271   N. Nishida, *Appl. Opt.* **14** (1970) 238.
4.272   L.H. Lin, K.S. Pennington, G.W. Stroke and A.E. Labeyrie, *Bell Syst. Tech. J.* **45** (1966) 659.
4.273   J.Upatnieks, J. Marks and R. Fedorowicz, *Appl. Phys. Lett.* **8** (1966) 286.
4.274   G.W. Stroke and A.E. Labeyrie, *Phys. Lett.* **20** (1966) 368.
4.275   L.H. Lin and C.V. Lo Bianco, *Appl. Opt.* **6** (1967) 1255.
4.276   P. Hariharan, *J. Opt. (Paris)* **11** (1980) 53.
4.277   P. Hariharan, *Optica Acta* **25** (1978) 527.
4.278   P.C Mehta, Unpublished Work.
4.279   H. Chen, A. Tai and F.T.S. Yu, *Appl. Opt.* **17** (1978) 1990.
4.280   A.D. Galpern, B.K. Rozhkov, V.P. Smaev and Yu.A. Vavilova, *Opt. Spectrosc.* **62** (1987) 810.
4.281   T.H. Jeong and E. Wesly, *Proc. SPIE* **1212** (1990) 183;

ibid. **1238 (1989)** 298, *Ibid.* **1183** (1990).

4.282 P.C. Mehta, to appear.

4.283 C. Solano, R.A. Lessard and D.C. Roberge, *Appl. Opt.* **24** (1985) 1189.

4.284 T. Mizuno, T. Goto, M. Goto, K. Matsui and T. Kubota, *Proc. SPIE* **1212** (1990) 40.

4.285 T.H. Jeong, B.J. Feferman and C.R. Bennett, *Proc. SPIE* **615** (1986) 2.

4.286 J. Blyth, *Holosphere* **8** (1979) 5.

4.287 P. Hariharan, *Opt. Commun.* **35** (1980) 42.

4.288 W. Spierings, *Holosphere* **10** (1981) 1.

4.289 P.C. Mehta and K.S.S. Rao, Unpublished Work.

4.290 S.L. Smith and T.J. Cretkovich, *Proc. SPIE* **462** (1984) 8.

4.291 S.L. Smith, *Proc. SPIE* **523** (1985) 42.

4.292 J.L. Walker and S.A. Benton, *Proc. SPIE* **1051** (1989) 192.

4.293 J.P. Waters, *Appl. Opt.* **11** (1972) 630.

4.294 J.P. Cathey, J.F. Hadwin and J.D. Pace, *Appl. Opt.* **12** (1973) 2683.

4.295 J.W. Goodman, W.H. Hutley, Jr., D.W. Jackson and M. Lehmann, *Appl. Phys. Lett.* **8** (1966) 311.

4.296 H.J. Caulfield, J.L. Harris and J.G. Cobb, *Proc. IEEE* **55** (1967) 1758.

4.297 W.T. Cathey, US Patent 3415587, Dec 10, 1968.

4.298 G.B. Brandt, *Appl. Opt.* **8** (1965) 1421.

4.299 M. Lang, G. Goldmann and P. Graf, *Appl. Opt.* **10** (1971) 168

4.300 P.C. Mehta, *Appl. Opt.* **13** (1974) 1279.

4.301 P.C. Mehta, *Appl. Opt.* **14** (1975) 2562.

4.302 P.C. Mehta, *Rev. d. Opt.* **7** (1975) 101.

4.303 D.J. De Bitteto and A.L. Dalisa, *Appl. Opt.* **10** (1971) 2292.

4.304 P.J. Van Heerden, *Appl. Opt.* **2** (1963) 387.

4.305 D. Armitage and A. Lohmann, *Appl. Opt.* **4** (1965) 399.

4.306 H.J. Caulfield, *Appl. Opt.* **9** (1970) 1218.

4.307 J.T. LaMacchia and J.C. Vincelette, *Appl. Opt.* **7** (1968) 1857.

4.308   P.C. Mehta and S. Swami, *Indian J. Pure Appl. Phys.* **12** (1974) 592.

4.309   K.M. Johnson, L. Hesselink and J.W. Goodman, *Appl. Opt.* **23** (1984) 218.

4.310   S.A. Collins, Jr. and H.J. Caulfield, *J. Opt. Soc. Am.* **6** (1989) 1568.

4.311   P.C. Mehta and S. Swami, *Indian J. Pure Appl. Phys.* **13** (1975) 347.

4.312   N. Nishida and M. Sakaguchi, *Appl. Opt.* **10** (1971) 439.

4.313   D. Anderson and D. Lininger, *Appl. Opt.* **26** (1987) 5031.

4.314   T.F. Krile, M.O. Hagler, W.D. Redus and J.F. Walkup, *Appl. Opt.* **18** (1979) 52.

4.315   J.T. LaMacchia and D.L. White, *Appl. Opt.* **7** (1968) 91.

4.316   P.C. Mehta and M. Singh, *Opt. Commun.* **7** (1973) 394.

4.317   E. Paek, J. Wullert and J. Patel, *Jap. J. Appl. Phys.* **29** (1990) 1332.

4.318   D. Psaltis, D. Brady and K. Wagner, *Appl. Opt.* **27** (1988) 1752.

4.319   E. Maniloff and K. Johnson, *Opt. Eng.* **29** (1990) 225.

4.320   Y. Taketomi, J.E. Ford, H. Sasaki, J. Ma, Y. Fainman and S. H. Lee, *Opt. Lett.* **16** (1991) 1774.

4.321   H. Bartelt and F. Sauer, *Opt. Commun.* **53** (1985) 296.

4.322   A. Lohmann and F. Sauer, *Appl. Opt.* **27** (1988) 3003.

4.323   N. Davidson, A.A. Friesem and E. Hasman, *Appl. Opt.* **31** (1992) 1810.

4.324   P.R. Haugen, H. Bartelt and S.K. Case, *Appl. Opt.* **22** (1983) 2822.

4.325   S.K. Case, P.R. Haugen and O.J. Loekberg, *Appl. Opt.* **20** (1981) 2670.

4.326   S.K. Case and P.R. Haugen, *Opt. Eng.* **21** (1982) 352.

4.327   H.O. Bartelt and S.K. Case, *Opt. Eng.* **22** (1983) 497.

4.328   M.S. Brown, *Optica Acta* **31** (1984) 507.

4.329   S.K. Case and V. Gerbig, *Opt. Commun.* **36** (1981) 94.

4.330   D. Casasent and J.-Z. Song, *Proc. SPIE* **523** (1985) 227.

4.331   S. Xu, G. Mendes, S. Hart and J.C. Dainty, *Opt. Lett.*

14 (1989) 107.
4.332 P.A. Newman and V.E. Rible, *Appl. Opt.* **5** (1966) 1225.
4.333 L.H. Lin, *J. Opt. Soc. Am.* **60** (1970) 714A.
4.334 J. Upatnieks and E. Leith, *Proc. SPIE* **883** (1988) 171.
4.335 W.J. Farmer, S.A. Benton and M.A. Klug, *Proc. SPIE* **1461** (1991) 215.
4.336 N.J. Phillips and W. Ce, *3rd IEE Conf. on Holographic Systems, Components and Applications* (1991) p. 8.
4.337 M. Miler, V. Morozov and A. Putitin, *Sov. J. Quant. Electron.* **16** (1989) 415.
4.338 W.T. Rhodes, in *Handbook of Optical Holography*, ed. H.J. Caulfield (Academic Press, 1979) p. 373.
4.339 V.A. Vanin, *Sov. J. Quant Electron.* **8** (1978) 809.
4.340 J.C. Palasis and J.A. Wise, *Appl. Opt.* **10** (1971) 667.
4.341 H.P. Herzig, *Opt. Commun.* **58** (1986) 144.
4.342 J. Oliva, A. Fimia and J.A. Quintana, *Appl. Opt.* **21** (1982) 2891.
4.343 N.J. Phillips and R Vander Werf, *J. Phot. Sc.* **33** (1985) 22.
4.344 B.L. Booth, *Appl. Opt.* **14** (1975) 593.
4.345 G.P. Wood, *Proc. SPIE* **615** (1986) 74.
4.346 T.J. Cvetkovich, *Proc. SPIE* **523** (1985) 47.
4.347 G.P. Wood, *Proc. SPIE* **44** (1989) 44.
4.348 K.J. Schell, *Proc. SPIE* **523** (1985) 331.
4.349 R.A. Bartolini, N. Feldstein and R.J. Ryan, *J. Electrochem. Soc. Solid State Sci. & Tech.* **120** (1973) 1408.
4.350 B.R. Clay and D.A. Gore, *Proc. SPIE* **45** (1974) 149.
4.351 J.R. Burns, *Proc. SPIE* **523** (1985) 7.
4.352 S.R. Pal and A.I. Carswell, *Appl. Opt.* **12** (1973) 1530.
4.353 M.L. Varshavchuk and V.O. Kobak, *Radio Eng. Electron. Phys.* **16** (1971) 201.
4.354 H. Ghandeharian and W.M. Boerner, *J. Opt. Soc. Am.* **68** (1978) 931.
4.355 P.C. Mehta and R. Hradayanath, *Appl. Opt.* **21** (1982) 4549.
4.356 G.V. Rozenberg and G.I. Gorchakov, *Atoms Oceanic*

Phys. **3** (1967) 400.

4.357 K. Gäsvik, *Optica Acta* **22** (1975) 189.

4.358 S.C. Som and R.A. Lessard, *Appl. Phys. Lett.* **17** (1970) 381.

4.359 A.W. Lohmann, *Appl. Opt.* **4** (1965) 1667.

4.360 O. Bryngdahl, *J. Opt. Soc. Am.* **57** (1967) 545.

4.361 C.N. Kurtz, *Appl. Phys. Lett.* **14** (1969) 59.

4.362 M. Born and E. Wolf, *Principle of Optics* (Pergamon Press, New York 1980).

4.363 O. Bryngdahl, in *Progress in Optics XI*, ed. E. Wolf (North-Holland, 1973) p. 169.

4.364 H. Nassenstein, *Phys. Lett.* **28A** (1968) 249; **29A** (1969) 175.

4.365 O. Bryngdahl, *J. Opt. Soc. Am.* **59** (1969) 1645.

4.366 T. Suhara, H. Nishihara and J. Koyama, *Opt. Commun.* **19** (1976) 353.

4.367 H. Nassenstein, *Optik* **30** (1969) 44, 201; **29** (1969) 456, 597.

4.368 H. Nassenstein, *Opt. Commun.* **1** (1969) 146; **2** (1970) 231.

4.369 W. Lukosz and A. Wüthrich, *Optik* **41** (1974) 191; *Opt. Commun.* **19** (1976) 232.

4.370 W.H. Lee and W. Streifer, *J. Opt. Soc. Am.* **68** (1978) 795, 802.

4.371 H.J. Caulfield, Q. Huang and J. Shamir, *Opt. Commun.* **86** (1991) 487.

4.372 P.C. Mehta, D. Mohan and A.K. Musla, Unpublished Work

4.373 J. Jahns and A. Huang, *Appl. Opt.* **28** (1989) 1602.

4.374 R.T. Chen, OE Reports, (March 1990) 13.

4.375 R.T. Chen, L. Sadovnik, T.M. Aye and T. Jannson, *Proc. SPIE*, **1212** (1990) 290.

CHAPTER 5

HOLOGRAPHIC INTERFEROMETRY

5.1. Introduction

Holographic interferometry produces interference between two or more waves, at least one of which is reconstructed by a hologram. The interfering waves may be separated in time. Holographic interferometry allows measurements, inspection and testing of not only optically flat surfaces but also of any three dimensional surface of arbitrary shape. It also allows interferometric examination of the object in different perspectives. Conventional optical interferometry can be used for optically polished and specularly reflecting surfaces only. If the object is considered to be made up of small mirror elements, then a single holographic interferogram can be equivalent to a large number of observations from a conventional optical interferometry.

The object wavefront can be stored in a hologram and compared with the object after a change that may have taken place. It is this storage or time delay aspect which gives the holographic method a unique advantage over conventional optical interferometry.

Horman [5.1] was the first to suggest the application of a hologram in the test arm of a Mach-Zehnder interferometer. The holographic interferometry of 3D objects was first introduced by Powell and Stetson [5.2, 5.3]. Different holographic interferometric techniques were developed by several researchers independently [5.1-5.9] almost simultaneously. In general, three forms of holographic interferometry are possible viz. double-exposure, single-exposure real-time and time-average.

5.2. Double-Exposure Holographic Interferometry

In double-exposure holographic interferometry, two holograms of the two object waves occurring sequentially in time are recorded on a single photographic plate. When such a hologram is reconstructed, two superimposed images of the same object are produced. The interference between these images produces interference fringes overlaid on the image of the object. The interference fringes are indicative of deformation, displacement, rotation and change in refractive index or thickness of the object. The fringes appear to be localized in space, not necessarily on the object. When viewing direction is altered, the fringes shift and change their form.

A hologram of the object in its initial unstressed state is recorded by exposing the photographic plate. Without removing the photographic plate from the setup, the object is stressed and a second exposure is made. The plate is then developed and reconstructed by the reconstruction wave to observe the interference. Let $U_o(x,y)$ and $U(x,y)$ represent the complex amplitudes of the object wave before and after stressing and $U_r(x,y)$ the complex amplitude of the reference wave. Then

$$U_o(x,y) = O_o(x,y) \exp[-j\phi(x,y)], \qquad (5.1)$$

$$U(x,y) = O_o(x,y) \exp[-j[\phi(x,y)+\Delta\phi(x,y)]]. \qquad (5.2)$$

The stressing introduces a small change in the phase $\Delta\phi(x,y)$.
The phase $\phi(x,y)$ is given by

$$\phi(x,y) = \frac{2\pi n d}{\lambda}, \qquad (5.3)$$

where n is the refractive index of the medium in which the light propagates and d is the physical path length from laser to the recording plate via object.

Both the states of the object are recorded as a double-exposure hologram using a reference wave of complex amplitude $U_r(x,y)$. When the plate is developed and reconstructed by $U_r(x,y)$, the complex amplitude of the reconstructed wave will be proportional to

$$[U_o(x,y) + U(x,y)].$$

The irradiance of the reconstructed wave will be proportional to

$$I(x,y) = |U_o(x,y) + U(x,y)|^2$$

$$= |O_o(x,y)|^2 \, 2\{1+\cos[\Delta\phi(x,y)]\}. \qquad (5.4)$$

Eq. 5.4 indicates that the object irradiance $|O_o(x,y)|^2$

is modulated by the fringe pattern $2\{1+\cos[\Delta\phi(x,y)]\}$. The change in phase $\Delta\phi$ may be related to displacement, rotation, tilt, bending, vibrational amplitude, temperature, pressure, strain, stress, electron density or mass concentration. The fringe represents the locus across the field which is associated with a particular optical path. The distance between two consecutive fringes represents a change equal to a wavelength in the optical path from the first exposure to the second. If no phase changes are induced between the two exposures, the fringes will be nonexistent. Bright fringes are the contours of $\Delta\phi$ values which are even-integer multiples of π, whereas the dark fringes correspond to the contours of constant $\Delta\phi$ values which are odd-integer multiples of π.

To calculate the value of $\Delta\phi$, we assume that a point on the object at P moves to P' (Fig. 5.1). For small displacements such that $|L|$ is much small than the optical path SO, $\Delta\phi$ is given by

$$\Delta\phi = L(k_2 - k_1),$$

$$= L \cdot K, \tag{5.5}$$

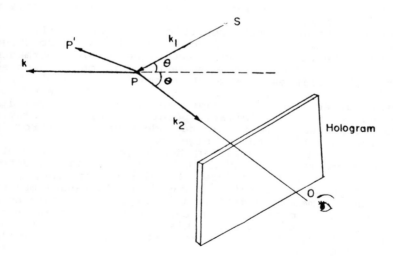

Fig. 5.1. Schematic diagram showing propagation vectors.

where $|\mathbf{k}_1|=|\mathbf{k}_2|=k=2\pi/\lambda$ are the magnitudes of the propagation vectors $\mathbf{k}_1$ and $\mathbf{k}_2$ of the incident and the scattered field. $\mathbf{K}=\mathbf{k}_2-\mathbf{k}_1$ is the holographic *sensitivity vector* which points along the bisector of the angle between the incident and viewing directions. If 2θ is the angle between the incident and viewing directions, then the magnitude of the sensitivity vector is given by

$$|\mathbf{K}| = 2k\cos\theta. \tag{5.6}$$

Eq. 5.5 forms the basis of quantitative interpretation of the holographic interferogram. In an interferogram, the fringe loci display constant values of L in the direction of the sensitivity vector. From the interferogram, the value of $\Delta\phi$ is calculated by counting the bright fringes and using

$$\Delta\phi = 2\pi N. \tag{5.7}$$

Eq. 5.5 then can be used to measure the displacement along the direction of **K**. However, the sense of the displacement is uncertain. As the fringes are described by $[1+\cos(\Delta\phi)]$ and since the cosine is an even function, the same fringes are produced by positive or negative deformations.

For most cases, **K** will be normal to the surface. Hence $\Delta\phi$ is insensitive to in-plane displacements and sensitive to normal displacements.

Double-exposure interferometric technique is simple to implement as no alignment is involved in the two reconstructed images. The double-exposure technique gives a permanent record of the change in the shape of the object between exposures. The fringes have good contrast and visibility as distortions in the emulsion and imperfections in the optical components affect both the images equally and the two diffracted wavefronts are similarly polarized. The intensity of the two reconstructed waves can be made equal by controlling the two exposure times to get high contrast fringes.

One difficulty in applying holographic interferometry to the measurement of deformation is the unambiguous determination of the fringe orders. In simple problems, prior knowledge of boundary conditions may help. In random motion, the problem becomes very difficult. Multiple-exposure holograms can also be recorded with each exposure for an incremental change in the object. This technique creates very sharp fringes which helps in accurate measurement of the fringe displacement.

5.3. Single-Exposure Real-Time Holographic Interferometry

In double-exposure technique the object can be studied at two fixed stressing conditions only unless multiplexing techniques are used to study the intermediate conditions [5.10, 5.11]. However, in many applications it is desirable to see the response of the object to a series of excitations. This can be accomplished by single-exposure real-time holographic interferometry which is also known as concomitant or live-fringe interferometry.

A single-exposure hologram of the object is recorded. The photographic plate is either developed *in situ* or removed from the setup, processed, dried and repositioned in its original recording position. The hologram is reconstructed by the reference beam. An observer looking through the hologram will see the virtual image of the object superimposed on the actual object which is still being illuminated by the laser beam. If the reconstruction has been done properly, the null condition results. If the object is now deformed, the observer will see the object superimposed with interference fringes which are produced due to interference between the reconstructed wave and the object wave. The fringe pattern will change in real-time if the object is subjected to different amount of stresses or deformations. The fringes will have a cosinusoidal intensity pattern. A vibrating object can also be studied through real-time holographic interferometry. The fringe pattern will be indicative of the vibrational amplitude [5.3] but the irradiance of the fringes will be weak due to the integrating time of the eye.

The instantaneous irradiance for a real-time hologram is given by [5.12]

$$I(x,y,t) = |U_o|^2 \, 2\{1-\cos[\Delta\phi(x,y,t)]\}. \tag{5.8}$$

The minus sign in Eq. 5.8 appears because the complex amplitude of the reconstructed object wave is negative due to a uniform phase shift of π rad.

The real-time holographic interferometry requires repositioning of the processed hologram with interferometric accuracy. Abramson [5.9] has recommended to use a three pin arrangement in which the plate is held by gravity. Repositioning is eliminated if the plate is developed *in situ*. For this a real-time plate holder known as liquid gate may be used which holds the plate in a rectangular vessel with optically flat windows and having inlet and outlet tubing. The vessel is filled with water and the photographic plate is loaded and exposed. The water is drained out and the developer is poured in. After the development and subsequent fixing the vessel is again filled with water and the hologram is reconstructed [5.13]. Development of the photographic plate without removing it can also be done by simply suspending the plate vertically down and raising beakers from below to submerge the plate in it [5.14, 5.15]. For *in situ* processing of the plate a mono bath [5.16] can considerably reduce the processing time.

Real-time recording materials such as photothermoplastic, photopolymer and photorefractive crystals are very convenient for real-time holography. A close circuit television camera may be useful to display the fringe pattern on the screen of a monitor.

5.4. Time-Average Holographic Interferometry

The multiple-exposure holography can be extended to a continuum of exposures resulting in time-average holographic interferometry. The method for recording a time-average hologram of a vibrating body is identical to that used for a conventional hologram of a stationary body. If the exposure time is larger than the periodic time of vibration of the body, the reconstructed image will be modulated by a system of fringes. These fringes are contour lines of equal displacement of the object surface.

An object vibrating in its normal mode will produce standing waves of vibration on its surface. At the nodes the object will be stationary whereas at antinodes, the amplitude of vibration will be large. The time-average hologram will show those portions of the object which remain stationary during the exposure (i.e. nodal points). For example, when the object vibrates sinusoidally, its velocity becomes zero at two positions of maximum displacement. Thus a time average holographic interferogram of a sinusoidally vibrating object is equivalent to a double-exposure interferogram showing the displacement between extreme positions.

Let us consider a body vibrating sinusoidally with amplitude $L(x,y,t)$. The complex amplitude of light from an object point in the hologram plane is

$$U_o'(x,y,t) = O_o(x,y)\exp\{j[\phi(x,y)+\Delta\phi(x,y,t)]\}, \qquad (5.9)$$

where $\phi(x,y)$ is the phase of light when the object is stationary and $\Delta\phi(x,y,t)$ is the change in phase of the light when the object is vibrating with angular frequency ω and is given by

$$\Delta\phi(x,y,t) = K.L(x,y)\sin\omega t, \qquad (5.10)$$

where K is the sensitivity vector. Eq. 5.9 can be written as

$$U_o'(x,y,t) = U_o(x,y)\exp[-jK.L(x,y)\sin\omega t], \qquad (5.11)$$

where
$$U_o(x,y) = O_o(x,y)\exp[-j\phi(x,y)] \qquad (5.12)$$

is the complex amplitude of light when the object is stationary.

When the hologram is recorded and replayed, the complex amplitude of the reconstructed wave will be proportional to the time average of $U(x,y,t)$ over the exposure time T and is equal to $U_o(x,y)C$, where

$$C = \frac{1}{T} \int_0^T \exp[-jK.L(x,y)\sin\omega t]dt \quad (5.13)$$

is known as the *characteristic function* for sinusoidal vibration. If the period of vibration is much less than the exposure time, $T \gg 1/\omega$, then

$$C = \lim_{T\to\infty} \frac{1}{T} \int_0^T \exp[-j(K.L(x,y)\sin \omega t]dt$$

$$= J_o[K.L(x,y)], \quad (5.14)$$

where $J_o[..]$ is the Bessel function of the first kind of order zero.

The instantaneous irradiance of the image is given by

$$I(x,y) = |O_o(x,y)|^2 J_o^2[K.L(x,y)]. \quad (5.15)$$

The image is modulated by fringes described by J_o^2. Dark fringes correspond to the points where the amplitude of vibration is such that the Bessel function is zero. Table 5.1 gives the zeros of the Bessel function J_o.

Table 5.1 Zeros of the Bessel function J_o.

Fringe order	True Zeros J_o, n	Asymptotic limit Eq. 5.16
1	2.405	2.356
2	5.520	5.495
3	8.654	8.639
4	11.792	11.781
5	14.931	14.923
6	18.071	18.064
7	21.212	21.206
8	24.352	24.489
9	27.493	27.489
10	30.635	30.631

For the case of a cantilever vibrating sinusoidally with amplitude a(x), K.L(x,y) is given by [5.12]

$$K.L(x,y) = \frac{2\pi}{\lambda} 2a(x). \tag{5.16}$$

The amplitude of vibration of an object point at any fringe, for example at the sixth dark fringe can be calculated by using the J_o value from Table 5.1. Thus

$$K.L = \frac{4\pi}{\lambda} a(x) = 18.071. \tag{5.17}$$

Therefore, $a(x)=1.439\ \lambda$. The amplitude of vibration will be increasing with the higher order of dark fringe.

The function J_o can be approximated by the asymptotic limit for large argument for third fringe onwards [5.17]. Thus

$$J_o(p) = \sqrt{\frac{2}{\pi p}} \cos(p - \frac{\pi}{4}). \tag{5.18}$$

Table 5.1 compares the values of true zeros with those given by Eq. 5.18. For the third fringe, the error in measurement will be 15 parts in about 8600 which shows that the asymptotic limit can be used.

Figure 5.2 shows the plot of J_o^2 against K.L. The first maximum is the brightest while the subsequent maxima decrease considerably. This may be compared with $\cos^2$ fringes. If one tries to photograph the interferogram, it would be difficult to record more than seven fringes due to large difference in the brightness of the maxima and limited dynamic range of the reconstruction film. Proper processing procedures can help in increasing the visibility of higher order fringes. The reconstruction negative can be treated in a bath of 2% solution of ammonium persulphate which dissolves away the silver from the dark portions. Development of the reconstruction film in a POTA developer (1-phenyl-3-pyrazolidone 1.5 g and sodium sulphite 30 g in 1 litre water) gives better results. This developer develops weakly exposed and over exposed films with almost equal density.

If a vibrating object is examined in a single exposure real-time setup, then the instantaneous reconstructed average intensity is given by

$$I = |O_o(x,y)|^2[1+J_o(\frac{4\pi a}{\lambda})]. \tag{5.19}$$

The fringes are similar to the time average hologram but

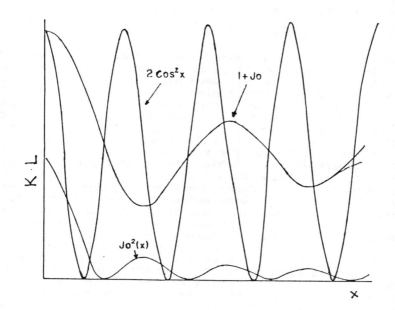

Fig. 5.2. Comparison of Bessel function with a $\cos^2$ function. For the curve $(1+J_o)$ the minima have non-zero values.

number is only half. The curve $(1+J_o)$ (Fig. 5.2) shows that the minima have non-zero values, therefore the fringe contrast is poor.

5.5. Stroboscopic Holographic Interferometry

The applications of time-average holographic interferometry is limited as this technique does not show the variation in the relative vibrational phase across the object surface and the fringe visibility decreases with the increase in the vibrational amplitude. To eliminate these deficiencies, stroboscopic holographic interferometry can be used [5.18-5.22].

In the stroboscopic holographic interferometry, a hologram of the vibrating object is recorded with a sequence of bright pulses of laser light. Usually, the photographic plate is exposed when the object is at its maximum positive

and negative displacements using brief laser pulses until
the plate reaches the desired level. The characteristic
function for the motion of the type $\mathbf{A}$ sinωt is $\cos^2(\mathbf{K}.\mathbf{A})$ if
the pulses are very short. This characteristic function
describes the fringes which have unit visibility for all
amplitudes.

The stroboscopic holographic interferometry needs the
generation of a sequence of pulses at appropriate frequency.
This can be achieved by mechanical or electro-optical
methods [5.20-5.24]. The simplest method is to use a perforated disk rotating with a synchronous motor and chopping
the laser frequency. Chopping frequencies as high as 13 kHz
have been achieved [5.18, 5.23] by mechanical methods.
Electro-optic crystals such as Pockels cell can also be used
for laser beam modulation for stroboscopic holography [5.19,
5.24]. Stroboscopic holographic interferometry can be implemented in real-time mode also [5.25].

Stroboscopic holography does not utilize the laser
energy fully, therefore long exposure times are needed. To
decrease the total exposure time, the photographic plate can
be biased by pre- or post-exposing with a laser beam of
uniform amplitude. Total exposure time can also be decreased
by preexposure the plate and further exposing it by two
[5.26] or four [5.27] symmetrically located pulses during
vibration period. Recording the time-average hologram with
time-dependent phase modulation of the object beam or the
reference beam can help when the reference-to-object beam
intensity ratio is low [5.28]. However, multiple exposures,
biasing and phase modulation decrease the diffraction efficiency if proper care is not taken [5.29].

Fringes with brightness nearly uniform in all orders
similar to stroboscopic technique can be obtained [5.30] if
a time-average interferogram of an object vibrating with
frequency ω is recorded with a laser light that is modulated
at 2ω.

5.6. Temporally Modulated Holography

Temporally modulated holography involves the recording
of a hologram when the object wave and/or the reference wave
are time dependent [5.31]. Time-average and stroboscopic
holographic interferometry are special cases of temporally
modulated holography. The temporal modulation of the reference wave allows vibration measurement with variable sensitivity and relative phase measurement across the vibrating
surface. The time-average interferometry gives the fundamental frequency whereas the frequency translated reference
beam can reveal any of the higher harmonics. The frequency
translated hologram can detect [5.32] small amplitude vibrations down to λ/100. Large vibrations can also be measured
by frequency translated holography. The amplitude modulated
reference wave holography maps the amplitude as well as

relative phase of a vibrating body [5.33]. The phase modulation of the reference wave is useful for object motion compensation in holography. It is also useful for recording holograms of moving objects [5.34].

5.7. Fringe Linearization Holographic Interferometry

Double-exposure holographic interferogram some times becomes difficult to interpret due to complicated interfence pattern produced. Fringe linearization interferometry (FLI) reduces the complicated fringe patterns in a double-exposure interferometry to a simple linear fringe pattern [5.35]. This technique is particularly useful for the detection of cuts and sub-surface cuts in a test sample.

The implementation of holographic FLI involves a simple modification of double-exposure holographic interferometry. The basic holographic FLI process is as follows [5.35].

Step 1   Expose a conventional hologram of the object at the initial stress state.

Step 2   Stress the object.

Step 3   Tilt slightly the object illumination beam and keep the reference beam fixed.

Step 4   Expose a second hologram on the same recording plate on the same area.

Step 5   Process and reconstruct the hologram.

In order to detect the defects, the frequency of the linear fringes should be increased to smooth out most of the gross deformations of the specimen but it should not be very high to smooth out the signature of the defects. The linearized fringe interferogram shows unique Fourier components. This property can be exploited for automatic analysis of the object deformations [5.35].

5.8. Desensitized Holographic Interferometry

The desensitized holographic interferometry is useful for flatness testing when the flatness of the test sample is of the order of several microns and r.m.s. surface roughness of the order of a wavelength or greater [5.36, 5.37].

The key element in the desensitized holographic interferometry is a holographic optical element (HOE) which performs simultaneously the functions of deflection, beam splitting, change of curvature and as a reference surface [5.38]. The principle of the interferometric is shown in Fig. 5.3. A HOE is made by recording the interference between two plane beams at an angle θ. The HOE is repositioned at its recording position and the test surface is placed behind it.

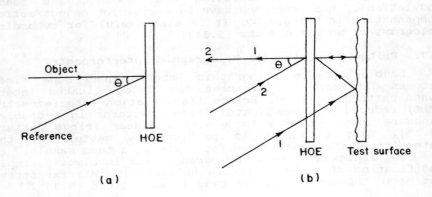

Fig. 5.3 Principle of desensitized holographic interferometer (b). The key component is a HOE which is recorded as shown in (a) [5.38].

To understand the working of the interferometer, we consider two rays 1 and 2. The 0 order of ray 1 is reflected by the test surface and falls on the HOE which diffracts it in the +1 order. The +1 order of ray 2 falls on the test surface. This beam is reflected back by the test surface and emerges through the HOE in the 0 order. The path difference between the two rays emerging through the HOE is a function of thickness w of the air gap between the HOE and the test surface and is given by [5.38]

$$\delta = 2w(1-\cos\theta). \tag{5.20}$$

The interference between the two beams produces fringes of equal thickness w given by

$$w = N\lambda/2(1-\cos\theta). \tag{5.21}$$

The desensitization factor is $1/(1-\cos\theta)$ compared to Fizeau interferometer which can be as high as 100 for small angles.

5.9. Digital Holographic Interferometry

Digital holographic interferometry is a technique for the measurement of phase with an accuracy of $2\pi/200$ as was shown by Hariharan et al. [5.39]. The technique is more suitable for measurements on real-time holographic interferometer fringes at a large number of points (100x100).

The image of real-time fringes is picked up by a camera consisting of 100x100 array of photodiodes. The measurement cycle consists of three successive scans, in which after the first scan, the phase of the reference beam is shifted by $+2\pi/3$ and then by $-2\pi/3$ with respect to the object beam. The intensity data at each detector are stored in the memory of a computer. If the first scan is made when the phase difference between the object and reference beam is $\Delta\phi$, then the intensity values at a point (x,y) due to each scan are given by

$$I_1 = I_o + I_c + 2(I_o I_c)^{1/2} \cos\Delta\phi, \tag{5.22}$$

$$I_2 = I_o + I_c - (I_o I_c)^{1/2} (\cos\Delta\phi + \sqrt{3} \sin\Delta\phi), \tag{5.23}$$

$$I_3 = I_o + I_c - (I_o I_c)^{1/2} (\cos\Delta\phi - \sqrt{3} \sin\Delta\phi), \tag{5.24}$$

where I_o and I_c are the intensities of the object and the reconstructed image respectively at the point (x,y). Eqs. 5.22-5.24 can be solved for $\Delta\phi$ giving

$$\tan\Delta\phi = \frac{\sqrt{3}(I_3 - I_2)}{(2I_1 - I_2 - I_3)}. \tag{5.25}$$

As I_1, I_2 and I_3 have been stored in the computer memory, the value of $\tan\Delta\phi$ can be calculated at each point.

The schematic diagram for digital holographic interferometer is given in Fig. 5.4

The phase of the reference beam is shifted by using a mirror mounted on a piezoelectric translator. The piezoelectric translator is translated by applying voltages by a computer controlled dc amplifier.

5.10. Fringe Localization

The holographic interference fringes formed by a diffuse reflecting object are localized where their contrast or visibility is maximum [5.40, 5.41]. The position of the plane of localization is decided by the type of displacement undergone by the object [5.5]. Generally, the fringes are localized on some curve in space. In order to record the fringes with high contrast, the imaging system is focussed on the plane of localization. Sometimes it becomes difficult to image both the object and the fringes when the fringes are localized far away from the object. Reducing the viewing aperture brings fringes into focus with the object.

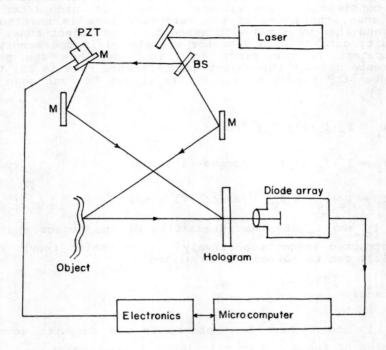

Fig. 5.4. Schematic diagram for digital holographic interferometer [after 5.39].

5.10.1. Pure Translation

In the case of pure translation the displacement L is constant at all the points of the object (Fig. 5.5). Waves from corresponding points such as P and P' on the wavefronts contribute to the formation of fringes. The phase difference $\Delta\phi$ between P and P' can be considered and a plane is found over which it remains constant. Therefore,

$$\Delta\phi = \mathbf{L}(\mathbf{k}_2-\mathbf{k}_1) = \text{constant}. \tag{5.26}$$

The fringe equation is

$$LK = 2\pi N. \tag{5.27}$$

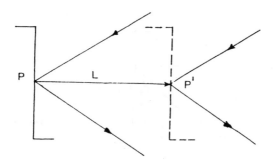

Fig. 5.5. Fringe localization for pure translation.

For plane wave object illumination k_1 is constant, hence from Eq. 5.26 k_2 must also be constant over the detector. This means that the fringes are localized at infinity. A lens placed at a distance equal to its focal length from the object surface shows parallel straight line fringes at its back focal plane.

If one takes a photograph of a double-exposure hologram of an object which is translated parallel to itself between the exposures, no fringes will be recorded on the object surface. If the object is translated normal to itself, again no fringes will be recorded on the object. The circular fringes are formed which are localized at infinity.

5.10.2. *Pure Rotation about an Axis in the Surface*

Let the object at P in the xy plane undergoes rotation about the y axis through a small angle θ and moves to the point P' as shown in Fig. 5.6. The object is illuminated by a plane beam of light. The direction of object illumination and observation lie in x plane. The phase difference ϕ between P and P' is

$$\phi = - \frac{2\pi}{\lambda} x\theta (\cos\theta_1 + \cos\theta_2) \qquad (5.28)$$

and the fringe equation is

$$2x\theta (\cos\theta_1 + \cos\theta_2) = N\lambda . \qquad (5.29)$$

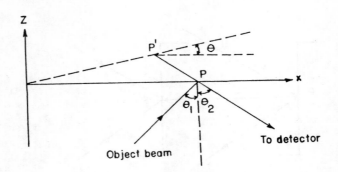

Fig. 5.6. Fringe localization for pure rotation about an axis in the surface.

where θ_1 and θ_2 are the angles of illumination and observation respectively. The fringe localization is found where the variation in the phase difference with the viewing direction θ_2 is equal to zero, i.e. $(d\phi/d\theta)=0$. This yields

$$z = \frac{(x\sin\theta_2\cos^2\theta_2)}{(\cos\theta_1+\cos\theta_2)}. \tag{5.30}$$

We can consider various directions of illumination and observation as given in Table 5.2

Table 5.2 Fringe equation and localization for different combinations of θ_1 and θ_2 angles.

θ_1, θ_2	Fringe Equation	Localization
$\theta_1 = 0°$, $\theta_2 = 0°$	$2\theta x = N\lambda$	$z = 0$
$\theta_1 = 45°$, $\theta_2 = 0°$	$[1+(1/\sqrt{2})]\theta x = N\lambda$	$z = 0$
$\theta_1 = 45°$, $\theta_2 = 45°$	$\sqrt{2}\,\theta x = N\lambda$	$z = (x/4)$
$\theta_1 = 0°$, $\theta_2 = 45°$	$[1+(1/\sqrt{2})]\theta x = N\lambda$	$z = x/[2(1+\sqrt{2})]$

When an object rotates about an axis lying in the plane of the object, straight fringes parallel to the axis of rotation are formed. The fringes are localized on the surface of the object when the observation is normal to the object. Table 5.3 gives the type of fringes and their localization for different displacements of an object.

Table 5.2 Type of fringes and their localization for some types of displacement.

A *Lateral translation* orthogonal to viewing direction.

Straight parallel fringes, normal to the direction of movement.

Localization in front of or behind the surface.

B. *Translation* in the viewing direction.

Circular fringes.

Localization not well defined.

C. *Tilt* about an axis lying in the surface.

Straight fringes parallel to the tilt axis.

Localization close to the surface.

D. *Rotation* about an axis parallel to the viewing direction.

Straight parallel fringes.

Direction varies according to the observation point.

Localization in front of or behind the surface.

The form of fringes which would be formed due to object translation or rotation can be known with the help of a holodiagram [5.42]. The holodiagram is useful for the qualitative understanding of the fringes which are obtained when the object undergoes different types of displacements. The holodiagram is also useful for optimizing the hologram recording setup when the object is in motion (section 4.10.8). Abramson [5.42] has introduced a moiré analogy to holographic interferometry for the prediction of fringe pattern due to simple motions.

5.11. Sandwich Hologram Interferometry

Sandwich hologram technique devised by Abramson [5.43] is an excellent way for elimination of ambiguities in interpretation by removing interference fringes generated by extraneous rigid body motion in double-exposure holographic interferometry. This technique is suitable for recording holograms in industrial environment [5.43].

Following steps shown in Fig. 5.7 are involved in making a sandwich hologram.

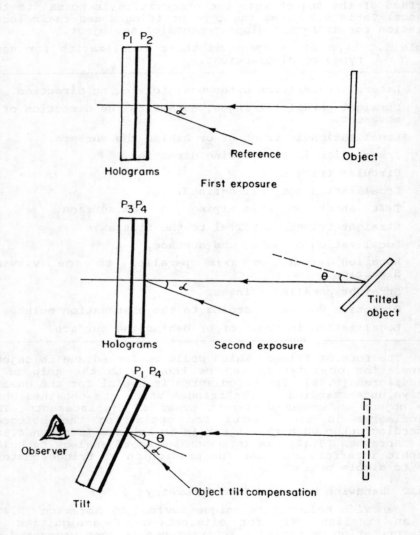

Fig. 5.7. Principle of sandwich holography [5.43].

1. Two plates P_1 and P_2 (without annihilation backing) are loaded in a plate holder with their emulsions facing the object. Using a spherical reference wave, the plates are exposed with the object in the initial unstressed state.

2. Similarly another pair of plates P_3 and P_4 are exposed with the object in the stressed state.

3. Develop all the four plates and make a sandwich by combining the holograms P_1 and P_4 (holograms P_3 and P_4 can also be sandwiched).

4. Reconstruct the sandwich hologram with original reference beam. Tilting the sandwich results in the change in the interference fringes. To remove the fringes due to rigid body tilt, the sandwich is tilted in the same direction as the actual object.

The technique can be implemented by only two plates cemented together with a spacer [5.44]. The technique has also been adapted to pulsed-laser holographic interferometry [5.45].

5.12. Applications

Holographic interferometry has proved to be an important tool for nondestructive testing of materials, medical and dental research, stress-strain analysis and optical component testing [5.46]. It has applications in almost all industries. An important application is in the nondestructive testing (NDT) of old paintings, icons and sculptures. Table 5.4 gives application fields of holographic interferometry.

5.12.1. Holographic Nondestructive Testing (HNDT)

Holographic nondestructive testing techniques are used to locate and evaluate cracks, disbonds, voids, delaminations, inhomogeneity and residual stresses in a test sample without destruction of the sample. The holographic interferometry techniques are applied for nondestructive testing of materials. The HNDT techniques can be used for the testing of laminated structures, turbine blades, solid propellant rocket motor casings, tyres and air foils. These techniques are also useful in medical and dental research.

In HNDT techniques, the test sample is interferometrically compared with the sample after it has been stressed (loaded). A flaw can be detected if by stressing the object it creates an anomalous deformation of the surface around the flaw. The holographic interferogram will show up the anomalous deformation by an abrupt change in the shape of the interference pattern. It is pointed out that the flaw is not located on the surface of the body. The success of

Table 5.4 Applications of holographic interferometry

Field	Applications
Aerospace	Defects in honeycomb plates, Testing of construction materials, Testing of welding methods, Inspection of rocket bodies, Flow visualization in wind tunnels, Vibration modes of turbine blades.
Automobiles	Testing of oil pressure sections, Testing of welding methods, Research in construction of automobile bodies, Construction of engines.
Machine tools and precision instruments	Measurement of deformations of machine parts, jigs and tools, Measurement inside cylinders, Measurements of stiffness (heat, static or dynamic), Analysis of construction of instruments and tools.
Electrical and electronic industries	Vibration modes of turbine blades, motors, transformers, loudspeakers, Testing of welding and adhesion, Testing of circuit parts, Analysis of audio equipments, etc. Leak test of batteries.
Heavy industry, ship building, civil engineering	Research in welding methods, Analysis of constructions, Vibration modes of turbine blades, Design of pipes, Research in concrete.
Chemical industry	Measurement of mixed fluids.
Tyre, rubber and plastics	NDT of tyres, Testing of molded products, Measurement of adhesion defects.
Medicine, dentistry	Measurement on living bodies, Chest deformation due to inhalation, Measurement on teeth and bones, Testing materials for dental surgery, Testing of urinary track, Measurement on eyes, ears, etc.
Musical instruments	Measurement of vibration modes.
Cultural articles and paintings	NDT and restoration.

HNDT of a material depends on the stressing technique adopted. The stressing should deform the body under test in such a manner that the 'good' areas are distinguished from the 'bad' areas simply by the visual inspection of the interferogram. In general, the flaw at a depth equal to its size can be detected by HNDT techniques.

Fringe Control Techniques

The object can be stressed by mechanical stressing, pressure or vacuum stressing, thermal stressing, vibrational stressing and magnetic stressing. The stressing of the object can create gross deformation and rigid body motion of the object. This will produce fine interference fringes in the interferogram if the test area is large. In such a situation, the interference fringes around the flaw will be very fine and it would not be detected by unaided eye.

By using fringe control methods, the effects of gross deformation and rigid body motion can be compensated [5.47, 5.48]. In the single-exposure real-time interferometry, this is accomplished by incorporating an adjustable mirror and a convex lens in the object beam path as shown in Fig. 5.8. A conventional hologram of the object is recorded and repositioned in the plate holder. The object is stressed and real-time fringes are observed. The gross deformations of the object are compensated by the manipulation of the lens and the mirror. A similar fringe control system can be adopted in double-exposure holographic interferometry. The hologram is recorded with separate reference beams angularly different for each exposure.

A fringe control technique which can be used not only to compensate the rigid body motion but also for the measurement of the deformation, has been demonstrated by Stimpfling and Smiglelski [5.49]. In this technique, the reference source, object illumination source and the hologram are fixed on a rigid stage as shown in Fig. 5.9a. After recording and repositioning the hologram the stage is moved to exactly superimpose the object and its reconstructed image (i.e. infinite fringe). In the case of the deformation of a bigger object, the compensation can be achieved only for a small area of the object. The displacement and deformation of the object can be quantified by measuring the displacement of the stage by a computer-aided displacement device so as to obtain the condition of infinite fringe. Such a measurement does not take fringe spacing and fringe localization into consideration. It is also possible to determine the sign of the local displacement.

The technique can also be used for motion compensation in double-exposure holographic interferometry as shown in Fig. 5.9b. The two holograms are recorded on two separate photographic plates. The first hologram is made on plate H_1

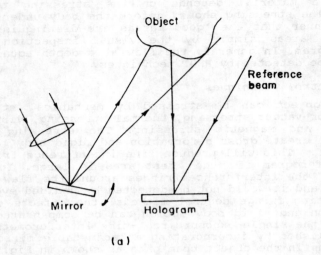

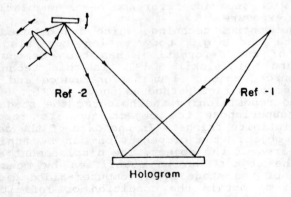

Fig. 5.8. Fringe control technique for real-time interferometer (a) and double-exposure interferometry (b) [5.47, 5.48].

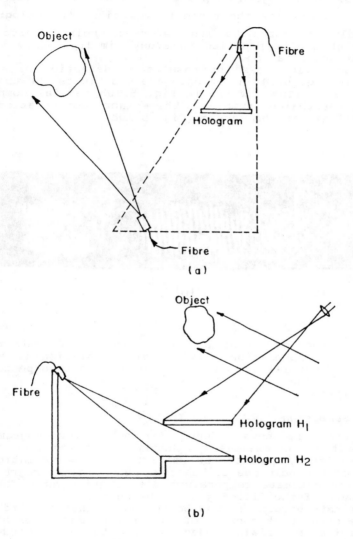

Fig. 5.9. Rigid body motion compensation for real-time interferometry (a) and double-exposure interferometry (b) [5.49].

and the second hologram on plate H_2 placed behind the first plate. To compensate the rigid body motion, the hologram H_2 is rigidly connected to its reference point source. The arrangement can be modified to study simultaneously a displacement and rotation.

Figure 5.10 shows the results of detection of a flaw [5.49]. In Fig. 5.10a no fringes are visible as the translation is large (1mm), while in Fig. 5.10b partial compensation has revealed fringes. The exact compensation has revealed flaws in the sample (Fig. 5.10c).

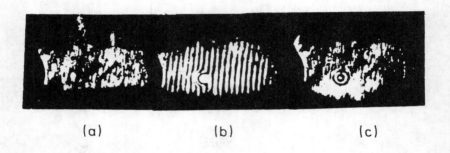

(a)　　　　　　　　　(b)　　　　　　　　　(c)

Fig. 5.10. Detection of a flaw using rigid body motion compensation [5.49]. (a) Translation 1 mm, (b) partial compensation and (c) exact compensation.

HNDT Examples

Thermal Stressing

Figure 5.11 shows a double exposure interferogram of a 100 W bulb as an example of thermal stressing. The interferogram clearly depicts the refractive index variations in the gas of the bulb due to heating. Such interferograms can be used to estimate temperature distribution as a function of pressure of the filled gas in the bulbs.

Laminate structures can be tested by thermal stressing. Laminate structures consist of two or more sheets of similar or dissimilar materials bonded together to form a sandwiched structure. These are bonded together to form a sandwiched structure. These structures are light weight but of high strength and are used as structural materials for aircrafts, missiles, jet-engine fan blades and turbine blades. The most common structural flaws present in laminates is the absence of a good bond line leaving disbands in some areas.

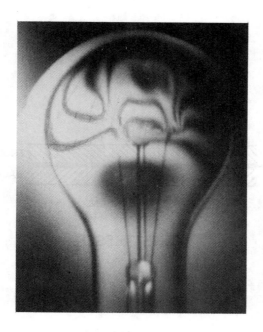

Fig. 5.11. Double-exposure interferogram of a 100 W bulb.

Fig. 5.12 shows a disbond in a two material structure bonded by an epoxy. When the upper sheet is exposed to thermal radiations, it will expand according to its thermal coefficient of expansion. But the bond line being a thermal insulator permits a smaller temperature change in the back sheet. Thus the two metallic sheets are at different temperatures and form a bimetallic structure. Around the disbond the structure is free to deform and hence the upper sheet will buldge out as shown in Fig. 5.12. A certain amount of deformation of the total structure will also result depending on the thickness and type of materials of the two sheets. But what is more important is the differential displacement between the disbond area and the surrounding area. This differential displacement which may be of the order of a wavelength can be detected by holographic interferometry.

Amadesi et al. [5.50] have used thermal stressing for testing of ancient Italian panel paintings. A representative

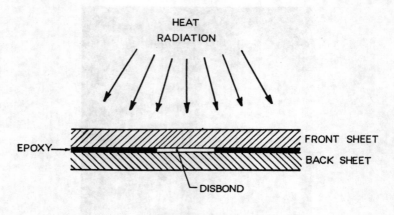

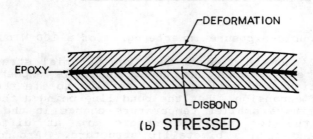

Fig. 5.12. Thermal stressing of a bimetallic strip.

result of the testing of a painting is shown in Fig. 5.13.

Electronic components and circuit boards can also be tested using thermal stressing [5.51, 5.52]. The HNDT techniques are gaining importance in the electronics industry as a sensitive and accurate method of locating manufacturing and assembly flaws in components and assembled modules.

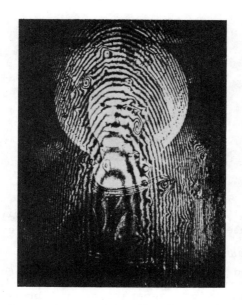

Fig. 5.13. Interferogram of an ancient Italian painting using thermal stressing [5.50].

Mechanical Stressing

Mechanical stressing can be used to detect cracks and observation of their growth in concrete [5.53] and plastic molding [5.54] and to locate fabric cuts and delaminations in flat sheets of fibre glass-reinforced plastics [5.55]. The HNDT techniques in study of concretes is useful as other methods of testing are not suitable due to coarse and heterogeneous structures of concrete. Plastic moldings are interesting candidates for HNDT techniques as the ultrasonic or radiological methods are not suitable because the ultrasound and gamma rays are either absorbed or reflected by such objects. The main source of flaws in plastic molding are due to structural nonuniformity and stored stresses. These stresses are released slowly over a period of time resulting in the development of cracks. Figure 5.14 shows holographic interferogram of a plastic body which shows the presence of a crack revealed by mechanical pressure [5.54]. Over the crack the fringes are broaken.

Fig. 5.14. Flaw detection in a plastic molding using double-exposure holographic interferometry.

Pressure and Vacuum Stressing

Pressure and vacuum stressing are useful for testing of aircraft and automotive tyres to detect a variety of flaws such as disbonds, voids, separations and broken belts. The inside surface of the tyre can be inspected by placing the tyre in a vacuum chamber [5.56, 5.57]. By changing the tyre pressure slightly between the exposures, or during real-time observation the exterior surface can be inspected [5.58]. The tyre air pressure may be varied between one twentieth to one fifth of atmosphere depending on tyre stiffness. Figure 5.15 shows the holographic interferograms of a new scooter tyre [5.54]. This result was later confirmed by the fact that the tyre, after running for about 2000 km started bulging out around 4 cm diameter at the same place.

Leak testing using vacuum stressing is an interesting application for the electronics industry [5.59]. A component hermetically sealed is first placed in a vacuum chamber and a hologram is recorded. The real-time observation at a different pressure will show the fringes which disappears at different rates depending on the size of the leak. Figure 5.16 shows pacemaker batteries undergoing leak testing. The leaking battery first forms fringes as it is deformed but these fringes disappear as the battery relaxes back to its original shape. The nonleaking battery forms fringes as they are deformed under changing pressure.

Internal pressurization is also useful for HNDT of pressure cylinders such as carbon composite cylinders and solid-fuel propellant grains for rocket motors [5.60-5.62].

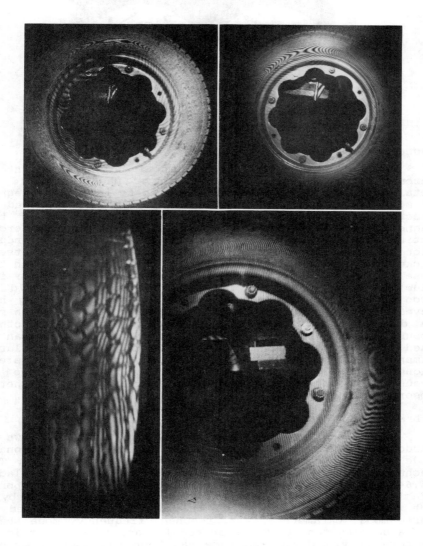

Fig. 5.15. Holographic interferograms of a scooter tyre using double-exposure holographic interferometry.

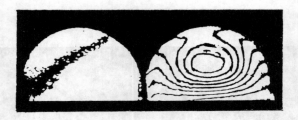

Fig. 5.16. HNDT of pacemaker batteries [5.59].

Magnetic Stressing

Mehta et al.[5.63] have applied holographic interferometry for observing the effect of magnetic field on metals. The effect of magnetic field is to induce net magnetism in the sample wherein each molecule may be supposed to undergo a process of alignment. The net effect of the magnet on the body is the aggregate of the attractive forces acting on different particles of the body. Magnetic induction is generally extremely feeble except in iron, steel, nickel and cobalt. Figure 5.17 shows that the fringes in the case of iron are much more closely spaced than those for aluminum confirming that the effect of magnetic field on iron is considerably greater. The magnetic stressing has revealed casting faults (Fig. 5.18) in an iron box which may be due to the presence of an air bubble or any other impurity. These results show that the HNDT techniques can be used to detect faults in ferromagnetic crystals of large dimensions. It can also be exploited to detect transient magnetic phenomena in ferrite rods such as passage of acoustic disturbances leading to variation in magnetic absorption.

Vibrational Excitation

Vibrational excitation is useful for revealing flaws in structural materials. The sample is vibrated in its resonant mode which creates nonuniform deformations near flaws. The technique can be used for the inspection of turbine blades [5.64] and fibre-reinforced plastics [5.65]. The response of the object to sinusoidal forces are characterized by a set of vibration modes, each of which is characterized by three modal parameters, viz. resonant frequency, mode shape and a damping factor. Of these modal shape i.e. geometrical pattern of vibration amplitude has been the most laborious to measure. Holographic vibration analysis provides a convenient method of displaying vibration mode shapes. A time-

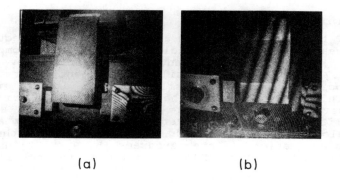

Fig. 5.17. Effect of magnetic field on an iron (a) and aluminium (b) sample [5.63].

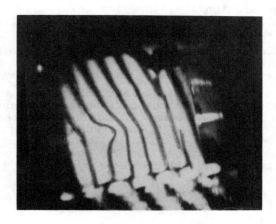

Fig. 5.18. Casting flaw revealed by magnetic stressing [5.63].

average hologram of the object vibrating in its one of the modes displays the mode shape. The time-average hologram can also exhibit vibration modes of musical instruments [5.66, 5.67].

Testing in Hostile Environment

HNDT techniques have been applied for the inspection of nuclear reactors [5.68] and nuclear power station components [5.69]. Holography has recently been applied by Parker [5.70, 5.71] in another hostile environment in recording hypervelocity projectiles similar to the armour piercing shells. This type of research has become possible mainly due to the development of reliable pulsed ruby lasers and the availability of recyclable thermoplastic recording system. The development of robust phase stepping algorithms for automatic evaluation of interferograms [5.39, 5.72] has also helped in taking the HNDT techniques to hostile and industrial environment.

5.13. Holographic Contouring

Holographic interferometry can be used to produce contour map of a three dimensional object. The contour map is a two dimensional image of the object modulated by a set of lines corresponding to the contour of constant elevation with respect to a reference plane [5.73]. Three techniques for contour generation are discussed below.

5.13.1. Two Wavelength Holographic Contouring

This method of holographic contouring involves [5.74-5.76] recording of two holograms of the object using light of two different wavelengths λ_1 and λ_2 and imaged by a telephoto lens. At the recording plane, the object beam of each wavelength interferes with the corresponding plane reference beam. The processed hologram is replaced at the recording position and reconstructed by reference beam at one of the wavelengths.

The separation between bright contour lines (contour interval) or contour height is given by

$$\Delta h = \lambda_1 \lambda_2 / 2n(\lambda_1 - \lambda_2), \tag{5.31}$$

where n is the refractive index of the medium surrounding the object. For n=1 and if λ_1 and λ_2 do not differ too much, then

$$\Delta h = \lambda^2 / 2\Delta\lambda, \tag{5.32}$$

where $\Delta\lambda = |\lambda_1 - \lambda_2|$. The two lines at 514 and 488 nm of argon ion laser produce contour height of about 4.75 μm.

5.13.2. Two Refractive Index Holographic Contouring

The object is placed in a tank with plane glass windows. The object is illuminated by a collimated beam using a beam splitter on the optical axis and imaged by a telephoto lens [5.77]. Two exposures are given to the photographic plate, the first with the tank filled with a liquid of refractive index n_1 and the second with the liquid of refractive index n_2.

The contour height is given by

$$\Delta h = \lambda/2|n_1 - n_2|. \tag{5.33}$$

By taking water and ethylene glycol in the tank for the two exposures, a contour height of 7.5 μm is obtained. If air is used with a change in pressure by 1 atmosphere for the two exposures, a contour height of 2 μm is obtained. A contour interval from 1 μm to 300 μm can be obtained using different combination of liquids [5.78]. The main drawback of this method is the requirement of the object to be immersed in a liquid.

5.13.3. Contouring by Change in the Illuminating Angle

Contours can also be generated by recording a double-exposure hologram of the object and in between the exposures the angle of illumination of the object is slightly changed. The contour surfaces generated are a set of hyperboloids of revolution.

The contour height is given by

$$\Delta h = \lambda/\sin(\Delta\theta/2), \tag{5.34}$$

where θ is the change in the illuminating angle [5.75]. By suitably selecting the direction of translation of the object between the two exposures [5.79] and displacement of the object source [5.80], contouring surfaces at any desired orientation can be produced. Sandwich holography has also been used for contouring [5.81]. Figure 5.19 shows a photograph of contours obtained by using a rainbow recording geometry [5.82].

5.14. Holographic Interferometry with Fibre Optics

The use of fibre optics in holography and holographic

Fig. 5.19. Contour generation by using a rainbow hologram recording geometry [5.82].

interferometry provides flexibility in optical setup and accessibility of remote information [5.83, 5.84]. Single mode fibres may be used for object illumination and for reference beam. Although multimode fibres can also be used but the output from these fibres contain spatial noise which makes these fibres less suitable for holographic applications. The holographic system using single mode fibres requires only the two ends of the fibres to be fixed. A coherent multimode imaging bundle may be used to transfer the object information to provide a remote holographic capability. It has also been demonstrated [5.85] that a coherent multimode bundle may be used to transfer the fringe structure (i.e. the hologram) from the object area to the recording station. The holographic fringes for this type of holography must, however, be very coarse. A resolution of 33 lines/mm in the reconstructed image has been reported using a 10 mm diameter multimode fibre optic bundle [5.85]. However, the need for ultra low spatial frequency of the hologram (a small angle reference beam) makes the probe head awkward. Experiments on real-time holographic interferometry using a flexible single mode image transmitting optical fibre of about 1 mm diameter has indicated the order of stability significantly better than that can be obtained with multimode bundles [5.83].

References

5.1 M.H. Horman, *Appl. Opt.* **4** (1965) 333.
5.2 R.L. Powell and K.A. Stetson, *J. Opt. Soc. Am.* **55** (1965) 1593, 612A.
5.3 K.A. Stetson and R.L. Powell, *J. Opt. Soc. Am.* **55** (1965) 1694, 1570A.
5.4 J.M. Burch, *Prod. Eng.* **44** (1965) 431.
5.5 K.A. Haines and B.P. Hildebrand, *Appl. Opt.* **5** (1966) 595.
5.6 R.E. Brooks, L.O. Heflinger and R.F. Wuerker, *Appl. Phys. Lett.* **7** (1965) 248.
5.7 L.O. Heflinger, R.F. Wuerker and R.E. Brooks, *J. Appl. Phys.* **37** (1966) 642.
5.8 R.J. Collier, E.T. Doherty and K.S. Pennington, *Appl. Phys. Lett.* **7** (1965) 223.
5.9 N Abramson, *Appl. Opt.* **13** (1974) 2019.
5.10 H.J. Caulfield, *Appl. Opt.* **11** (1972) 2711.
5.11 P. Hariharan and Z. Hegedus, *Opt. Commun.* **9** (1973) 152.
5.12 C.M. Vest, *Holographic Interferometry* (John Wiley & Sons, New York, 1979).
5.13 K Biedermann and N.E. Molin, *J. Phys. E: Sci. Instrum.* **3** (1970) 669.
5.14 J.D. Bolstad, *Appl. Opt.* **6** (1967) 170.
5.15 W.F. Fagon, *Opt. Laser Technol.* **4** (1972) 167.
5.16 P. Hariharan, C.S. Ramanathan and G.S. Kaushik, *Appl. Opt.* **12** (1973) 611.
5.17 J.E. Sollid, *Proc. SPIE* **25** (1971) 171.
5.18 E. Archbold and A.E. Ennos, *Nature* **217** (1968) 942.
5.19 P. Shajenko and C.D. Johnson, *Appl. Phys. Lett.* **13** (1968) 22.
5.20 B.M. Watrasiewicz and P. Spicer, *Nature* **217** (1968) 1142.
5.21 A.N. Zaidel', L.G. Malkhasyan, G.V. Markova and Yu. I. Ostrovskii, *Sov. Phys. Tech. Phys.* **13** (1969) 1470.
5.22 P.A. Fryer, *Prep. Prog. Phys.* **33** (1970) 489.
5.23 A.E. Ennos and E Archbold, *Laser Focus* **4** (1968) 58.
5.24 G.M. Mayer, *J. Appl. Phys.* **40** (1969) 2863.

5.25 D.S. Elinevskii, R.S. Bekbulatov, Yu.N. Shaposhnikov, V.A. Eryshev, A.M. Burenkin and Yu. G. Yurtaev, *Strength Mater.* **8** (1976) 95.

5.26 C.S. Vikram, *Nouv. Rev. Opt. Appl.* **4** (1973) 147; *Opt. Commun.* **7** (1973) 347.

5.27 K.N. Chopra and G.S. Bhatnagar, *Appl. Opt.* **13** (1974) 2468.

5.28 C.S. Vikram and R.S. Sirohi, *Optica Acta* **19** (1972) 39.

5.29 P.C. Mehta, *Optica Acta* **21** (1974) 529.

5.30 K.A. Stetson, *J. Opt. Soc. Am.* **60** (1970) 1378 **62** (1972) 698.

5.31 C.C. Aleksoff, *Appl. Opt.* **10** (1971) 1329.

5.32 M.H. Zambuto and W.K. Fischer, *Appl. Opt.* **12** (1973) 1651.

5.33 N. Takai, M. Yamada and T. Idogawa, *Opt. Laser Technol.* **8** (1976) 21.

5.34 F.M. Mottier, *App. Phys. Lett.* **15** (1969) 44.

5.35 G.O. Reynolds, D.A. Servaes, L.R. Izquierdo and J.B. DeVelis, *Proc. SPIE* **523** (1985) 160.

5.36 N. Abramson, *Optik* **30** (1969) 56.

5.37 W. Jaerich and G. Makosch, *Appl. Opt.* **17** (1978) 740.

5.38 P. Jacquot and P.M. Boone, *Proc. SPIE* **1212** (1990) 207.

5.39 P. Hariharan, B.F. Oreb and N. Brown, *Opt. Commun.* **41** (1982) 393; *Appl. Opt.* **22** (1983) 876.

5.40 K.A. Stetson, *Optik* **31** (1970) 576; *J. Opt. Soc. Am.* **64** (1974) 1.

5.41 W.H. Steel, *Optica Acta* **17** (1970) 873.

5.42 N. Abramson, *Appl. Opt.* **9** (1970) 97; *Appl. Opt.* **10** (1971) 2155; *Optik* **37** (1973) 337.

5.43 N. Abramson, *Appl. Opt.* **13** (1974) 2019; **14** (1975) 981; **16** (1977) 252.

5.44 P. Hariharan and Z.S. Hegedus, *App. Opt.* **15** (1976) 848.

5.45 H. Bjelkhagen, *Appl. Opt.* **16** (1977) 1727.

5.46 R.K. Erf, *Holographic Nondestructive Testing* (Academic Press, New York 1974).

5.47 E.B. Champagne, *Proceeding of Symp. on Engineering Applications of Holography* (Feb 1972, Los Angeles), Society of Photo-Optical Instrumental Engineers, Redando Beach, CA, 1972) p. 133.

5.48 L.A. Kersch, in *Optical and Acoustical Holography*, ed. E Camatini (Plenum Press, 1972) p 277.
5.49 A. Stimpfling and P. Smiglelski, *Proc. SPIE* **600** (1985) 194.
5.50 S. Amadesi, F Gori, R. Grella and G. Guattari, *Appl. Opt.* **13** (1974) 2009.
5.51 W.J. Harris and D.C. Woods, *Mater. Eval.* **32** (1974) 50.
5.52 J.R. Crawford and R. Benson, *Appl. Opt.* **15** (1976) 24.
5.53 A.D. Luxmoore, *Nondestr. Test.* **6** (1973) 258.
5.54 P.C. Mehta, Unpublished Work.
5.55 K. Grünewald, W. Fritzsch, A.V. Harnier and E. Roth, *Polym. Sci. Eng.* **15** (1975) 16.
5.56 G.M. Brown, *Proc. SPIE* **34** (1973) 75.
5.57 G.M. Brown, in *Holographic Nondestructive Testing*, ed. R.K. Erf, (Academic Press, New York 1974) p. 355.
5.58 R.M. Grant and G.M. Brown, *Mater. Eval.* **27** (1969) 79.
5.59 D. Rosenthal and R. Garza, *Photonics Spectra* (Dec 1987) 105.
5.60 O.J. Burchett, *Mater. Eval.* **30** (1972) 25.
5.61 M.D. Mayer and T.E. Katayanagi, *J. Test Eval.* **5** (1977) 47.
5.62 J.P. Waters, *App. Opt.* **10** (1971) 2364.
5.63 P.C. Mehta, D. Mohan, C. Bhan, P. Lal and R. Hradaynath, *Opt. Laser Technol.* (Oct. 1982) 269.
5.64 L.A. Kersch, *Mater. Eval.* **29** (1971) 125.
5.65 K. Grünewald, W. Fritzsch, A.V. Harnier and E. Roth, *J. Compos. Mater.* **9** (1975) 394.
5.66 A.E. Ennos, in *Optical and Acoustical Holography*, ed. E. Camatini (Plenum Press, New York 1972)p. 61.
5.67 P.T. Ajith Kumar, J.P. Thomas and C. Purusthotham, *Appl. Opt.* **29** (1990) 2841.
5.68 B. Tozer, *Opt. Eng.* **24** (1985) 746.
5.69 A. Clothier, R.C. Forsey, R. Glanville, A.L. Gordon, C.A. Mead, G.T. Thompson and B.A. Tozer, *Proc. Electro-Optics/Lasers Conf.* Brighton UK (Butterworths Scientific, 1984).
5.70 I.C. Cullis, R.J. Parker and D. Swell, *Proc. SPIE* **1358** (1990) 52.
5.71 R.J. Parker, in *3rd IEE Conf. on Holographic Systems,*

Components and Applications (1991) 234.
5.72 R. Dandliker and R. Thalman, Opt. Eng. **24** (1985) 824.
5.73 J.R. Varner, in Holographic Nondestructive Testing, ed. R.K. Erf (Academic Press 1974) p. 105.
5.74 K.A. Haines and B.P. Hildebrand, Phys. Lett. **19** (1965) 10.
5.75 P. Hildebrand and K.A. Haines, J. Opt. Soc. Am. **57** (1967) 155.
5.76 J.S. Zelenka and J.R. Varner, Appl. Opt. **7** (1968) 2107.
5.77 J.S. Zelenka and J.R. Varner, Appl. Opt. **8** (1969) 1439.
5.78 E.S. Marrone and W.B. Ribbens, Appl. Opt. **14** (1975) 23.
5.79 N. Abramson, Appl. Opt. **15** (1976) 1018.
5.80 P. De Mattia and V.F. -Bellani, Opt. Commun. **26** (1978) 17.
5.81 N. Abramson, Appl. Opt. **15** (1976) 200.
5.82 P.C. Mehta, C. Bhan and D. Mohan, J. Opt. (Paris) **11** (1980) 119.
5.83 J.A. Gilbert and T.D. Dudderar, SPIE Inst. Series, **IS 88** (1990) 146.
5.84 T.H. Jeong, Proc. SPIE **747** (1987) 25.
5.85 T.D. Dudderar, J.A. Gilbert and A.J. Boehnlein, Appl. Opt. **24** (1985) 628.

CHAPTER 6

HOLOGRAPHIC OPTICAL ELEMENTS

Holograms can be made to work as gratings, lenses, aspherics or any other type of optical element. Holographic optical elements (HOEs) are able to duplicate most of the functions provided by glass optics over a narrow spectral bandwidth. A large size holographic lens recorded on a film base can be rolled into a small piece which can be unrolled when required. Large optical apertures, light weight and lower cost are the main features of HOEs.

6.1. What is a HOE ?

A HOE can be considered as a generalized grating structure. A holographic lens is a hologram of a point source [6.1] or in other words it is a photographic recording of interference pattern between a plane wave and a spherical wave, or between two spherical waves. The HOEs have following features:

(a) HOEs are diffractive elements unlike glass lenses which are reflective or refractive. HOEs are therefore wavelength sensitive. The focal length, aberrations, image orientation, etc. of a holographic element vary with wavelength. This dispersive property is undesirable in broadband systems.

(b) A single HOE can produce the function of multiple lenses.

(c) A single HOE can serve multiple functions. It can be used as a lens, beam splitter and spectral filter simultaneously.

(d) The HOE itself is a thin film (thickness a few μm) of recording material on a substrate. The optical power of a HOE is independent of the shape of the substrate. This means that the directions of the entering and exiting beams are not controlled by the substrate normal or its refractive index.

(e) In a conventional glass lens at least one surface is

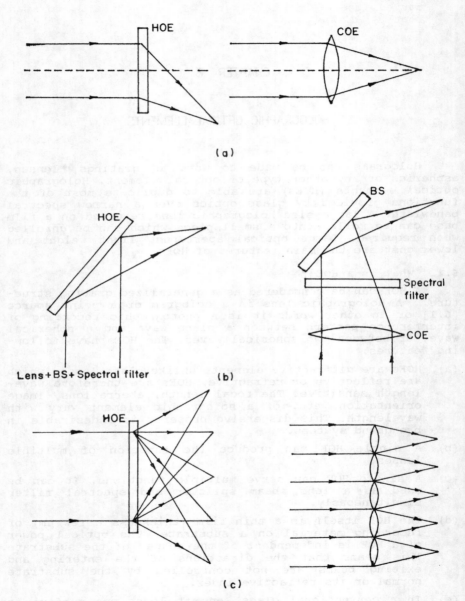

Fig. 6.1. Characteristics of HOEs.

curved, whereas in a holographic lens there is no such requirement.

(f) In the case of a reflective HOE, the substrate may not be of high quality unlike glass lenses which require glass to be highly isotropic, homogeneous and free from any strains.

(g) The HOE is light in weight.

(h) The fabrication and replication of HOEs are relatively easy because no grinding and polishing are required.

(i) A holographic element can produce multiple image waves corresponding to different diffraction orders. A lens of primary focal length F has a focal length of F/n for the nth diffraction order.

(j) With the availability of real-time recyclable recording media, any desired system function can be recorded, studied and erased as and when desired.

Figure 6.1 shows schematically some of the features of HOEs.

6.2. Hologram of a Point

We consider the recording and reconstruction of a hologram of a point source. A point source produces a diverging spherical wave given by

$$O = O_o \exp[(j\pi/\lambda F)(x^2+y^2)], \qquad (6.1)$$

where F is the distance of the point source to the hologram recording plane.

The hologram is recorded with a plane reference beam at an angle θ given by

$$R = R_o \exp[-j\alpha x], \qquad (6.2)$$

where $\alpha = (\sin\theta)/\lambda$ is the carrier frequency of the hologram.

The hologram is recorded and reconstructed by a plane beam. The zero order beam is a plane beam of constant amplitude travelling in forward direction. The finite size of the hologram aperture produces diffraction in the zero order beam. The image forming beams for

virtual image, $O_o R_o \exp[(j\pi/\lambda F)(x^2 + y^2) + j\alpha x]$ (6.3)

and

real image, $O_o R_o \exp[(-j\pi/\lambda F)(x^2 + y^2) + j\alpha x]$ (6.4)

520 LASERS AND HOLOGRAPHY

are the quadratic function of coordinates, producing a phase shift similar to spherical lenses. The first phase factor $\exp[(j\pi/\lambda F)(x^2+y^2)]$ is similar to a diverging lens, while the second phase factor $\exp[j\alpha x]$ shows the effect of a virtual prism giving an upward deflection. The net result is an off-axis diverging spherical wave whose focus (virtual image) is located at a distance F behind the hologram. Similarly the converging spherical wave $\exp[(-j\pi/\lambda f)(x^2+y^2)]$ produces a real image at a distance F in front of the hologram.

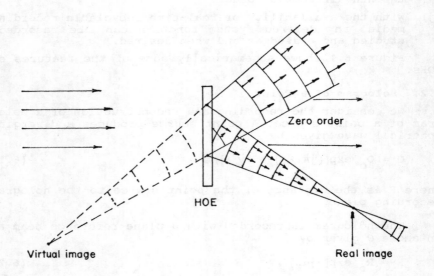

Fig. 6.2. Reconstruction of the hologram of a point.

Figure 6.2 shows the reconstruction of the hologram of a point. From the figure it is clear that the hologram of a point source produces a holographic lens with converging and diverging properties.

6.3. Resolution of a HOE

The resolution of a HOE may be determined by considering the angular spreads of the reference, object and the reconstruction wavefronts. The angular resolution is limited by these spread functions. The angular spread function of a source such as a pinhole is approximately given by the ratio of the pinhole size to its distance from the hologram. The limiting resolution (LR) is given by [6.2, 6.3]

$$LR = D/(D\delta_o + 2F\delta_r), \qquad (6.5)$$

where δ_o and δ_r are the sizes of the pinholes used for generating the two interfering wavefronts, D is the aperture and F is the focal length of the HOE. From this relation it is clear that for high resolution (a) pinholes of the smallest possible sizes should be used, (b) the HOE should be as large as possible and (c) the hologram should have the smallest possible f/number.

Figure 6.3 shows the variation of resolution with the pinhole size for different values of f/number [6.3]. It is observed that the resolution of the HOE decreases with the increase in the f/number and the pinhole size. For a conventional glass lens, the resolution is inversely proportional to the f/number. If we use a pinhole of 2 μm, the resolution will be limited to 310 lines/mm for a lens of f/no.1. Figure 6.4 shows an image of a resolution chart recorded by a holographic lens.

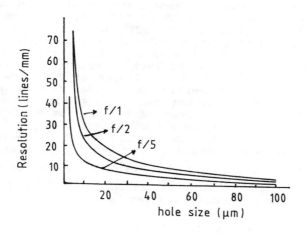

Fig. 6.3. Variation of resolution of a holographic lens with the pinhole size for different values of f-number [6.3].

Fig. 6.4. Image of a resolution chart produced by a holographic lens [6.3].

6.4. Design Aspects

The design of a holographic optical element is a much difficult task than that of an optical lens. Any design of HOEs has to consider several factors such as construction beam parameters, reconstruction beam parameters, diffraction efficiency, fringe spacing and index modulation of the recording medium. Further the aberrations of HOEs are required to be minimized. Most of the HOEs and HOE based optical systems lack in symmetry which make the design more complex. The main difficulty in the design of HOEs is the absence of established starting points and lack of previous experience.

The two major factors in the design of HOEs are fringe spacing and diffraction efficiency. The fringe spacing determines the image location and the optical power. The diffraction efficiency is governed by the shape of fringes and their distribution in the recording medium.

Let us first consider the recording of a generalized grating using two beams $O\exp(j\phi_o)$ and $R\exp(j\phi_r)$, where O and R are the amplitudes and ϕ_o and ϕ_r are the phases of the two waves. The recorded grating is reconstructed by the wave $C\exp(j\phi_c)$. The image terms are given by

$$COR\exp[j\{\phi_c \pm (\phi_o - \phi_r)\}]. \tag{6.6}$$

The position and shape of the fringes in the HOE are given by the phase difference between the two recording waves $(\phi_o - \phi_r)$. The image point can be calculated by knowing

$$\phi_i = \phi_c \pm (\phi_o - \phi_r) \tag{6.7}$$

from the system geometry. The aberrations are obtained by comparing the calculated phase with a similar expression of the phase of a spherical wave [6.4].

Although computerized ray tracing is the best way for designing HOE based optical systems, other approaches provide useful guidelines. The lens analogy of HOE helps in ray tracing HOEs without consideration of the recording beams [6.5].

In many applications it is desired to use HOEs at a wavelength that is different than the construction wavelength [6.6]. This is because the recording materials such as dichromated gelatin and photoresist are most sensitive to shorter wavelengths whereas the applications lie in near infrared region. The HOEs recorded at one wavelength and if used at other wavelengths will exhibit large aberrations. It is, however, possible to provide corrections at the recording stage. An angular adjustment can be used to compensate for wavelength change. At the new reconstruction wavelength λ_1, maximum efficiency will result if

$$\lambda/\sin\theta = \lambda_1/\sin\theta_1. \tag{6.8}$$

This condition can be satisfied at every point in a plane recording.

Rao and Mehta [6.7] have recorded a multiplexed holographic lens at 633 nm which can be used simultaneously at 633, 514 and 488 nm with proper correction of focal lengths so that when it is illuminated with a plane beam consisting of 633, 514 and 488 nm wavelengths, it will produce focal spot for each wavelength in the same plane (Fig. 6.5). The angular deviation can be corrected by using a holographic grating with a spatial frequency that is the average spatial frequency of the multiplexed holographic lens (section 4.12.7). Such a multiplexed holographic lens can be used for image processing of photographic colour transparencies and pseudocolouring.

If the HOEs are not fabricated and used with care, they will exhibit many more aberrations than may be present in conventional glass lenses. The higher order aberrations become important for HOEs of large apertures for which alignment tolerances are very low [6.8]. The condition that will eliminate all the aberrations simultaneously is to duplicate one of the construction beams in the reconstruction process [6.9]. This is, however, difficult to maintain when HOEs are to be used as components in instruments.

The third, fifth and seventh order aberrations in an off-axis holographic lens of f/no.1 are shown [6.7] in Fig. 6.6 with the parameters: reference and object beam angles

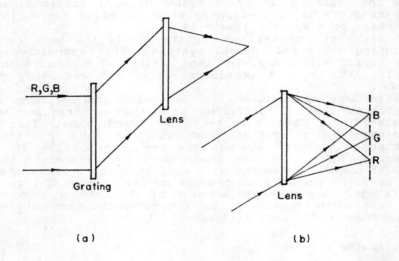

Fig. 6.5. (a) Combination of a holographic grating and a multiplexed HOE to obtain single focal point for three wavelengths, (b) reconstruction of a multiplexed HOE recorded at 633 nm with proper correction of focal lengths for three wavelengths [6.7].

with the normal to the HOE 40°, focal length 10 mm and the aperture 10 mm. The magnitudes of the third, fifth and seventh order aberrations are in decreasing order. The seventh order aberration is quite high even for small deviations in the reconstruction angle.

If we allow permissible aberrations to be less than 5 μm, the alignment tolerances for cometic (ϵ_c) and astigmatic (ϵ_a) aberrations for reference beam angle=10°, aperture=25 mm and focal length=500 mm are [6.3]

$$\epsilon_c \leq 7.2 \times 10^{-5} \text{ rad} \simeq 15 \text{ sec of arc}$$

$$\epsilon_a \leq 5.6 \times 10^{-6} \text{ rad} \simeq 1.2 \text{ sec of arc}$$

The upper limits for alignment tolerances are very low. The

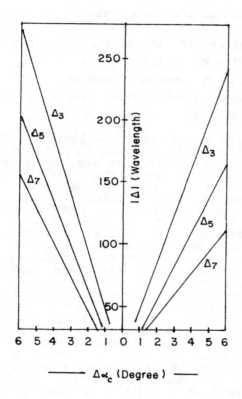

Fig. 6.6. The third, fifth and seventh order aberrations in an off-axis holographic lens of f/no.1 [6.7].

HOE must, therefore, be mounted in rugged mounts specifically designed to withstand vibrations, shocks and other environmental variations.

6.5. Fabrication

The fabrication of holographic optical elements requires the basic facilities of hologram recording. Special attention is needed with regard to vibration isolation, quality of substrate and recording material. The full potentialities of HOEs depend on the availability of high quality recording materials to achieve high diffraction efficiency and image quality. The required characteristics of the recording materials are

(a) Peak diffraction efficiency above 90% at the working wavelength.
(b) High spectral sensitivity at the recording wavelength.
(c) High resolution capability.
(d) Resistance to environmental changes.
(e) Low scattering and absorption losses.
(f) Adequate thickness.

Photoresist is a good recording material for the surface relief holograms and dichromated gelatin and photopolymer for the volume holograms. The swelling property of DCG is very helpful in fabricating HOEs at any desired wavelength with narrow band or wide band operation. Very narrow band reflectivities can also be obtained.

Testing of HOEs is a difficult task as no test optics is available as in the case of glass optics [6.10]. The focal length, efficiency and spectral characteristics can be measured. The fringe profile can be studied by an electron microscope. The complete testing of HOEs is possible by putting the HOE in the actual system and studying the performance of the system.

Figure 6.7 shows the flow chart for the fabrication of holographic gratings (or any HOE). For Lippmann gratings reflection coating is not needed. The photoresist is spin coated on an optical substrate to produce a uniform layer of resist of about 0.5 μm thickness. To achieve higher diffraction efficiencies, the grating is blazed by controlling the groove profile to a triangular shape. For this the photoresist coated substrate is placed in the interference pattern at a small angle such that one of the two beams is incident through the back of the substrate [6.11]. The surface profile, after development, depends on the shape of the insoluble layers near the surface. The method requires substrates of optical quality as one of the beams pass through it. Triangular shape profile can also be produced by first exposing the photoresist coated plate by a sinusoidal pattern and then modifying its profile by ion-beam etching [6.12].

Holographic optical elements can also be made by computers. The computer generated holographic technology has produced various HOEs, the most notable ones are aspherics and those for optical interconnects. The efficiency of the binary HOEs are poor and may be improved either by copying them in DCG or using multiphase quantization.

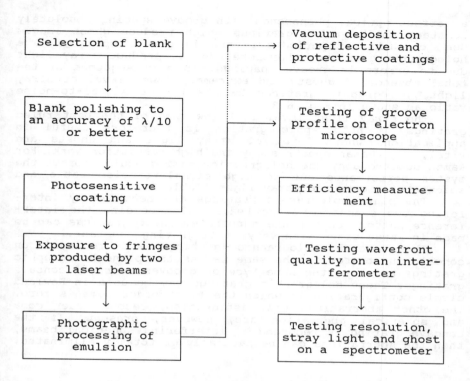

Fig. 6.7 Flow chart for making HOEs.

6.6. Holographic Gratings/Mirrors

Holographic diffraction gratings are recorded by interfering two plane or spherical beams. The grating constant a i.e. the distance between two successive fringes is given by

$$a = \frac{\lambda}{2\sin\theta} \quad (6.9)$$

where 2θ is the angle between the two symmetric interfering beams. The angle θ and the wavelength of recording λ fix the groove spacing of the grating. Since the maximum value for 2θ can be $180°$, the minimum groove spacing obtainable in this manner is equal to $\lambda/2$. Use of different wavelengths can produce very small groove spacings, resulting in holographic gratings with as many as 6000 lines/mm.

A comparison may made between the holographic gratings and classical ruled gratings [6.13]. The holographic gratings do not produce ghosts as their fabrication involves a perfect optical phenomenon with groove spacing absolutely constant unlike ruled gratings which produce ghost images due to random or periodic errors in groove spacings. The holographic gratings, therefore, do not produce stray light due to errors in groove spacings. As a consequence of the total absence of ghosts and extremely lower level of stray light, holographic gratings have a higher signal-to-noise ratio as shown in Fig. 6.8.

The stray light level is below that of the best ruled gratings. A holographic grating is therefore useful in applications where level of stray light and ghosts are vital, e.g. Raman spectroscopy and high resolution work. For Raman spectroscopy the holographic grating would improve the system performance due to high signal-to-noise ratio and absence of ghosts and stray light [Table 6.1].

The plane holographic gratings are recorded by interfering two symmetric plane beams and recording the interference pattern on a plane substrate. These gratings can be made in large sizes with very small grating constant.

The concave holographic gratings are recorded on concave substrates in the same way as the plane holographic gratings and have the same type of grooves as ruled concave gratings. These holographic gratings may be used in Rowland circle configuration in which the first order coma is zero. The other aberrations viz. astigmatism, second order coma and spherical aberration are, however, present. If the concave grating is recorded by interfering spherical beams, then the aberrations can be partially or totally eliminated.

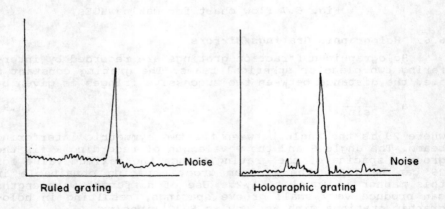

Fig. 6.8. Noise levels in ruled and holographic gratings.

Table 6.1 Comparison of holographic gratings with ruled gratings [6.13].

Physical property	Ruled grating	Holographic grating
Size of grating	Small size	Much larger size than the maximum size of ruled gratings.
Number of lines per millimeter	Generally up to 3600 lines/mm	As high as 6000 lines/mm
	Density of grooves is dictated by mechanical limitations.	Availability of different wavelengths is the limiting factor for groove density.
Free spectral range	Depends on number of lines/mm	Depends on number of lines/mm
	Free spectral range increases as the density of lines increases.	High densities are available for holographic gratings, hence larger free spectral ranges are possible.
Blaze wavelength	Can be blazed at any angle for maximum efficiency.	The groove shape and depth are controlled to maximize the efficiency at any wavelength.
	A grating blazed at λB in 1st order is also blazed at $\lambda B/2$ in 2nd order.	No such thing available.
Efficiency	Very high	Very high
Ghost and stray light	Present	May be totally absent. The stray light level is below that of the best ruled gratings.
Signal-to-noise ratio		Higher than ruled gratings.
Resolution	Depends on the width of the grating, the working angles and the wavelength.	Similar to ruled grating. Due to very low stray light the actual resolution attained is higher.

Table 6.1 contd.

Physical property	Ruled grating	Holographic grating
Aberrations	No aberrations when used with parallel beams.	Similar to ruled gratings.
	Aberrations will exist, when using a plane grating in non-parallel beam.	Possibility of aberration correction when used in non-parallel beam configuration.
	Concave gratings have very large astigmatism, 2nd order coma and spherical aberration.	Complete aberration correction is possible (perfect stigmatic) for 3 wavelengths.

It is possible to record complete stigmatic holographic gratings for three wavelengths.

The holographic grating recorded in photoresist is metalized to convert it into a reflective grating. It can be replicated by embossing methods. These gratings have large bandwidths due to their dispersive nature. The diffraction efficiency can be 90% for linearly polarized beams. The grating performance is polarization dependent. The polarization selectivity increases as the grating frequency increases, thereby reducing efficiency. The grating is a surface relief pattern, therefore it should be protected against damage during handling.

The behavior of a volume reflective holographic grating is different than a metalized surface relief grating. The volume gratings are recorded in silver halide, photopolymer or dichromated gelatin.

The bandwidth of the volume reflection gratings are very narrow due to their selective dispersive nature. However, special processing techniques can be used to make a wideband grating, if desired. Efficiency of the order of 99% is possible. The volume reflection holographic gratings are insensitive to the polarization of the incident beam unlike surface relief gratings [6.14]. The fringe structure is within the volume of the emulsion, hence they can be easily cleaned unlike surface relief gratings. These gratings may be termed as holographic mirrors and can be tuned to reflect any wavelength.

A flat reflection HOE can be made to function as a plane, spherical, hyperboloidal and ellipsoidal mirror

Table 6.2 Fabrication of different types of holographic mirrors [after 6.17].

Holographic mirror type	Wavefronts type	Source location
Plane	Plane, plane or Spherical, spherical	
Spherical	Diverging spherical, converging spherical	
Hyperboloidal	Diverging spherical, diverging spherical	
Ellipsoidal	Diverging spherical, converging spherical	
Paraboloidal	Diverging spherical, plane	
	Converging spherical, plane	
	Diverging spherical, diverging cylindrical	

[6.15-6.17]. These mirrors are fabricated by taking different combinations of the two interfering wavefronts as given in Table 6.2.

When a holographic element is recorded with a divergent beam, the angle of incidence varies across the hologram surface. The laser beam after passing through the emulsion is reflected back by a front surface mirror (Fig. 6.9) and produces a standing wave pattern in the emulsion. Since the angle of incidence varies, the fringe spacing increases along the length of the hologram. This results in a variation in the peak reflection wavelength λ_x across the hologram aperture for a fixed measurement angle according to the relation [6.18]

$$\lambda_x = \frac{n\lambda}{[R^2\cos^2\theta/(R^2+x^2+2Rx\sin\theta)+n^2-1]^{1/2}}, \qquad (6.10)$$

where n is the refractive index of the medium, λ is the construction wavelength, R and x are as shown in Fig. 6.9.

The peak reflectivity can be shifted even to longer wavelengths by increasing the angle of incidence. The angle of incidence more than the critical angle may be achieved by using a prism which is index matched to the substrate as shown in Fig. 6.10. The peak reflection wavelength λ for normal incidence operation is given by [6.18]

$$\lambda = \frac{s\lambda}{\cos[\theta_p+\sin^{-1}[\sin(\theta-\theta_p)/n]]}, \qquad (6.11)$$

where s is the swelling factor of the emulsion and θ_p is the

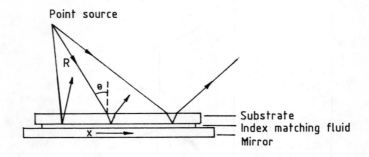

Fig. 6.9. Recording of a holographic mirror [6.18].

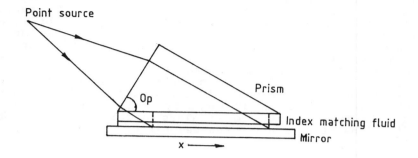

Fig. 6.10. Shifting of peak reflectivity by increasing the angle of incidence using a prism [6.18].

prism angle. The peak reflectivity can be shifted to near infrared region of the spectrum by using a prism with large prism angle and increasing the angle of incidence.

The peak reflection efficiency is controlled by the index of modulation and the emulsion thickness. The DCG reflection HOEs can have a very broad bandwidth (>300 nm) and a high diffraction efficiency (>99.5%) over the entire waveband [6.19, 6.20]. These gratings can neither be classified as Bragg gratings since their bandwidth is too wide nor as Raman-Nath since their efficiency is too high. They can be considered as a new class of Bragg gratings with chirp normal to the surface [6.19, 6.21]. The vertical nonuniformity of the grating constant and refractive index modulation can be achieved in DCG gratings. A large number of grating layers M can be created holographically by using the relation [6.21]

$$M = 2\bar{n}t/\lambda, \qquad (6.12)$$

where $\bar{n}$ is the average refractive index of the holographic material, t is the hologram thickness and λ is the Bragg wavelength.

For a hologram thickness of 20 μm and refractive index of 1.55, 124 layers can be recorded at 0.5 μm. The refractive index profile can be modified from sinusoidal to quasi-rectangular shape. This can be compared to the step-by-step deposition of layers by vacuum deposition techniques by which it is difficult to obtain more than 35 layers.

6.7. Applications of HOEs

As mentioned earlier, HOEs are able to duplicate most

of the functions provided by glass optics only if optical system operates over narrow spectral bandwidth or requires chromatic dispersion. In general, HOEs will not replace glass optics except where their unique characteristics are desired in some system application. Some of the unique functions that HOEs provide are multiple function optics, unusual configuration, narrow spectral response, etc. Beam combiners, laser scanner, spectral filters, interferometers, gratings, laser collimators, computer interconnects, solar concentrators and wavelength division multiplexers/ demultiplexers are the main areas where typical characteristics of HOEs provide advantages over glass optics. Some of the applications of HOEs are discussed below.

6.7.1. Spectral Filters

HOEs can be used as long pass, short pass or notch filters (Fig. 6.11). These have helped in realizing compact spectrometers. The short pass filter may be used for anti-Stokes applications, long pass for Stokes applications and notch filter for both of these applications [6.22]. The holographic spectral filters can also be used as narrow band beam splitters for confocal microscopy.

A notch filter rejects a narrow band of selected frequencies and allows maximum transmission outside the rejected band. The increase in the number of planes will narrow down the rejected band of frequencies. The requirement of maximum transmission outside the rejected band is accomplished by the elimination of side lobes by small index modulation between planes. Holographic notch filters are ideal since their behaviour can be made almost independent of the angle of incidence for specific geometries. These filters are emerging as unique laser eye protection filters. They block the laser source line and pass the remainder of the visible spectrum with high transmission. The holographic filters can give optical densities of more than 5.0 in the block region and high transmission of 75 to 88% in the pass band region [6.22]. The filters can be tuned by simply tilting them with respect to the incident light. Figure 6.12

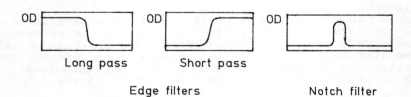

Fig. 6.11. Holographic spectral filters [6.22].

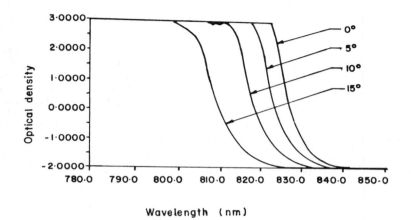

Fig. 6.12. Tuning of a holographic edge filter by tilting [6.22].

shows the tuning of a holographic edge filter by tilting. The optical density scale of -2.0 to 3.0 has been obtained by putting a neutral density filter of 2.0 in the reference beam of the spectrophotometer [6.22].

Multiple holographic notches can be superimposed on the same substrate for protection against more than one frequency. Holographic notch filters for 300 nm to 3 μm have been produced with plane or curved substrates [6.20].

6.7.2. Application in Optical Communication

Holographic gratings and lenses have promising applications in fibre optics and integrated optics as couplers, multiplexers and demultiplexers. This is because of the fact that a single HOE can perform all the functions of collection, wavelength separation and focussing. The HOE can be fabricated easily to match the numerical aperture of the fibre [6.23]. Mode conversion holocouplers can be made for converting the output of a fibre to a form suitable for an integrated optic circuit. A multiplexed holocoupler can couple one-to-many, many-to-one or many-to-many fibres. In the multiplexed holocoupler for connecting the output from many fibres to one or multiple fibres, it is difficult to completely eliminate cross-talk. One possibility is to use a stack of N holograms, each carrier multiplexed, for connection of N fibres to N fibres [6.24, 6.25].

Wavelength division multiplexers (WDMs) or wavelength division demultiplexers (WDDs) are useful in increasing the bandwidth capacity of fibre optic communication and local area network systems by sending simultaneously signals having different wavelengths. Most of the existing WDMs are limited by the number of channels and/or high insertion loss. In the holographic WDM, the number of channels is limited only by the bandwidth of the holographic grating [6.26]. The bandwidth can be increased by recording the HOE in dichromated gelatin with a vertical nonuniformity of the grating constant, refractive index or refractive index modulation [6.21]. The function of HOE is to split the incident radiation into channels of different wavelengths.

Figure 6.13 shows the holographic WDM design for local area network communication application [6.26]. The fibre array is attached to an aspheric lens by an adhesive. The fibre spacing b is related to the channel spectral bandwidth $\Delta\lambda$ and the focal length f of the aspheric lens by [6.26]

$$b = (2f\tan\theta)(\Delta\lambda/\lambda), \qquad (6.13)$$

where θ is the Littrow angle. A fibre separation of about 143 μm is possible with a hologram bandwidth of 150 nm for a Littrow angle of $6.9°$ at 810 nm and channel-to-channel separation of 30 nm.

Computer generated HOEs have also been considered for making WDMs [6.27]. The reflection HOE based WDMs can be constructed in a small size (length 3.5 cm, diameter 1 cm) [6.26]. Using superimposed HOEs several spectral windows at 800, 1300 and 1550 nm can be multiplexed or demultiplexed.

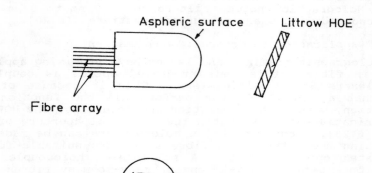

Fig. 6.13. Holographic WDM design using a volume Littrow HOE [6.26].

6.7.3. HOEs in Compact Disks

The most intriguing application of HOEs has recently been made in optical heads for compact audio and video disk players and computer disk drives [6.28-6.31]. The optical head has to perform three optical functions besides focussing the light on the disk. These functions are focussing error detection optics, tracking error detection optics and splitting the back reflected beam on the disk. Conventional optical head is very bulky and heavy. The HOEs can be used in optical heads for producing tracking error signals along with retrieval of the signal. They can also correct the lens aberrations, reshape the laser beam and compensate for the wavelength shift of the laser diode [6.31, 6.32].

In the optical pickup for compact audio and video disk player, a holographic grating is used which splits the laser beam into three. The zero order beam is used for signal retrieval, while the two first order beams produce the error signals. During alignment the grating is manipulated such that the two diffracted spots straddle across a track so that the signals in the detectors are balanced. During play if the centre spot becomes off the track, the difference in the detector signals will be the tracking error signal.

The device has been made in a compact form by integrating the laser chip and the detector chip in a thick film ceramic board [6.33]. The holographic grating is mounted on the top of the hybrid device. The device can also be used to monitor the focussed position of a microscope objective.

6.7.4. Holographic Laser Beam Attenuator

High efficiency volume holographic gratings can be used for attenuation of visible and near infrared laser beams. Fig 6.14 shows the working of a holographic laser beam attenuator designed by Heriot-Watt University and manufactured by Edinburgh Instruments. The holographic grating has a maximum efficiency at an angle of incidence of $5°$ from normal. The efficiency drops to zero at normal incidence. The maximum transmission (~97%) therefore occurs at normal incidence and minimum transmission (<2%) at $5°$ from normal. The diffracted light is blocked by an iris diaphragm which allows only the zero order beam. The device can withstand 20 W of input laser power and can produce a continuously variable attenuation up to 18 dB.

6.7.5. HOE based Fibre Optic Gyroscope

For fibre optic sensors 3dB couplers are the most crucial. HOEs can be recorded in dichromated gelatin or photopolymer with beam splitting ratio approaching 1:1 with a total higher diffraction efficiency. Fig. 6.15 shows the arrangement [6.34] of using a double-exposure HOE in a fibre optic gyroscope. The HOE acts differently in forward pass

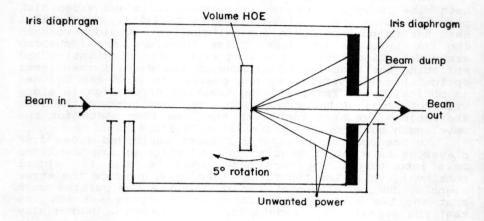

Fig. 6.14. Holographic laser beam attenuator.

and return pass. It operates as two lenses and one beam splitter (1:1) for the incident light on the fibre, while it acts as two lenses and one beam combiner for the light emitted by the fibre. On the screen the fringes will be observed which will move when the fibre ring spins. Knowing the fibre length and the change in fringes, the angular velocity of the spinning ring can be determined.

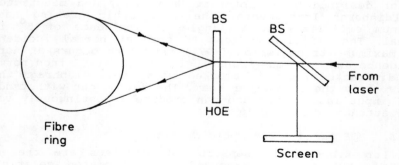

Fig. 6.15. Principle of a HOE based fibre optic gyroscope [6.34].

6.7.6. Holographic Scanner

Laser beam scanning is one of the important applications of HOEs which has commercially been exploited. Holographic scanners have already established their usefulness for printing and copying. A laser beam scanner consists of a laser source, a device to deflect the beam, a detector of light and an electronic device for interpretation [6.35, 6.36]. The scan path can be in one, two or three dimensions of space consisting of continuous or discrete points. It can be irregular or of uniform shape and orientation, bright or dim, monochromatic or multicoloured depending on the type of deflection and the light source.

The use of HOEs in laser beam scanner offers number of advantages as compared to rotating mirrors and acousto-optic scanners. The same HOE can be used as deflector and for collecting the scattered light by the object. Thus the location of the return image is invariant with the position of the scan path in the object field. The size and weight of the HOE are very less making the drive system and packaging easier requiring a low drive power. The HOEs also serve as ambient light filter. These are two dimensional devices with three dimensional scan possibility with increased flexibility. The holographic scanners do not produce any beam displacement and nonlinear deflection exhibited by some mirror scanners.

Holographic scanners are mechanically moving deflectors like rotating mirrors but have smooth surfaces unlike polygons. These devices are highly wobble-insensitive and have extreme scan flexibility. The scan path is usually achieved by translating or rotating the HOEs. Both transmission and reflection type of HOEs can be used depending on the required scan line uniformity. The reflection HOEs will introduce cross-scan errors due to their mirror like behaviour. The transmission HOEs offer wobble-free characteristics but require good optical substrates.

A single HOE or a combination of HOEs can be used in a holographic scanner depending on the required application and accuracy. Single hologram scanners involve rotating or translating holographic gratings or lenses. Single rotating grating scanner can produce circular scan lines. If the grating is replaced by a rotating holographic lens no additional conventional glass optics is required and the scanner geometry becomes simplest. A tandem holographic scanner consisting of two faceted holograms provides more scan flexibility, higher scan capacity and any three dimensional scan paths. However, the light throughput is limited due to the use of two holograms. Far superior pixel rates and scan capacity than the conventional scanner have been achieved by holographic scanners. A scan rate of several million pixels/s can be achieved with a scan angle of $5-360°$, scan efficiency of 100%, scan linearity of 0.02% and 20,000 spots

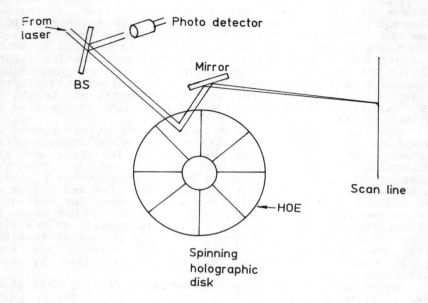

Fig. 6.16. Principle of a typical holographic scanner.

per line. Figure 6.16 shows the principle of a typical holographic scanner.

6.7.7. Diffractive-Refractive Telescope

In telescopes, microscopes, and binoculars doublet or triplet objective systems are needed since the longitudinal chromatic aberrations for a singlet objective are extremely large and difficult for a conventional eyepiece to correct. These objectives are limited by stronger surface curvatures and larger weight. Large singlet objectives can be used if holographic lenses are used in the eyepieces. The large longitudinal chromatic aberrations of singlet refractive objectives are corrected by negative dispersion of the diffractive eyepieces. By splitting the holographic eyepiece into two elements [6.37, 6.38] paraxial lateral dispersion may also be corrected.

Figure 6.17 compares the longitudinal chromatic aberration of a thin refractive objective lens with that of a diffractive eyepiece lens [6.37]. The dispersion in HOEs is

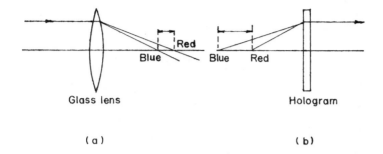

Fig. 6.17. Longitudinal chromatic aberrations of a glass lens and a holographic lens [6.37].

very large which results in longitudinal chromatic aberrations which are many times larger than those of glass lenses. Thus by setting the chromatic aberrations of the refractive objective and diffractive eyepiece equal and opposite, longitudinal chromatic aberration is eliminated. The correction of lateral colour is achieved by splitting the holographic eyepiece into two separated lenses. Figure 6.18a shows a schematic diagram of a refractive-diffractive hybrid telescope proposed by Stone [6.37]. Figure 6.18b shows the details of the diffractive eyepiece which consists of two holographic lenses.

In the diffractive-refractive telescope only one optical material is needed, whereas in the conventional glass achromatic telescope two different materials are required. The objective of the telescope of the hybrid system is not achromatic, hence the required surface curvatures are reduced. This will result in less optical material requirement, less weight, larger maximum apertures and low monochromatic aberrations.

6.7.8. Applications in Architecture

Holography can assist in changing the architectural environment in a number of ways. Display holograms have been incorporated in many offices. Special holographic gratings can be used for day lighting applications in offices and houses. The need has arisen due to the requirement of a certain minimum quantity of light energy for health and for psychological reasons [6.39].

Holographic diffractive structures are designed to redirect sunlight entering through the windows towards the ceiling and upper surfaces at the back of the room as shown

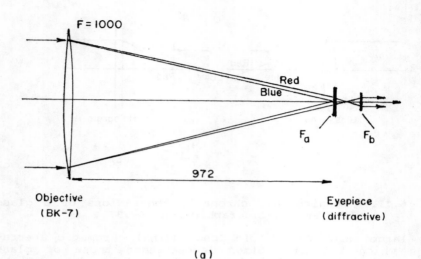

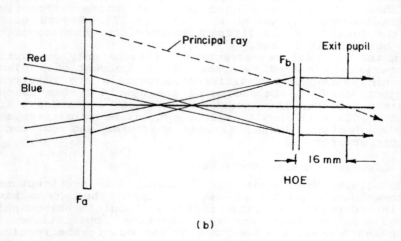

Fig. 6.18. (a) Schematic diagram of a refractive-diffractive hybrid telescope, (b) details of the diffractive eyepiece [6.37].

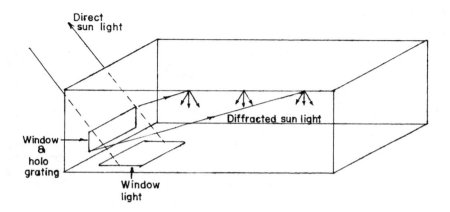

Fig. 6.19. Use of a holographic grating for day lighting application [after 6.39].

in Fig. 6.19. The effect produced is equivalent to skylight with the full sunlight spectrum or any desired portion of it. Thus the architect can allow efficient use of light from smaller apertures and direct light into places where it would not normally go. Holographic gratings and hollow light pipes can guide light into the rooms bringing daylight without windows. The holographic windows will also block the ultraviolet rays reaching the room. The holographic gratings may be multiplexed to diffract efficiently for different solar altitudes.

Another interesting architectural application of holograms is in reflecting light into streets or other public places which would otherwise be blocked from the sun due to tall buildings.

6.7.9. Beam Combiners

A holographic beam combiner is a dual function optical element which simultaneously acts as an optical window and as a lens or mirror. Beam combiners are important components for various display systems such as head-up display (HUD), helmet mounted display and night vision goggle systems. Because of their multiple component capability viz. imaging, beam splitting and spectral selection, HOEs provide increased flexibility in the selection of image forming geometry. Conventional display systems are limited by lower field-of-view, poor see-through capability and requirement of higher cathode ray tube (CRT) power levels. The HOEs fabricated in dichromated gelatin and photopolymers can overcome all these difficulties.

Holographic Head-up Display System

A head-up display system is used in aircrafts to provide various flight and weapon aiming data to the pilot while he is viewing outside the aircraft. The data is generated on a small CRT. The light from the CRT is collimated and folded towards the pilot's eyes through a beam splitter which is about 25% reflective and 75% transmitting. The data on the CRT is thus projected to infinity to allow better integration of the visual message with the outside real world. The field-of-view is limited by the size of the collimating lens and by the distance of the lens from the eyes. The field can be increased by employing a large aperture lens system but that is very difficult to manufacture. The CRT light is also not fully utilized as only 25% light is reflected by the beam splitter. The see-through capability is reduced as only 75% of the incident light is transmitted to the pilot.

The holographic combiner increases the field-of-view in two ways. Firstly, the collimating lens can be brought nearer to the eyes by making the combiner itself as a reflective lens (or concave mirror). Secondly, the size of the combiner can be made as large as 300 mm. The holographic combiner can have a diffraction efficiency of the order of 99% in reflection at the CRT peak wavelength, therefore the light from the CRT can efficiently be utilized. This will prolong the useful life of the CRTs as these can be operated at low voltages with the holographic combiners. The transmittance of the holographic combiner can also be very high (>85%).

The holographic HUD can be designed to produce a wide field-of-view ($30^\circ \times 20^\circ$) display. However, to produce a high resolution image free from any aberration, the hologram aberrations are required to be compensated in the relay lens. The relay lens is a multi-component lens which involves tilted, decentred, split and aspheric components for full compensation of the HOE aberrations.

Figure 6.20 shows a diagram of a holographic HUD used in F-16 aircraft [6.40]. The system consists of three HOEs. The HOE-1 reflects the beam from the relay lens to the HOE-2 which acts as a thin beam splitter. The HOE-2 reflects the beam to the HOE-3 which is a spherical reflecting collimator. A single HOE can also be used but in that case the incident light on the HOE will be at a very large angle. This will produce large amount of aberrations in the image. The use of three HOEs ensures a very small angle of incidence to the image forming HOE-3.

The helmet mounted display provides a very compact

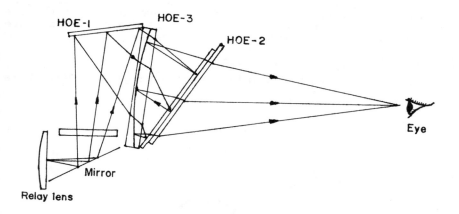

Fig. 6.20. A holographic head-up display system using three HOEs [6.40].

design. The CRT and the relay optics is mounted on the helmet and the HOE is fabricated on the visor [6.41].

The concept of head-up displays has been used for modern motor cars for efficient driving as shown in Fig. 6.21. While in the aircraft HUD the image is at infinite distance, in the automobile HUD the image is optically formed at a distance of 2 m in slightly downward direction

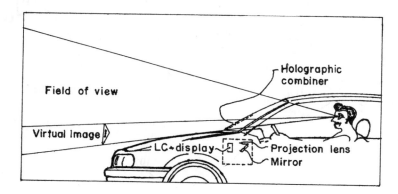

Fig. 6.21. Holographic combiner for a car [6.42].

in front of the driver [6.42]. The aim of the HUD on automobiles, is to monitor speeds, stopping distances and warning signals without head down viewing of panels. The combiner hologram is made on the windshield and the data are produced on a liquid crystal display system.

Holographic Night Vision Goggles

With the conventional night vision goggles (NVGs) there

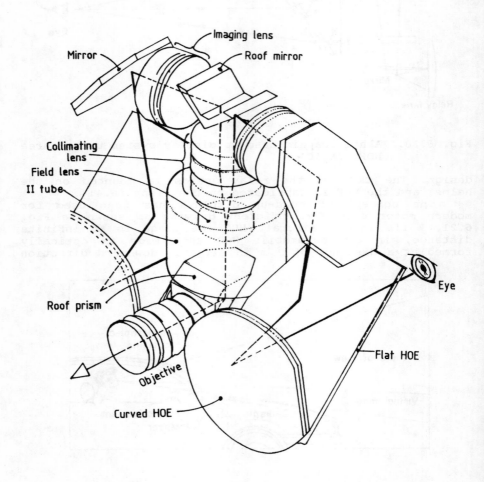

Fig. 6.22.   A holographic night vision goggle [6.44].

is no permanent see-through capability. In the event of a flare or flash of light, the image intensifier tube of the goggle gets saturated and the wearer is temporally blinded. The wearer has to flip-off the goggle to operate any instrument or to take an action. Holographic NVGs eliminates these drawbacks of the conventional goggles.

The main component of a holographic NVG is a holographic eyepiece which is a reflective holographic lens [6.43, 6.44]. The holographic eyepiece forms the image in reflection by the light received from the image intensifier tube. As the HOE is transparent, the viewer can see-through it in the event of a flash of light. The system layout of a holographic night vision goggle is shown in Fig. 6.22. The objective of the system receives some photons under night conditions which are amplified by an image amplifier tube. The light from the phosphor of the image intensifier tube is collimated and split into two branches by a roof mirror. A relay lens in each branch produces an aerial image of the phosphor screen which forms the input to the image forming HOE. To reduce the angle of incidence to the HOE, a plane HOE is inserted in the path.

6.7.10. Fingerprint Sensor

A HOE can be used as a fingerprint sensor [6.45]. The

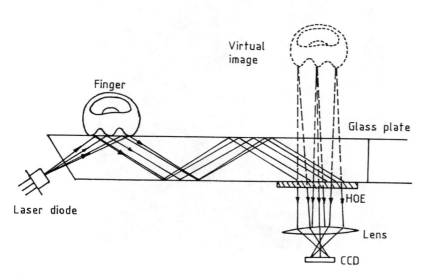

Fig. 6.23. The principle of a HOE based fingerprint sensor [6.45].

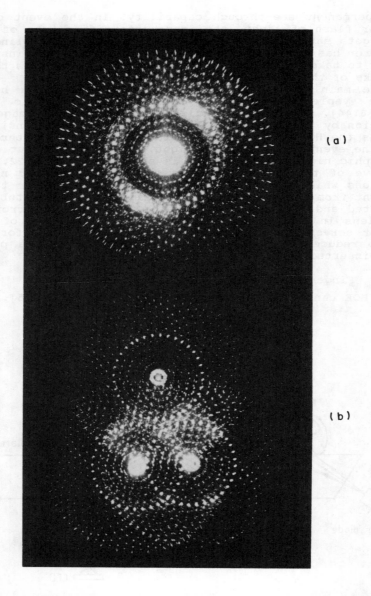

Fig. 6.24. Patterns generated by multiplexed holographic gratings.

basic principle of the sensor is shown in Fig. 6.23. A 10 mm glass plate which contains a holographic grating is illuminated from below by a diode laser. When the plate is touched, the phenomenon of frustrated total reflection occurs, thereby guiding the light scattered by ridges of the fingerprint within the glass plate. The information about the grooves is not guided as there is a small gap between the finger print grooves and the glass plate. The holographic grating at the other end of the glass plate forms a virtual image of the ridges of the fingerprint. A lens and a CCD are used to form the real image of the fingerprint. The system produces a high contrast image of the fingerprint as the signal (light from ridges) is separated from the noise (light from grooves).

The total reflection geometry ensures the operator's eye safety requirements. The setup can be used for personal identification for security purposes.

6.7.11 HOEs in Art

An interesting application of HOEs can be in the creation of artistic patterns and laser shows. Specific holographic multiplexed gratings can be recorded or computer generated which can transform a laser beam into a two or three dimensional pattern on a screen. Figure 6.24 shows two patterns created by optically recorded holographic multiplexed gratings. The pattern of Fig. 6.24a has been produced by a single carrier multiplexed holographic grating, while that shown in the Fig. 6.24b is the result of a sandwich of two multiplexed gratings when illuminated by a He-Ne laser.

Reference

6.1 M.J.R. Schwar, T.P. Pandya and F.J. Weinberg, *Nature* **215** (1967) 239.

6.2 E.B. Champagne and N.G. Massey, *App. Opt.* **8** (1969) 1879.

6.3 P.C. Mehta, S. Swami and V.V. Rampal, *Appl. Opt.* **16** (1977) 445.

6.4 E.B. Champagne, *J. Opt. Soc. Am.* **57** (1977) 51.

6.5 W.C. Sweatt, *J. Opt. Soc. Am.* **67** (1977) 803.

6.6 M.R. Latta and R.V. Pole, *Appl. Opt.* **18** (1979) 238.

6.7 K.S.S. Rao and P.C. Mehta *Proc. SPIE* **1319** (1990) 283.

6.8 P.C. Mehta, K.S.S. Rao and R. Hradaynath, *Appl. Opt.* **21** (1982) 4553.

6.9 J.N. Latta, *Appl. Opt.* **10** (1971) 599, 609.

6.10 D.H. Close, *Opt. Eng.* **14** (1975) 408.

6.11 M.C. Hutley, *Diffraction Gratings,* (Academic Press, 1982).

6.12 Y. Aoyagi, K. Sano and S. Namba, *Opt. Commun.* **29** (1979) 253.

6.13 Jobin Yvon Catalogue.

6.14 B. Moslehi, P. Harvey, J. Ng and T. Jannson, *Opt. Lett.* **14** (1989) 1088.

6.15 G.D. Mintz, D.K. Morland and W.M. Boerner, *Appl. Opt.* **14** (1975) 564.

6.16 P.C. Gupta and A.K. Aggarwal, *Appl. Opt.* **16** (1977) 472.

6.17 S. Anand Rao and S.V. Pappu, *Rev. Sc. Instr.* **51** (1980) 809.

6.18 S.S. Duncan, J.A. McQuoid and D.J. McCarthey, *Proc. SPIE* **523** (1985) 196.

6.19 J. Jannson, T. Jannson and K. Yu, *Sol. Energy Mat.* **14** (1986) 289.

6.20 J.R. Magarinos and D.J. Coleman, *Opt. Eng.* **24** (1985) 769; *Proc. SPIE* (1985) 523.

6.21 J. Jannson, I. Tergara, Y. Qiao and G. Savant, *J. Opt. Soc. Am.* **A8** (1991) 201.

6.22 A. Rizkin and K. Rankin, *Photonic Spectra* (Jan 1992) 113.

6.23 O.D.D. Soares, *Opt. Eng.* **20** (1981) 740.

6.24 G. Goldman and H.H. Witte, *Opt. Quant. Elect.* **QE-14** (1972) 75.

6.25 A.M.P.R. Leite, *Holographic Elements for Optical Transmission Systems, Ph.D Thesis*, Univ. of London (1979).

6.26 B. Moslehi, P. Harvey, J. Ng and T. Jannson, *Opt. Lett.* **14** (1989) 1088.

6.27 M. Kato and K. Sakuda, *Appl. Opt.* **31** (1992) 630.

6.28 Y. Kimura, S. Sugama and Y. Ono, *Appl. Opt.* **27** (1988) 668.

6.29 Y. Kimura, S. Sugama and Y. Ono, *Jap. J. Appl. Phys.* **26 Suppl. 26-4** (1987) 131.

6.30 C.S. Ih, L.Q. Xiang, B.H. Zhuang, C.W. Yang, K.Q. Lu, R.S. Tian and Z.M. Wang, *Proc. SPIE* **1052** (1989) 176.

6.31 C.S. Ih, R. Tian, B. Zheng and K. Lu, *Proc. SPIE* **1316** (1990) 390.

6.32 Y. Yoshida, T. Miyake, Y. Kurata and T. Ishikawa,

Proc. SPIE **1401** (1990) 58.

6.33   W.H. Lee, *Opt. Eng.* **28** (1989) 650.

6.34   Z. Yunlu, H. Dahsiung, W. Ben and T. Huiying, *Proc. SPIE* **1238** (1989) 452.

6.35   L. Beiser, *Holographic Scanning* (John Wiley, 1988) p. 234.

6.36   *Laser Scanning and Recording,* eds. L. Beiser and B.J. Thompson, SPIE Milestone Series **378** (1985).

6.37   T.W. Stone, *Proc. SPIE* **1212** (1990) 257.

6.38   T.W. Stone and N. George, *Appl. Opt.* **15** (1988) 542.

6.39   S. Weber, E. King and R. Ian, *Proc. SPIE* **747** (1987) 96.

6.40   B.M. Woodcock, *Proc. SPIE* **399** (1983) 333.

6.41   R.J. Withrington, *Proc. SPIE* **147** (1978) 161.

6.42   M.-A. Beeck, T. Frost and W. Windeln, *Proc. SPIE* **1507** (1991).

6.43   L.G. Cook, *Proc. SPIE* **193** (1979) 153.

6.44   E. Schweicher, *Proc. SPIE* **1032** (1988) 349.

6.45   S. Igaki, S. Eguchi, F. Yamagishi, H. Ikeda and T. Inagaki, *Appl. Opt.* **31** (1992) 1794.

CHAPTER 7

INTERCONNECTS

7.1. Introduction

There is a growing requirement of increasing the throughput of computers and data processors for several areas which need real-time calculations of high complexities. These areas include remote sensing, seismic data interpretation, defence early warning systems, simulations for weather prediction, aerodynamic modeling and recognition. Computer architects are working on the design of parallel computers to realize more powerful machines. A connection machine has as high as 65,000 small processing elements interconnected in a parallel network. More complex interconnections are required for neural networks where each connection has an associated weighting factor.

To meet the demand, an increasing number of high speed components are densely packed in a device. But this creates several difficulties. Goodman et $al.$ [7.1] have dealt with the problems of electronic chips and computers and offered various possible optical schemes which can be incorporated in an electronic computer. As the number of elements in a single chip increases due to scaling of feature size, the number of interconnects also increases but the perimeter of the chip where the interconnection pads are made, does not grow proportionately [7.1]. For example, a circuit involving 10^5 gates would require about 2000 interconnections, for which connection pads will be separated only by 20 μm on a chip size of 10 mm×10 mm. Thus, the communications between different chips may be limited by the number of pins available on the chips. This is true because the available space for connections is limited to two dimensional planes.

The scaling of feature size also results in an increase in the current density which leads to greater electromigration effects and ultimately in the breaking of conductor lines [7.1].

Another interconnection problem arises due to the need for the transfer of a timing signal, known as clock, to various parts of the chip and to different devices for synchronous operation of multiple devices and circuits. The difficulty arises in designing circuits for high speed operation where different parts of the circuit receive the same clock at different times. This is known as clock skew and is a result of interconnect delay. Transistors based on GaAs have been fabricated which can switch states as fast as a few ps, however, actual processing time in computers is much higher due to interconnect delays.

Extensive efforts have been made for minimizing the interconnection lengths and improving the transmission speed. Linking different chips and realizing intra-chip connections become limiting factors in the circuit design. As the number of interconnections increases, the power needed to drive the transmission lines increases to 50% or more than the total power for high clock rate. Thus, the performance of high speed electronic circuits is limited by their signal transmission capacity, power consumption, cross-talk, reliability and clock skew.

7.2. Optical Interconnects

Optical interconnects have the potential of increasing the throughput of computing systems. Unlike electrical interconnections, optical interconnects are insensitive to mutual interference effects. The velocity of electrical signal propagation in a transmission line depends on the capacitance per unit length. The propagation of optical signals on the other hand, is free from capacitive loading effects leading to faster point-to-point communication of several gigabit/s. The optical signals can propagate through free space or through waveguides without significant cross-coupling. In free-space configuration propagation paths may share the same space thereby permitting a more compact design and crossbar type connections. The use of free space permits the density of interconnects to increase to the fundamental diffraction limit of optics with minimal cross-talk.

The electronic interconnections are fixed, while the optical interconnects can be reprogrammable by using dynamic components. Such a capability will have significant impact on the design of new computers. The optical interconnects will reduce the number of electrical pins on each chip.

The optical interconnects will have applications for long distances between data processing systems and also for short distances within data processing systems. The optical interconnect in data processors will handle high data rates, allow parallel communication, eliminate ground-loop problems and dissipate less heat.

The optical interconnects will have major applications not only for electronic connections in computers but also in two dimensional digital optical computing [7.2, 7.3]. Considerable efforts are being made to realize optical interconnects with desired characteristics. Various devices based on standard optical apparatus [7.4], lenslet arrays [7.5], optical fibres [7.1, 7.6-7.8], integrated optic waveguide [7.1] and holograms [7.1, 7.9, 7.10] have been suggested. The integrated optic waveguide interconnects have limitations of high attenuation, high bending loss and constraint of a plane.

An important step in optical interconnect is the requirement of transformation of electrical signals into photonic signals. Diode lasers or light emitting diodes are used to generate photons and semiconductor detectors to transform photons back into electrical signals. LEDs can be used for interconnects requiring lower speeds (100 MHz) and lower fanouts, while diode lasers are used for interconnects requiring higher speeds and large fanout [7.11].

7.3. Classification of Holographic Interconnects

The holographic optical elements (HOEs) determine the fanin, fanout and distribution of signals and implement the interconnection pattern between sources and detectors. The use of HOEs allows great flexibility in designing the interconnection pattern. Collins and Caulfield [7.10] have categorized different types of holographic interconnects. The interconnection pattern can be represented by a matrix whose components are given by [7.10]

$$m_i = \begin{cases} 1 \text{ if source i is connected to receiver j} \\ 0 \text{ otherwise.} \end{cases} \quad (7.1)$$

A holographic lens, for example, has a connection matrix M=1 for one-to-one connection. For one-to-many interconnections, the matrix may be represented by

$$M = [1,0,1,1,0,1]. \quad (7.2)$$

Eq. 7.2 shows that a single source is connected to detectors 1, 3, 4 and 6. This is implemented by a multiplexed HOE or a multifaceted HOE.

For many-to-one connections the connection matrix may be

$$M = \begin{bmatrix} 1 \\ 0 \\ 1 \\ 1 \\ 0 \\ 1 \end{bmatrix}. \quad (7.3)$$

Eq. 7.3 implies that the sources 1,3,4 and 6 are required to be connected to the same detector. This is also implemented by a multiplexed HOE. The function of the interconnect given by Eq. 7.3 is that of a beam combiner. For many-to-many connections, the matrix M is of the form

$$M = \begin{bmatrix} 1 & 0 & 1 & 1 & 0 & 1 \\ 0 & 1 & 0 & 0 & 0 & 1 \\ 1 & 0 & 1 & 0 & 1 & 0 \\ 0 & 0 & 0 & 0 & 1 & 0 \\ 1 & 1 & 0 & 1 & 0 & 1 \\ 0 & 0 & 1 & 0 & 0 & 0 \end{bmatrix}. \qquad (7.4)$$

Matrix given in Eq. 7.4 can be constructed for regular or random positions of sources and detectors.

7.4. HOE Size

The HOE may be of transmission (thin or volume) or reflection type. Let us estimate the required volume of the HOE for connection of large input and output elements. The system should be able to connect any element of the input plane to any element of the output plane [7.12, 7.13]. A thin interconnect can be made by using a multifaceted hologram [7.14]. Fig. 7.1 shows the configuration for the transmission hologram interconnection. The input and output planes contain NxN elements. In the case of a thin interconnect, the hologram is a n x n multifaceted element whose

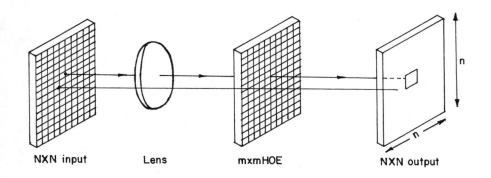

(a)

Fig. 7.1a

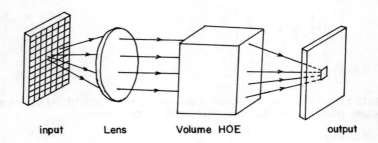

   (b)

Fig. 7.1. Holographic transmission interconnect configuration. (a) Thin interconnect, (b) volume interconnect [7.13].

function is just to deflect the light beam. Therefore, each facet of the HOE contains a grating structure. The number of interconnect channels depends on the number of facets. The desired spot size at each detector decides the size of each facet.

Assuming square elements and following Destine and Guha [7.13], the amplitude u(x) of the light incident on the output plane is given by the sinc function

$$u(x) = A \operatorname{sinc} \frac{a(x) - ps_1}{\lambda f}, \qquad (7.5)$$

where a is the size of the input element, p is the element number, s_1 is the centre-to-centre separation between the elements at the input and f is the distance between the planes. If s_2 is the centre-to-centre distance between the elements at the output plane, then the minimum separation between the planes is given by [7.13]

$$f = \frac{N(S_1 + S_2)}{2}. \qquad (7.6)$$

Eq. 7.5 gives the distance d between the first zeros of the output spots as

$$d = \frac{2f\lambda}{a} \qquad (7.7)$$

Eqs. 7.6 and 7.7 give

$$\frac{1}{m\lambda} = \frac{S_1 + S_2}{ad},$$

$$= \frac{k_1}{d} + \frac{k_2}{d}, \qquad (7.8)$$

where $k_1 = S_1/a$ and $k_2 = S_2/d$.

For close packing $k_1 = k_2 = 1$ which gives the relationship for smallest elements as

$$S_1 = a, \quad S_2 = d. \qquad (7.9)$$

Smallest system volume results when

$$a = d = N\lambda. \qquad (7.10)$$

For NxN elements, the dimensions of input plane, output plane and HOE, and the distance between them are all $N^2\lambda$. Therefore, the total volume will be $N^6\lambda^3$ and the volume per interconnect would be

$$\text{Volume/interconnect} = \frac{N^6\lambda^3}{N^2}$$

$$= N^4\lambda^3. \qquad (7.11)$$

Thus for a plane of 1000x1000 and $\lambda=1$ μm, the sizes of input and output planes will be 1 m^2 and the system volume will be 1 m^3. The input-output interconnection time t is

$$t = \frac{N^2\lambda}{c}$$

$$= 3.3 \text{ ns}, \qquad (7.12)$$

where c is the speed of light.

If we use a volume HOE (Fig. 7.1b), the volume required per interconnect will be much smaller as given by [7.15]

$$\text{Volume/interconnect} = N^3\lambda^3. \qquad (7.13)$$

The size of the HOE plane can thus be reduced if we

consider the storage capability of the hologram. The number of interconnects that can be supported are then about $10^8/cm^2$ [7.16]. However, the realization of this number of interconnections are limited by several factors such as aperture effect, dynamic range of the recording medium, low diffraction efficiency, granular noise, input-output angle restrictions, medium resolution limitation and cross-talk.

It has been shown by Collins and Caulfield [7.10] that an emulsion of 15 μm thickness can record only about five superimposed holograms each with 100% efficiency. However, 100 holograms are possible in a recording material of 1 mm thickness. With the presently available materials it is impossible to have many superimposed holograms as the thick materials have smaller refractive index modulation capability. A possible solution to the material limitation is to use resonated holograms in which the interaction between light and the hologram is increased optically by inserting the hologram in an optical resonator [7.10]. With the development of suitable recording materials, it would be possible to record a large number of superimposed holograms with very high equal efficiencies to achieve extremely high number of interconnections, of the order of $10^{15}-10^{18}$ per second.

Any design of a holographic interconnect should take into consideration the number of fanouts, hologram-IC mask separation, hologram size and the spot size of the light at the detector plane. The resolution should be maintained over all the detector locations. The hologram diameter and its separation from the IC mask should be minimized [7.17].

7.5. Desirable Characteristics

A holographic optical element to work as an optical interconnect should have the following characteristics.

a) The HOE should provide suitable f-number optics to match the divergence of laser diodes. The output from the laser diodes consists of ellipsoidal expanding wavefronts. The position and direction of the incident light to the HOE will be dependent on the geometrical arrangement of the sources. The pattern of each signal distribution may be different. The HOE must be able to perform well under these conditions.

b) The efficiency of the HOE should be high so that most of the source light is transferred to the detectors. Highly efficient HOE will reduce the source power consumption. This will reduce the optical noise and heat dissipation [7.18].

c) The HOE should produce well focussed spots at the detector locations. Ideally the spot size should closely match the size of the detectors (10 μm).

d) The optical noise produced by the hologram should be

minimum. The undiffracted zero order beam from the hologram will produce a background signal at the detectors. The scattering from the hologram also creates noise. Spurious higher order diffraction orders must be reduced. Computer generated HOEs may produce quantization noise which must be reduced.
e) The HOE should be tolerant to misalignment due to thermal variation and laser output drift. It should also provide a mechanism for its alignment with the sources and detectors. The alignment tolerance is of the order of ±5 μm.
f) It should preferably be dynamic and programmable.

While designing a holographic interconnect the above mentioned requirements are considered. Generally it is preferred to have sources and detectors in a single plane due to packaging requirements.

7.6. Configurations

Various configurations using transmission or reflection holographic optical elements have been suggested for point-to-point (one-to-one), broadcast system (one-to-many), interrupt line (many-to-one) and the bus line (many-to-many) interconnections. Most of the work has been confined to free-space interconnection schemes. However, a few schemes based on integrated optics and waveguide holography have been demonstrated recently. Both optically generated and computer generated holograms have been used for optical interconnects. Recent work has also concentrated on the realization of reconfigurable dynamic interconnects [7.19].

7.6.1. Free-Space Focussed Spot Interconnects

Optical free-space interconnects based on holographic optical elements can be used for short distance interconnections e.g. chip-to-chip or board-to-board and for optical permutation networks [7.1]. The main function of HOE in such interconnects is to image the source onto the detector sites simultaneously. Fig. 7.2 shows the schematic diagram of an optical interconnect which can be realized by using a transmission HOE. The HOE is a multiplexed holographic lens which produces multiple focussed spots. The scheme is highly flexible as any desired interconnection pattern can be realized with efficiency in excess of 50%. The interconnect of this type is suitable for clock distribution.

A free-space interconnect based on multifaceted transmission HOE is shown in Fig. 7.3. Each hologram facet focusses the light at a particular detector. One of the critical requirements of free-space focussing type interconnect is the high degree of alignment precision so that each spot strikes at the correct location on the chip. The location of the focussed spot can change due to drift in

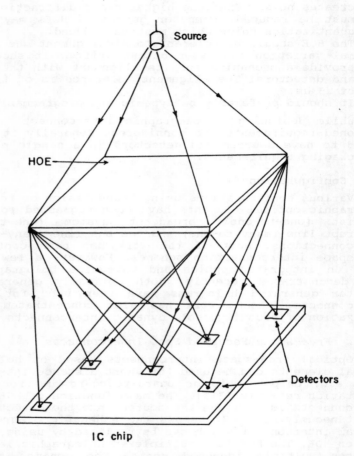

Fig. 7.2. Transmission holographic interconnect for clock distribution [7.1].

the laser wavelength. The diode lasers produce a wavelength shift of 0.25 nm/°K. Temperature stabilization within ±1°C is therefore necessary [7.20].

The wavelength stability requirements on the laser can be relaxed if collector lenses are used in the system [7.21-7.23]. The first HOE (Fig. 7.4) collimates the laser beam and directs it to another faceted HOE which brings the beams

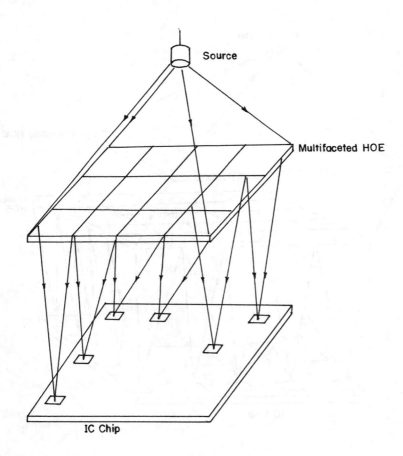

Fig. 7.3. Multifaceted transmission HOE interconnect.

to sharp focus on the IC chip [7.20, 7.21]. This doublet architecture allows much larger temperature variations and alignment tolerances. The doublet configuration can be made aberration free up to third order [7.21].

The interconnect density can be increased by increasing the number of facets and introducing fanout capability in the HOE. Such an architecture is depicted [7.20] in Fig. 7.5 which allows approximately 40,000 interconnects/cm^2. How-

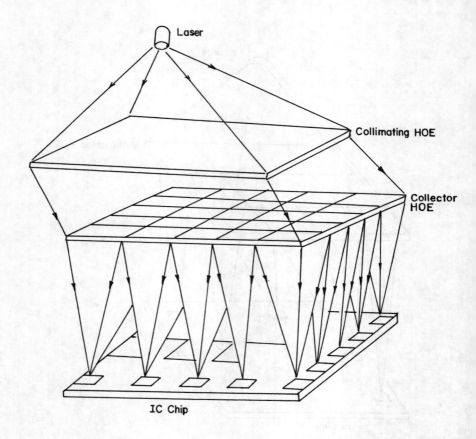

Fig. 7.4. Interconnection architecture employing collimator and collector HOEs [7.21].

ever, this will require the increase in the laser power by a factor equal to the fanout by each facet of the reflective HOE.

As the interconnect density increases, the fabrication and alignment become more complex. A holographic doublet configuration may be used for point-to-point interconnection with high density [7.24]. The basic unit of the interconnect is a holographic doublet which is a combination of a

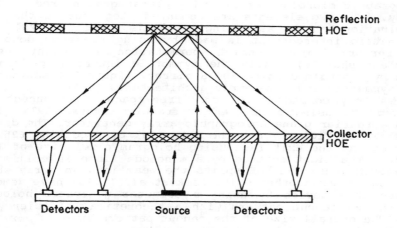

Fig. 7.5. Holographic interconnect architecture employing a collector HOE and a reflection HOE with fanout capability [7.20].

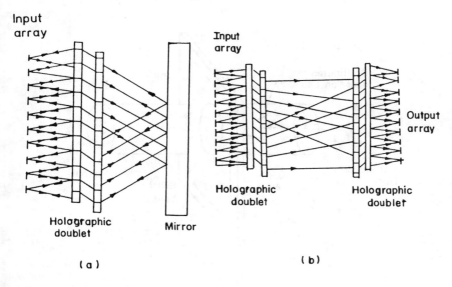

Fig. 7.6. Symmetric (a) and asymmetric (b) interconnects for point-to-point connections [7.24].

holographic microlens array and a planar grating redirection array. The two elements are cemented together so that the combination behaves as a single on-axis element [7.25]. An asymmetric interconnect is made by using two such doublets to perform point-to-point permutation of light beams (perfect shuffle), while the symmetric interconnect can be made by a single doublet and a mirror. Figure 7.6 shows both the symmetric and asymmetric architectures [7.24].

A very compact (0.01 cm^3) free-space interconnect can be made by using a gradient refractive index (GRIN) lens with a Fourier plane holographic array generator. The device as shown in Fig. 7.7 has the potential of producing 625x625 beamlets at the exit face using hologram pixel size of 1 μm [7.26]. The light emitted by a monomode fibre is collimated by a GRIN lens. The HOE splits the beam into an array which is focussed by another GRIN. Kirk et al. [7.26] have demonstrated this device by using a computer generated hologram giving 5x5 fanout pattern with a theoretical efficiency of 72%. The overall size of the fanout pattern was 70 μmx70 μm and the spot separation was about 17 μm.

The optical sources and detectors may be fabricated [7.1] on GaAs and Si chips, respectively and connected in a hybrid approach such that the light is generated along the edges of the Si chip as shown in Fig. 7.8. The reflection

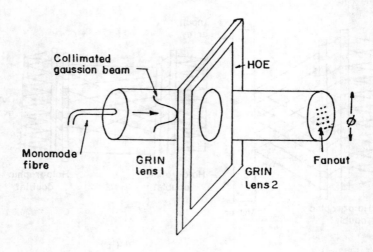

Fig. 7.7. Compact interconnect using GRIN lenses and a HOE [7.26].

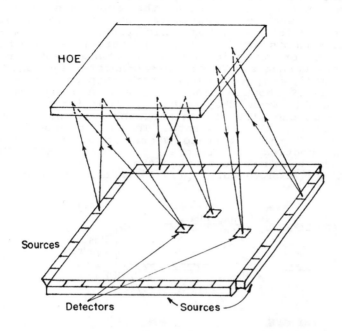

Fig. 7.8. Hybrid GaAs/Si approach for interconnection [7.1].

HOE routes the optical signal to appropriate locations. The detector-amplifier combinations can be fabricated on a Si chip.

The HOEs required for optical interconnects have been fabricated in dichromated gelatin [7.27, 7.28]. Computer generated holograms have also been used for multifaceted interconnects [7.20, 7.29]. If the HOEs are computer generated, then the scattered light must be minimized to avoid cross-talk.

7.6.2. Waveguide Hologram Interconnects

The free-space interconnects are limited by the problems connected with the packaging of optical systems. The optomechanical packaging is at best suitable for simple systems but inadequate in terms of size and robustness for complex systems [7.30]. The performance of the free-space holographic interconnect is sensitive to alignment, wavelength drift of the source and the image spreading effects.

566 LASERS AND HOLOGRAPHY

These problems may be solved by using folded optical systems with light propagation inside the substrate. The micro-optical components are fabricated on the glass substrate with holographic or lithographic techniques [7.29-7.36]. Such interconnects may be termed as substrate mode hologram interconnects or waveguide hologram interconnects or planar optics interconnects. These interconnects are useful for board-to-board, chip-to-chip or intra-chip communications.

Figure 7.9 shows a diagram of a planar optical interconnect. Light from the source is collimated by a HOE fabricated on one side of a glass substrate and directed at an angle within the substrate. The light reaches at the other face of the substrate where it is reflected by a reflecting coating and reaches to another HOE. The HOE focusses the beam onto the detectors.

Another architecture [7.37] which integrates HOEs,

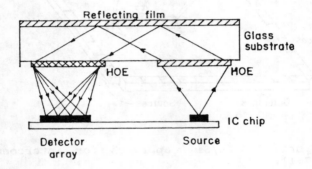

Fig. 7.9. Planar interconnect configuration [7.21].

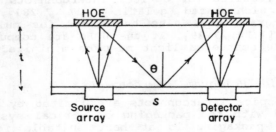

Fig. 7.10. Integration of HOEs, sources and detectors in a planar configuration [7.37].

sources and detectors on a glass plate is shown in Fig. 7.10. For a single reflection between the two HOEs, the distance S between the source and the detector is given by

$$S = 2t \tan\theta. \tag{7.14}$$

The glass thickness and the number of reflections between the two sides of the glass plate can be properly chosen to increase the distance S. It would be useful for chip-to-chip communications in a VLSI or in optical computers to integrate the passive optics with devices made of GaAs or other optoelectronic materials. Jahns et al. [7.37] have given a conceptual layout of a hybrid integration of an optical system on a silicon substrate (Fig. 7.11). The laser

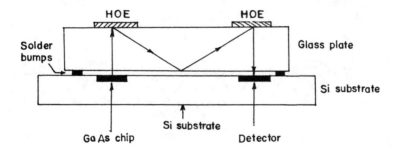

Fig. 7.11. Integration of optical system for interconnection on a silicon substrate [7.37].

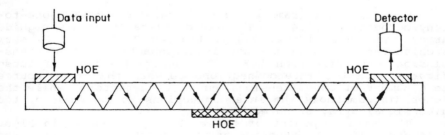

Fig. 7.12. Waveguide mode holographic interconnect [7.28].

chip and the detector array are recessed into the Si substrate.

In the waveguide mode or substrate mode optical interconnect the light is guided in a glass plate by total internal reflection [7.28, 7.33, 7.35, 7.38, 7.39]. The HOEs carry out collimation, focussing and coupling. An intermediate field lens (Fig. 7.12) allows some freedom from light tolerances on laser wavelength drift due to change in temperature [7.28]. It is possible to use polarization selective substrate mode holographic interconnects for implementing a reconfigurable optical bus with more than 500 nodes [7.40]. Figure 7.13 shows the architecture for a two directional holographic interconnect.

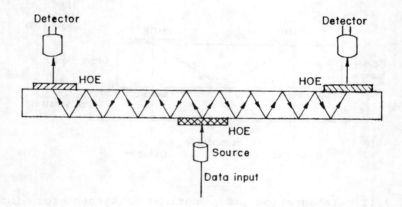

Fig. 7.13. Waveguide mode two directional holographic interconnect [7.28].

Waveguide holograms can also be used to provide one-to-many connections [7.41]. The device is based on a waveguide array generator [7.42, 7.43]. Figures 7.14 and 7.15 show architectures for one-to-many interconnections using transmission and reflection HOEs. The upper diagrams in these figures show the interconnections between the input source to a number of one dimensional array of detectors, while the lower diagram shows the interconnections to a number of two dimensional array of detectors.

The main component of the architecture shown in Figs. 7.14 and 7.15 is a waveguide array generator. The light from the source is guided into the glass plate by a HOE. The glass plate contains a second HOE (grating) which diffracts a part of the incident guided beam at each reflection.

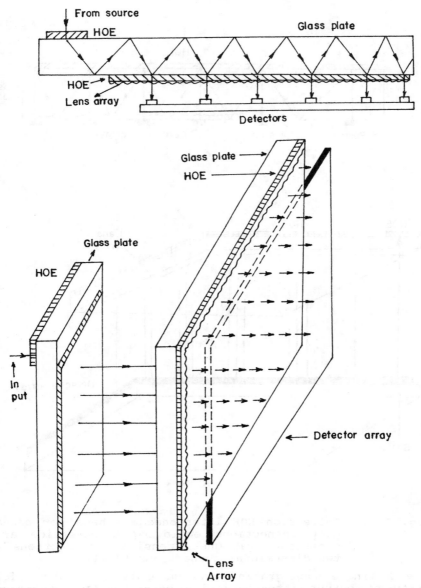

Fig. 7.14. Transmission HOE interconnect scheme for one-to-many connections employing a waveguide array generator. (a) One dimensional connections, (b) two dimensional connections [7.41].

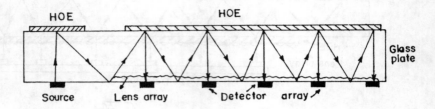

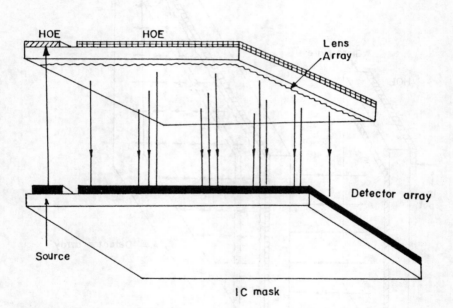

Fig. 7.15. Reflection HOE interconnect scheme for one-to-many connections employing a waveguide array generator. (a) One dimensional connections (b) two dimensional connections [7.41].

A single such grating will generate a one dimensional array of output interconnection beams. A two dimensional array of interconnection beams can be generated by using two one dimensional devices with their lines normal to each other as shown in Fig. 7.14. The first device generates an array of one dimensional beams each of which enters the second device and produces a two dimensional array of interconnection beams.

A second approach is to fabricate a double-exposure holographic grating such that two gratings have their lines normal to each other. A lens array is placed before the detector array to focus the beams onto the detectors. In principle, the holographic recording technique and photolithographic technique should be able to give a practicable interconnect, however, obtaining beams of uniform intensity is a challenging task. It may be difficult to fabricate a grating with a control over its diffraction efficiency in two dimensions. Discrete HOEs at predetermined locations with required diffraction efficiency can produce an optical interconnect with focussed spots of equal intensity [7.44].

Takeda and Kubota [7.45] have suggested to use a waveguide array generator as a key element for free-space optical interconnect. The grating used for such an application is a surface relief type so that array of beams are generated on both sides of the waveguide. As shown in Fig.

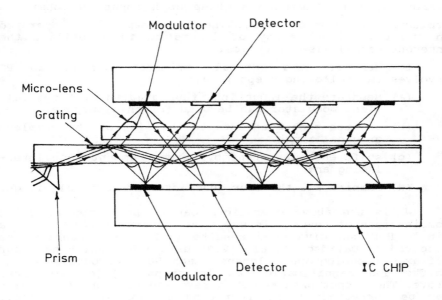

Fig. 7.16. Interconnection architecture employing a waveguide array generator which produces array of beams on both sides of the waveguide [7.45].

7.16, the array generator is sandwiched by two IC chips. At the two outer surface of the array generator microlens arrays are fixed at appropriate locations. The laser beams are focussed on the reflective type of laser modulators which are fabricated on the chip surface. The reflected modulated beams are collimated and focussed by microlens arrays on the photodetectors.

7.6.3. Perfect Shuffle Interconnects

Perfect shuffle (PS) transform and its inverse (PS^{-1}) are two important interconnect schemes. The PS operation divides a given group into two halves and then interlaces them. The neighbouring pairs may be interchanged if desired [7.46]. The PS can be described by [7.47]

$$n' = \begin{cases} 2n & \text{if } 0 \leq n < N/2 \\ 2n-N+1 & \text{if } N/2 \leq n < N \end{cases}, \qquad (7.15)$$

where n' is the new location of the nth input. The PS can be performed $3\log_2 N$ times on a group of N input to obtain an arbitrary permutation [7.9]. The transform can be operated on two dimensional array of inputs to fully utilize the inherent parallelism of optics.

In general, the optical implementation of 2D PS involves the following steps [7.9]:

(a) Separate the input into four quadrants. A spatial light modulator can be used for this purpose,

(b) stretch each quadrants to the size of the complete original input,

(c) shift each quadrant by proper amount for interlacing and

(d) combine all the four quadrants into a single image.

While the above operations can be performed by using combinations of several lenses and prisms [7.47, 7.48], a single HOE can give good results [7.9]. The technique as proposed by Davidson et al. [7.9] uses a HOE which has four off-axis subholographic lenses. Each HOE introduces proper shifts and magnification as in steps (b)-(d) mentioned above. The output obtained is a reversed version of PS which can be corrected by incorporating a lens in the system. Figure 7.17 shows the result of the PS operation on an input [7.9]. It is also possible to implement PS^{-1} using a HOE.

7.6.4. Dynamic Holographic Interconnects

Dynamic reconfigurable interconnections are desirable

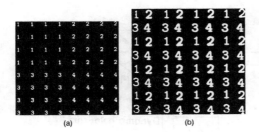

Fig. 7.17. Perfect shuffle using a HOE. (a) Input, (b) output [7.9].

for providing connections between processing elements and memories or to other processing elements. In a small computer, connection between each processing element can be accomplished by fixed interconnects. But for the futuristic parallel computers this scheme is not viable because of astronomical number of connections required. The solution lies in dynamic interconnection network which will establish interconnections only when they are needed. The network may be a single bus type or of a complex crossbar type. The dynamic interconnections will also ease out the responsibility of software for reconfiguration in the case when all fixed type connections are made as is done in present electronic computers. The dynamic holograms will reconfigure the circuits and handle the computational problem at hand at any instant. The interconnection pattern can be changed as per requirement during computation. The dynamic interconnects will also be used in learning machines.

Real-time hologram recording materials are useful for dynamic interconnects. The electro-, acousto- or magneto-optic spatial light modulators form the active switching elements. Figure 7.18 shows the use of two photorefractive crystals for realizing a dynamic holographic interconnect [7.49]. The two crystals are situated above an array of IC chips. The bottom crystal is used as a hologram storage medium and the top crystal as a phase conjugate mirror. The device works in two steps:

Step 1: Source 1, in the low light level mode, illuminates the phase conjugate mirror after transmitting through the lower photorefractive crystal. The phase conjugator returns an amplified phase conjugate beam back towards the lower crystal.

Step 2: Source 1 is turned off. Source 2 is then turned on to a high light level. The light from the source 2 will interfere with the decaying phase conjugate beam. This

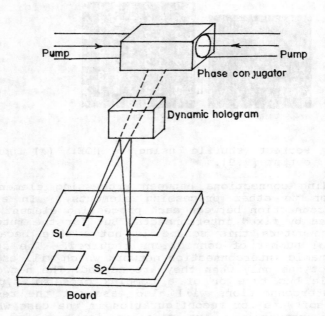

Fig. 7.18. A scheme for dynamic holographic interconnect [7.49].

interference pattern is recorded as a reflection hologram in the crystal, thus creating the required connection between the two points.

The scheme requires that all sources must be mutually coherent. A single source with fibre distribution arrangement can serve as different sources. It is also required that the detectors are located in close proximity to the sources. It may be possible to design a single element which can serve as a source and a detector.

Large interconnections can be achieved by recording multiplexed holograms in photorefractive crystals. The difficulty with the photorefractive materials is their erasure during recording and reading if the holograms are not fixed. This problem can be somewhat tackled by using exponentially decreasing exposures [7.50]. This method, however, is not good for making large interconnections. The erasure can be minimized by recording spatially separated holograms in the crystal [7.51]. Rastani and Hubbard [7.52] have suggested an architecture that uses spatially multiplexed beams and angularly multiplexed input beams to record multiplexed holographic gratings in BSO crystal.

Figure 7.19 shows the schematic diagram for a large interconnect architecture. An array of NxN input beams is generated by using Fresnel lens array or Dammon grating. A spatial light modulator (SLM-1) controls the configuration of the input beams. The beams are collimated and allowed to interfere with reference beam in BSO. The spatial light modulator SLM-2 masks most of the beams to reduce the crystal erasure problem. The reference beam is an array of N xN array of spatially separated collimated beams passed by the spatial light modulator (SLM-3).

The recording is done by turning on a single pixel on SLM-2 and a single pixel on SLM-3. Thus N^2 gratings are recorded in the crystal due to the interference between the angularly multiplexed input beams and the single reference beam. A readout beam which can be allowed by the spatial light modulator SLM-4 can reconstruct the phase conjugate of the input beams which can be focussed on detector array by using a beam splitter and a lens.

Different pixels on SLM-2 and SLM-3 are activated and the corresponding gratings can be recorded on different

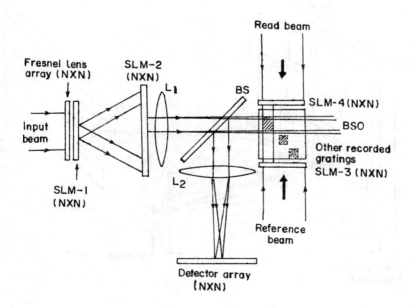

Fig. 7.19. Large dynamic interconnect [7.52].

portions of the crystals. Thus SLM-2, SLM-3 and SLM-4 provide a shield to the photorefractive crystal to protect it from undesirable exposures and thereby eliminating the erasure problem.

Another approach is to use switchable masks before a multifaceted HOE. The light beam from an emitter is spread over the entire HOE. Any connection can be established by switching the suitable mask so that the corresponding hologram can be used for connection. Electro-optical or magneto-optical switches can be used as masks [7.53].

Schulze [7.54] has produced dynamic holograms whose interference patterns can be changed dynamically by using very high resolution spatial light modulators based on liquid crystal layers or E-O crystal (like KDP) layers. The layer is sandwiched between two parallel electrodes, one of which is transparent and homogeneous while the other carries the structure of the desired interference pattern. When electronic field is applied between the electrodes, the field shows the inhomogeneity similar to the electrode

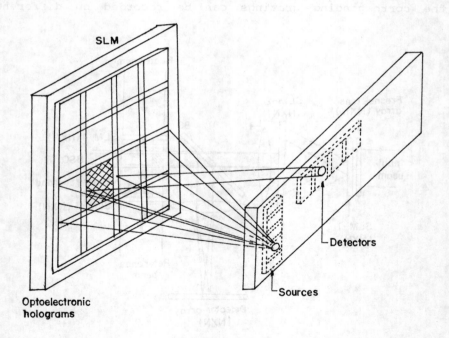

Fig. 7.20. Optical crossbar switch employing opto-electronic HOEs [7.54].

structure. An alternate approach is to introduce a photoconductor (amorphous silicon) layer with the optically active layer (liquid crystal or E-O layer) in between two homogeneous electrodes. When the photoconductor is illuminated with suitable interference pattern, then its conductivity and electric field change accordingly [7.54]. The holograms recorded in such a spatial light modulator are known as optoelectronic holograms because their interference patterns are controlled optically and electronically.

Figure 7.20 shows an optical crossbar switch using optoelectronic holograms [7.54]. Light from the source is expanded to illuminate one row of holograms of the matrix. The light beam can be directed to the desired detector by turning on the appropriate hologram. If a hologram of size 1 mm×1 mm is taken, then a 32×32 matrix would require a hologram of size 32 mm×32 mm. All the connections can be controlled dynamically.

Figure 7.21 shows the schematic diagram of a device which may be used for reconfigurable holographic interconnect [7.55]. Basically it consists of a spatial light modulator as a shutter array which can be programmed to operate a desired element of a HOE. The prerecorded multi-

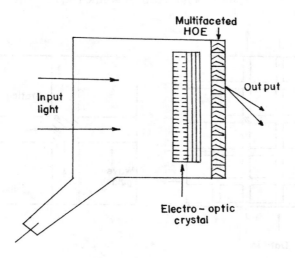

Fig. 7.21. Device for reconfigurable holographic interconnect [7.55].

faceted HOE is mated to the shutter array.

An integrated optic dynamic switch array based on photorefractive effect has been realized by Aronson and Hesselink [7.56] for chip-to-chip communication. This type of interconnect has potential for large scale integration of the order of 100x100 switches/cm^2. Fig. 7.22 shows the principle of the switch. A two dimensional array of single mode waveguides is formed on a lithium niobate crystal by Ti indiffusion technique. At the intersections of crossing waveguides, iron (Fe) is locally doped to achieve the photo-refractive effect in $LiNbO_3$ [7.57]. The grating is written at the crossing waveguides by guiding two short wavelength beams from outside (Fig. 7.22b). The signal is at longer wavelength at which the material is not photorefractive. The resulting gratings switch signals from guide-to-guide. A spatial light modulator can control the connections by switching on the desired intersection.

The main advantage of this architecture is that straight waveguides (without any bends) are used for the interconnect. The interconnection density can be very high as the geometry makes use of the substrate area effectively.

The integrated optic dynamic switch array should use a material with large electro-optic coefficient, fast response time and a low loss waveguide technology. The materials

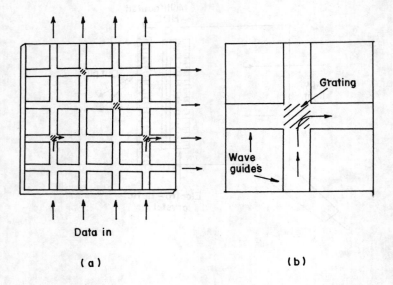

Fig. 7.22. Integrated optic dynamic switch array [7.56].

such as GaAs or GaAs/AlGaAs systems satisfy these requirements. GaAs requires only a writing time of 20 μs at intensities of 4 W/cm^2 [7.58].

Pinhole hologram (section 4.14.5) have the potential for programmable holographic interconnect. Several interconnection patterns can be recorded, each through a separate pinhole [7.59]. A pixelated spatial light modulator can be used in the plane of the reconstructed image of the pinholes. The pixels can be opened in a programmable manner to achieve a particular interconnection.

Jenkins and Sawchuk [7.60] showed a 3D dynamic interconnection architecture for connecting 2D NxN arrays in parallel without interconnection weights. Caulfield [7.61] was the first to give an architecture for n^4 interconnects, each of which can be associated with proper weights. The scheme consisted of a NxN reflective SLM, a beam splitter and a lens. Since then many variations of this architecture have been proposed.

7.7. Challenges

The present developments in the area of optical interconnects show that optics will become an integral part of future high throughput computers. Holographic optical elements will be important constituents of optical interconnects as routing elements as these can provide 10^{15}-10^{18} interconnections per second. To achieve the goal, improvements in HOE, both optically recorded and computer generated, are needed. The present recording materials limit the diffraction efficiency and angular resolution of superimposed exposures in the same hologram. Improved holographic materials with better thickness and refractive index modulation are needed [7.10, 7.62]. Photorefractive materials appear to be the most suited for making reconfigurable interconnects. The dynamic interconnect will be especially important for neural networks where learning in interconnect must be associated with a weighting factor. Research needs to focus on materials to achieve faster response (<1 ms), greater sensitivity and control over erasure.

Besides above requirements in the area of HOEs, the development in other areas are also needed in order to realize the full advantages of optical interconnects. The lower power consumption, high-speed, reliability and low cost will be achieved by monolithic integration of electronic and optoelectronic devices. The optoelectronic *packaging* compatible with high-speed electronics is another challenge. The requirement is to incorporate both active (e.g. optical source) and passive (e.g. HOEs) components into a small package to ensure short and long term stability and robustness of the device [7.63].

References

7.1 J.W. Goodman, F.I. Leonbergers, S. Kung and R.A. Athale, *Proc. IEEE* **72** (1984) 850.

7.2 B.K. Jenkins, P. Chavel, R. Forcheimer, A.A. Sawchuck and T.C. Strand, *Appl. Opt.* **23** (1984) 3465.

7.3 H.-I. Jeon and A.A. Sawchuk, *Appl. Opt.* **26** (1987) 261.

7.4 K.-H. Brenner and A. Huang, *Appl. Opt.* **27** (1988) 135.

7.5 I. Glasser and L. Perelmutter, *Opt. Lett.* **11** (1986) 53.

7.6 R. Arrathoon and S. Kozaitis, *Opt. Lett.* **12** (1987) 956.

7.7 J.A. Fried, *Opt. Eng.* **25** (1986) 1132.

7.8 M.E. Marhic, Y. Birk and F.A. Tobagi, *Opt. Lett.* **10** (1985) 629.

7.9 N. Davidson, A.A. Friesem and E. Hasman, *Appl. Opt.* **31** (1992) 1810.

7.10 S.A. Collins, Jr. and H.J. Caulfield, *J. Opt. Soc. Am.* **A 6** (1989) 1568.

7.11 L. Bergman, A. Johnston, R. Nixon, S. Esener, C. Guest, P. Yu, T. Drabik, M. Feldman and S.H. Lee, *Proc. SPIE* **625** (1986) 117.

7.12 A.A. Sawchuk and T.C. Strand, *Proc. IEEE* **72** (1984) 758.

7.13 M.W. Derstine and A. Guha, *Opt. Eng.* **28** (1989) 434.

7.14 H. Bartelt and S.K. Case, *Appl. Opt.* **24** (1985) 2051.

7.15 D. Psaltis and J. Hong, *Opt. Lett.* **11** (1986) 812.

7.16 P.R. Haugen, S. Rychnovsky, A. Hussain and L.D. Hutcheson, *Opt. Eng.* **25** (1986) 1076.

7.17 R.K. Kostuk, J.W. Goodman and L. Hesselink, *Appl. Opt.* **25** (1986) 4362; **26** (1987) 3947.

7.18 D.A.B. Miller, *Opt. Lett.* **14** (1989) 146.

7.19 A.E. Chiou and P. Yeh, *Appl. Opt.* **31** (1992) 5536.

7.20 E. Bradley, P.K.L. Yu and A.R. Johnston, *Opt. Eng.* **28** (1989) 201.

7.21 D. Prongue and H.P. Herzig, *Proc. SPIE* **1281** (1990) 113.

7.22 J.W. Goodman, F. Leonberger, S.-Y. Kung and R. Athale, *Proc. IEEE* **72** (1984) 850.

7.23 E. Bradley and P.K.L. Yu, *Proc. SPIE* **835** (1988) 298.

7.24 E.J. Restall, B. Robertson, M.R. Taghizadeh and A.C. Walker, in *3rd IEE Conf. on Holographic Systems,*

Components and Applications (1991) p. 127.

7.25 K.S.S. Rao, C.Bhan and P.C. Mehta, in *Optoelectronic Imaging*, eds D.P. Juyal, P.C. Mehta and M.K. Sharma (Tata McGraw-Hill, India 1987) p. 460.

7.26 A.G. Kirk, H. Imam and T.J. Hall, in *3rd IEE Conf. on Holographic Systems, Components and Applications* (1991) p. 161.

7.27 E.J. Restall, I.R. Redmond and A.C. Walker, in *3rd IEE Conf. on Holographic Systems, Components and Applications* (1991) p. 40.

7.28 H. Kobolla, N. Lindlein, O. Falkenstörefer, S. Rosner, J. Schmidt, J. Schwider, N. Streibl and R. Völkel, in *3rd IEE Conf. on Holographic Systems, Components and Applications* (1991) p. 123.

7.29 C. Sebillotte and T. Lemoine, in *3rd IEE Conf. on Holographic Systems, Components and Applications* (1991) p. 52.

7.30 J. Jahns and A. Huang, *Appl. Opt.* **28** (1989) 1602.

7.31 F. Sauer, *Appl. Opt.* **28** (1989) 386.

7.32 K.-H. Brenner and F. Sauer, *Appl. Opt.* **27** (1988) 4251.

7.33 R.K. Kostuk, M. Kato and Y.-T. Huang, *Appl. Opt.* **28** (1989) 4939.

7.34 J. Jahns, *Appl. Opt.* **29** (1990) 1998.

7.35 R.K. Kostuk, Y.-T. Huang, D. Metherington and M. Kato, *J. Opt. Soc. Am.* **A 28** (1989) 4939.

7.36 M. Kato, Y.-T. Huang and R.K. Kostuk, *J. Opt. Soc. Am.* **7** (1990) 1441.

7.37 J. Jahns, Y.H. Lee, C.A. Burrus Jr. and J.L. Jewell, *Appl. Opt.* **31** (1992) 592.

7.38 H.J. Haumann, H Kobolla, F. Sauer, J. Schwider, W. Stork, N. Streibl and R. Völkel, *Proc. SPIE* **1319** (1990).

7.39 J. Jahns and W. Däschner, in *Optical Computing* (OSA Technical Digest Series, 1991) **6** p. 29.

7.40 R.K. Kostuk, M. Kato and Y.-T. Huang, *Appl. Opt.* **29** (1990) 3848.

7.41 P.C. Mehta, Unpublished Work.

7.42 T. Kubota and M. Takeda, *Opt. Lett.* **14** (1989) 651.

7.43 A.N. Kaul, D. Mohan and P.C. Mehta, in *Proc. Emerging Optoelectronic Technologies*, eds. A. Selvarajan, B.S. Sonde, K. Shenai and V.K. Tripathi (Tata-McGraw Hill,

India, 1992) p. 209.

7.44. P.C. Mehta, in *Optoelectronics: Technologies and Applications* eds. A. Selvarajan, K. Shenai and V.K. Tripathi (SPIE Optical Engineering Press, Washington, 1993) p. 280.

7.45 M. Takeda and T. Kubota, *Appl. Opt.* **30** (1991) 1090.

7.46 K.H. Brenner and A. Huang, *Appl. Opt.* **27** (1988) 135.

7.47 H.S. Stone, *IEEE Trans. Comput.* **C-20** (1971) 153.

7.48 A.W. Lohmann, *Appl. Opt.* **25** (1986) 1543.

7.49 J. Wilde, R. McRuer, L. Hesselink and J.W. Goodman, *Proc. SPIE* **752** (1987) 200.

7.50 K Blotekjaer, *Appl. Opt.* **18** (1979) 57.

7.51 A. Marrakchi, W.M. Hubbard, S.F. Habiby and J. S. Patel, *Opt. Eng.* **29** (1990) 215.

7.52 K. Rastani and W.M. Hubbard, *Appl. Opt.* **31** (1992) 598.

7.53 W.E. Ross, D. Psaltis and R.H. Anderson, *Proc. SPIE* **341** (1982) 191.

7.54 E. Schulze, *Proc. SPIE* **862** (1987) 50.

7.55 J.A. Neff, *Photonics Spectra* (Aug 1989) 135.

7.56 L.B. Aronson and L. Hesselink, *Proc. SPIE,* **1212** (1990) 304.

7.57 V.M. Shandarov and S.M. Shandarov, *Sov. Tech. Phys. Lett.* **12** (1986) 20.

7.58 M.B. Klein, *Opt. Lett.* **9** (1984) 350.

7.59 S. Xu, G. Mendes, S Hart and J.C. Dainty, *Opt. Lett.* **14** (1989) 107.

7.60 B.K. Jenkins and A.A. Sawchuk, *J. Opt. Soc. Am.* **A 2** (1985) 26.

7.61 H.J. Caulfield, *Appl. Opt.* **26** (1987) 4039.

7.62 S.A. Collins, Jr., S.F. Habiby and P.C. Griffith, *Proc. SPIE* **625** (1986) 167.

7.63 B.K. Jenkins, S H. Lee and C.L. Giles, *Appl. Opt.* **31** (1992) 5423.

CHAPTER 8

INFORMATION PROCESSING

Lasers and holograms play an important role in coherent optical information processing. The ability to record both amplitude and phase of an optical wave has made holograms ideal spatial filters for performing various operations such as matched filtering, inverse filtering, differentiation, addition, subtraction, convolution and correlation. Both linear, space invariant and non-linear, space variant processing techniques employ principles of holography. The holographic lenses can be used as elements for Fourier transformation in a coherent image processor. Holographic information storage techniques promise high density memory devices. The current trends show that holograms will be used as key elements in optical computers. This chapter discusses various applications of holograms in optical data processing, image sharpening, image subtraction, pattern recognition, image multiplication and memories.

8.1. Optical Data Processor

Optical processors process the information encoded in light beams. The analogue optical processor utilizes the ability of a lens to obtain Fourier transform (FT) and convolution to achieve advanced linear algebra operations. The lenses have the simultaneous capability of imaging and processing. The Fourier transform operation is carried out in 2D simultaneously for all input points with the speed of light. The lens may be considered to have stored in it the required 2D FT processing programme [8.1, 8.2].

Figure 8.1 shows the basic system for a general optical image processor. P_1 is the input plane consisting of 2D light amplitude distribution $f(x_1,y_1)$, whose Fourier transform $F(u,v)$ is displayed by the lens L_1 at the plane P_2 and is given by

$$\mathcal{F}[f(x_1,y_1)] = F(u,v)$$

$$= \iint f(x_1,y_1)\exp[-2\pi j(ux_1 + vy_1)]dx_1 dy_1, \qquad (8.1)$$

where (u,v) are the spatial frequencies of the input object given by

$$(u,v) = \left(\frac{x_2}{\lambda f}, \frac{y_2}{\lambda f}\right), \qquad (8.2)$$

and f is the focal length of the FT lens. Eq. 8.1 is the familiar FT relationship. The optical FT has the following properties/characteristics:

(a) The amplitude of the spectrum is independent of the input location.

(b) The spectrum is not changed by the inversion $(x,y) \to (-x,-y)$.

(c) The location of the spectrum is independent of the location of the corresponding spatial frequency in the input.

(d) The amplitude of a spatial frequency in the FT spectrum gives information about the amount of that particular spatial frequency present in the input.

(e) Optical signals that change gradually (i.e. gross features, background etc.) have low spatial frequency spectrum, whereas those signals that change drastically over short distances (i.e. fine details, edges etc.) have higher spatial frequencies.

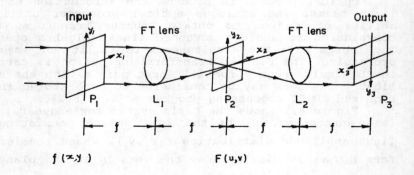

Fig. 8.1. Schematic diagram of an optical image processor.

It is thus clear that the spectrum in the plane P_2 gives valuable information about the light amplitude distribution in the input plane P_1.

We can place a filter function $T(u,v)$ at P_2. The lens L_2 displays the FT of the product $F(u,v)\,T(u,v)$ at the plane P_3. The plane P_3 will show the original or modified object depending on the filter function. $T(u,v)$ can be a function affecting amplitude or phase or both of the light distribution in the plane P_2. If, for example,

$$T(u,v) = G^*(u,v), \qquad (8.3)$$

one can obtain the optical correlation at the output plane P_3.

$$\mathcal{F}[F(u,v)G^*(u,v)] = f * g, \qquad (8.4)$$

where $*$ denotes correlation operation and $G^*(u,v)$ is the complex conjugate of $G(u,v)$. This equation is useful for pattern recognition as discussed in section 8.4.

It is evident from the above considerations that optical data processing is a mathematical process which consists in a complex multiplication of the spectrum of the input by the filter function.

Holograms can be used in two ways in an optical data processor. Firstly, they can be used as complex spatial filters which may be optically recorded or computer generated [8.2]. Secondly, holographic lenses can be used as Fourier transform lenses in the image processing setup [8.3, 8.4].

The holographic lenses for use as FT elements may be recorded by a plane wave and a spherical wave. The diameter of the collimated wave limits the aperture of the lens. If the emulsion side of the hologram is aluminized, the reconstructed wavefront is reflected from the emulsion without passing through the glass substrate of the hologram. This allows some relaxation on the material and quality of the glass base of the holographic lens. The holographic lenses can also be of Lippmann type. The emulsion shrinkage should be carefully compensated for to avoid any changes in the angle of the diffracted beams.

The astigmatism in the images due to emulsion shrinkage may be minimized by recording the hologram such that the object and the reference beams make equal angles with the normal to the photographic plate. The HOEs will not show any chromatic aberration if the light source for the optical data processor is the laser of the wavelength for which the

HOEs are optimized. The other aberrations can be minimized by accurately aligning the HOEs in the data processor.

Figure 8.2 shows the FT spectra of a circular aperture, a benzene ring structure, muslin cloth and coarse cloth produced by a reflective holographic lens [8.5]. Figure 8.3 compares the Fourier spectra of annular apertures obtained in the first diffraction order with the diffraction pattern of the objects obtained in the zero order of the holographic

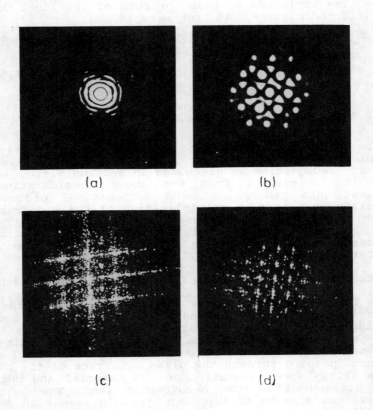

Fig. 8.2. FT spectra of a circular aperture (a), a benzene ring structure (b), a muslin cloth sample (c) and a coarse cloth sample (d) produced by a reflective holographic lens [8.5].

Fig. 8.3. Comparison of Fourier spectra of annular apertures with the diffraction pattern obtained in the zero order of a holographic lens [8.5].

lens [8.5]. A comparison of the low exposure and overexposed photograph of the FT spectrum reveals the extent of halos and flare light (Fig. 8.4). Figures 8.2-8.4 show that a holographic lens can produce Fourier spectra of high quality with few halos and little flare light.

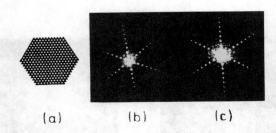

Fig. 8.4. Comparison of a low exposure (b) and an overexposed (c) photograph of the FT spectrum of the object (a) [8.5].

The schematic diagram of an optical data processor using holographic lenses are shown in Fig. 8.5. The input object transparency to be processed is illuminated by a collimated beam of coherent light. First holographic lens L_1 is positioned at a distance equal to its focal length from the object which produces the Fourier transform spectrum of the object at the focal plane (filter plane). The holographic lens L_2 is placed at a focal length from the filter plane and produces an inverted image of the object at the image plane. The filtering is carried out by inserting the desired filter function in the filter plane. The optical data processor employing transmission holographic lenses (Fig. 8.5a) requires a distance of 4f between the input plane and the image plane, similar to glass lens based optical data processor (Fig. 8.1). The holographic reflection lenses make the processor compact (Fig. 8.5b) as it requires one-half of the length (2f) of the optical bench as required for the data processor using glass lenses.

The HOE based optical data processors have been used for various filtering experiments [8.3-8.5], viz. raster/halftone removal, frequency doubling, separation of two signals, contrast enhancement, directional filtering etc. Figures 8.6-8.8 show the image processing results obtained by an optical data processor employing holographic reflective lenses [8.5]. Figure 8.6 shows the effect of a binary filter on an object containing periodic structure. The spectrum consists of discrete components of the fundamental frequency of raster and its harmonics. A binary filter, in this case simply an aperture, blocks all but one of the frequency components and removes the undesired raster pattern. It is observed that the raster pattern has been completely removed.

Two superimposed signals can be separated by filtering out the Fourier spectrum of the undesired signal. The object consisted of a photograph of a baby superimposed with a

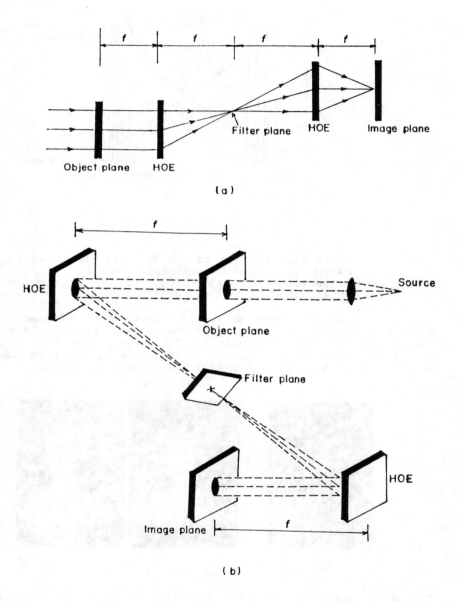

Fig. 8.5a,b

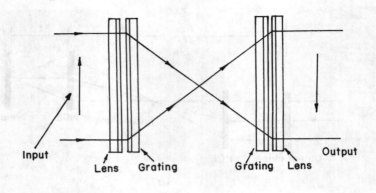

(c)

Fig. 8.5. Schematic diagram of an optical data processor using HOEs. (a) Transmission geometry [8.3], (b) reflection geometry [8.5], (c) on-axis geometry [8.8].

text. The Fourier spectrum of the binary text was recorded on a film and accurately positioned in the filter plane to obtain the image of the baby in the image plane (Fig. 8.7). It is observed from Fig. 8.7 that there is some residual noise that is a result of misalignment of the filter.

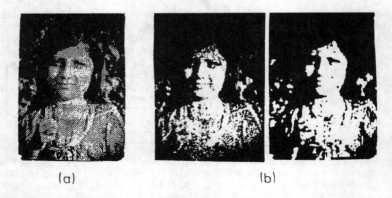

Fig. 8.6. Removal of raster pattern. (a) input (b) output [8.5].

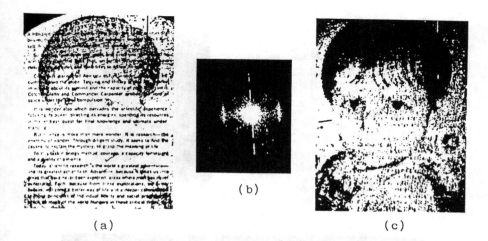

Fig. 8.7. Filtering of undesired signal from two superimposed signals. (a) Input image, (b) spectrum, (c) filtered image [8.5].

The image contrast can be improved or even reversed (Fig. 8.8) by suitably attenuating or removing the low order frequencies of the spectrum of the object. The filter for this is simply an amplitude filter having a maximum density at the centre and a gradual decrease in density towards the higher frequency region. The Fig. 8.8 also shows the effect of directional filtering.

The results of spatial filtering experiments show that an optical data processor employing HOEs can perform various analogue image processing operations. Because the reflective HOE based data processor requires only one-half of the space required by a processor using glass lenses or transmission HOEs, it may be used for onboard data processing on spacecrafts and satellites. The most distinguishing characteristics of such a processor is its extreme light weight. Since the angular alignment tolerances for holographic lenses are very low [8.5, 8.6], the processor must be made rugged, designed and mounted to withstand vibrations, shocks and other environmental problems associated with spacecrafts and satellites [8.7]. Parabolic mirrors can also fold the system but they are quite heavy and the weight of those to be used in a spacecraft laboratory would exceed 0.9t [8.7].

A HOE based optical data processor offers the advantage of processing in several channels. This can be achieved in two ways. Firstly, it is possible to achieve simultaneous processing of the same input signal by using different

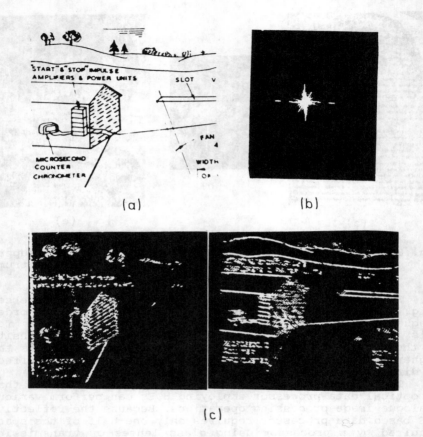

Fig. 8.8. Contrast reversal and directional filtering. (a) Object, (b) spectrum, (c) filtered image [8.5].

diffraction orders of the holographic lens [8.5]. The higher orders in a volume holographic lens are, however, absent. Secondly, the holographic lens can be multiplexed using carrier multiplexing technique.

A holographic lens can be combined with a holographic grating of equal spatial frequency to obtain on-axis HOE. Two such combinations can be used to form an optical data processor [8.8] as shown in Fig. 8.5c. The lens-grating combination is not fully corrected for chromatic aberration. Hence such a HOE would produce wavelength dependent on-axis Fourier planes, providing flexibility in handling various spatial frequencies of different colour components of an input colour transparency. The arrangement of Fig. 8.5c can also be used for producing colour zooming.

A multiplexed holographic lens can be recorded [8.9] which would have equal focal lengths at 633, 514 and 488 nm. Such a HOE can be used for restoration of information from faded colour photographic transparencies and pseudocolouring of black and white photographs.

8.2. Image Sharpening (Deblurring)

Holographic techniques can be used to sharpen blurred photographs [8.10]. The blurring may be accidental or due to instrumental imperfections. The accidental blurring results from defocussing of the imaging instrument, linear motion of the object and/or camera and the presence of atmospheric turbulence between the object and the camera.

The Fourier transform of a δ-function is a constant for all spatial frequencies. In the case of defocussing, the Fourier transform of the spread function is no more a constant for all spatial frequencies. The effect of this would be a decrease in the contrast of the image and a decrease of resolution at higher frequencies.

The blurring process results in an attenuation, suppression or dislocation of certain spatial frequencies. By looking at the photograph, it may appear that there is a irretrievable loss of information. A perfect image can be recorded by an imaging system only if its point-spread function (impulse response) is a Dirac delta function. In other words, a perfect imaging system images a point object as a sharp point. The presence of motion, defocus and atmospheric turbulence modifies the point-spread function. In the case of a defocus, a sharp point on the object is imaged as a blurred point. Therefore all the points of the object are recorded as blurred points. The net result of this is to produce a defocussed image. In a linear motion, the sharp points are imaged as sharp lines which result in a blurred image. It is thus clear that the object information is available in a blurred or defocussed image in a coded form. The holographic techniques decode this information to retrieve a sharp image. Two holographic techniques are described below to achieve this aim.

8.2.1. Fourier Transform Division Filter Method (Inverse Filter Method)

The imaging operation can be described as a convolution integral given by

$$g(x',y') = \int\int_{-\infty}^{\infty} f(x,y) h(x'-x, y'-y) \, dx \, dy, \qquad (8.5)$$

where $g(x',y')$ is the amplitude transmittance of the blurred photograph, $f(x,y)$ is the amplitude transmittance of the object and $h(x'-x, y'-y)$ is the point-spread function of the imaging system.

Eq. 8.5 can be written in the symbolic form as

$$g = f * h, \qquad (8.6)$$

where * denotes the convolution operation.

If we take the Fourier transform of Eq. 8.6, the convolution integral takes the form of the product of the Fourier transforms

$$\mathcal{F}[g] = \mathcal{F}[f * h]$$

$$G[u,v] = F[u,v] \, H[u,v], \qquad (8.7)$$

where G, F and H are the Fourier transforms of the function g, f and h respectively, and are given by

$$\mathcal{F}[g] = G(u,v) = \iint g(x',y') \exp[-2\pi j(ux + vy)] dxdy,$$

$$\mathcal{F}[h] = H(u,v) = \iint h(x,y) \exp[-2\pi j(ux + vy)] dxdy,$$

$$\mathcal{F}[f] = F(u,v) = \iint f(x,y) \exp[-2\pi j(ux + vy)] dxdy. \qquad (8.8)$$

Eq. 8.7 reveals that the true image F can be retrieved by

$$F(u,v) = \frac{G(u,v)}{H(u,v)}. \qquad (8.9)$$

Therefore if a filter with the transmittance function $T(u,v)$ given by

$$T(u,v) = 1/H(u,v) \qquad (8.10)$$

is inserted in the Fourier plane of the image processing system, then output plane will show the sharp image.

The filter $T(u,v)$ is known as Fourier transform division filter. This can be realized as a two-part sandwich filter

$$T(u,v) = \frac{H^*(u,v)}{|H(u,v)|^2} \qquad (8.11)$$

$$= T_1(u,v) T_2(u,v).$$

The two components of the filter, $H^*(u,v)$ and $|H(u,v)|^{-2}$ are realized separately starting from the point-spread function of the system.

The blurred transparency to be used as the input in the image processing setup should be linear in amplitude transmittance. This can be achieved by preparing the input as a positive transparency, processed such that the product of the values of γ for the negative and positive transparencies is equal to 2. The γ in each step should be measured by a densitometer.

The next step is to infer the point-spread function $h(x,y)$ of the imaging system. This can be done by looking at the blurred photograph carefully and searching for the cause of blurring, whether it is due to defocus or motion blur. Always look for the image of a point in the scene (*a priori* information about the scene is helpful). A line image would imply a motion blur, the direction of line showing the direction of motion. A blurred point image will be due to defocus.

Mathematically, the function H can be represented as

$$H(\psi) = 1 - a\psi^2/\psi_{max}^2, \text{ for blurred image} \qquad (8.12)$$

and

$$H(\psi) = 2J_1(b\psi)/(b\psi), \text{ for defocussed image,} \qquad (8.13)$$

where $\psi = (u^2+v^2)^{1/2}$. In Eqs. 8.12 and 8.13, a and b are constants whose values depend on the magnitude of imperfections.

The function $h(x,y)$ is also processed by negative-positive procedure, similar to that used for preparing $g(x',y')$. The transparency representing $h(x,y)$ is placed in the input plane of the processor. A photographic plate is exposed in the Fourier plane and processed to obtain the negative with a γ equal to 2. The transmittance of this plate will be equal to

$$T_1(u,v) = [|H|^2]^{-\gamma/2}$$

$$= |H|^{-2} \quad (\text{ for } \gamma = 2). \qquad (8.14)$$

The $H^*(u,v)$ component of the sandwich filter can be realized by recording a Fourier transform hologram of the spread function $h(x,y)$ i.e. by recording linearly the interference between $H(u,v)$ and a plane reference wave. The transmittance of this hologram is given by

$$T_2(u,v) = 1 + |H(u,v)|^2 + H(u,v) + H^*(u,v). \qquad (8.15)$$

The last term on the right hand side of Eq. 8.15 is the required transmittance. The illumination of the sandwich filter with the field G(u,v) will produce the required field

$$G(u,v) \cdot H^*(u,v) |H(u,v)|^{-2} = F(u,v). \qquad (8.16)$$

Taking the Fourier transform of this field will generate f(x,y).

The filter fabrication is, in fact, an art. The filter $|H|^{-2}$ is required to be made linear over $10^4:1$. The dynamic range of the photographic material is only 30:1 over γ equal to 1. The dynamic range can somewhat be extended by using POTA developer. A dynamic range of $10^6:1$ can be achieved by masking method [8.11]. Figure 8.9 shows the example of image restoration.

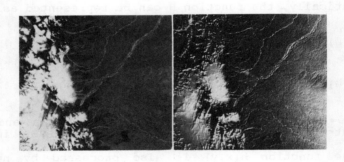

(a) (b)

Fig. 8.9. Image restoration. (a) Original, (b) restored.

Improvements in this technique requiring only a single filter have also been suggested.

8.2.2. Correlative Holographic Decoding Method

In this technique [8.12], the function G(u,v) is multiplied by the complex conjugate $H^*(u,v)$ of H(u,v) to obtain

$$G(u,v)H^*(u,v) = F(u,v)H(u,v)H^*(u,v). \qquad (8.17)$$

Eq. 8.17 is valid if

$$H(u,v)H^*(u,v) = 1. \qquad (8.18)$$

Eq. 8.18 is the decoding condition [8.13], which may be written in the form of spatial autocorrelation operation in the (x,y) domain as

$$\int\!\!\!\int_{-\infty}^{\infty} h(x,y)h^*(x+x',y+y')\,dx\,dy = \delta(x',y'),$$

$$h * h^* = \delta, \qquad (8.19)$$

where $\delta(x',y')$ is the Dirac delta function. The implementation of this method can be done in three ways.

Scheme A

A lensless Fourier transform hologram of the function $g(x,y)$ is recorded. The recorded pattern is given by

$$I(u,v) = [1 + G(u,v)][1 + G(u,v)]^*$$
$$= 1 + |G(u,v)|^2 + G(u,v) + G^*(u,v). \qquad (8.20)$$

The hologram is developed and replaced at its recording position and illuminated by the wave H. The irradiance of the hologram is given by

$$I(u,v)H(u,v) = H(u,v) + H(u,v)|G(u,v)|^2$$
$$+ H(u,v)G(u,v) + H(u,v)G^*(u,v). \qquad (8.21)$$

Substituting the value of $G(u,v)$ from Eq. 8.7 gives

$$I(u,v)H(u,v) = H(u,v) + H(u,v)|G(u,v)|^2$$
$$+ H(u,v)F(u,v)H(u,v)$$
$$+ H(u,v)F^*(u,v)H^*(u,v). \qquad (8.22)$$

The last term on the right hand side of Eq. 8.22 gives

$$F^*(u,v)H^*(u,v)H(u,v) = F^*(u,v) \quad (\text{if } HH^* = 1). \qquad (8.23)$$

F^* is the complex conjugate of the image wave F. Taking the Fourier transform of this irradiance will produce $f^*(x,y)$.

Thus the method consists in recording a lensless Fourier transform hologram of the blurred photographic transparency (positive) and reconstructing it with an extended point source describing h(x,y).

Scheme B

In this scheme a lensless Fourier transform hologram of the function g(x,y) is recorded using h(x,y) as the extended reference source. The recorded pattern is given by

$$I(u,v) = [G(u,v) + H(u,v)][G(u,v) + H(u,v)]^*$$
$$= |G(u,v)|^2 + |H(u,v)|^2 + G(u,v)H^*(u,v)$$
$$+ G^*(u,v)H(u,v). \tag{8.24}$$

The hologram is developed and reconstructed with a plane wave of unit magnitude. The irradiance of the hologram is given by

$$I(u,v) = |G(u,v)|^2 + |H(u,v)|^2 + F(u,v)H(u,v)H^*(u,v)$$
$$+ F^*(u,v)H^*(u,v)H(u,v). \tag{8.25}$$

The third term on the right hand side of Eq. 8.25 gives

$$F(u,v)H(u,v)H^*(u,v) = F(u,v) \quad (\text{if } HH^* = 1). \tag{8.26}$$

Taking the Fourier transform of this irradiance will produce f(x,y). Thus the method consists in recording a lensless Fourier transform hologram of the blurred photographic transparency (positive) using an extended point source describing h(x,y) and reconstructing it with a plane wave of unit magnitude.

Scheme C

In this scheme a lensless Fourier transform hologram of the function h(x,y) is recorded using a reference beam derived from a point source (i.e. a δ-function). The hologram recording is given by

$$I(u,v) = [1 + H(u,v)][1 + H(u,v)]^*$$
$$= 1 + |H(u,v)|^2 + H^*(u,v) + H(u,v). \tag{8.27}$$

The hologram is reconstructed with G(u,v). The term of interest will be

$$G(u,v)H^*(u,v) = F(u,v)H(u,v)H^*(u,v)$$

$$= F(u,v) \quad \text{(if } HH^* = 1\text{).} \tag{8.28}$$

Taking the Fourier transform of this irradiance will produce $f(x,y)$. This method is suitable for restoring any photograph recorded with the same imaging system at the same condition [8.14], once a lensless Fourier transform hologram of the spread function of the imaging system is recorded. Schemes A and B, on the other hand require recording of separate holograms for each blurred image recorded with the same instrument.

In all the three extended source schemes discussed above, the functions $g(x,y)$ and $h(x,y)$ are required to be prepared with proper photographic care mentioned in the inverse filter method.

8.3. Image Subtraction

The problem of extraction of small differences between two scenes is of considerable interest for earth resources studies, urban development, automatic tracking, surveillance and robotics. The image subtraction may also find application in video communication since only the difference between images in successive cycles rather than the entire image may be transmitted to compress the bandwidth.

The principle of optical image subtraction involves modulation of the two images by waves which differ in phase by π radian. The positive and negative transparencies of an image are out of phase by π radian, hence the superposition of these two will produce the difference between them [8.15].

Several holographic schemes have been developed for image subtraction. In general, the holographic techniques involve amplitude subtraction. Some techniques involve modulation of the scene information by interference fringes or colours. Most of the methods need the recording of a hologram, hence real-time implementation is difficult.

8.3.1. Single-Exposure Method

A hologram (image, Fourier transform or Fresnel type) of one of the transparencies P_1 is recorded, developed and repositioned at its recording place. The second photographic transparency P_2 is now placed in the same position of the first transparency. The hologram of P_1 is illuminated by the reference beam. If the two images in the output plane are in exact registration, subtraction will occur [8.16] since the recorded hologram of P_1 is a negative and is thus π radian out of phase. For perfect cancellation, the amplitude of the light transmitted by P_2 should be equal to the amplitude of the reconstructed image of P_1. This can be ensured by keeping the ratio of reference-to-signal beam intensity equal to the ratio of the reconstruction-to-the image beam intensity. Once a photographic transparency is stored in the hologram, it can be compared with any information.

8.3.2. Double-Exposure Method

The method involves [8.17] the recording of a Fourier transform hologram of the first transparency P_1. The phase of the reference beam is then shifted by π radian. The object is replaced by the second transparency P_2 and a second exposure is given to the photographic plate. The photographic plate after processing is illuminated by the reference beam to obtain the difference between the two transparencies. The π phase shift can be introduced by inserting a π phase plate in the path of the reference beam. This can also be achieved by using a glass plate in the reference beam and suitably changing its orientation before the second exposure.

8.3.3. Holographic Beam Splitter Method

The method involves [8.18] a special beam splitter recorded by holographic technique using three beam interferometry. For recording the beam splitter, two beams, A and B, which lie in the same plane are incident at angles θ_1 and θ_2 on a photographic plate. The third beam C lies in a plane orthogonal to the plane of the beams A and B and makes an angle θ_3 with the axis as shown in Fig. 8.10. The photographic plate is exposed and developed linearly. The hologram records three sets of interference fringes with their orientation depending on the angles between A, B and C. Such a hologram when illuminated with the beams A and B simultaneously will reconstruct ten different beams. For getting the difference, the two transparencies P_1 and P_2 are inserted into beams A and B respectively and the beam C is blocked. Out of ten diffracted beams, one in the direction of C will show the difference, provided a phase change of π radian has been introduced into the path of either A or B.

This method has the potential of real-time capability as no recording of a hologram is needed for each pair of object transparencies. The information (optical or electrical) can be introduced by using spatial light modulators in beams A and B.

8.3.4. Cross-Polarization Method

The holographic image subtraction can also be implemen-

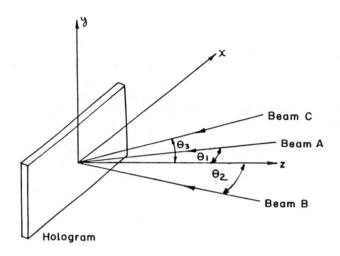

Fig. 8.10. Recording of a holographic beam splitter for image subtraction [8.18].

ted by using different polarizations for the two object transparencies. If the holographic images of the two transparencies are reconstructed with their polarization at right angles to each other, then the superposition of the two will show only their difference. The method can be implemented by recording the hologram of the first transparency P_1 by using a circularly polarized reference beam and vertically polarized object beam. The second transparency P_2 is recorded with horizontally polarized object beam. This may be accomplished [8.19] by using the object illumination with light circularly polarized in the opposite sense to the reference beam. A linear polarizer is used in the vertical position for the transparency P_1 and in the horizontal position for the transparency P_2. This results in a π radian phase difference between the two object illuminations.

The method can also be implemented in single-exposure [8.19]. Both the transparencies P_1 and P_2 are positioned symmetrical with the optical axis. The transparencies P_1 and P_2 are illuminated with a vertically polarized and a horizontally polarized beam respectively. A Fourier transform hologram of both the transparencies are recorded in a single-exposure by using a circularly polarized reference beam.

8.3.5. Holographic Shear Lens Method

This method utilizes a double frequency holographic lens [8.20] as shown in Fig. 8.11. The two photographic transparencies P_1 and P_2 are positioned symmetrically with respect to the optical axis and simultaneously illuminated with a coherent plane beam. The holographic shear lens may be recorded employing double-exposure technique with carrier frequencies separated by $\Delta\nu$. The holographic lens shears and recombines the images of P_1 and P_2 at the output plane. The shear lens is tilted at a suitable angle such that the light path through P_1 and P_2 differs by desired phase to establish condition for subtraction. The technique is similar to a double grating shearing interferometer for optical subtraction.

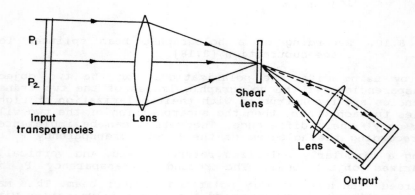

Fig. 8.11. Holographic shear lens technique for image subtraction [8.20].

8.3.6. Rainbow Holographic Method

This technique for identifying the difference between two images is based on the principle of double-exposure holographic interferometry rather than subtraction of amplitude. The technique produces both the signals superimposed with common portions modulated with interference fringes [8.21].

The optical setup is that of a typical rainbow hologram recording system. The two exposures are given to the photographic plate, one for each transparency P_1 and P_2. Before the second exposure, however, the photographic plate is displaced slightly, say by 20 μm, in a direction normal to the direction of propagation of light. The recorded hologram is reconstructed by a white light source. The similar portions in the two transparencies interfere and produce fringes, while the uncommon portions do not show any interference fringes. The method does not require any critical alignment once the hologram is recorded. Any white light source can be used for reconstruction. The method can also be implemented for knowing the difference in two 3D scenes.

8.3.7. Dynamic Hologram Method

Dynamic hologram method can produce image subtraction using a piezoelectric mirror in the optical set up [8.22] as shown in Fig. 8.12. The object transparency P_1 is placed in beam 1. A reflection hologram is formed in a photorefractive crystal (BGO) whose crystallographic axis which is normal to the crystal surface, is oriented parallel to the bisector of the reference beam and the object beam. After recording the hologram of the transparency 1, the piezoelectric mirror is translated by a distance of λ/4 and the hologram of the transparency P_2 is recorded. Due to the slow response and

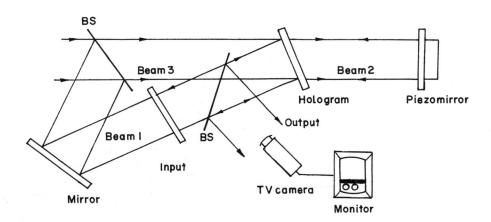

Fig. 8.12. Dynamic hologram method of image subtraction [8.22].

erase time of BGO, the two images can both exist for a short time interval and show the difference on the monitor.

8.3.8. Colour Coding Method

This method gives a comparison of two transparencies rather than giving their subtraction. The similar parts of the two transparencies are produced coded in white while the distinct parts of each transparency in different colour. Figure 8.13 gives a schematic diagram of the setup which uses two holographic lenses [8.23]. The photographic transparency P_1 is placed close to the HOE-1 and back illuminated with an expanded beam of He-Ne laser producing a red image on the screen. The second photographic transparency P_2 is illuminated with an incandescent lamp. The HOE-2 placed

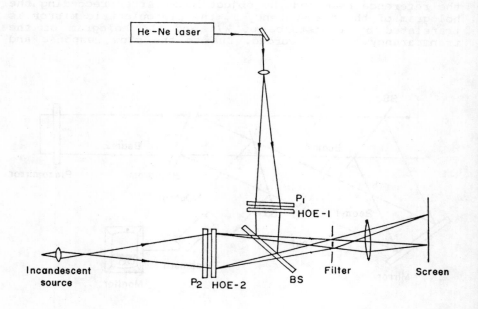

Fig. 8.13. Comparison of two transparencies using colour coding technique employing HOEs [8.23].

near the transparency P_2 produces colour spectrum in its focal plane where an aperture rejects the red colour to produce a greenish image of P_2 on the screen. The beam splitter BS and the transparency P_2 are adjusted to obtain perfect registration of the two images. The output will show the similar parts of P_1 and P_2 in white, distinct parts of P_1 in red and those of P_2 in green colours. This pseudo-colour image subtraction technique can be implemented in real-time [8.24].

8.4. Character and Pattern Recognition

One of the most significant applications of holography is in the character and pattern recognition. The use of holograms as complex filters in optical data processing was realized quite early. This is because of the capability of the holograms to record both the amplitude and phase of a wave. The use of a matched filter for the recognition of characters and patterns is a direct consequence of the associative nature of the holograms.

The aim of an optical pattern recognition system is to determine the presence, location and class of objects present in a scene. The optical recognition system is attractive for automated inspection, robotics and missiles due to its capability of handling large space bandwidth data at high frame rate.

8.4.1. Vander Lugt Frequency Plane Correlator

The optical pattern recognition system that is being widely used is the frequency plane correlator of Vander Lugt [8.25]. To recognize an object (character, pattern etc.) in a given scene, a matched spatial filter (MSF) of the object is first recorded. The filter after development is repositioned at its recording position and the given scene is entered at the input. If the object at the input scene and that recorded in the filter are similar, a sharp point of light will appear at the output. Thus the presence of a light point at the output will indicate the presence of a similar object in the input scene and the location of the point will indicate the location of the object at the input. If the two objects differ, the output will not be a sharp point.

The recognition mechanism can be understood as follows. The matched filter is a Fourier transform hologram of the object to be recognized. This is therefore a record of the interference pattern between the Fourier transform of the object and a point source. If this hologram is reconstructed with a point source, the image of the original object is produced. If, however, the hologram is reconstructed by the object wave, an image of the point source is recreated. If there are two similar objects in the input scene, two images of the point source will be produced with their locations corresponding to the locations of the objects in the scene. If the object in the scene differs slightly from the recorded information in the filter, the hologram will reconstruct a less intense point image.

Figure 8.14 shows a general setup for making a matched filter. At the input plane two amplitude distributions $p(x,y)$ and $q(x,y)$ are located at the front focal plane of a Fourier transform lens. At the back focal plane a photographic transparency is exposed and linearly developed. The amplitude transmittance of the resulting hologram is given by

$$\begin{aligned} I(x,y) &= |P(u,v) + Q(u,v)\exp[-2\pi j\alpha y_1]|^2 \\ &= |P(u,v)|^2 + |Q(u,v)|^2 + P(u,v)Q^*(u,v)\exp[2\pi j\alpha y_1] \\ &\quad + P^*(u,v)Q(u,v)\exp[-2\pi j\alpha y_1], \end{aligned} \quad (8.29)$$

where $\alpha = (\sin\theta)/\lambda$, $P(u,v)$ and $Q(u,v)$ are the Fourier transforms of $p(x,y)$ and $q(x,y)$ respectively; and u and v are spatial frequencies.

This filter is accurately repositioned at the filter plane. The functions $p(x,y)$ and $q(x,y)$ are removed from the input plane and the function $r(x,y)$ is entered. The filter is thus illuminated by the Fourier transform $R(u,v)$ of

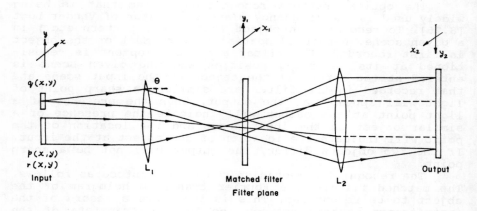

Fig. 8.14. General setup for optical correlation.

$r(x,y)$. The amplitude distribution after the filter becomes

$$U = R(u,v) \cdot I(x,y). \tag{8.30}$$

This amplitude distribution is Fourier transformed by lens L_2 giving the amplitude distribution at the output plane as

$$\mathcal{F}[U] = \mathcal{F}[R(u,v)\{|P(u,v)|^2 + |Q(u,v)|^2\}]$$
$$+ \mathcal{F}[R(u,v)P(u,v)Q^*(u,v)\exp\{2\pi j\alpha y_1\}]$$
$$+ \mathcal{F}[R(u,v)P^*(u,v)Q(u,v)\exp\{-2\pi j\alpha y_1\}]. \tag{8.31}$$

If $q(x,y)$ is a δ-function, then the second term in the right hand side of Eq. 8.31 is the convolution of $p(x,y)$ and $r(x,y)$.

$$p * r = p(x_2,y_2) * r(x_2,y_2) * \delta(x_2, y_2 + \alpha\lambda f). \tag{8.32}$$

If the functions $p(x,y)$ and $r(x,y)$ are the same then the second term in the right hand side of Eq. 8.31 gives the autocorrelation of $p(x,y)$ and a sharp peak (maximum intensity point) is produced at $(0,-\alpha\lambda f)$ in the output plane.

The third term in Eq. 8.31 is the cross-correlation of $p(x,y)$ and $r(x,y)$

$$p * r = p^*(-x_2, -y_2) * r(x_2, y_2) * \delta(x_2, y_2 - \alpha\lambda f) \tag{8.33}$$

The value of this correlation has a small peak centered at $(0,\alpha\lambda f)$ in the output plane.

The first term on the right hand side of Eq. 8.31 is centered at the origin of the (x_2, y_2) plane. If angle θ is of sufficient large value, then the convolution and correlation terms will be diffracted in opposite direction sufficiently separated to be viewed independently.

The angle θ can be fixed by considering the widths of various output terms in Eq. 8.31. If W_p and W_r are the widths of p and r in y direction, then θ is given by

$$\sin\theta \geq \frac{1}{f}[\frac{3}{2} W_p + W_r]. \tag{8.34}$$

8.4.2. Experimental Techniqe

Figure 8.15 shows a schematic diagram of a practical system for optical correlation. The one arm of the correlator produces the Fourier transform of the object at plane P_2

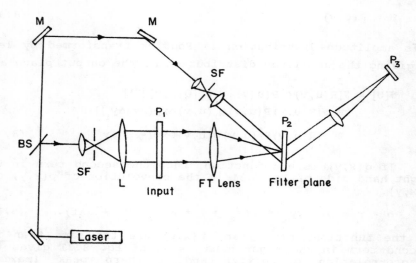

Fig. 8.15. Practical system for optical correlation.

using a Fourier transform lens L_1. The matched filter is recorded by interfering the Fourier transform of the object with a plane reference beam.

The processed matched filter is rotated by 180° and repositioned at plane P_2. The correlation appears at $+\theta°$ with respect to the path of signal beam. This makes the system compact as otherwise the correlation would have appeared at $-\theta°$ along the path of the reference beam. The second Fourier transform lens L_2 is placed one focal length from the filter plane along the path of correlation beam.

The reference-to-object beam intensity ratio ρ for the filter fabrication is normally chosen to be equal to 1. The important point in the filter fabrication is to select the proper spatial frequency band in which the beam balance ratio is set. An input containing urban scene will show the Fourier transform with more energy in the higher frequency components. A rural imagery will have more energy in the lower frequency components. If a scene contains both urban and rural information, then with $\rho=1$ set at higher frequency band, a higher correlation intensity will be obtained for urban image as compared to rural image [8.26]. Figure 8.16 compares the correlation intensity of urban and rural images versus the spatial frequency band for $\rho=1$. It has been observed that the correlation intensity value is pro-

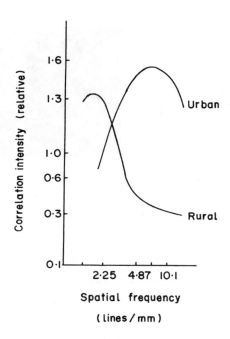

Fig. 8.16. Correlation intensity of urban and rural images versus spatial frequency band for $\rho=1$ [8.26].

portional to the square of the area of the image while the SNR varies as area of the image [8.26].

Figure 8.17 shows the results of matched spatial filtering. A spatial filter for the character A was made. The input pattern is shown in Fig. 8.17a and the output pattern is shown in Fig. 8.17b. The location of correlation peaks (intense points) are proportional to the locations of the character A in the input. The other points which are less intense are due to cross-correlations and are indicative of the limitation of optical character and word recognition. The cross-correlation signals can be suppressed by thresholding the output.

8.4.3. Applications

The correlation techniques have mainly been applied for the recognition of character, word and pattern. The correlation technique has also been exploited for the nondestructive testing of materials. It has been used to detect nonuniformity in small welds [8.27], to measure

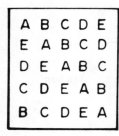 

Fig. 8.17. Results of matched spatial filtering using a thermoplastic film. (a) Input, (b) output.

strain in metals [8.28], to examine material surfaces for microdeformations [8.29], to monitor the aging process of rubber [8.30], to evaluate ceramic materials [8.31] and to study blood cells and malignant cancer tissues.

One of the important applications suggested for pattern recognition using the Vander Lugt filter is the recognition of fingerprints [8.32]. The main limitation of the method is the requirement that the fingerprint must be in the form of a transparency. Frustrated internal reflection has been used to provide real-time input of the fingerprint to the recognition system [8.33]. The use of frustrated internal reflection and a reflective matched spatial filter can lead to a simple, inexpensive and compact fingerprint correlator which permits the fingerprint verification in real-time [8.34]. Figure 8.18 shows the schematic diagram of such a correlator. The system employs a 90° isosceles prism. The finger is pressed against the face of the prism which frustrates the internal reflection along the ridges of the finger. The finger positioning can be accomplished through a slot arrangement on the surface of the prism. The system is quite sensitive to the misalignment [8.35]. The reflective matched spatial filter is embossed on the personnel security card or credit card. The detection of the correlation is achieved by using two detectors. The detector A detects the correlation radiation which can be measured against the background radiation detected by detector B. The system can be used for ownership verification or personnel identification.

A nonlinear element may be incorporated in the system for decision making [8.36]. A digital optical system has also been suggested for fingerprint analysis and classification [8.37]. This system uses a magneto-optic spatial light modulator for real-time generation of filters.

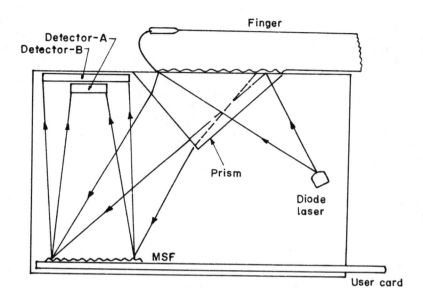

Fig. 8.18. Schematic diagram for the recognition of fingerprints using frustrated internal reflection and a reflection matched filter [8.34].

An interesting application of optical correlation technique has been made for the detection and measurement of leak rates in hermetically sealed packages [8.38]. The technique depends on the correlation of the stressed object with its matched filter as a function of time. The method consists in recording a matched filter of the object enclosed in a vacuum chamber. The matched filter is placed at its recording position and correlation signal is measured. As the vacuum is drawn on the chamber, the package distorts due to the pressure differential and the correlation signal will drop to zero. If there is a leak in the package, the pressure differential will soon disappear and the package will return to its original shape, recovering the correlation signal. The leak rate can be calculated by the relation [8.38]

$$\text{Leak rate} = \frac{V_i}{\tau} \left[\frac{P_e}{P_a} - 1 \right], \tag{8.35}$$

where V_i is the internal free volume of the package, P_e is the external applied pressure, P_a is the atmospheric pressure and τ is the time constant. τ is a function of leak geometry, temperature and the internal free volume of the package. The relationship between the recovery time and time constant can be found using a nonleaking package and controlling the external applied pressure.

The technique can detect leaks in the range of 10^{-1} to 10^{-6} atm cc/s. The main advantage of this technique is the elimination of the classic problem of fringe analysis to determine the quality of the sample. The system can be automated by using a real-time recording material for the filter fabrication. The method has been tested on pacemaker batteries [8.38]. It has been observed that the technique has the advantage over conventional leak testing as it requires no backfilling with tracer gas.

Recently Psaltis et al. [8.39] have used optical memory disks for entering data in the input of an optical correlator. The optical disk offers the advantage of reading the blocks (pages) of data in parallel by simply rotating the disk. The data acquisition is greatly simplified because the beam steering is needed only in one direction. The system once perfected would accomplish a correlation rate for 100x 100 pixel images of 40,000/s [8.39] which would be difficult by an electronic system.

8.4.4. Fresnel Hologram Correlator

Fresnel holograms, instead of Fourier transform holograms, can be used for optical correlation which would avoid the use of a lens in the setup. The matched Fourier filter technique requires the exact repositioning at the Fourier plane and the input in the form of a transparency. Using a Fresnel hologram extended objects can be correlated.

Figure 8.19 shows a schematic diagram of an optical correlator using a Fresnel filter. The object surface represented by the function f(x,y) is placed at the input plane P_1 at a distance f from the hologram recording plane P_2. The reference point source is also placed at a distance f from the recording plane. The complex amplitudes of the object and reference arriving at the recording plane are denoted by O(u,v) and R(u,v) respectively and are represented in the Fresnel approximation as

$$O(u,v) = \iint f(x,y)\exp\{(j\pi/\lambda f)[(x-u)^2 + (y-v)^2]\}dxdy \quad (8.36)$$

and

$$R(u,v) = \exp[(j\pi/\lambda f)(u^2+v^2)]\exp[(j\pi/\lambda f)(\mu u + \nu v)], \quad (8.37)$$

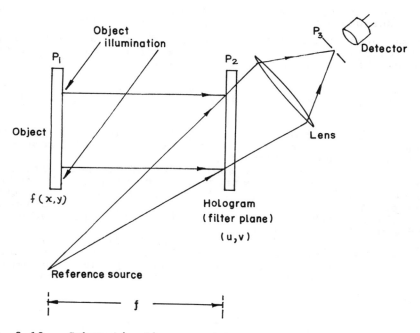

Fig. 8.19. Schematic diagram of an optical correlator using a Fresnel filter.

where μ and ν are the cosine angles with respect to the recording plane. In Eqs. 8.36 and 8.37 a constant phase multiplication factor has been omitted. The hologram has an amplitude transmittance I(u,v) proportional to the incident intensity pattern. The term of interest with regard to correlation measurement is $O^*(u,v)R(u,v)$ which may be expressed as

$$O^*(u,v)R(u,v) = \iint f^*(x,y)\exp\{(-j\pi/\lambda f)[(x-u)^2+(y-v)^2]\}dxdy$$

$$\cdot \exp[(j\pi/\lambda f)(u^2+v^2)]\exp[(j\pi/\lambda f)(\mu u+\nu v)].$$

(8.38)

The recorded hologram is repositioned at its original position and illuminated by the original object wavefront. The wavefront reconstructed is given by $O(u,v)O^*(u,v)R(u,v)$, which is proportional to original reference wave. A lens placed behind the hologram will focus this wavefront onto a photodetector which will measure the correlation signal.

In the input plane the new object is inserted which can be described by g(x,y). The light distribution at plane P_2 is measured by scanning the plane and integrating the field amplitude of the reconstructed wavefront over the entire plane. The correlation intensity is given by

$$I = \left| \iint f^*(x,y) g(x,y) dx dy \right|^2. \tag{8.39}$$

Maximum correlation intensity will be obtained when the function f(x,y) and g(x,y) are identical. This represents the case of autocorrelation.

In order to show the application of Fresnel correlator, we discuss the nondestructive evaluation of ceramic materials [8.40]. A Fresnel filter of a small test area on the surface of a silicon nitride ceramic was recorded and correlated with the thermally stressed sample which may cause microstructural changes.

Figure 8.20 shows the variation in autocorrelation signal intensity when the sample was displaced in planes parallel and perpendicular to the filter plane. The in-plane displacements rapidly decreases the correlation signal

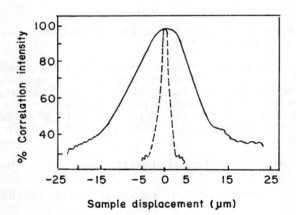

Fig. 8.20. Autocorrelation signal intensity as a function of sample displacement parallel and perpendicular to the filter plane [8.40].

intensity suggesting that each in-plane displacement provides a new input scene to the filter. In contrast, the out-of-plane displacement gradually decreases the correlation intensity, indicating that the input in each displacement provides a scaled version of the original. An out-of-plane displacement of about 12 μm produces a 50% signal loss, whereas an in-plane displacement of only 1.8 μm produces the same signal loss. These experiments show that the correlation technique is capable of detection of surface changes into submicron range. This also imposes the critical requirement of the object stability for the correlation technique [8.40].

Figure 8.21 shows the recovery time for initial autocorrelation value after the ceramic sample had been heated for 180 s. The technique is useful for the measurement in the differences between identically prepared samples and in studying the thermal response for ceramics containing different minority constituents [8.40].

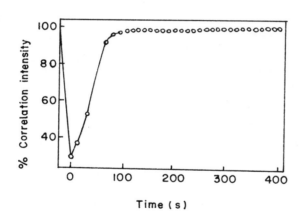

Fig. 8.21. Recovery time for initial correlation value after sample heating [8.40].

It may be mentioned that the matched spatial filter will show a decrease in correlation intensity as $(1 - a)^{-4}$ for a 2D image when there is a scale change 'a' between the input and reference function. The correlator based on the Fresnel hologram is rather insensitive to the scale change [8.41] as it produces a decrease in correlation intensity as $(1 - a)^{-2}$. The Fresnel hologram technique can also be used when the object to be detected is illuminated with a spatially extended band-limited source [8.41]. It means that there is no need of incoherent-to-coherent conversion as in the case of a complex amplitude correlator using matched spatial filter. The degradation of the correlation peak is less sensitive to a change in scale or rotation.

Table 8.1 gives various transformations of information when different functions are used as p, q and r in Fig.8.14.

Table 8.1 Some transformations of information

Operation	Functions		
	p	q	r
Diffraction grating	δ	δ	δ
Multiple images	$\sum \delta(x+a)$	δ	r
Fourier transform hologram	p	δ	δ
Correlation (pattern recognition)	p	δ	r
Correlation (pattern recognition)	p	q	δ
Deblurring (decoding, Scheme A)	f*h	δ	h
Deblurring (decoding, Scheme B)	f*h	h	δ
Deblurring (decoding, Scheme C)	h	δ	f*h
Deblurring (inverse filter) with filter 1/H	h	δ	f*h

8.5. Holographic Information Storage

The concept of holographic information storage was first proposed about 30 years ago [8.42]. The main advantage of holographic storage is its low access time as large blocks of information can be accessed simultaneously. However, poor performance of available recording materials and beam deflectors were the main obstacles in realizing holographic memories. Efforts were made to realize read/write memories in photorefractive crystals [8.43,

8.44]. These attempts were limited by the speed, capacity, efficiency and poor crystal quality. The interest in holographic memories has renewed in recent years due to new advances in recording materials, associative memory and neural net systems [8.45].

In general, the theoretical storage capacity in two dimension is

$$S_{2D} = 1/\lambda^2 \quad \text{(thin hologram)} \tag{8.40}$$

and that in three dimension is

$$S_{3D} = n^3/\lambda^3 \quad \text{(thick hologram)}, \tag{8.41}$$

where n is the refractive index of the recording material [8.42].

For $\lambda=514.5$ nm and $n=1.5$,

$$S_{2D} = 3.78 \times 10^6 \text{ bits/mm}^2$$

and

$$S_{3D} = 2.48 \times 10^{10} \text{ bits/mm}^3.$$

In practical situations these values are considerably reduced due to the finite size of the hologram, grain noise of the recording medium, detector noise, limitations of beam deflectors and cross-talk.

Figure 8.22 shows the basic components of a block-organized random access holographic information storage system [8.46]. It consists of a page composer, a Fourier transform lens, a recording medium, a beam deflector and a detector array. A Fourier transform hologram of each of the page (encoded in bit or image format) is recorded on the recording medium as a subhologram. The diffraction between boundaries of each subhologram determines the minimum distance Δ between two subholograms. The distance Δ can be given by the Rayleigh criterion as

$$\Delta = 1.22\lambda b/a, \tag{8.42}$$

where a/b is the numeric aperture of the hologram assumed to be circular of diameter a and b is the distance of the hologram from the output plane [8.47]. The capacity of the storage is limited by the finite aperture of the subholograms and image projection effects.

The information capacity of a block-organized holographic memory is equal to the number of bits that can be read

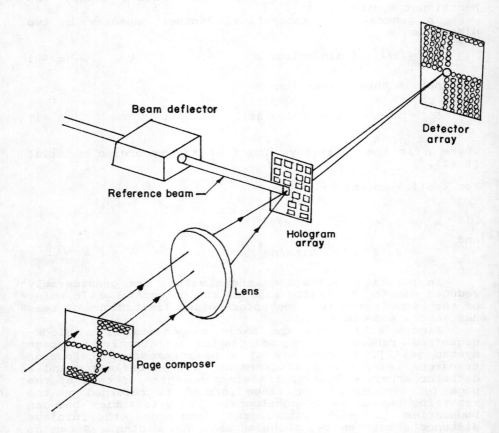

Fig. 8.22. Basic components of a block-organized random access holographic information storage system [8.46].

out in parallel. If we assume that all bits of data are simultaneously read from a single subhologram, then the capacity C is given by

$$C = l_x l_y m_x m_y ,$$

$$= M_h^2 M_d^2 , \qquad (8.43)$$

where $M_h^2 = l_x l_y$ is the number of holograms and $M_d^2 = m_x m_y$ is the detector elements. Assuming l_i and m_i of the order of 100, capacity of the storage plate becomes

$$C = (100)^2 (100)^2 = 10^8 \text{ bits.}$$

The detector response time is limited by the time required to change the detector to a threshold voltage level. The optical power P needed to switch each of the 10^4 detectors is given by [8.48]

$$P = 1.45 \times 10^{-5} \nu \; (\mu m), \qquad (8.44)$$

where ν is the operating frequency. Assuming $\nu = 0.1$ MHz and 20% diffraction efficiency of each subhologram, the required power is 5 µW/detector.

A photorefractive crystal can be used to store 2D information to realize a read/write memory system. A page oriented system based on cerium doped SBN fibres [8.49] is currently under development at Stanford University with the performance parameters as 10,000 bit pages, 1-10 µs latency and a capacity of 200 to 1000 Mb.

The holographic memory system based on 3D hologram in a photorefractive crystal in which each page is stacked, shortens the data retrieval time drastically [8.50]. For example, the amount of data that can be transferred by a holographic 3D memory device in 1 second will need 5 hours by a fastest rotating magnetic disk device.

Figure 8.23 shows a schematic diagram of a holographic 3D memory system [8.50]. It consists of a frequency doubled Nd:YAG laser, a two dimensional spatial light modulator (SLM), a scanner for page selection and a CCD camera. The two dimensional light patterns (called as pages) to be stored are created by the spatial light modulator and are recorded in the storage crystal as Fourier transform holograms. To retrieve the data the correct location of the reference beam is selected by the scanner. The emerging bit pattern is received by a CCD detector. The system performance has been improved by replacing the relatively large single photorefractive crystal (1x1x0.5 cm) with an array of many smaller crystals [8.50]. The array of crystals are less sensitive to cross-talk and also improves the storage capacity and access time.

For achieving a storage capacity at least 100 times the storage capacity of today's optical disks for computer applications, a new architecture is being used in the

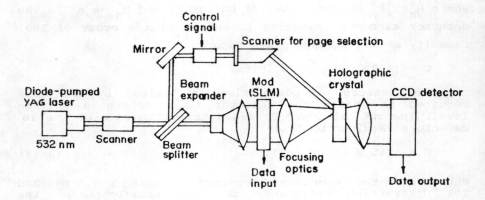

Fig. 8.23. Schematic diagram of a holographic 3D memory system [8.50].

experimental Sparta's holographic 4D memory system [8.51] (Fig. 8.24). The principle of this extremely high storage capacity is based on a technique called spectral hole burning in which the holograms are written in a dye-polymer medium at different wavelengths. Each hologram recorded as a

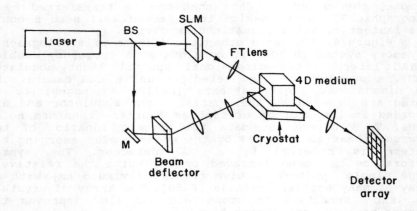

Fig. 8.24. Schematic diagram of a holographic 4D memory system [8.51].

megabit image can only be reconstructed by the laser wavelength that recorded it. This is an important development in the architecture as it has added a fourth dimension in the recording medium. As about 1000 different laser wavelengths are possible, the storage capacity of such a holographic memory device could be increased by three orders of magnitude to about 10^{12} bits.

The architecture of 4D memory system [8.51] is in principle similar to that shown in Fig. 8.23 for a 3D holographic memory system except that a tunable laser system has been used. The recording material needs a very low temperature which can be achieved by using liquid helium.

8.5.1. Holographic Associative Memory

Holographic associative memory is another important achievement in realizing a very high speed processing system useful for an artificial neural computer [8.52, 8.53]. The electronic computers have reached a stage of very high processing speeds leading to the construction of supercomputers. However, it is not like that computers can perform all the tasks given to it with ease. There is a persistent difficulty for performing tasks like simple pattern recognition problem. Such problems appear simple for humans but very cumbersome for conventional computers in spite of the fact that computer's processing power can tackle hundreds of problems that are beyond the capability of a human brain which operates at a comparatively low speed (at a clock speed of 100 Hz compared with more than 25 MHz for even a PC). It means that there is something wrong in the basic approach of conventional computers. Natural systems of intelligence in biological systems are therefore being examined for building new computers.

It is known that very primitive organisms have very sophisticated perceptual processing which is necessary for their survival. Biological systems show adaptivity and learning capability although brain has no logic or arithmetic circuits, registers, addresses, decoders and memory units typical to computers. As we know, brain is not as good in arithmetic operations as a computer, but it outperforms the most powerful modern computers in association, generalization, categorization, classification, feature extraction, recognition and identification. Brain can also supplement missing information (Fig. 8.25). It can interpret fuzzy or incomplete data, understand distorted speech and recognize people from momentary glimpses.

The human brain contains a large number of processing elements known as neurons. There are 10^{10}-10^{11} neurons in brain each making about 10^3-10^4 synaptic interconnects, i.e. there are 10^{13}-10^{14} interconnections. Neural computing is fundamentally different from the conventional Von Neumann computer. A neuron never operates alone. Each neuron

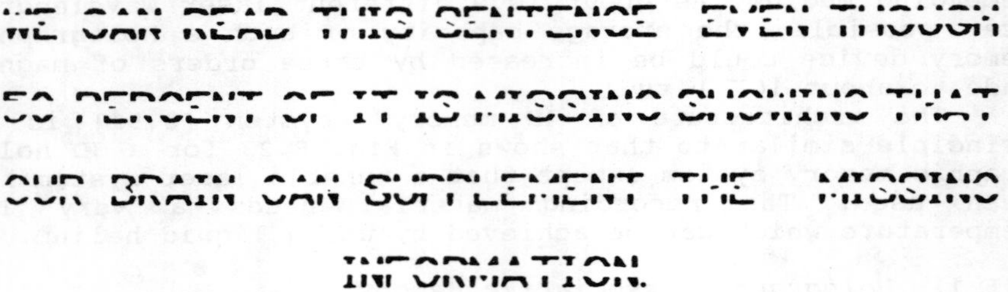

> WE CAN READ THIS SENTENCE EVEN THOUGH
> 50 PERCENT OF IT IS MISSING SHOWING THAT
> OUR BRAIN CAN SUPPLEMENT THE MISSING
> INFORMATION.

Fig. 8.25. Brain can supplement missing information.

independently evaluates its state and decides to change it if the sum of its synaptic inputs exceeds a given threshold. Thus a highly nonlinear operation is carried out. A set of neurons collectively work on the same problem. The global communication in neural nets allows each neuron to change its state in response to the state of entire network. The type of connectivity that exists in a neural network allows thousands of neurons to collectively and simultaneously influence the state of each neuron. Neural nets modify their structure to fit the requirements of the problem.

As a consequence of the associative aspect of the neural nets, the memory remembers the stored information and produces the output even when it is stimulated by a partial or distorted input of the stored information. Even if some of the neurons are destroyed (it always happens in brain) the associative nature is not affected. The collectivity, adaptivity and fault tolerance are the main features of neural nets which provide a new way of viewing signal processing and computing problems.

The main components of neural nets are

(a) Neurons : These are the main processing elements. They perform a nonlinear operation, mainly thresholding.

(b) Weights : Strength of connections between different neurons determine the weight.

(c) Connections : The memory resides in connections. It should be modifiable.

Hopfield Model of Neural Network

Holography has to play a major role in realizing an

optical associative memory similar to that of the brain. The holographic associative memory schemes have been implemented based on the Hopfield model of neural network. In this model, inputs and outputs are considered as vectors [8.54]. The system consists of a vector S of N processing elements interconnected by weighted links. The values of weights form a matrix T. The weights of the interconnection matrix are formulated by summing up the outer product of each of the memory vectors

$$T_{ij} = \sum_{m=1}^{M} s_i^m s_j^m \qquad (8.45)$$

$$T_{ii} = 0. \qquad (8.46)$$

In Eq. 8.45 the summation is over M memories. Multiplying T_{ij} with any stored vector S reproduces a noisy estimate of the same vector

$$\sum_{i}^{N} T_{ij} s_i^1 = (N-1) s_i^1 + \sum_{m \neq 1}^{M} \sum_{i \neq j}^{N} s_i^m s_j^1 . \qquad (8.47)$$

The first term on the right hand side of Eq. 8.47 is the desired term, whereas the second term is the unwanted interference which can be eliminated by thresholding.

The output is re-entered into the system. The iteration of the information around the system converges to a stable state only if that vector has already been stored in the memory. It means that an input vector is associated with one particular memory vector which becomes the output. Therefore the input vectors which are partial or distorted result in the stable state.

The associative storage and retrieval process can be explained by a mathematical example [8.55]. Let the two vectors be

$$S(1)^T = (\ 1 \quad 1 \quad -1 \quad -1\ ) \qquad (8.48)$$

and

$$S(2)^T = (\ 1 \ -1 \quad 1 \ -1\ ). \qquad (8.49)$$

The vectors are stored by T_{ij} matrix

$$T_{ij} = S(1).S(1)^T + S(2).S(2)^T$$

$$= \begin{bmatrix} 2 & 0 & 0 & -2 \\ 0 & 2 & -2 & 0 \\ 0 & -2 & 2 & 0 \\ -2 & 0 & 0 & 2 \end{bmatrix} \qquad (8.50)$$

For retrieval of vector $S(1)$ it is multiplied by T_{ij}

$$T_{ij}\, S(1) = \begin{bmatrix} 2 & 0 & 0 & 2 \\ 0 & 2 & -2 & 0 \\ 0 & -2 & 2 & 0 \\ -2 & 0 & 0 & 2 \end{bmatrix} \begin{bmatrix} 1 \\ 1 \\ -1 \\ -1 \end{bmatrix}$$

$$= 4 \begin{bmatrix} 1 \\ 1 \\ -1 \\ -1 \end{bmatrix} \xrightarrow{\text{Thresholding}} S(1). \qquad (8.51)$$

Holographic implementation

It is observed from the above discussion that there are three steps involved in the association of the input with a memory. These steps are multiplication of the input vector with the memory matrix, thresholding of the resultant vector and iteration of the information in the system. The memory can be stored as weighted interconnections in a hologram [8.56]. It is, in principle, possible to achieve 10^{15}-10^{18} parallel weighted interconnections in one second [8.57]. The output from the holographic interconnects are imaged on a nonlinear device which performs the thresholding. The device should not give any output until the incident light intensity reaches the threshold level.

Both digital and analogue associative memories can be implemented using holograms. In digital holographic associative memory, computer-generated holograms can be used to form weighted interconnects [8.56]. In analogue holographic associative memory, the incident information is compared with the stored information in the hologram like an optical correlator [8.58, 8.59]. The correlation output is thresholded and is fed back for iteration which terminates the output to a single solution.

A single hologram memory element can be used between two phase conjugate mirrors (PCMs) [8.60, 8.61]. The PCMs provide the thresholding operation. The holographic memory elements may be Fresnel or Fourier transform holograms. The memory elements can also be multiplexed. Figure 8.26 shows an optical arrangement of an analogue memory system using a Fourier transform hologram [8.56].

Different signals are recorded as superimposed Fourier transform holograms using different reference beams. When

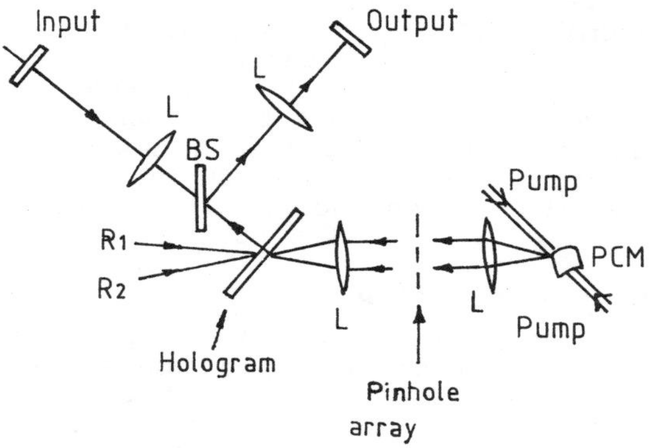

Fig. 8.26. Holographic analogue associative memory system using a FT hologram [8.56].

any partial information is available in the input, a correlation signal will be generated. A phase conjugate mirror will provide the feedback and thresholding necessary to generate an output. Another phase conjugate mirror can be placed in the output plane which would provide the iteration [8.56]. Figure 8.27 shows the result of obtaining complete

Fig. 8.27. (a) Input in the memory system, (b) output when reference beam is blocked and 'HOL' is in the input, (c) output when reference beam is blocked and 'OL' is in the input [8.62, 8.63].

information from the holographic memory when only a partial information is available in the input ('OL'). Figure 8.27 also shows the single iteration output when the complete information ('HOL') is placed in the input and blocking the reference beam [8.62, 8.63].

Keller and Gmitro [8.64] have analyzed a neural network architecture of multifaceted interconnection holograms and optoelectronic neurons. Their technique based on Gerchberg-Saxton [8.65] algorithm followed by a random-search error minimization has shown that the system has the capacity to connect 5000 neuron outputs to 5000 neuron inputs with bipolar synapses with the encoded synaptic weights accuracy of 5 bits.

8.6. Holographic Image Multiplication

Holographic image multiplication has potential applications in the generation of multiple images of integrated circuits, multiple holographic matched filtering in pattern recognition and in optical interconnects. The formation of multiple images from a single master can be achieved by a well corrected step-and-repeat camera. This method is, however, very time consuming, for example, 100 separate exposures are required for an array of 10 by 10 images. The multiple imaging can also be achieved by using fly's eye lenses [8.66], by an array of pinholes [8.67] and birefringence [8.68].

Lu [8.69] and Groh [8.70] have made specialized holographic optical elements for image multiplication. The multiple imaging is achieved by forming the convolution of the object with an array of δ functions in a spatial filtering setup. The complex filter is made by Fourier transform or lensless Fourier transform holograms of the array of point sources. Figure 8.28 shows the recording of such a holographic element and its application in image multiplication.

Kalestynski [8.71] has recorded a hologram of the object to be multiplied by using a number of reference beams. The hologram on reconstruction gives multiple images of the object corresponding to each reference beam. The multi-beam reference field is produced by an array of microlenses distributed in a programmed way in a mask. The fixed separation between the lenses restricts its use. Mehta and coworkers [8.72, 8.73] has recorded lensless Fourier transform hologram of the master object using a holographic grating in the reference beam. The number and location of the multiple images depend upon the spatial distribution of the reference point sources generated by the holographic grating. The multiple images produced by this method are shown in Fig. 8.29.

A multifocus holographic lens may be recorded by using a specially designed contact screen [8.74] or a crossed

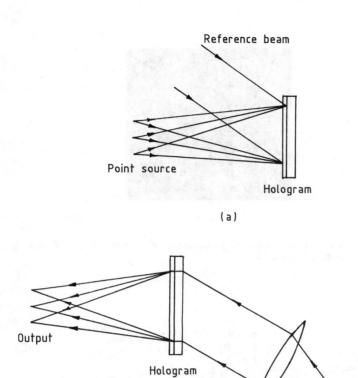

Fig. 8.28. (a) Recording of a hologram of an array of point sources. (b) multiple image generation.

holographic grating [8.75] in a coherent optical filtering system as shown in Fig. 8.30. The Fourier transform of the 2D contact screen S produces an array of spectra islands of nearly the same intensity at P_1. The hologram is recorded at plane P_2 by introducing a collimated reference beam. An

Fig. 8.29. Multiple images produced by lensless FT hologram method [8.73].

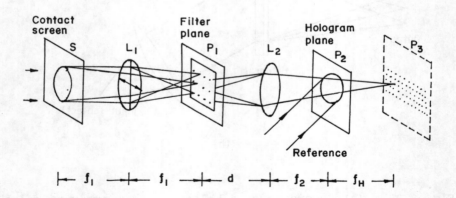

Fig. 8.30. Recording of a multifocus holographic lens for image multiplication [8.74].

aperture at the plane P_1 selects any desired portion of the spectra to vary the number of foci and their positions. The recorded multifocus lens is used in an imaging system for image multiplication.

8.7 Coherent Noise Reduction

The images obtained from an optical data processor and hologram reconstruction are greatly affected by various

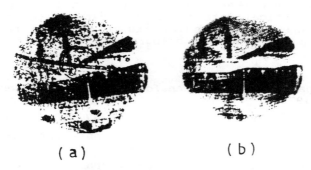

Fig.8.31. Noise removal by using a frequency modulated holographic lens. (a) original object, (b) improved

reflecting and scattering points present in the optical elements of the processor or in the hologram recording and reconstruction optics. This is because the highly coherent light of the laser produces multiple interference of light bouncing off different scatterers. The over-all effect is to produce speckle, i.e. to introduce random intensity fluctuations in the image. The properties of speckle depends on the coherence of the incident light and the nature of the scatterers. The images containing speckle may be extremely noisy with a lower resolution. It may be difficult to extract information from such images if the range of spatial frequencies of the speckle is the same as that of the information [8.76, 8.77].

Speckle noise has been a serious problem for coherent imaging systems and holographic memories as even by taking utmost care in cleaning the optics, it cannot be completely eliminated. Many methods have been suggested for reducing the noise although the subject is still open. The methods can be classified as [8.78]: (a) rotation of lens, diffuser, source, etc. (b) diffuse wavefronts, (c) incoherent addition of wavefronts, (d) multiple coherent and incoherent wavefronts, (e) random and periodic phase modulation with or without sampling, (f) spatial filtering and (g) multiple wavelengths of illumination.

Frequency modulated holographic lenses [8.78] and holographic gratings [8.79. 8.80] have also been used for coherent noise elimination. While in the grating method the averaging of the noise takes place, in the frequency modulated phase plate using a HOE method, interference between neighbouring points of the object are reduced. The phase plate incorporates two complementary HOEs so that half of the signal is diffracted into one place and the remaining half into another place. Since the properties of the Fourier transforming lens have been introduced in the phase plate itself, the optical data processor system becomes simpler [8.78] Figure 8.31 shows the result of noise removal by using a frequency modulated holographic lens.

Reference

8.1 H. Lipson, *Optical Transforms* (Academic Press, New York, 1972)

8.2 P.C. Mehta, *Hyperfine Interaction* **37** (1987) 325.

8.3 A.K. Richter and F.P. Carlson, *Appl. Opt.* **13** (1974) 2924.

8.4 H.R. Manjunath and S.V. Pappu, *Appl. Opt.* **14** (1975) 2562; **15** (1976) 849.

8.5 P.C. Mehta, S. Swami and V.V. Rampal, *Appl. Opt.* **16** (1977) 445.

8.6 E.B. Champagne and N.G. Massey, *Appl. Opt.* **8** (1969) 1879.

8.7 G.D. Mintz, D.K. Morland and W.M. Boerner, *Appl. Opt.* **14** (1975) 564.

8.8 K.S.S. Rao, C. Bhan and P.C. Mehta, in *Opto-electronic imaging,* D.P. Juyal, P.C. Mehta and M.K. Sharma eds. (Tata-McGraw Hill, New Delhi 1987) p. 460.

8.9 K.S.S. Rao and P.C. Mehta, *Proc. SPIE* **1319** (1990) 283.

8.10 G.W. Stroke and R.G. Zech, *Phys. Lett.* **25A** (1967) 89.

8.11 G.W. Stroke, F. Furrer and D. Lamberty, *Opt. Commun.* **1** (1969) 141.

8.12 G.W. Stroke, *Phys. Lett.* **27A** (1968) 252, 405.

8.13 G.W. Stroke, *An Introduction to Coherent Optics and Holography* (Academic Press, New York 1966).

8.14 G. Groh and G.W. Stroke, *Opt. Commun.* **8** (1970).

8.15 J.F. Ebersole, *Opt. Engg.* **14** (1975) 436.

8.16 K. Bromley, M.A. Monahan, J.F. Bryant and B.J. Thompson, *Appl. Opt.* **10** (1971) 174.

8.17 D. Gabor, G.W. Stroke, R. Restrick, A.F. Khouser and D. Brumm, *Phys. Lett.* **18** (1965) 116.

8.18 K. Matsuda, N. Takeya, J. Tsujiuchi and M. Shinoda,

Opt. Commun. **2** (1971) 425.

8.19 E. Marom, Opt. Commun. **6** (1972) 86.

8.20 V.V. Rao, C. Joenathan and R.S. Sirohi, Appl. Opt. **21** (1982) 3864.

8.21 A.K. Musla, M.Tech Dissertation, University of Poona, (1990).

8.22 Y.H. Ja, Opt. Commun. **42** (1982) 377.

8.23 Y.G. Jiang, Appl. Opt. **21** (1982) 3138.

8.24 D. Zhago and H.K. Liu, Opt. Lett. **8** (1983) 99.

8.25 A.B. Vander Lugt, IEEE Trans. Inf. Theory **IT-10** (1964) 139.

8.26 D. Casasent and A. Furman, Appl. Opt. **16** (1977) 1662.

8.27 J.W. Wagner, Appl. Opt. **20** (1981) 3605.

8.28 E. Marom, Appl. Opt. **9** (1970) 1385.

8.29 I.V. Kiryuschcheva and V.A. Rabinovich, Meas. Tech. USA **24** (1981) 275.

8.30 K. Hinch and K. Brokopf, Opt. Lett. **7** (1982) 51.

8.31 F. McLysaght and J.A. Slevin, Appl. Opt. **30** (1991) 780.

8.32 V.V. Horvath, J.M. Holeman and C.Q. Lemmond, Laser Focus **6** (1967) 18.

8.33 L.M. Frye, F.T. Gamble and D.R. Grieser, Bull. Am. Phys. Soc. **35** (1990) 1866.

8.34 F.T. Gamble, L.M. Frye and D.R. Grieser, Appl. Opt. **31** (1992) 652.

8.35 F.T. Gamble, P.C. Griffith and K. Mersereau, Bull. Am. Phys. Soc. **32** (1987) 11293.

8.36 G. Khitrove, L. Wang, V.Esch, R. Feinleib, H.M. Chou, R.W. Sprague, H.A. MacLeod, H.M. Gibbs, K. Wagner and D. Psaltis, Proc. SPIE **881** (1988) 60.

8.37 D. Farrant, B.F. Oreb, J.P. Modde and P. Hariharan, Proc. SPIE **1053** (1989) 110.

8.38 C. Fitzpatrick and Ed. Mueller, Proc. SPIE **1332** (1990) 185.

8.39 D. Psaltis, M.A. Neifield and A. Yamamura, Opt. Lett. **14** (1989) 429 ; Appl. Opt. **29** (1990)

8.40 F. McLysaght and J.A. Slevin, Appl. Opt. **30** (1991) 780.

8.41 M. Guoguang, W. Zhaoqi, C. Dongqing and W. Faxiang,

Optik **75** (1987) 97.

8.42   P.J. Van Heerden, *Appl. Opt.* **2** (1963) 393.
8.43   G.A. Alphonse and W. Phillips, *RCA Rev.* **37** (1976) 184.
8.44   A.L. Mikaeliane, in *Optical Information Processing,* E.S. Barrekette, G.W. Stroke, Yu.E. Nesterikhin and W.E. Kock eds. (Plenum Press 1978) **2** p. 217.
8.45   G. Pauliat and G. Roosen, *Int. J. Opt. Computing, Special issue on Photorefractive Materials* (1993).
8.46   A. Vander Lugt, *Appl. Opt.* **12** (1973) 1675.
8.47   H. Kiemle, *Appl. Opt.* **13** (1974) 803.
8.48   R.K. Kostuk, *Proc. SPIE* **1316** (1990) 150.
8.49   L. Hesselink and S. Redfield, *Opt. Lett.* **13** (1988) 887.
8.50   G. Marcinkowski, *Photonics Spectra* (Jan 1992) 113.
8.51   D. Lytle, *Photonics Spectra* (Oct. 1991) 135.
8.52   G. Eichmann and H.J. Caulfield, *Appl. Opt.* **24** (1985) 2051.
8.53   H. Mada, *Appl. Opt.* **24** (1985) 2063.
8.54   J.J. Hopfield, *Proc. Natl. Acad. Sci. USA* **79** (1982) 2554.
8.55   R. A. Athale, *Short Course, Annual Tech. Symp. SPIE,* Aug. 1987.
8.56   H.J. White and N.B. Aldridge, *Opt. Eng.* **27** (1988) 30.
8.57   H.J. Caulfield, *Appl.Opt.* **26** (1987) 4039.
8.58   N.H. Farhat, D. Psaltis, A.Prata and E. Pack, *Appl. Opt.* **24** (1985) 1469.
8.59   R.A. Athale, H.H. Szu and C.B. Friedlander, *Opt. Lett.* **11** (1986) 482.
8.60   B. Macukow and H. Arsenault, *Appl. Opt.* **26** (1987) 924.
8.61   B.H. Soffer, G.J. Dunning, Y. Owecjko and E. Marom, *Opt. Lett.* **11** (1986) 118.
8.62   C. Bhan and P.C. Mehta, *Proc. SPIE* **1319** (1990) 279.
8.63   P.C. Mehta and C. Bhan, *Def. Sc. J.* **42** (1992) 107.
8.64. P.E. Keller and A.F. Gmitro, *Appl. Opt.* **31** (1992) 5517.
8.65. R.W. Gerchberg and W.O. Saxton, *Optik* **35** (1972) 237.
8.66   W.E. Rudge, W.E. Harding and W.E. Mutter, *IBM J. Res. Dev.* **2** (1963) 146.
8.67   J. J. Murray and R. E. Maurer, *Semicond. Prod.* **5**

(1962) 30.
8.68 W.J. Tabor, *Appl. Opt.* **6** (1967) 1275.
8.69 S. Lu, *Proc. IEEE* **56** (1968) 116.
8.70 G. Groh, *Appl. Opt.* **7** (1968) 1643; **8** (1969) 967.
8.71 A. Kalestynski, *Appl. Opt.* **12** (1973) 1946; **16** (1977) 3027.
8.72 P.C. Mehta, *Opt. Commun.* **13** (1975) 70.
8.73 P.C. Mehta, C. Bhan and R. Hradaynath, *J. Opt. (Paris)* **10** (1979) 133.
8.74 Y.Z. Liang, D. Zho and H.K. Liu, *Appl. Opt.* **22** (1983) 3451.
8.75 P.C. Mehta and K.S.S. Rao, Unpublished work.
8.76 A. Kozma and C.R. Christensen, *J. Opt. Soc. Am.* **66** (1976) 1257.
8.77 P C. Mehta, C. Bhan and R. Hradaynath, *Opt. Laser Technol.* (Aug 1984) 193.
8.78 P.C. Mehta and R. Hradaynath, *Appl. Opt.* **18** (1979) 2394 and references cited therein.
8.79 P.K. Katti and P.C. Mehta, *Appl. Opt.* **15** (1976) 530.
8.80. P.C. Mehta, D. Mohan and R. Hradaynath, *Opt. Commun.* **28** (1979) 171.

CHAPTER 9

MEDICAL APPLICATIONS OF LASERS AND HOLOGRAPHY

9.1. Medical Applications of Lasers

9.1.1. Introduction

Attempts to use lasers in medicine and biology started shortly after the invention of laser in 1960. Continued interest established the use of lasers in ophthalmology and cardiovascular systems. Retinal photocoagulation, treatment of skin disorders and laser surgery still form the hardcore of laser applications in medicine. To explore potential new applications, studies are vigorously pursued to improve understanding of basic mechanisms involved in laser interaction with cells and tissues. Basic work in the area of short pulse photothermal effects, subnanosecond photodisruption and tissue specific photosensitizer-carrier systems, is expected to provide more precise laser applications in medicine and biology. Laser angioplasty and corneal surgery are examples of clinical application, which have been studied using a variety of laser systems, often using different laser tissue interaction mechanisms.

In addition to ophthalmology and cardiology, laser has also found applications in a wide variety of medical practice i.e. in diagnosis as well as surgical procedures, particularly in radiology, gastroenterology and oncology [9.1-9.6]. The clinical use of a laser is based on a compromise between many factors such as laser tissue interaction mechanism, the penetration depth of laser light in tissue and of course the availability of laser and the suitable optical fibre for transmission of radiation to the desired point of interaction. Laser-tissue interaction mechanism is a function of wavelength, power, and pulse duration. The penetration depth of laser light depends upon the optical properties (e.g. reflection, absorption, scattering) of the tissue at the selected wavelength.

The lasers used for medical applications include Ar (514nm), Nd:YAG (frequency doubled with KTP giving 532 nm) Er:YAG, Ho:YAG, CO_2, tunable dye, excimer (Ar, Kr and XeCl) for a variety of uses including dermatology, ENT, gynaecology, urology, general surgery and neurosurgery, angioplasty, pulmonology, tissue welding and stone fragmentation. The wavelength region covered spans the spectrum from UV (~ 200 nm) to mid IR (up to 10 μm).

The CO_2 laser radiation (10.6 μm) is most highly absorbed in soft tissue, which makes CO_2 laser useful for cutting and vaporizing. The argon laser (at 514 nm) and the Nd:YAG: KTP (at 532 nm) are conventionally considered as superficial coagulators due to absorption by red or black pigment. When the beam from either of the two lasers is delivered through fibre and the tip of the fibre is dragged through tissue it will cut. Argon lasers have been primarily used in ophthalmology as retinal photocoagulators. In dermatology it is used for triggering photocoagulation of skin masks caused by over growth of blood vessels such as portwine stains, angiomas and spider vessels. These lasers have also been used in otology, microlaryngoscopy and neurosurgery. The Nd:YAG laser is considered as a deep coagulator. It is also used for extreme precision in cutting. The xenon chloride excimer laser at 308 nm has proved effective in bone cutting. Ophthalmological applications of excimer lasers include photorefractive keratectomy (PRK), astigmatism correction and glaucoma filtering. The 630 nm radiation from dye laser is almost exclusively used in photodynamic therapy (PDT) and the 577 nm radiation is used for precise destruction of cosmetic vascular lesions. The 504 nm radiation from pulsed dye laser is used for lithotripsy, the fragmentation of kidney stones by shock waves set up by the laser pulse.

The solid state tunable lasers with their compact size and longer lifetimes, are now becoming popular. Diode lasers are replacing the argon lasers in ophthalmic applications. Alexandrite lasers tunable for 730-780 nm, are establishing a foothold in lithotripsy. Ti:sapphire's tunable qualities also make it an attractive alternative for many applications. The pulsed Ho:YAG at 2.1 μm performs clean vaporization of plaque in angioplasty without thermal damage to surrounding tissue. It has also application in arthroscopic surgery to repair damaged cartilage in knees and other joints. It can be used in refractive surgery of eye and ENT applications. The Er:YAG laser at 2.9 μm has superior cutting properties in orthopedics. It is predicted to make the most impact in dentistry because Er:YAG radiation can both drill holes in hard tissue and remove the soft decay of caries.

The development of newer lasers is again expected to bring their use in medicine in the years to come. Greater coverage of wavelength, better efficiency, smaller size and cost effectiveness are likely to be the considerations.

9.1.2. Laser-tissue Interaction Mechanisms

Laser treatment depends basically on the following mechanisms:

Photothermal

It is based on volumetric rate of heat production that occurs due to absorption of laser light by tissue chromophores like blood, pigment, cholesterol etc. The heat produced is used to thermally damage tissue irreversibly, as in cauterization of cancerous tissue or control of haemorrhage. The heat production can also be utilized for welding separate tissues together (e.g blood vessels, nerves). At higher rates of heat deposition, ablation occurs and is used to vaporize tumors or plaque in arteries.

Photochemical

Photosensitizing chemicals, when injected intravenously into the human body, can react with laser irradiation and produce products that lead to irreversible damage of tissues in which photosensitizer is accumulated. Photosensitizers such as haematoporphyrin derivative (HPD), which accumulate in tumors, are used for treatment of cancer with a technique called photodynamic therapy. Low level irradiation (~25 mW/cm^2) at 0.6 µm for about an hour results in photochemical reaction of HPD and destroys the tumor cells.

Photoacoustical

Photoacoustic interactions use shock waves to fragment or disrupt tissue. The high laser intensities result in a plasma which produces shock waves. This mechanism finds use in ophthalmology (cataract treatment) and urology (fragmentation of ureter stones).

Photoablation

This requires sufficient heat generation to vaporize the irradiated tissue. The depth of ablation is a function of the penetration depth of laser light and duration of the pulse. The technique enables removal of plaque from arteries and corneal refractive surgery. The removal has to be very precise, of the order of tens of µm, to prevent perforation of arteries and a fraction of µm for corneal refractive surgery. The ArF excimer (193 nm) and Er:YAG (2.9 µm) are the preferred lasers respectively.

9.1.3. Delivery of Laser Radiation through Fibres

The laser radiation has to be transmitted from the source to the target through the fibre light guide. For CW lasers, adequate fibre materials are available for the UV, visible and near IR (up to 2.5 μm). Pure silica large-core (200-1000 μm) fibres have become the standard in medical laser delivery systems because of their low cost, high laser damage threshold, biocompatibility, strength and reliability. The zirconium fluoride and sapphire fibres may extend the range to 5 μm. This would permit endoscopic application of the Er:YAG laser at 2.94 μm. For pulsed lasers, electric breakdown in fibre material at high intensities of nanosecond and picosecond pulses needs to be overcome. The fibre coupling of 193 nm laser pulses from excimer laser poses problems. There are still difficulties in finding materials for CO_2 laser at 10.6 μm though first generation fibres with higher loss are available.

For laser delivery through fibres, the sharp fibre tips are undesirable as they perforate arteries even when the laser is not on. Efforts to eliminate sharp tip have resulted in several tip designs (Fig. 9.1) such as

(a) hot tip, in which the sharp end is covered by a rounded metal cap heated by the laser radiation,

(b) ball tip, in which the end of the fibre is melted to form a ball,

(c) window tip, in which a transparent rounded sapphire or silica window is mounted at a fixed distance from the end of the fibre,

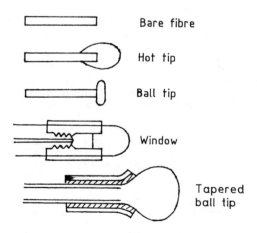

Fig. 9.1. Optical fibre tips commonly used for laser angioplasty.

(d) a tapered ball tip, which produces a bell shaped radiant exposure profile on the tissue surface in contact with the ball tip.

The fail safe design of the fibre delivery system ensures that when laser intensity is too high the input end is destroyed before the intensity gets high enough to damage the output end, thus preventing an undesirable fibre failure inside a patient. If the fibre beaks, it is held together by the teflon coating and does not break off.

The fibre optic transmission of laser energy may pose difficulties when high peak power and short duration pulses approach the damage threshold of fibre or cladding medium. The use of mid infrared wavelengths requires infrared fibres (made of chalcogenide and silver halide materials) which are yet not perfected for low loss transmission. The excimer laser radiation at around 300 nm can however be transmitted by the fused silica fibres.

Laser light can directly reach the eye, retina, and skin. With endoscopy fibres, laser irradiation can reach the mouth, larynx, pharynx, trachea, oesophagus, stomach, common bile duct, pancreas, caecum, sigmoid colon, rectum, urethra, bladder, ureter, vagina, cervix and uterus. With a minor invasive surgical procedure, laser light through fibres can also reach arteries and veins, coronaries, the heart ventricles and atria, and the inner most parts of the brain.

9.1.4. Study of Tissue Fluorescence as a Diagnostic Aid

During the last few years, laser induced fluorescence has found increasing use in biomedical studies, both at fundamental cell level as well as for tissue diagnostics (e.g. detection of malignant tumors). Fluorescence characterization of tissue relies on different spectral properties of normal and diseased tissue. Apart from using the natural fluorescence of the tissue (autofluorescence) certain fluorescent agents with enhanced affinity to certain tissue types can be injected systematically to enhance the discrimination. Investigations of tissue based on laser induced fluorescence use a pulsed or continuous laser, mostly in UV region. For capturing the fluorescence, a diode detector matrix, along with a low light level image intensifier, is used. Gating the intensifier rejects the ambient light, allowing faint fluorescence signal to be recorded in full daylight

The removal of plaque in obstructed vessels by laser radiation administered through a fibre optic light guide (laser angioplasty) has recently attracted great attention. In this operation, reliable identification of the target under irradiation is needed to avoid vessel perforation. It has been suggested that fluorescence characterization of the vessel wall could be performed via the same fibre as that used for the delivery of high power laser pulses for plaque removal. Using nitrogen laser radiation at 337 nm, very prominent differences between atherosclerosis lesions and normal artery wall can be found in the fluorescence spectra, both for the large arteries such as aorta, arteria femoralis and pulmonalis, as well as for coronary arteries. The influence of blood can be overcome by observing that blood only acts as a light attenuator and does not change the decay characteristics of observed fluorescence. Time resolved techniques provide an alternative approach to increase the sensitivity and selectivity of fluorescence based diagnosis. This is so because, even though the spectra of different fluorophores may overlap, their fluorescence decays usually differ appreciably. Fluorescence techniques are important in ophthalmology for early diagnosis of diseases such as cataract and diabetic retinopathy. Studies have ben performed on the pigments (melanin and lipofuscin) of the retinal pigment epithelium to evaluate age-related changes in their fluorescence emission.

To collect tissue fluorescence spectra using fibre optics, a multifibre laser catheter like the one shown in Fig. 9.2 is used [9.3]. The device consists of a central fibre, used to illuminate the tissue, and six collection fibres, used to collect resulting fluorescence. A trans-

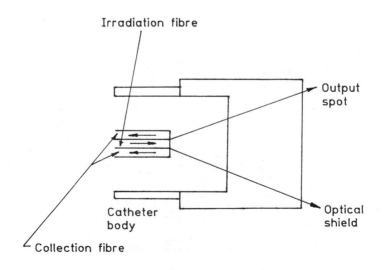

Fig. 9.2. Diagnostic laser catheter.

parent optical shield encases all of the fibres and is used to displace any intervening fluid between the tissue and the device, as this is a contact system. The optical fibres are fixed within the shield and positioned so that the collection fibres view only the irradiated portion of the tissue. By setting the optical fibres at a preselected distance from the optical shield, a well defined spot diameter or irradiated area of tissue is produced. The device is designed to be placed in contact with the tissue to be analyzed. The delivery/collection geometry of this device is then well defined, and the intensities of the fluorescence return signals can be calibrated.

Output spot diameters range from 0.5 mm to 1.5 mm with catheter of outside diameters of 1 to 2.5 mm. The number, dimensions, and arrangement of the optical fibres can also be varied to maximize collection efficiency. Typical fibre core diameter range from 100 to 200 μm. The optical fibres and shield are made of UV grade quartz, thus allowing for a variety of operating wavelengths of laser, both CW and pulsed. These diagnostic laser catheters are constructed to be flexible, using biocompatible materials, and incorporate radiologically opaque markers for angioscopic guidance for *in-vivo* use. Collection of spectra with the laser catheter is shown in the block diagram of Fig. 9.3. The collection fibres of the device are imaged at the entrance slit of a single monochromator. Other detection systems can also be used.

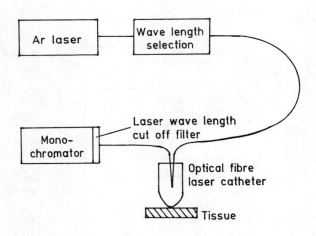

Fig. 9.3. Probe collection system.

9.1.5. Medical Applications of Solid State Lasers

A number of solid state lasers have recently appeared which have the advantages of compactness, portability, temporal mode versatility and a large number of wavelengths. These are proving viable competitors to dye lasers and argon lasers in the visible region. They provide tunability, as in the case of Alexandrite, and some newer wavelengths in the near IR around 2-3 μm. Thulium laser provides 2.01 μm, Er:YAG 2.9 μm, Co:MgF 1.75-2.5 μm and Raman shifted Nd:YAG 1.9 μm. The use of these solid state lasers in medicine is briefly summarized as follows:

Nd:YAG

There are different modes of operating this versatile laser e.g. CW, frequency doubled and frequency tripled, free running pulsed and Raman shifted to 1.9 μm.

The new delivery accessories for CW version include dual effect sculpted fibres (Fig. 9.4) that are truncated to allow a higher proportion of delivered energy in the forward direction but not as much as a bare flat ended fibre would. Compared to bare fibres (Fig. 9.4), these dual effect fibres provide improved surgical precision. For highest precision in contact cutting however, standard sharpened fibres (Fig. 9.4) are preferred. It is desirable to monitor and control the temperature of sculpted fibre tips in order to reduce tip fracture and loss of tip sharpness.

The latest versions of frequency doubled Nd:YAG surgical products provide significantly increased power capability, up to 36W at 532 nm and 100W at 1064 nm. Higher power enhances their utility for general surgery applica-

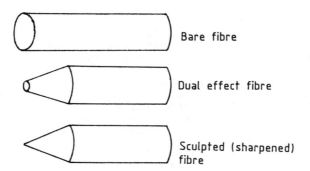

Fig. 9.4. Comparison of fibre optic tip shapes. Dual effect fibre represents a compromise between a bare fibre and a sculpted fibre.

tions that involve extensive tissue cutting or debulking of larger tissue masses. The applications include percutaneous lumbar diskectomy and treatment of vascular skin lesions using scanning delivery systems. Frequency doubled Nd:YAG lasers may also be used in place of argon lasers for some ophthalmic procedures and as potential modalities for refractive surgery.

Frequency tripled Nd:YAG systems operating at 355 nm have potential applications in coronary laser angioplasty where they may replace 308 nm XeCl excimer lasers and holmium lasers.

Nd:YAG lasers in free running pulsed mode (pulse energies of 100 to 700 mJ with pulse duration 150-800 μs) have gained popularity for dental applications. They are used for soft tissue cutting procedures, such as gingivectomy, frenectomy and drainage of abscesses. Hard tissue procedures on bone and teeth include caries removal, etching of dentine, pit and fissure sealing, and preparation of dental implant sites in bone. In addition to dentistry, free running pulsed Nd:YAG lasers are also being used for ophthalmic procedures such as transcleral cyclophotocoagulation and trabeculoplasty. Recently the use of this laser for arthroscopic surgery has also been suggested.

Nd:YLF

Mode locked Nd:YLF lasers operating at 1053 nm and 527 nm (frequency doubled) are used for ophthalmic applications. The picosecond pulses (typically 60 ps wide at 300 μJ energy and 1 kHz repetition rate) allow one to perform retinal photocoagulation and photodisruptive ophthalmic procedures with high degree of precision. Nd:YLF lasers with diode pumping have been developed for this purpose.

Er:YAG

The operating wavelength (2.936 μm) of erbium doped YAG laser is strongly absorbed by water component of soft tissue and by the water, organic and calcium phosphate components of hard, calcified tissue such as bone and teeth. Use of this laser in pulsed mode improves the precision in ablating the tissue and reduces thermal injury at cut tissue edges. Hemostasis is correspondingly poor. The transmission of 2.9 μm through optical fibre poses problems. Future infrared transmitting fibres (such as zirconium fluoride) are expected to remove this difficulty.

Ho:YAG

Introduced in 1990, surgical holmium lasers are rapidly finding a place in the medical field. These are free running pulsed devices which typically emit 1 to 2 J of pulse energy, with pulse durations of 250 to 350 μs. Pulse rates

10 to 20 Hz and average power of 20 to 30 W are typical. The operating wavelength 2.1 μm is absorbed strongly by the water component of tissue, resulting in shallow tissue penetration and good surgical precision. Compared to CO_2 laser, it provides better hemostasis in most tissues.

Alexandrite

Tunable over 720 to 820 nm, Alexandrite lasers offer deep tissue penetration capability at high peak power and high pulse energy. When Q-switched, these lasers are useful for lithotripsy. Pulsed duration of about one μs are typical with pulse energy of 200 mJ. Q-switched Alexandrite lasers provide alternative to pulsed dye laser lithotripters. Dual wavelength versions of Alexandrite laser (frequency doubled 375 nm initiating shockwave for fragmentation, followed by 750 nm pulse to break up the stone) may be able to efficiently fragment a wide variety of stones with less total pulse energy and power than single wavelength systems. Frequency doubled Alexandrite lasers also find use in dental applications. Pulsed free running Alexandrite lasers can be used for tattoo removal, as well as treating benign pigmented lesions. For this application, Alexandrite laser pulse duration of several hundred μs and pulse energies of several joules are employed. Raman shifted (anti Stokes) Alexandrite lasers could become an alternative to dye lasers for applications in dermatology, ophthalmology and photodynamic therapy.

Thulium Laser

Laser wavelength at 2.01 μm or 1.96 μm from thulium lasers are closer to the water absorption peak at 1.93 μm. They may, therefore, provide improved cutting and vaporization compared to holmium lasers:reduced thermal injury at cut edges and enhanced surgical precision. They are also likely to replace holmium lasers for coronary angioplasty.

$Co:MgF_2$

Cobalt lasers can be tuned continuously throughout a nominal range of 1.75 μm to 2.5 μm, completely through the water absorption peak at 1.93 μm. They may therefore provide a means to continuously vary the surgical precision between that of a holmium laser and a CO_2 laser. Cobalt lasers are also being considered as an alternative to holmium lasers and thulium lasers for angioplasty.

AlGaAs Diode Lasers

Semiconductor diode lasers that provide watts of power at room temperature are currently available. AlGaAs diodes giving radiation in the 790-860 nm range have already found therapeutic applications in ophthalmology for trans-pupillary, endo-ocular, and transcleral photocoagulation procedures. The deeply penetrating 800 nm wavelength, absorbed strongly by tissues that contain melanin, is claimed to offer advantages over blue-green argon laser wavelengths where intervening blood or tissue must be penetrated in order to treat targeted structures. Ophthalmic diode lasers are very compact and lightweight, and are air cooled devices (convection or forced air). They can operate from a battery pack and are portable. Table 9.1 lists some important terms used in medicine.

Table 9.1 Some medical terms relevant to the use of lasers in medicine and surgery.

Term	Meaning
Endophotocoagulation	Photocoagulation of the retina or other intraocular structure.
Endoscope	A thin, rigid, or flexible device that relays images of objects at its distal end to a viewing eye piece at the proximal end.
Frenectomy	Removal of a portion of the web like piece of soft tissue attached to the middle portion of the upper lip and upper gum line.
Gingivectomy	Removal of gum tissue.
Haemostasis	Control of bleeding by cauterization (heat sealing) of blood vessels at cut tissue edges.
Hyperthermia	Raising an entire body, or certain body parts, to temperatures a few degrees Celsius above normal body temperature.
Laparoscopic Cholecystectomy	A surgical procedure in which gall-bladder is removed through a small abdominal incision.
Angioplasty	Removal of arterial plaque that partially or completely blocks arteries.
Lithotripsy	Fragmentation of stones (calculi) in the body.
Photoemulsification	To soften or break up, opacified lens into small pieces and to remove them

Table 9.1 contd.

Term	Meaning
Thermal Keratoplasty	through a small eye incision. Reshaping of cornea by gently heating the cornea without coagulating or vaporizing corneal tissue.
Percutaneous lumbar diskectomy	Partial removal of herniated vertebral discs in order to decompress the spinal cord.
Posterior capsulotomy	Transpupillary photodisruption of the posterior lens capsule in order to restore clear vision. The capsule is normally a clear membranous bag that encloses the lens of the eye. Clouding of the posterior capsule occurs in cataract surgery.
Refractive surgery	Surgical methods to permanently change the curvature/refractive power of the cornea to eliminate the corrective eye wear.
Sclerostomy	Creation of a small hole in the sclera (white part of the eye) to facilitate drainage of aqueous fluid and control of intraocular pressure. This is the treatment for glaucoma.
Cyclophoto-coagulation	Treatment of glaucoma.
Vitrectomy	Removal of opacified vitreous humour to restore unobstructed vision.

9.1.6. Use of Lasers in Ophthalmology

Photocoagulation by laser in ophthalmology was the first important medical application of laser and still remains in the forefront of laser usage in ophthalmic surgery.

Lasers have been used in eye surgery for more than twenty five years. They can cut the tissue (with Nd:YAG laser) or coagulate (with Ar laser). We list below four important usages which have gained acceptance over a period of time.

Retinal Photocoagulation

One of the most familiar and earlier laser surgery in the eye is photocoagulation of retinal tissues. It is a thermal interaction. If sufficient heat is generated by the absorption of laser light in the tissue, photocoagulation occurs and a lesion is formed.

The retina is the thin sensory tissue that rests on the inner surface of the back of the eye and is the actual transducer of the eye, converting light energy into electrical pulses. On occasion, this layer will detach from the wall, resulting in the death of the detached area from lack of blood and leading to partial, if not total, blindness in that eye. A laser can be used to 'weld the retina into its proper place on the inner wall. Laser coagulation of clear retinal tissue is accomplished by heat conduction. Laser light is absorbed in the pigment epithelium and conducted to the retina.

A disease associated with diabetics is diabetic retinopathy where the retina is not receiving enough nutrition. If left untreated, blindness will result from haemorrhage and traction detachment. This is treated by destroying the peripheral retina, by placing strategically hundreds of lesions on the retina, and allowing only the small foveal area to survive. This improves the retinal metabolic balance (i.e. nutrition fed to the useful area is sufficient to keep the area alive). The patient loses night and peripheral vision but maintains the critical central vision. Retinal photocoagulation can also be used to cauterize small blood vessels growing into inappropriate areas.

The parameters to be controlled in photocoagulation based treatment are the beam energy and wavelength, and the size and position of the spot focussed on the retina.

Corneal Refractive Surgery

Recently interest has grown in the use of lasers for corneal reshaping (radial keratotomy). In this procedure, a UV laser (ArF excimer at 193 nm or 4th or 5th harmonic of Q-switched Nd:YAG) makes incision in the cornea by ablative photodecomposition in order to change the corneal curvature to compensate for myopia. This of course requires proper selection of laser parameters such as wavelength, pulse duration, energy density etc. and a practical beam delivery system to handle very short wavelengths. Additional parameters to control are the depth of incision (it must not perforate the cornea) and eye motion. Q switched Er:YAG (wavelength 2.94 μm) has also been used for recontouring the cornea surface by using axicon beam shaping. Astigmatic errors could be corrected by controlling the symmetry of the tissue removed.

Cataract Capsulotomy

With enough photons in a small volume, electric fields become so intense as to strip electrons off atoms, ionizing them and producing a plasma. This phenomenon is the basis for photodisruptors in the eye. With a very short focus (f/3, 0.2 mm waist) and energy delivered in less than 30 ns, it is possible to blow holes in the nearly transparent membranes, which sometimes obscure vision after cataract surgery. For this purpose, Nd:YAG lasers at 1064 nm deliver typically 10^{12} W/cm^2. The retina is protected both by the reflecting plasma and by the beam divergence, which increases the area by a factor of 400. Beam aiming and focus, as well as pulse shaping and timing, are critical.

Glaucoma Surgery

In Glaucoma, pressure in the eye is raised. Both photocoagulation (thermal) and photodisruption (ionizing) are used to treat glaucoma by reducing the amount of aqueous humour in the chamber between the cornea and the lens or iris of the eye. Laser light, in this method, makes a drainage hole through the iris (iridectomy), or seals off some of the fluid generating trabecular meshwork (trabeculoplasty). In this treatment, aiming and focus control are both important. argon, Nd:YAG and CO_2 lasers have been used.

Diagnostic Aids

The use of lasers in diagnostic instrumentation is quite varied. Interference fringes or laser speckle formed at the retina are used to assess contrast sensitivity, a complete measure of visual acuity. Laser Doppler velocimetry measures the speed of red blood cells in the vessels visible inside the eye. Imaging with a scanning laser ophthalmoscope allows use of lower light intensities than conventional instruments and extends to a wide range of psychophysical tests the ability to observe the retinal locus where the patients response is being measured. Scatter, fluorescence and Raman shifted light are all collected to evaluate properties of the eye.

Laser's spatial coherence is used to form interference fringes on the retina. As the spacing of the fringes is varied the patient indicates whether or not they are visible. Fringe contrast is also varied to yield retinal MTF. Laser measurement of contrast sensitivity is a typical psychophysical phenomenon; patients respond to changes in the stimulus presented, and these responses form the data. Stimulus parameters (e.g. fringe spacing, orientation, contrast) are manipulated by the experimental setup (Fig 9.5). Fringes are formed at the retina by the overlapping of two laser beams. Only the components with similar polarizations interfere, so the contrast is variable without change in irradiance. The laser used is generally a He Ne and the retinal irradiance is in the range of 1 $\mu m/cm^2$.

A mat surface illuminated with laser (coherent) light appears speckled by dark and bright flecks. There are local

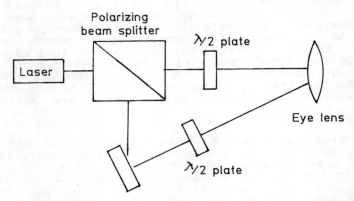

Fig 9.5. Measurement setup for contrast sensitivity of eye.

interference areas caused by the scatter of coherent light by a surface whose roughness is at least half a wavelength in depth. If the surface is moved and the speckle is seen as moving with it, then the eye is myopic. A hyperopic eye sees motion in the contrary sense. The actual instrumentation requires only a rotating cylinder with a rough surface and an expanded laser beam for retinal irradiances of less than 1 $\mu W/cm^2$.

The transparency of the lens changes normally with age, and sometimes degenerates entirely with the formation of a cataract. These changes are evaluated in scattering studies and by fluorescence and Raman emission, using high radiance laser light. Fluorescein dye may be safely introduced into the eye, where its rate of transport is an important diagnostic indicator. Fluorophotometers assess the dye concentration at locations within the transparent media.

The retina image is the most commonly used diagnostic procedure in ophthalmology. We can use the lasers high radiance to see the retina more efficiently with an electro-optical imaging device called the scanning laser ophthalmoscope.

9.1.7. Angioplasty with Lasers

Laser angioplasty is a method of opening arteries that are obstructed by atherosclerotic plaque. It has advantages over conventional surgery, balloon angioplasty and other forms of vascular interventions. Laser radiation may be introduced into arteries via small optical fibres, thus avoiding major surgery. The radiation can remove plaque rather than displacing it, thus reducing the high rate of restenosis that occurs with balloon angioplasty. Radiation can be preferentially absorbed by plaque, thereby adding an element of safety that may not exist with mechanical devices. Problems like frequent perforations, dissections and damage to the underlying normal artery wall can be avoided by careful selection of laser parameters. The massive thermal injury that occurs with continuous wave lasers can be minimized by using pulse durations that are short enough to make thermal diffusion negligible. The pulse duration should be much shorter than the tissue thermal relaxation time. Calcified plaque can be removed with high intensity radiation and at a wavelength where plaque absorption is much greater than normal artery absorption. Precise ablation is achieved at wavelengths where plaque absorption is strong or can be enhanced with exogenous chromophores.

The principal appeal of cardiovascular laser therapy, in which pathologic cardiovascular tissues are ablated with laser energy, revolves around the fact that laser energy can be transmitted via optical fibres and thus allows percutaneous solution to interventions that previously required open heart surgery. The concept of percutaneous laser angioplasty is derived from percutaneous balloon angioplasty. The latter employs a balloon mounted on the tip of a catheter which can be advanced through a needle puncture of the skin via the peripheral arterial system to the coronary arteries using fluoroscopic guidance. The balloon is then positioned within the atherosclerotic lesion and when inflated, typically reduces the degree of narrowing caused by the lesion. Laser angioplasty offers the potential to physically eliminate, rather than simply remodel, a portion of the atherosclerotic lesion.

To satisfy the requirement of laser angioplasty, the laser beam must be interfaced or coupled to the proximal face of the fibre and the fibre must be delivered in appropriate fashion to the coronary arterial lesion. Cardiovascular laser systems, therefore, differ markedly from those developed for dermatologic, ophthalmologic and other medical applications by virtue of the fact that the laser power source represents only one of four major components of the clinical system, the other three being the fibre, the coupler and the delivery catheter.

Much of the earlier work on angioplasty was centered around argon, Nd:YAG, and CO_2 lasers. However, use of these lasers has caused perforations, aneurysms, pain and arterial spasm emanating from injury to normal arterial structures adjacent to or underlying the plaque. The excimer laser operating in the UV can make well defined cuts in biological material with very little injury to the adjacent tissue. Also, the strong UV absorption by arterial tissue leads to a shorter penetration depth of the radiation; smaller etch depth per pulse may provide greater operator control of the ablation process.

Laser endarterectomy, a method of reopening occluded arteries, has been limited by difficulties in ablating calcified plaque. A study has demonstrated that pulsed visible radiation (482 nm, 1 µs duration, 40-80 mJ/pulse) can ablate calcified plaque at intensities (50 MW/cm^2) that are readily transmitted down flexible (320 µm diameter) optical fibres and at flouences which are below the threshold for ablating normal artery. The ablative process removes material primarily by formation of a plasma and fracturing of the plaque, rather than by thermal vaporization, leading to ejection of fine debris and a luminescent plume. The ablation efficiency is as high as 100 mg/J.

Images of the calcified regions of arterial samples can be produced based on the observation that the autofluorescence between 480 and 630 nm is considerably more intense for calcified tissue than for noncalcified tissue. This allows for a unique identification of calcified areas in the tissue. The clinical application of this technique are very promising. Low intensity visible laser radiation may be transmitted through an optical fibre advanced through an intra-arterial catheter. Detection of the autofluorescence peak intensity as a function of the position of the fibre tip would yield complete images of the inner arterial wall for diagnostic purposes. The development of a laser angioplasty system permitting alternative fluorescence imaging and ablation can improve the safety of laser angioplasty. Multicolour imaging systems for tissue fluorescence diagnostics have also been proposed leading to a major impact in diagnostic medicine. Fluorescence imaging systems for tumour localization have been constructed.

9.1.8. Other Medical Applications

Laser radiation is now routinely used in medical applications to cut, shape, treat and remove soft tissues of the body. Hard (calcified) tissues have been removed. Calcified plaque in vitro have been ablated utilizing a pulsed XeCl excimer laser operating at λ=308 nm. Kidney stones *in vivo* have been removed utilizing flash lamp pumped dye laser light delivered through an optical fibre. High irradiance pulsed Nd:YAG laser radiation at 1064, 532 and 266 nm cuts through bone and calcified plaque almost as easily as it cuts through the soft tissue of normal arterial walls. Wavelengths at 2.1, 2.9, 3.0 and 10.6 µm from Ho:YSGG, Er:YAG, free electron and CO_2 lasers have also been used for studying hard tissue ablation. The near-IR tissue ablation is associated with the very strong liquid water absorption peak centered at λ=2.9 µm. All three major tissue constituents of bone, hydroxyapatite, collagen and water have absorption peaks in the 3 µm region. The HF laser, tunable from 2.6 to 3.1 µm, with well shaped clean pulses of variable width and good spatial quality beam, is very well suited to the study of tissue ablation. It is effective for ablating both hard and soft biological tissue with clean cutting and very little heat transfer to the surrounding tissue.

The use of infrared lasers for removing tissue is hampered by the nonavailability of optical fibres for energy transmission at these wavelengths. Therefore lasers operating at wavelengths shorter than 2.5 μm have been investigated. The pulsed holmium laser operating at 2.1 μm has been shown to produce reasonably clear and rapid removal of a variety of tissues inluding liver, stomach, colon, bone, and cardiac valves. Typically it produces zones of residual thermal injury extending from 0.4 to 1.0 mm from the ablation site. In comparison, the CW Nd:YAG laser produces thermal injury zones of several millimeters. Tissue is approximately 80% water and water has a strong absorption band at 1.94 μm. Thus, use of a laser at a wavelength closer to this peak of absorption band is likely to produce cleaner incisions than the holmium laser. A thulium laser operating at 2.0 μm has been used; at this wavelength the absorption coefficient of water is 65 cm^{-1} as compared to 24 cm^{-1} at 2.1 μm.

In dermatology, lasers can be used to remove portwine stains and birth marks. Argon laser is also used for this purpose. Skin incision can be made with CO_2 laser. Tatoos can be treated using either the carbon dioxide or argon laser by the removal of skin down to the layer of pigment.

Carcinoma of the lung gives extreme distress due to suffocation caused by obstruction of one of the main air passages by tumour. Significant palliation can be achieved by removal of this tumour using either the Nd:YAG or CO_2 laser. The laser is used with a bronchoscope under anaesthesia.

Argon and Nd:YAG lasers have been used to control the bleeding of the ulcers of the upper gastrointestinal tract. The laser radiation is transmitted through a flexible fibre introduced through the biopsy channel of a flexible fibre-optic gastroscope. The acute bleeding can be controlled in 70 to 100% of cases. In cases of cancer of the oesophagus, which causes inability to swallow, the disorder can be relieved by removal of obstructing tumour using the Nd:YAG laser.

In gynaecology, the premalignant condition of the cervix at the neck of the uterus can be treated with a CO_2 laser to vaporize lesions precisely and without significant bleeding. In large majority of cases, the use of anaesthetic is not required. In addition, there is no scarring of the cervical canal.

In neurosurgery, lasers offer the true no-touch surgery

without any mechanical interference to vital areas of the brain and without even a temporary interruption of the circulation. CO_2 laser can be used to incise and remove tumours and to create precise lesions of the spinal cord for the relief of pain. A three dimensional computer model of a tumour is created from CT scans and the laser is controlled by computer, ensuring complete removal of deep-seated lesions. The Nd:YAG laser can be used to debulk haemorrhagic tumours with good control of bleeding. It is also used to control both normal and abnormal blood vessels. The argon laser is also used for the no-touch precise control of normal blood vessels and vascular malformation of the brain.

Surgery to the air passages can be performed in otolaryngology with the CO_2 laser with precision, no bleeding, no postoperative swelling and with reduced scar formation and contracture. There is suggestion that early cancer of the vocal cords may be treated by laser excision, rather than by radiation therapy, giving an equally high cure rate of 80 to 90%. Lesions can be removed from the mouth and tongue with great precision, minimal bleeding, rapid healing, and a remarkable lack of postoperative pain. The Nd:YAG laser is used in urology to destroy tumours of the bladder via a cystoscope, without a general anaesthetic. Both the tumour within the bladder and tumour within the bladder wall can be destroyed without the risk of perforation. The use of lasers in photodynamic therapy has been established. It has been shown that after intravenous injection, haematoporphyrin derivative (HPD) is widely distributed in the body, but is then selectively retained by malignant tissues. Three days after injection, HPD will have been cleared from the majority of normal tissues, but will still be present in any area of malignancy. If a tumour containing HPD is exposed to blue violet light, it will fluoresce. Studies have ben carried out on the diagnosis of early lung cancer using a fibre optic bronchoscope that incorporates a krypton laser as the light source and an image intensifier to detect the areas of tumor fluorescence.

If the HPD in a tumour is exposed to red light, singlet oxygen is produced which is cytotoxic to the cells which contain HPD, but sparing surrounding normal tissues. Early lung cancer has been diagnosed and cured by this technique.

9.2. Medical Applications of Holography

Recent development in holographic techniques have shown that they are gaining importance in biomedical research also. The holographic techniques have been used with a certain degree of success in dentistry, urology, otology and orthopedics [9.7-9.9]. Most of the applications use holographic interferometric techniques. The radiological applications of holography include the synthesis of three dimensional images from a series of two dimensional images using the techniques of holographic cylindrical stereogram, holographic conical stereogram and multiplex hologram [9.10] (section 4.12). While none of the medical applications of holography has matured into a commercial instrument, the research work that is being carried out shows such a promise. Holographic endoscopy is emerging as a powerful instrument for medical applications.

9.2.1. Holographic Endoscopy

Endoscopic holography combines the features of holography and endoscopy. It has opened up the possibility of non contact high resolution 3D imaging and nondestructive measurements inside natural cavities of the body or in any difficult to access environment [9.11 - 9.21]. The ability to record a 3D large focal depth and high resolution image of internal organs and tissues greatly enhances the detection capability. The holographic endoscopy may be of two types. In the one form the hologram is recorded inside the endoscope, while the other form uses an external recording device.

The internal hologram recording endoscope requires a miniaturized holographic setup inside the instrument and records a reflection hologram. It mainly consists of three parts; a film cartridge, a diaphragm and a single mode optical fibre (core diameter 4 μm) cable. The three parts are assembled in three adjustable stainless steel or brass tubes. The film is placed at 10° to the normal to the endoscope. Figure 9.6 shows the diagram of a holographic endoscope [9.17].

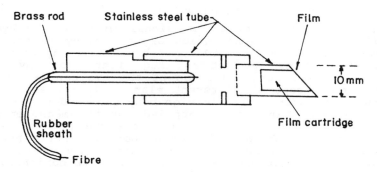

Fig. 9.6 Schematic diagram of a holographic endoscope [9.17].

The holograms are viewed under a powerful microscope allowing for the observation of individual cells. A lateral resolution of 7 μm has been obtained [9.20] in the reconstructed image which shows that the technique can be used for cellular structure analysis and may even substitute biopsy in tumour diagnosis. Specific dyes can be used to enhance the contrast of the tissue before recording the holograms as has been used extensively in gynecology and gastrointestinal tract.

The external holographic endoscope records the hologram outside the endoscope using an external reference beam. Endoscopes with extremely small outer diameter can be used but this results in a loss of parallax and a small entrance pupil producing speckles in the reconstructed image. However image plane holograms can be recorded to reconstruct the image with a white light source reducing the problem of speckles [9.21].

In order to obtain a high signal-to-noise ratio, the holographic endoscope should use imaging systems with a high numerical aperture and a low f/no.. This can be achieved by using gradient-index (GRIN) rod lenses. The speckle noise is reduced by illuminating and imaging the object by the same GRIN lens [9.13, 9.14, 9.19]. Panoramic imaging may be used by incorporating a panoramic lens in the holographic endoscope [9.12]. Photorefractive crystals can be used as holographic storage devices in a holoendoscope [9.21]. These will make a new class of instruments for use not only in medical diagnostics but also in industrial testing.

Holographic endoscope may be attached to a salpingoscope for fallopian tube investigations or to otoscope for the inspection of outer and middle ear via an acoustic system to generate vibrations of the tympanic membrane [9.20].

9.2.2. Holography in Ophthalmology

Recording of a three dimensional image of the eye was one of the earliest application of holography in the field of ophthalmology [9.22, 9.23]. A fibre bundle may be used to illuminate the eye to get an image free from any shadow [9.24]. The reconstructed image of the eye can be examined under a microscope at different planes [9.25]. Any retinal detachment or intraocular foreign body can be detected.

Holography can also be applied for the measurement of corneal topography and crystalline lens changes and for the study of surface characteristics of both the nerve head and the cornea. Current methods of determining the shape of the central surface miss the central part and its periphery. The major advantage of holographic technique is the ultra high precision (sub-μm range) with which such measurements are possible [9.26, 9.27]. Figure 9.7 shows the interferogram of a cornea with 3 diopter of astigmatism [9.27]. The number of fringes in x-direction are more than in the y-direction showing a plus axis of astigmatism in the retinal plane.

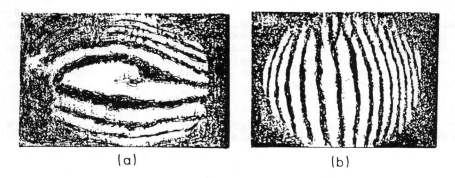

Fig. 9.7 The interferogram of a cornea in x-direction (a) and y-direction (b) [9.27].

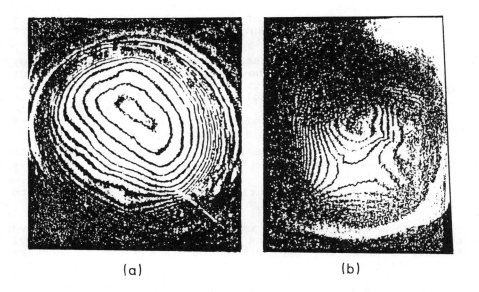

Fig. 9.8 Elastic expansion of the cornea of two fresh enucleated bovine eyes revealed by holographic interferometry [9.28].

The elastic expansion of the cornea can also be measured by holographic interferometry. This information is vital for corneal surgery. Förster et al. [9.28] have examined the expansion of the cornea of fresh enucleated bovine eyes as a result of a small increase in interocular pressure. Their first investigations have revealed that each bovine cornea has its own typical expansion as shown in Fig. 9.8. Differences between the central and perilimbal cornea have also been observed.

9.2.3. Holography in Dentistry

Holography can be applied for dental research in a variety of forms. Some of these are:

(a) Holographic toothprints can be stored for identification and forensic purposes,

(b) Detection of defective gold solder joints used in dental bridges [9.29],

(c) Contactless measurement of *in-vivo* tooth mobility and movement in three dimensions [9.30],

(d) Prosthetic application such as measurement of dimensional changes of the tissue bearing surfaces of maxillary full dentures due to deformation of dentures material by oral fluids [9.30, 9.31],

(e) Holograms can be used for storing orthodontic study models which can be retrieved by a laser beam or a white light source for accurate 3D measurements. More than 1000 holograms could be stored in the space of 16 dental models [9.32]. The holographic images are clinically reliable and random errors are not clinically significant [9.33],

(f) Holograms can be employed as training aids in the disciplines of dental anatomy and operative dentistry and

(g) Holographic contouring techniques can be used to reveal the topography of teeth [9.30].

Both continuous wave and pulse laser holography have been used for applications of holography in dentistry. Another interesting application of lasers is to measure luster changes from teeth *in-vivo* to determine effectiveness of dentifrice abrasivity [9.34]. Laser speckle techniques have been used to evaluate procedures for tooth colour restoration [9.35].

Altshuler et al. [9.36-9.38] have recently considered human tooth as an optical device and used their theory for calculation of tooth transparency and optical topography of tooth. The theory shows that human tooth is a natural waveguide which transports the optical energy and information to the soft tissues.

9.2.4. Holography in Otology

Otology has also attracted the attention of the scientists as many of the structures related to ear act as vibration transmitters. The holographic interferometric techniques are powerful for studying vibrational analysis of these structures [9.39-9.42]. The holography has been applied for each of the three parts of the human peripheral hearing organ viz. the outer ear, the middle ear and the inner ear. The ear is embedded in the temporal bone which form a part of the skull base. The following holographic investigations have been carried out:

(a) Vibration patterns of macerated human skull have been studied using time-average holographic interferometry [9.39]. The sample was excited using a bone conduction vibrator.

(b) The vibrational behaviour of inner ear parts such as an unrolled cochlear and a coiled basilar membrane have been studied [9.39, 9.43].

(c) Time-average holographic interferometry has been used for the study of vibrational analysis of incudo-mallar joint with forces applied to the middle ear muscles [9.44]. The results demonstrated that the incus and malleus move like a lever around a frequency dependent axis.

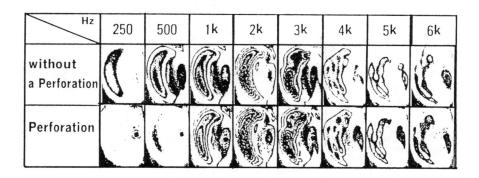

Fig. 9.9 Vibrational patterns of human tympanic membrane obtained by time-average holography [9.39].

(d) Vibrational analysis of human tympanic membrane has been done with time-average holography [9.39]. The tympanic membrane has also been studied with a ventilation tube and a perforation [9.41]. The study has revealed that after the insertion of the ventilation tube, the overall vibrational pattern of the tympanic tube remains unchanged except a decrease in the amplitude of vibration at 500 Hz. The vibrational pattern of perforated tympanic membrane also does not change, except a decrease in the vibrational amplitude at low frequency. This has been shown in Fig. 9.9 [9.41].

9.2.5. Miscellaneous Studies

Holographic interference microscope has been used to investigate the changes in stable cell line LSTC-SF1 (size 50 μm) infected with a virus causing Aueski disease [9.45, 9.46]. The holographic interference microscope is shown schematically in Fig. 9.10. The microscope utilizes a

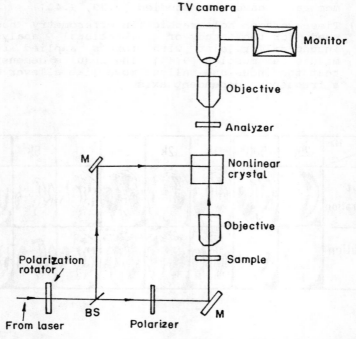

Fig. 9.10 Schematic diagram of a holographic interference microscope [9.46].

magnification of the sample. The hologram is recorded in a photorefractive crystal. Figure 9.11a shows the holographic image of the cell 4 hours after infection with the virus. The black lines are the contours of the cell. A fringe structure superimposed on the image of the cell is due to a slight wedge in the photorefractive crystal. Fig 9.11b which shows in fact the difference between the first hologram and the hologram recorded 10 hours after the infection of the cell. The results reveal that the virus strain causes relatively rapid change in the cell system [9.46].

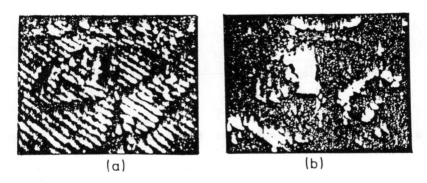
(a)　　　　　　　　　　(b)

Fig. 9.11 Two holographic images of the cell LSTF-SF14 [9.46].

Holographic interferometric techniques have received attention in the area of orthopedics. Osteosynthesis with external fixator which is used for some types of long bone fractures, can be studied [9.47] holographically for its fixation strength. Holographic double-exposure interferometry has been applied to study biomechanical characteristics of tibia fixed with external fixator. Dry bone in cantilever bending mode has been studied by heterodyne holographic interferometry to determine the piezoelectric coefficient in the bone [9,48].

9.2.6. Biological Holography with X-Ray Laser

We end this chapter by making a brief introduction to a recent experiment conducted [9.49] to produce X-ray holograms of microscopic test objects that could lead to the recording of holograms of living biological specimens, such as cells in their natural environment. X-ray holography has the potential of examining the samples in aqueous solution with very high resolution.

Figure 9.12 shows the recording geometry of a X-ray hologram. The arrangement is similar to that proposed by

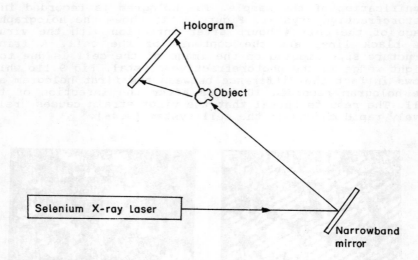

Fig. 9.12 Optical system for the recording of a hologram with X-rays [9.49].

Gabor. The X-ray beam from a selenium X-ray laser (λ=20 nm) falls on a narrow band X-ray mirror which reduces the broadband X-ray background produced by the Se laser. The mirror has a flatness and roughness of better than 2 nm which preserves the coherence of the X-ray laser beam. The image can be reconstructed optically. The arrangement shown in Fig. 9.12 could record the holograms of 8 μm diameter carbon fibre and a 3D object made of gold bars 10 μm thick.

Further work is needed to obtain the X-ray wavelength near 4.4 nm with improved coherence, where the contrast between the water and the protein is maximized.

References

9.1 R. Bimgruber, ed., *IEEE J. Quant. Electron.* **26** (1990) 2146-2305 (Special issue on Lasers in Biology & Medicine).

9.2 T.F. Dewsch, C.A. Puliafito eds., *IEEE J. Quant. Electron* **23** (1987) 1701-1845, (Special issue on Lasers in Biology & Medicine).

9.3 M.J.C. Van Gemert and A.J. Wetch, *IEEE Eng.in Medicine & Biology* **4** (1989) 10-20.

9.4 J.A.S. Carruth, *IEEE Eng. in Medicine & Biology* (March 1986) **37**.

9.5 R.H. Webb, *IEEE Eng. in Medicine & Biology* (Dec. 1985) 12.
9.6 Laser Surgery:Advanced characterisation, Therapeutics, and systems, *Proc. SPIE* **1066** (1989).
9.7 *Holography in Medicine and Biology*, ed. G. Von Bally (Springer-Verlag 1979).
9.8 *Holography, Interferometry and Optical Pattern Recognition in Biomedicine*, eds. H. Podbielska and A.Katzir *Proc. SPIE* **1425** (1991).
9.9 *Optics in Biomedical Science* eds. G. Von Bally and P. Greguss (Springer Verlag 1982).
9.10 J. Tsujiuchi, *Proc.SPIE* **1429** (1991) 326.
9.11 W. Roesch, *Endoscopy* **6** (1974) 190.
9.12 P. Greguss, in *Optics in Biomedical Sciences*, eds. G Von Bally and P.Greguss (Springer Verlag 1982) p. 96.
9.13 G. Von Bally, W. Schmidthaus, H. Sakowski and W. Mette, *Appl. Opt.* **23** (1984) 1725.
9.14 G. Von Bally, E. Brune and W. Mette, *Appl. Opt.* **25** (1986) 3425.
9.15 G. Raviv, M. Marhic and M. Epstein, *Opt. Commum.* **55** (1985) 261.
9.16 M.D. Friedman, H.I. Bjelkhagen and M. Epstein, *J.Laser Appl.* **1** (1988) 40.
9.17 H.I. Bjelkhagen, M.D. Friedman and M. Epstein, *Proc. SPIE* **1136** (1989) 186.
9.18 G. Von Bally, *Proc. SPIE* **1136** (1989) 952.
9.19 G. Von Bally, D. Dirksen, Y. Zou and E. Kraetzig, *Proc. SPIE* **1136** (1989) 158.
9.20 G. Von Bally, *SPIE Institute Series* **IS 8** (1990) 335.
9.21 H. Podbielska, *Proc.SPIE* **1429** (1991) 207.
9.22 R.F. Van Ligten, *Am. J. Optom.* **43** (1966) 351.
9.23 J.L. Calkins and C.D. Leonard, *Invest. Opthalmol.* **9** (1970) 458.
9.24 T. Kawara and H. Ohzu, *Jap. J. Opthalmol.* **21** (1977) 287.
9.25 R.L. Wiggins, K.D. Vaughan and G.B. Friedmann, *Arch. Opthlmol.* **88** (1972) 75.
9.26 L.D. Bores, *Proc. SPIE* **1429** (1991) 217.
9.27 M.H. Fridlander, M. Mulet, K. Buzard, N. Granet and P. Baker, *Proc. SPIE* **1429** (1991) 229.

9.28　W. Förster, H. Kasprzak, G. Von Bally and H. Busse, *Proc. SPIE* **1429** (1991) 146.

9.29　L. Wictorin, H. Bjelkhagen and N. Abramson, *Acta Odontol. Scand.* **30** (1972) 659.

9.30　H.I. Bjelkhagen, in *Holography in Medicine and Biology*, ed. G. Von Bally (Springer Verlag 1979) p. 157.

9.31　T. Matsumoto, T. Fujita, R. Nagata, T. Sugimura and Y. Kakudo, in *Holography in Medicine and Biology*, ed. G. Von Bally (Springer Verlag 1979) p. 170.

9.32　S.T. Higgins, *J. Audio Visual Media Med.* **7** (1984) 59.

9.33　P.J. Keating, R.A. Parker, D. Keane and L. Wright, *British J. Orthodontics* **11** (1984) 119.

9.34　G. Redmalm, G, Johannsen and H. Ryden, *Swedish Dent. J.* **9** (1985) 29.

9.35　R.A.J. Groenhuis, W.L. Jongebloed and J.J. Bosch, *Caies Res.* **14** (1980) 333.

9.36　G.B. Altshuler and V.N. Grisimov, *Proc. SPIE* **1353** (1989) 97.

9.37　G.B. Altushuler and V.N. Grisimov, *Dokl. USSR Ser.* **310** (1990) 1245.

9.38　G.B. Altushuler, V.N. Grisimov, V. Ermolaev and I. Vityaz, *Proc. SPIE* **1429** (1991) 95.

9.39　G. Von Bally, in *Holography in Medicine and Biology*, eds. G. Von Bally (Springer Verlag 1979) p. 183, 198.

9.40　W. Fritze, in *Holography in Medicine and Biology*, ed. G. Von Bally (Springer Verlag 1979) p. 206.

9.41　M. Maeta, S. Kawakami, T. Ogawara and Yu. Masuda, *Proc. SPIE* **1429** (1991) 152.

9.42　S. Kawakami, *J. Otoaryngol. Jap.* **90** (1987) 536.

9.43　G. Edlund, G. Wikander and N.-E. Molin, *J. Sound and Vib.* **59** (1978) 299.

9.44　K. Høgmoen and T. Gundersen, *Proc. Symp. Electrocochleography and Holography in Medicine, Münster*, **1** (1976) 247.

9.45　D. Tonchev, S. Zhivkova and M Miteva, *Appl. Opt.* **29** (1990) 4753.

9.46　D. Tonchev, S. Zhivkova, M. Miteva, I. Grigoriev and I. Ivanov, *Proc. SPIE* **1429** (1991) 76.

9.47　A. Kojima, R. Ogawa, N. Izuchi, M. Yamamoto, T. Nishimoto and T. Matsumoto, *Proc. SPIE* **1429** (1991) 162.

9.48   B. Ovryn and E.M. Haacke, *Proc. SPIE* **1429** (1991) 172.
9.49   J.E. Trebes, *Photonics Spectra* (Jan. 1989) 133.

NAME INDEX

A

Abramson, N., 307,465, 483,495,513,514,516
Aggarwal, A.K., 385,468, 471,550
Ajith Kumar, P.T., 515
Akiyama, I., 472
Albe, F., 471
Aldridge, N.B., 632
Aleksoff, C.C., 514
Alferness, R., 464,467
Allen, L., 90
Alphonse, G.A., 581
Altshuler, G.B., 656,662
Amadesi, S., 515
Amodei, J.J., 470
Anderson, D., 476
Anderson, R.H., 582
Andrade, A.A., 467
Andrews, J.R., 473
Angell, D.K., 350,468
Aoyagi, Y., 550
Arai, A., 249
Arakawa, Y., 248
Archbold, E., 513,514
Arecchi, F.T., 250
Armitage, D., 475
Armstrong, J.A., 142
Armstrong, M., 465
Aronson, L.B., 578,582
Arrathoon, R., 580
Arsenault, H., 632
Ashkin, A., 470
Athale, R.A., 580,582
Auston, D.H., 142
Aye, T.M., 478

B

Babu, S.S.C., 467
Baez, V.A., 252,463
Bagby, R., 472
Bahuguna, R.D., 394,471
Baker, P., 662

Balan, N.F., 468
Ballman, A.A., 470
Barrekette, E.S., 632
Bartelt, H., 476
Bartolini, R.A., 469,477
Basov, N.G., 250
Basu, M.K., 142
Bazargan, K., 406,429, 470,473,474
Beaulieu, A.J., 250
Beauregard, A., 394,471
Beek, M.-A., 551
Beiser, L., 551
Bekbulatov, R.S., 514
-Bellani, V., 516
Ben, W., 551
Benlarbi, B., 464
Bennett, C.R., 475
Bennett Jr., W.R., 91
Benson, R., 515
Benton, S.A., 386,392, 429,465,471-473,475,477
Bergman, L., 580
Bernal, E., 469
Bertelt, H., 580
Bhan, C., 394,464,471-473,515,516,581,630, 632,633
Bhar, G.C., 249
Bhatnagar, G.S., 514
Bhatnagar, V.S., 142
Bialkowski, S.E., 467
Biedermann, K., 464-466, 513
Biran, B., 464
Birk, Y., 630
Birner, S.M., 473
Bismuth, G., 470
Bjelkhagen, H., 514,661, 662
Blackie, R., 472
Blacksley, K., 473
Bloembergen, N., 250
Bloom, A.L., 91

NAME INDEX

Blotekjaer, K., 582
Blyth, J., 427,467, 475
Bodunov, E.N., 468
Boehnlein, A.J., 516
Boerner, W.M., 477,550, 630
Boivin, A., 474
Boj, P.G., 468
Bolstad, J.D., 513
Bonmati, A., 468
Boone, P.M., 514
Booth, B.L., 468,477
Bores, L.D., 661
Born, M., 463,477
Bosch, J.J., 662
Bouts, F.A.J., 333,466
Boyd, D.D., 470
Boyd, G.D., 90
Bradlay, E., 631
Brady, D., 476
Brandt, C.B., 472
Brau, C.A., 250
Brenner, K.-H., 580-582
Bridges, W.B., 249
Brienza, M.J., 141
Brock, P.J., 469
Brokopf, K., 631
Bromley, K., 630
Brooks, R.E., 513
Brown, G.M., 469,515
Brown, K.C., 337,466
Brown, M.S., 476
Brown, N., 514
Brumm, D., 464,630
Brune, E., 661
Bryant, J.F., 630
Bryngdahl, O., 465,478
Bryskin, V.Z., 470
Burch, J.M., 513
Burchett, O.J., 515
Burenkin, A.M., 514
Burke, W., 470
Burkhardt, C.B., 465,473
Burns, G., 248
Burns, J.R., 477
Burrus Jr., C.A., 631
Busse, H., 662
Button, K.J., 250
Buzard, K., 661
Byer, R.L., 248

C

Cai, L.Z., 466
Caimi, F., 470
Calixto, S., 355,467,468
Calkins, J.L., 661
Camatini, E., 515
Campbell, M., 469
Carlson, F.P., 630
Carlson, R.O., 248
Carrig, T.J., 248
Carruth, J.A.S., 660
Carswell, A.I., 477
Casasent, D., 470,476,631
Case, S., 348,467
Case, S.K., 476,580
Casperson, L., 90
Cassey Jr., H.C., 248
Cathey, J.P., 475
Cathey, W.T., 465,474,475
Caulfield, H.J., 433,465, 475-478,513,554,558,579, 580,582,632
Ce, W., 466,477
Champagne, E.B., 268,515, 549,643
Chang, B.J., 350,465,468
Changkakoti, R., 467,469
Chau, H.M., 470
Chavel, P., 580
Chen, F.S., 367,470
Chen, H., 392,471,474
Chen, R.T., 478
Chesler, R.B., 142
Chester, A.N., 249
Cheung, N., 470
Chhachhia, D.P., 471
Chiou, A.E., 580
Chopra, K.N., 514
Chou, H.M., 551
Christensen, C.R., 633
Clay, B.R., 477
Clothier, A., 515
Cobb, J.G., 475
Colburn, W.S., 468
Coleman, D.J., 467,550
Collier, R.J., 465,474, 513
Collins Jr., S.A., 433, 476,554,558,580,582
Cook, B.D., 274,464
Cook, L.G., 551

Cooke, D.J., 466
Cotter, D., 250
Couture, J.J.A., 357,469
Crawford, J.R., 515
Credelle, T.L., 469
Crenshaw, M.M., 466,470
Crespo, J., 468
Cretkovich, T.J., 475
Cross, L., 398,401,472
Cullis, I.C., 516
Curran, R.K., 346,467
Cutler, C.C., 141
Cvetkovich, T.J., 477

D

Da, X.-Y.,394,472
Daehlin, T.,469
Dähne, S., 466
Dahsiung, H., 551
Dainty, J.C., 476,581
Dalisa, A.L., 475
Dandliker, R., 516
Danielmeyer, H.G., 91
Das, T.K., 371,470
Däschner, W., 581
Davidson, N., 476,572,580
Deacon, D.A.G., 250
De Bitteto, D.J., 404, 474,473-475
Decker, A.J., 474
Demaria, A.J., 141
DeMattia, P., 516
Denisyuk, Y.N., 464,465
Derstine, M.W., 580
Deutschbein, O.K., 248
De Velis, J.B., 464,514
Dewsch, T.F., 660
DiDomenico, M., 141
Dingle, R., 248
Dirksen, D., 661
Doherty, E.T., 513
Dongqing, C., 631
Donnelly, J.P., 248
Dover, S., 472
Drabik, T., 580
Drinkwater, J., 472
Dubois, F., 278,464
Dudderar, T.D., 516
Duncan, S., 513,550
Duncon Jr., R.C., 469
Dunn, M.H., 90

Dunning, G.J., 632
Durrant, A. V., 90
Dutta, K., 472
Dziedzic, J.M., 470

E

Eastes, J.W., 468
Ebersole, J.F., 630
Edlund, G., 662
Eguchi, S., 551
Eichmann, G., 632
Einstein, A., 90
Elias, L.R., 250
Elinevskii, D.S., 514
El-Sum, H.M.A., 252,463
Embach, J.T., 471
Enloe, L.H., 473
Ennos, A.E., 513-515
Epstein, M., 661
Erf, R.K., 514-516
Erko, A.I., 468
Ermolaev, V., 662
Eryshev, V.A., 514
Esch, V., 631
Evtuhov, V., 142
Ewing, J.J., 143

F

Fagon, W.F., 513
Fagot, H., 465,474
Fainman, Y., 476
Falkenstörefer, O., 581
Fan, T.Y., 248
Fargion, D., 473
Farhat, N.H., 632
Farrant, D., 631
Faughnan, B.W., 469
Faxiang, W., 631
Fedorowicz, R.J., 475
Feferman, B.J., 475
Feinberg, J., 470
Feinlib, R., 631
Feldman, M., 580
Feldstein, N., 477
Fenner, G.E., 248
Fimia, A., 468,477
Fischer, W.K., 514
Fisher, R.A., 464
Fitzpatrick, C., 631
Forcheimer, R., 580

NAME INDEX

Förster, W., 656,662
Forsey, R.C., 515
Fowler, V.J., 141
Fowles, G.R., 143
Fox, A.G., 38,39,90
Foy, P.W., 248
Frazer, D.B., 470
Fried, J.A., 580
Friedlander, C.B., 632
Friedlander, M.H., 661
Friedman, M.D., 661
Friedmann, G.B., 661
Friesem, A.A., 317,464, 465,471,472,474,476
Fritze, W., 662
Fritzsch, W., 515
Frost, T., 514
Frye, L.M., 631
Fryer, P.A., 514
Fujita, T., 662
Fukazawa, K., 472
Funkhouser, A., 464
Furman, A., 631
Furrer, F., 630
Furukawa, S., 472
Fusek, R.L., 472

G

Gabor, D., 251,252,254,256, 463,630
Galpern, A.D., 474
Gamble, F.T., 631
Garza, R., 515
Gäsvik, K., 478
Gaylord, T.K., 464
Geogiou, E.T., 248
George, N., 385,465,471, 551
Georgekutty, T.G., 467
Gerbig, V., 476
Gerchberg, R.W., 632
German, K.R., 248
Geusic, J.E., 141,142,247
Ghandeharian, H., 477
Gibbs, H.M., 631
Gilbert, J., 250
Gilbert, J.A., 516
Giles, C.L., 582
Giordmaine, J.A., 142
Gladden, J.W., 468
Glanville, R., 515

Glaser, I., 472
Glass, A.M., 470
Glasser, I., 580
Glauber, R.J., 91
Glenn, W.H., 141
Goldman, G., 550
Goldmann, G., 475
Gomi, M., 472
Goodman, J.W., 465,475, 476,552,580
Gorchaakov, G.I., 477
Gordon, A.L., 515
Gordon, J.P., 90
Gore, D.A., 477
Gori, F., 515
Goto, M., 475
Goto, T., 475
Graf, P., 475
Granet, N., 661
Grant, R.M., 515
Graver, W.R., 350,468
Greguss, P., 661
Grella, R., 515
Grieser, D.R., 631
Griffith, P.C., 582,631
Grigoriev, I., 663
Grisimov, V.N., 662
Groenhuis, R.A.J., 662
Groh, G., 626,630,633
Grover, C.P., 394,471
Grube, A., 349,466,467
Grünewald, K., 515
Guattari, G., 515
Gudzenko, L.I., 250
Guenther, B.D., 467
Guest, C., 580
Guha, A., 580
Gundersen, T., 662
Guo, C.S., 466
Guoguang, M., 631
Gupta, P.C., 550
Gustafson, T.K., 142

H

Haas, W,E., 473
Habiby, S.F., 582
Hachett, P.A., 142
Hacke, E.M., 663
Hadwin, J.F., 475
Hagler, M.O., 476
Haines, K.A., 468,513,516

Hale, M.M., 250
Hall, R.N., 143,248
Hall, T.J., 581
Halle, M.W., 472
Halstead, A.S., 249
Hanbury-Brown, R., 91
Hanna, D.C., 249,250
Hänsch, T.W., 200,249
Harding, W.E., 632
Hariharan, P., 420,427,
   465,466, 471,474,475,
   490,513,514,631
Harnier, A.V., 515
Harper, J.S., 468
Harrey, P., 550
Harris, J.L., 475
Harris, S.E., 141,250
Harris, W.J., 515
Hart, S., 472,476,582
Hasman, E., 476,580
Haugen, P.R., 476,580
Haumann, H.J., 581
Havranek, J., 473
Hayashi, J., 248
Heflinger, L.O., 513
Hegedus, Z., 471,474,513,
   514
Hegedus, Z.S., 475
Hellworth, R.W., 142
Herman, G.T., 473
Herriau, J.P., 470
Herriot, D.R., 143,249
Herzig, H.P., 477,580
Hesselink, L., 465,476,
   578,580,582,630
Hicki, R., 470
Higgins, S.T., 662
Higuchi, K., 473
Hildebrand, B.P., 513,
   516
Hinch, K., 631
Hodges, D.T., 250
Hogmoen, K., 662
Holeman, J.M., 631
Honda, K., 467
Honda, T., 472,473
Hong, J., 580
Hopfield, J.J., 632
Horman, H.M., 479,513
Horvath, V.V., 631
Hradaynath, R., 463,515,
   549,633
Hsieh, J.J., 248
Huang, A., 478,580-582
Huang, Q., 473
Huang, Y.-T., 581
Hubbard, W.M., 574,582
Hubel, P.M., 474
Huff, L., 472
Huignard, J.P., 470
Huiying, T., 551
Hussain, L., 580
Hutcheson, L.D., 580
Hutley, M.C., 550
Hutley, W.H., 475

I

Ian, R., 551
Idogawa, T., 514
Igaki, S., 551
Ih, C.S., 550
Ikeda, H., 551
Imam, H., 551
Inagaki, T., 551
Ishikawa, T., 550
Ishizuka, T., 468
Ivanov, I., 662
Izquierdo, L.R., 514
Izuchi, N., 662

J

Ja, Y.H., 631
Jackson, D.W., 475
Jacobson, R.E., 337,466
Jacquot, P., 514
Jaerich, W., 514
Jaffy, S.M., 472
Jahns, J., 478,581
Jain, S.C., 141
Jain, T.C., 473
Jakob, M., 250
James, T.H., 465
Jannson, J., 550
Jannson, T., 478,550
Javan, A., 143,249
Jenkins, B.K., 579,580,
   582
Jeon, H.-I., 580
Jeong, M.H., 466
Jeong, T.H., 470,471,474,
   475,516

NAME INDEX

Jewell, J.L., 581
Jiang, Y.G., 631
Joenathan, C., 471,631
Johannsen, G., 662
Johnson, C.D., 513
Johnson, K., 476
Johnson, K.M., 465,476
Johnson, L.F., 143
Johnston, A., 580
Johnston, A.R., 580
Johnston, G., 473
Joly, L., 466
Jongebloed, W.L., 662
Joshi, L., 473
Joshi, U.C., 141
Juyal, D.P., 581,630

K

Kakudo, Y., 662
Kalestynskii, A., 626,633
Kalinkin, V.V., 468
Kamala, R., 471
Kanaya, M., 462
Kandilarov, P., 467
Kang, D.-K., 473
Karr, M.A., 142
Kasper, F.G., 464,465
Kasper, J.V., 250
Kasprazak, H., 662
Katayanagi, T.E., 515
Kato, M., 581
Katsuma, H., 472,473
Katti, P.K., 633
Katzir, A., 661
Kaul, A.N., 581
Kaura, S.K., 385,468,471
Kaushik, G.S., 465,466, 513
Kawabe, T., 472
Kawacz, L., 473
Kawakami. S., 662
Kawara, T., 661
Keane, D., 662
Keating, P.J., 662
Keller, P.E., 632
Kersch, L.A., 515
Kestigian, M., 470
Kakudo, Y., 662
Keys, D.E., 468
Khajuria, R.D., 473
Khare, A. K., 250

Khitrove, G., 631
Khouser, A.F., 630
Kiemle, H., 632
Kimura, Y., 550
King, E., 551
King, M.A., 472
King, M.C., 471,472
Kingsley, J.D., 248
Kirkpatrick, P., 252,463
Kirk, A.G., 581
Kiryuschcheva, I.V., 631
Kishino, K., 249
Kiss, Z.J., 247,469
Klein, M.B., 582
Klein, W.R., 274,464
Klug, M.A., 477
Knight, G.R., 465
Knight, P.L., 90
Kobak, V.O., 477
Kobayashi, K., 249
Kobolla, H., 581
Kock, W.E., 632
Koecher, W., 141
Kogelnik, H., 90,91,248, 249,278,311,464
Kojima, A., 662
Komar, V.G., 414,474
Konyukhov, V.K., 250
Korzinin, V.L., 468
Kosar, J., 466
Kostuk, R.K., 466,580,581, 632
Koyama, J., 478
Kozaitis, S., 580
Kozma, A., 317,465,633
Krätzig, E., 470,661
Krile, T.F., 476
Krylov, V.N., 470
Kubota, T., 467,473,475, 571,581,582
Kumar, S., 466
Kung, S., 580
Kung, S.-Y., 580
Kunugi, K., 471
Kurata, Y., 550
Kurtz, C.N., 478
Kurtz, R.L., 308,465
Kuwayama, T., 468

L

Labeyrie, A.E., 474

LaMacchia, J.T., 470,475, 476
Lamb Jr., W.E., 90.91, 107
Lamberts, R.L., 465
Lamberty, D., 630
Laming, F.P., 468
Lang, M., 475
Lankard, J.R., 143,249
Lanzl, F., 469
Lashkov, G.I., 468
Latta, J.N., 268,269,463, 549
Latta, M.R., 549
Lau, E., 466
Launay, J.C., 470
Laurie, K.A., 250
Lee, S.H., 476,580,582
Lee, T.C., 469
Lee, T.P., 248
Lee, W.H., 478,551
Lee, Y.H., 581
Lehmann, M., 475
Leite, A.M.P.R., 550
Leith, E., 252,258,477
Leith, E.N., 463
Lemaire, P., 464
Lemmond, C.Q., 631
Lemoine, T., 581
Lempicki, A., 249
Lengyl, B.A., 142
Leonard, C.D., 464-467, 661
Leonberger, F., 580
Leonberger, F.I., 580
Lessard, R.A., 394,467, 469,471,474,475,478
Levenson, M.D., 469
Levinstein, H.J., 470
Li, T., 38,39,90
Liang, Y.Z., 633
Lima, C.R.A., 469
Lin, G.J., 469
Lin, J.W., 469
Lin, L.H., 311,465,466, 474,477
Lindlein, N., 581
Linford, G.T., 248
Lipson, H., 630
Liu, H.K., 467,631,633
Lo Bianco, C.V., 474

Logan, R.A., 248
Lohmann, A., 252,463,465, 475,476,478,582
Losevsky, N.W., 468
Luckett, A., 470
Lu, K., 550
Lu, S., 626,633
Lukosz, W., 478
Luxmoore, A.D., 515
Lytle, D., 632

M

Ma, J., 476
Mack, M.E., 141
MacLeod, H.A., 631
MacQuigg, D.R., 464
Macukow, B., 632
Mada, H., 632
Madey, J.M.J., 143,250
Maeta, M.,662
Magarinos, J.R., 467,550
Magnusson, R., 464
Maiman, T.H., 247
Maitland, A., 90
Makosch, G., 514
Malkhasyan, L.G., 513
Maloletov, S.M., 468
Malov, A.N., 467,468
Malyutin, A.A., 142
Mandel, L., 91
Maniloff, E., 476
Manjunath, H.R., 630
Mansharamani, N., 142
Marcinkowski, G., 632
Marcos, H.M., 91,247
Marcuse, D., 90
Marhic, M.E., 661,580
Markova, G.V., 513
Marks, J., 474
Marling, J.B., 249
Marom, E., 471,630-632
Marrakchi, A., 582
Marrone, E.S., 516
Marshall, T.C., 250
Massey, N.G., 268,463, 549,630
Masuda, Yu., 662
Mathews, D., 143
Matsuda, K., 630
Matsui, K., 475
Matsumoto, K., 468

Matsumoto, M., 468
Matsumoto, T., 662
Mauer, P.B., 247
Maurer, R.D., 247
Maurer, R.E., 632
Mayer, G.M., 514
Mayer, M.D., 515,515
Mazakova, M., 467
McCarthey, D.J., 550
Mcclug, F.J., 142
McFarland, B.B., 249
McGrew, S.P., 467
McLysaght, F., 631
McQuoid, J.A., 550
McRuer, R., 582
Mead, C.A., 515
Mehta, P.C.,268,269,385, 394,404,463,464,471- 473,475,476,514-516, 549,581, 630,632,633
Meier, R.W., 463
Meiyue, L., 471
Mendes, G., 582
Mersereau, K., 631
Metherington, D., 581
Mette, W., 661
Meyerhofer, D., 347,466
Meyers, R.A., 142
Micheron, F., 470
Mickish, D., 468
Mikaeliane, A.L., 632
Mikami, O., 249
Miler, M., 477
Mileski, R., 473
Miller, D.A.B., 580,581
Miller, R.A., 248
Mintz, G.D., 550,630
Miteva, M., 662
Miyake, T., 550
Mizuno, T., 423,475
Modde, J.P., 631
Mohan, D., 473,515,516, 478, 581,633
Molin, N.E., 513
Molin, N.-E., 662
Mollenauer, L.F., 248
Monahan, M.A., 630
Monroe, B.M., 468
Morland, D.K., 550,630
Morozov, V., 477
Moruzzi, V.L., 249

Moslehi, B., 550
Mottier, F.M., 514
Moulton, P.F., 248
Mulet, M., 661
Muller, Ed., 631
Muller, G., 473
Murata, K., 386,471
Murphy, J.A., 473
Murray, J.J., 632
Musla, A.K., 478,631
Mutter, W.E., 632

N

Nagata, R., 662
Namba, S., 550
Nassau, K., 143
Nassenstein, H., 458,478
Nathan, M.I., 248
Neeland, J.K., 142
Neifield, M.A., 551
Nesterikhin, Yu. E., 551
Neumann, D.B., 464
Newell, J.C., 467
Newman, P.A., 477
Ng, J., 550
Ng, W.K., 248
Nichols, R.W., 248
Nishida, N., 465,475,476
Nishihara, H., 478

O

Ohimura, A., 472
Ohyama, N., 472,473
Ohzu, H., 661
Okada, K., 400,472
Oliva, J., 477
Olness, D., 248
Ono, A., 466
Ono, Y., 550
Oraevskii, A.N., 250
Oreb, B.F., 514,631
Orlowski, R., 470
Ose, T., 467
Ostrovaskii Yu, I., 513
Owecjko, Y., 632
Owen, H., 467

P

Pace, J.D., 475

Paek, E., 476
Palasis, J.C., 477
Pal, S.R., 477
Pampolne, T.R., 468
Pancheva, M., 467
Pandya, T.P., 549
Panish, M.B., 248
Pappu, S.V., 467,469,550
Pardo, M., 468
Parker, J.V., 249
Parker, R.A., 662
Parker, R.J., 465,516
Parrent, G.B., 463
Pastor, C., 468
Patel, C.K.N., 33,142,249
Patel, J., 476
Patel, J.S., 582
Pauliat, G., 632
Pautrat, C.C., 248
Pennington, K.S., 350, 468,474,513
Peppalardo, R., 249
Perelmutter, L., 580
Perry, L.M., 308,465
Peterson, O.G., 249
Petterson, G.E., 470
Phelan, R., 466
Phillips, N.J., 466,477
Phillips, W., 469,470, 632
Pimentel, G.C., 250
Piper, J.A., 250
Podbielska. H., 661
Pollock, C.R., 248
Porter, D., 466
Powell, R.L., 479,513
Prata, A., 632
Pressley, R.J., 247
Prokhorov, A.M., 33,90, 250
Prongue, D., 580
Psaltis, D., 476,580,582, 612,631,632
Puliafito, C.A., 660
Purusthotham, C., 515
Putitin, A., 477

Q

Qiao, Y., 550
Quintana, J.A., 468,477

R

Rabinovich, V.A., 631
Rainsdon, M.D., 473
Rallison, R.D., 467
Ramanathan, C.S., 465, 466,513
Ramian, G.J., 250
Rampal, V.V., 141,142, 249,630
Rankin, K., 550
Rao, K.S.S., 463,464,475, 523,549,581,630,633
Rao, S. Anand, 550
Rao, V.V., 631
Rastani, K., 574,582
Raviv, G., 661
Rayner, D.M., 142
Rebartdao, J.M., 467
Redfield, S., 632
Redmond, I.R., 581
Redus, W.D., 476
Redzikowski, M., 466
Reinhart, F.K., 142
Rentzepis, P.M., 142
Restall, E.J., 580,581
Restrick, R., 630
Reynolds, G.O., 464,514
Rhodes, C.K., 250
Rhodes, W.T., 477
Ribbens, W.B., 516
Rible, V.E., 477
Richter, A.K., 630
Rigden, J.D., 249
Rizkin, A., 550
Roberge, D.C., 475
Robertson, B., 580
Rodor, V., 469
Roesch, W., 661
Rogers, G.L., 252,463
Rojas, D., 470
Roosen, G., 632
Rosenthal, D., 515
Rosner, S., 581
Ross, W.E., 582
Rossi, J.A., 142
Roth, E., 515
Rotz, E.B., 464
Rozenberg, G.V., 477
Rubinstein, C.B., 473
Rudge, W.E., 632
Rudolf, P., 470

NAME INDEX

Russel, P.St.J., 278,464
Ryan, R.J., 477
Rychnovsky, S., 580
Ryden, H., 662

S

Sadovnik, L., 478
Saito, T., 472
Sakaguchi, M., 476
Sakowski, H., 661
Sakuda, K., 550
Samelson, H., 249
Sano, K., 550
Sargent, M., 90
Saroyan, R.A., 248
Sasaki, M., 467
Sassat, J.M., 470
Sato, K., 414,472,473
Sato, R., 471
Sauer, F., 476,581
Savant, G., 550
Sawchuck, A.A., 579,580
Saxton, W.O., 632
Schafer, F.P., 249
Schawlow, A.L., 33,90,143, 247
Schelev, M.Y., 142
Schell, K.J., 477
Schellenbrg, F.M., 469
Schlafer, J., 141
Schmidt, J., 581
Schmidt, W., 249
Schmidthaus, W., 661
Schneider, I., 469
Schryver, F.De, 464
Schulz, E.O., 250
Schulze, E., 576,582
Schulze, R., 469
Schwab, J., 465
Schwar, M.J.R, 549
Schweicher, E., 551
Schwettman, H.A., 250
Schwider, J., 581
Scrivener, G.E., 469
Scully, M.O., 90,91
Sebillotte, C., 581
Selvarajan, A., 581,582
Servaes, D.A., 514
Shajenko, P., 513
Shamir, J., 478

Shandarov, S.M., 582
Shandarov, V.M., 582
Shankoff, T.A., 339,346, 466,467
Shan, Q., 394,471
Shapiro, S.L., 142
Shaposhnikov, Yu.N., 514
Sharma, M.K., 581,630
Shaw, W.G., 469
Shawchuk, A.A., 582
Sheel, D., 467
Shelepin, L.A., 250
Shenai, K., 581
Shen, Y.R., 250
Sherstyuk, V.P., 468
Shimura, K., 473
Shinoda, M., 630
Shirakura, A., 473
Siegman, A.E., 90
Silfvast, W.T., 143,250
Singer, J.R., 142
Singh, K., 371,466,470
Singh, M., 476
Sirohi, R.S., 468,471, 514,631
Skogen, J., 469
Slevin, J.A., 631
Smaev, V.P., 474
Smigielski, P., 309,415, 474
Smiglelski, P., 499,515
Smith, H.M., 465,469, 470
Smith, M.V., 91
Smith, P.W., 141
Smith, R.C., 249
Smith, R.E., 141
Smith, R.G., 470
Smith, S.L., 475
Smith, T.I., 250
Smothers, W.K., 468
Snavely, B.B., 249
Snitzer, E., 247,248
Soares, O.D.D., 550
Soffer, B.H., 249,632
Solano, C., 475
Sollid, J.E., 513
Soltys, T.J., 248
Solymer, L., 465,467
Som, S.C., 478
Somekh, S., 249

Sonde, B.S., 581
Song, J.B., 466
Song, J.-Z., 476
Sorokin, P.P., 91,143,249
Sperley, K.M., 469
Spicer, P., 513
Spierings, W., 466,475
Spong, F.W., 469
Sprague, R.W., 631
Srivastava, K.P., 142
Staebler, D.L., 469,470
Staselko, D.I., 470
Steel, W.H., 471,514
Steier, W.H., 142
Stetson, K.A., 464,479, 513,514
Sthel, M.S., 359,469
Stimpfling, A., 465,499, 515
Stone, H.S., 582
Stone, T.W., 551
Stork, W., 581
Strand, T.C., 580
Streibl, N., 581
Streifer, W., 478
Strin, B.A., 471
Stroke, G.W., 464,474, 630,632
Suematsu, Y., 249
Sugama, S., 550
Sugimura, T., 662
Suhara, T., 478
Sukhanov, V.I., 468
Suzuki, H., 466
Suzuki, M., 472
Suzuki, T., 470
Swami, S., 476,549
Sweatt, W.C., 549
Swell, D., 516
Szeto, L.H., 250
Szu, H.H., 632

T

Tabor, W.J., 632
Taghizadeh, M.R., 580
Tai, A., 474
Takai, N., 514
Takeda, M., 571,582
Takeya, N., 630
Taniguchi, N., 468
Teitel, M.A., 472

Tergara, I., 550
Thalman, R., 516
Thareja, R.K., 250
Thaxter, J.B., 470
Thomas, J.P., 515
Thomas, S.J., 249
Thompson, B.J., 463,551, 630
Thompson, G.T., 515
Thornton, W.A., 474
Tian, R.S., 550
Tien, P.K., 42, 90
Tobagi, F.A., 580
Tonchev, D., 662
Townes, C.H., 33,90,143, 247
Tozer, B.A., 515
Trebes, J.E., 662
Treholme, J.B., 248
Tremblay, R., 471
Tripathi, V.K., 582
Trolinger, J., 463
Trotler, J.C., 470
Trout, T.J., 468
Tsujiuchi, J., 400,472, 473,630,661
Tubbs, M.R., 469
Tuccio, S.A., 249
Tufte, D.N., 469
Twiss, R.Q., 91
Tyagi, R.K., 249

U

Uchida, T., 141
Underkoffler, J.S., 472
Upatnieks, J., 252,258, 463,464,466,471,474 ,477
Urbach, J.C., 469
Utaka, K., 249

V

Vander Lugt, A.B., 605, 631,632
Van Driel, H.M., 471
Van Gemert, M.J.C., 660
Van Heerden, P.J., 475, 631
Van Ligten, R.F., 661
Van Renesse, R.L., 333, 466
Van Uitert, L.G.,247

NAME INDEX

Vaneslow, W., 465
Vanin, V.A., 477
Varner, J.R., 516
Varshavchuk, M.L., 477
Vaughan, K.D., 661
Veki, A., 141
Verdeyen, J.T., 142
Vest, C.M., 513
Vikram, C.S., 514
Vincelette, J.C., 475
Vityaz, I., 662
Völkel, R., 631
Von Bally, G., 661,662
Vuylsteke, A.A., 142

W

Wagner, J.W., 661
Wagner, K., 476,631
Wagner, W.G., 142
Waidelich, W., 469
Walker, A.C., 580,581
Walker, J.L., 429,475
Walkup, J.F., 476
Wang, L., 631
Wang, Q.-Q., 394,472
Wang, Z.M., 550
Ward, A.A., 420,466,467, 474
Waters, J.P., 475,515
Watrasiewcz, B.M., 513
Webb, R.H., 661
Weber, A.M., 468
Weber, M.J., 248
Weber, S., 551
Wecht, K.W., 142
Weide, H.-G., 466
Weinberg, F.J., 549
Werf, R.V., 477
Wesly, E., 424,474
Wetch, A.J., 660
Whinnery, J.R., 90
White, A.D., 249
White, D.L., 476
White, H.J., 632
Wictorin, L.,. 662
Wiener-Avnear, E., 471
Wiggins, R.L., 661
Wikander, G., 662
Willis, C., 142
Windeln, W., 581
Winick, K., 350,468

Wise, J.A., 477
Withrington, R.J., 581
Witte, H.H., 550
Wolf, E., 91,463,478
Wolfe, R., 469
Woodcock, B.M., 551
Wood, D.C., 515
Wood, G.P., 477
Wood II, O.R., 250
Wopschall, R.H., 468
Wright, L., 472,662
Wuerker, R.F., 513
Wullert, J., 476
Wüthrich, A. 478
Wyant, C.B., 472

X

Xiang, L.Q., 550
Xu, S., 476,582
Xueqang, C., 471

Y

Yagishita, T., 468
Yamada, M., 514
Yamagishi, F., 551
Yamagishi, Y., 468
Yamaguchi, M., 472,473
Yamamoto, M., 661
Yamamura, A., 581
Yang, C.W., 550
Yariv, A., 90,91,248
Yeh, P., 580
Yingming, X., 471
Yoshida, Y., 550
Young, C.G., 247,248
Yu, F.T.S., 392,471, 474
Yu, K., 550
Yu, P.K.L., 580
Yunlu, Z., 551
Yuratitch, M.A., 250
Yurtaev, Yu.G., 514

Z

Zaidel', A.N., 513
Zambuto, M.H., 514
Zech, R.G., 630
Zelenka, J.S., 317,465, 516
Zhago, D., 630,633

Zhaoqi, W., 631
Zheng, B., 550
Zhimin, Q., 471
Zhivkova, S., 662
Zhou, G.-P., 472
Zou, Y., 661

SUBJECT INDEX

A

aberrations
    third order 268
    fifth order 268
    seventh order 269
active mode locker 101-103
Alexandrite laser 160
amplitude transmittance 256,320,321
autocorrelation 614

B

beam combiners 543
biological holography 658
bleaching process 330-334
Bragg wavelength 188
broadening
    homogeneous 30,31
    inhomogeneous 30,31
buried heterostructure 184

C

cavity coupling 59
cavity dumping 123
character recognition 605
chemical lasers 218
chromaticity diagram 416
coherence 74-83
coherent noise reduction 628,629
colour centre lasers 168
colour holography 416
convolution 607
correlation 586,614,615
cross-correlation 607

D

DCG plates preparation 340,341
deblurring 593
de Broglie wavelengths 3
depolarization effects 450

developers 327,329,330
dichromated gelatin 338
    cross-linking 339
    mechanism of recording 339
    optimization 344
    preparation 340,341
    self-developing 350
diffraction efficiency 309-311
diffractive-refractive telescope 540
dispersion compensation 404
distributed feedback lasers 188
dye lasers 193
dynamic hologram 371

E

efficiency 309
Einstein coefficients 24
electromagnetic radiation 9
electronic interconnections 553
embossed holograms 449
evanescent wave holography 453
eye safe lasers 164
excimer lasers 221

F

far infrared lasers 234
Fermi level 176,179
fingerprint correlator 610
fingerprint sensor 547
Fourier transform 583-587, 593
Fourier transform division filter 593
Fourier transform hologram 291
Fraunhofer hologram 286
free electron laser 238

Fresnel correlator 614
Fresnel hologram 285
Fresnel filter 612
fringe control techniques 499
fringe localization 491
fringe stabilization 299
fringe visibility 297
frustrated internal reflection 610

G

gain saturation 29
Gaussian beams 39
gas lasers 206

H

harmonic generation 241
harmonic oscillator 7
health hazards 373
Heaviside function 186
Heisenberg's uncertainty principle 3
hermetically sealed packages 611
heterojunction lasers 180
HOEs in art 549
hole burning 66
holodiagram 308
hologram
    aberrations 268
    amplitude 2721
    archival storage 336
    characteristics 254
    classification 272
    cylindrical 255
    edge-lit 443
    Fourier transform 291
    Fraunhofer 286
    Fresnel 285
    image 293
    imaging equations 263
    lensless Fourier transform 287
    Lippmann
    local reference beam 429
    moving object 306
    multifaceted 555, 436
    multiplexed 432
    of a point 519
    phase 272
    pinhole 440
    plane 273
    rainbow 386-393
    reconstruction
    recording geometries 301
    recording materials 314
    reflection 281,303
    replication 445
    thin 273
    $360°$ 380
    storage 336,338
    transmission 301
    volume 277
holographic associative memory 620
holographic cinematography 414
holographic combiner 544, 545
holographic contouring 510
    change in illumination angle 511
    two refractive indices 511
    two wavelengths 510
holographic endoscopy 653
holographic gratings 527-533
holographic head-up display 544,545
holographic image multiplication 626
holographic information storage 616
holographic interconnect 552
    challenges 579
    classification 554
    configuration 559
    desirable characteristics 558
holographic interferometry 479
    application 497
    digital 490
    double-exposure 479, 480

SUBJECT INDEX

desensitized 489
fibre optics 512
fringe linearization 489
sandwich 595
stroboscopic 487
time-average 484
single-exposure 482,483
holographic lithography 459
holographic mirrors 527-533
holographic night vision goggles 546,547
holographic nondestructive testing 497-499,502-510
holographic optical elements
    aberrations 523
    applications 533
    design aspects 521
    fabrication 525
    flow chart for fabrication 527
    resolution 520
holographic scanner 539
holographic stereogram 398
holographic television 414
holography
    colour 416
    display 378
    dynamic 371
    evanescent wave 453
    Gabor 256
    in-line 256
    off-axis 258
    Leith-Upatnieks 258
    multiple-exposure 430
    polarization 450
    practical 294
    temporally modulated 488
    waveguide 453
homogeneous broadening 30,31
Hurter and Driffield curve 318-320

I

image
    blur 396
    deblurring 593
    hologram 293
    latent 322
    magnification 267
    multiplication 626
    restoration
    sharpening 593
    subtraction 599
        colour coding method 604
        cross polarization method 600
        double-exposure method 600
        dynamic hologram method 603
        holographic beam splitter method 600
        holographic shear lens method 602
        rainbow holographic method 602
        single-exposure method 599
imaging equations 263
information processing 583
information storage 616
interaction Hamiltonian 12

L

Lamb dip 67
lasers
    carbon monoxide 220
    chemical 218
        HF/DF 219
    colour centre 168
    diode 178
    diode laser pumping 165
    DFB 188
    dye 193
    excimer 221
    eye safe 164
    far infrared 234
    free electron 238
    gas
      Argon ion 209
      CO_2 210
      gas dynamic CO_2 215
      He-Ne 206
      TEA CO_2 212
    heterojunction 180
    metal vapour 226
      Copper vapour 229
      He-Cd 228

Hg-Br 228
  plasma recombination 234
  nitrogen 224
  neodymium 150
  quantum well 184
  semiconductor 173
  slab 167
  tunable 159
  x-ray 658
laser ions 143
laser noise 84
latent image 322
lensless Fourier hologram 287
lens-pinhole spatial filter 300
Lippmann 281
local reference beam hologram 429

M

magnification
  lateral 267
  longitudinal 267
  angular 268
magnetic stressing 508
Maxwell's equations 5
mechanical stressing 505
medical applications 634
metal vapour lasers 226
mode competition 64
mode locking 93,121
mode pulling 67
mode stabilization 108
modulation transfer function 315,316
multifaceted hologram 436, 555
multiple-exposure holography 430

N

neodymium lasers 150
neurons 621, 622
night vision goggles 546, 547
nitrogen laser 224
noise-to-signal ratio
nonlinear recording 316

O

operators
  creation, annihilation 16
  non commuting Hermitian 7
ophthalmology 645,654
optical communication 535
optical gain 26
optical data processor 583
optical interconnects 553
optical resonators 31,52
open resonator cavity 45
otology 657
orthoscopic image 255,269-273

P

perfect shuffle 572
phase modulation 118
photochromic materials 364
photodichroic materials 367
photoelectric effect 4
photopolymer 352
photorefractive crystals 367
photoresist 358
photothermoplastics 361
pinhole hologram 440
Plank's energy distribution formula 2
p-n junction laser diode 178
polarization effects 297
polarization holography 450
POTA developer 486
practical holography 294
pressure stressing 506
pseudocolouring 426
pseudoscopic image 255,269-273
pulse compression 118

Q

Q of resonator 52
Q-switching 123,125,134
quantum well lasers 184

R

Rayleigh Jeans law 1

recording materials
    dichromated gelatin 338
    photochromic 364
    photodichroic 367
    photopolymer 352
    photorefractive 367
    photoresist 358
    photothermoplastic 361
    silver halide 324
    summary 372
refractive index modulation 311
relaxation oscillation 136
replication 445
resonant modes 33
reversal bleach 332
ruby laser 145

S

sandwich hologram interferometry 495
saturable absorber 99
scanning object beam holography 430
self pulsing 107
semiconductor laser 173
sensitivity vector 482
signal-to-noise ratio 312
silver halide emulsion 318
    bleaches 330-334
    developers 327-330
    development 327-330
    H & D curve 318-320
    response 318
    sensitivity 322
    speed 322
silver halide sensitized gelatin 350
slab laser 167
solarization 319
solid state lasers 144
spectral filter 539
spontaneous radiation 20
stimulated radiation 20
stimulated scattering 242, 247
stressing
    magnetic 508
    mechanical 505
    pressure 506
    thermal 502
    vacuum 506
synaptic interconnects 621
synchronous mode locking 116

T

thermal lensing 167
thermal stressing 502
threshold condition 57
transition probability 11
triplet quenchers 193
tunable solid state lasers 158

U

unstable resonator 54
ultrashort pulses 115, 121

V

vacuum stressing 506
Vander Lugt correlator 605
vibration isolation
    pneumatic 298
    electronic 298
vibrational excitation 508
viewing zone 402

W

waveguide hologram 453
WDM 535, 536
Wien's law 2
Wiggler 239

X

x-ray laser 658
x-ray hologram 658